ORGANIC CHEMISTRY STUDY GUIDE

ORGANIC CHEMISTRY STUDY GUIDE: KEY CONCEPTS, PROBLEMS, AND SOLUTIONS

ROBERT J. OUELLETTE

Professor Emeritus, Department
of Chemistry, The Ohio State University

AND

J. DAVID RAWN

Professor Emeritus,
Towson University

ELSEVIER

AMSTERDAM • BOSTON • HEIDELBERG • LONDON • NEW YORK • OXFORD
PARIS • SAN DIEGO • SAN FRANCISCO • SINGAPORE • SYDNEY • TOKYO

Elsevier
Radarweg 29, PO Box 211, 1000 AE Amsterdam, Netherlands
The Boulevard, Langford Lane, Kidlington, Oxford OX5 1GB, UK
225 Wyman Street, Waltham, MA 02451, USA

Notice
Knowledge and best practice in this field are constantly changing. As new research and experience
broaden our understanding, changes in research methods, professional practices, or medical
treatment may become necessary.

Practitioners and researchers must always rely on their own experience and knowledge in evaluating
and using any information, methods, compounds, or experiments described herein. In using
such information or methods they should be mindful of their own safety and the safety of others,
including parties for whom they have a professional responsibility.

To the fullest extent of the law, neither the Publisher nor the authors, contributors, or editors,
assume any liability for any injury and/or damage to persons or property as a matter of products
liability, negligence or otherwise, or from any use or operation of any methods, products,
instructions, or ideas contained in the material herein.

ISBN: 978-0-12-801889-7

Library of Congress Cataloging-in-Publication Data
Ouellette, Robert J., 1938-
    Organic chemistry study guide : key concepts, problems, and solutions / Robert J. Ouellette,
J. David Rawn.
        pages cm
    ISBN 978-0-12-801889-7
1. Chemistry, Organic–Problems, exercises, etc. I. Rawn, J. David, 1944- II. Title.
    QD257.O94 2014
    547.0076–dc23

                                                                    2014027632

British Library Cataloguing in Publication Data
A catalogue record for this book is available from the British Library

For Information on all Elsevier publications
visit our website at http://store.elsevier.com/

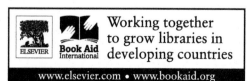

CONTENTS

Companion book: Robert J. Ouellette and J. David Rawn; Organic Chemistry: Structure, Mechanism, and Synthesis; ISBN 9780128007808; June 2014.

Companion website: http://booksite.elsevier.com/9780128007808/

1 STRUCTURE AND BONDING IN ORGANIC COMPOUNDS

KEYS TO THE CHAPTER

Atomic Structure and Properties

Two periodic trends are important to understanding the physical and chemical properties of organic compounds. They are electronegativity and atomic radius.

The electronegativity scale is an index of the attraction of an atom for an electron. It increases from left to right in a period and from bottom to top in a group of the periodic table. The order of electronegativities for the three most common elements in organic molecules, excluding hydrogen, is C < N < O. Their electronegativity values differ by 0.5 between neighboring elements in this part of the second period. There is a more pronounced difference between second and third period elements. Thus, fluorine and chlorine differ by 1.0, as do oxygen and sulfur. The order of the electronegativity values of the halogens is I < Br < Cl.

Ionic and Covalent Bonds

There are two main classes of bonds. Ionic bonds predominate in inorganic compounds, but covalent bonds are much more important in organic chemistry. When positive and negative ions combine to form an ionic compound, the charges of the cations and anions must be balanced to give a neutral compound. For ionic compounds, the cation is named first and then the anion. Thus, ammonium sulfide contains $(NH_4)_2$ and S^{2-}. Two ammonium ions are required to balance the charge of one sulfide ion, so the formula of ammonium sulfide is $(NH_4)_2S$. Parentheses enclose a polyatomic ion when a formula unit contains two or more of that ion, and the subscript is placed outside the parentheses.

A covalent bond forms when two nuclei are simultaneously attracted to the same pair of electrons. Carbon usually forms covalent bonds to other elements. The stability of Lewis structures is attributed to the octet rule that states that second row elements tend to form associations of atoms with eight electrons (both shared and unshared) in the valence shell of all atoms of the molecule.

One or more pairs of electrons can be shared between carbon atoms. Single, double, and triple bonds are linked one, two, and three pairs of electrons, respectively. In applying the octet rule, the bonding electrons are counted twice. That is, each atom "owns" the bonding electrons, so they count toward the total of eight for each atom.

With the exception of bonds to carbon and to hydrogen, carbon forms polar covalent bonds to other elements. The degree of polarity depends on the difference in the electronegativity values of the bonded atoms. The direction of the bond moment is indicated by an arrow with a cross at the end opposite the arrow head. The symbols δ^+ and δ^- indicate the partially positive and partially negative atoms of the bonded atoms.

Strategy for Writing Lewis Structures

When we write a Lewis structure, we first need to know how many electrons are in a molecule based and where they are located.

Consider vinyl chloride, C_2H_3Cl, which is used to produce polymers for commercial products such as PVC pipes. It contains a total of 18 electrons. Hydrogen forms only one bond in all compounds. Chlorine also forms one bond to carbon. The basic skeleton of the molecule is shown below.

Organic Chemistry Study Guide: Key Concepts, Problems, and Solutions http://dx.doi.org/10.1016/B978-0-12-801889-7.00001-7

The molecular skeleton accounts for eight electrons; two per single bond. Each carbon atom still needs two more electrons to complete its octet, and the chlorine atom needs six. The six electrons on chlorine form three lone pairs. Each carbon contributes one electron to the single bond. Each carbon has four electrons, and each donates one more to form a double bond.

Formal Charge

We determine formal charges in several steps.

1. Count the total number of valence electrons for each atom in the molecule.
2. Each atom "owns" its nonbonded electron pairs.
3. Electrons in bonds are shared equally between the bonded atoms; in a single bond each atom gets one electron, in a double bond it gets two, and so forth.
4. If an atom has more electrons in the bonded structure than it would have if neutral, it has a formal negative charge; if it has fewer electrons than it would have as a neutral atom, it has a formal positive charge.

A few simple rules make it easy to determine the formal charge in most cases by inspection. For example, if nitrogen has three bonds—regardless of the combination of single, double, or triple bonds—and a pair of electrons, then it has no formal charge. If there are four bonds to nitrogen—regardless of the combination of single, double, or triple bonds—the nitrogen atom has a formal +1 charge. Similarly, if oxygen has two bonds—regardless of the combination of single or double bonds—and two pairs of electrons, then it has no formal charge. If there are three bonds to oxygen—regardless of the combination of single or double bonds—the oxygen atom has a formal +1 charge. The structure shown below contains an oxygen atom with a +1 formal charge; the entire species has a net +1 charge.

Resonance Theory

For most compounds, one Lewis structure describes the distribution of electrons and the types of bonds in a molecule. However, for some species a single Lewis structure does not provide an adequate description of bonding. Resonance structures provide a bookkeeping device to describe the delocalization of electrons, giving structures that cannot be adequately described by a single Lewis structure. Such bonding is described using two or more resonance contributors that differ only in the location of the electrons. The positions of the nuclei are unchanged. The actual structure of a molecule that is pictured by resonance structures has characteristics of all the resonance contributors.

Curved arrows are used to show the movement of electrons to transform one resonance contributor into another. The electrons move from the position indicated by the tail of the arrow toward the position shown by the head.

The degree to which various resonance forms contribute to the actual structure in terms of the properties of the bonds and the location of charge is not the same for all resonance forms. The overriding first rule is that the Lewis octet must be considered as a first priority. After that, the location of charge on atoms of appropriate electronegativity can be considered.

Valence-Shell Electron-Pair Repulsion Theory

Like charges repel each other, so the electron pairs surrounding a central atom in a molecule should repel each other and move as far apart as possible. We use valence-shell electron-pair repulsion (VSEPR) theory to predict the shapes of molecules. VSEPR theory allows us to predict whether the geometry around any given atom is tetrahedral, trigonal planar, or linear.

Using VSEPR theory requires that regions of electron density be considered regardless of how many electrons are contained in the region. Thus, a single-bonded pair or two pairs of electrons in a double bond are considered as "equal." The following rules cover most cases.

1. Two regions containing electrons around a central atom are 180° apart, producing a linear arrangement.
2. Three regions containing electrons around a central atom are 120° apart, producing a trigonal planar arrangement.
3. Four regions containing electrons around a central atom are 109.5° apart, producing a tetrahedral arrangement.

The electron pairs around a central atom may be bonding electrons or nonbonding electrons, and both kinds of valence-shell electron pairs must be considered in determining the shape of a molecule. When all of the electron pairs are arranged to minimize repulsion, we look at the molecule to see how the atoms are arranged in relation to each other. The geometric arrangement of the atoms determines the bond angles.

Consider the structure of an isocyanate group in methylisocyanate.

$$\text{H—}\overset{\overset{\displaystyle H}{|}}{\underset{\underset{\displaystyle H}{|}}{\text{C}}}\text{—}\ddot{\text{N}}\text{=C=}\ddot{\text{O}}\text{:}$$

methyl isocyanate

The nitrogen atom has three regions containing electrons around it. They are a single bond, a double bond, and a nonbonded pair of electrons. So, these features will have a trigonal planar arrangement, and the R—N=C bond angle is 120°. The isocyanate carbon atom has two groups of electrons around it—two double bonds—so they will have a linear arrangement. The N=C=O bond angle is 180°.

Dipole Moments

The polarity of a molecule is given by its dipole moment. The dipole moment depends upon both the polarity of individual bonds and the arrangement of those bonds in the molecule. In some molecules, the dipole moments are pointed in opposite directions so that they cancel one another. As a result, there is no net resultant dipole moment. In other molecules, the dipole moments may reinforce each other or partially cancel, causing a net dipole moment.

Atomic and Molecular Orbitals

Atomic orbitals are mathematical equations that describe the discrete, quantized energy levels of atoms. They are described as 1s, 2s, 2p, and so forth. Each atomic orbital can contain a maximum of two electrons with opposite spins. The square of the equation for an atomic orbital gives the probability of finding an electron within a given region of space.

The concepts developed for atomic orbitals can be extended to molecular orbitals that extend across a molecule. Molecular orbitals are linear combinations of atomic orbitals, which represent the distribution of electrons over two or more atoms. The important concepts are summarized below.

1. The number of molecular orbitals must equal the number of atomic orbitals used to generate them.
2. Molecular orbitals, as Well as atomic orbitals, are represented by wave functions whose value may be positive or negative and is a function of geometry.
3. There are two types of bonding molecular orbitals to hydrogen and to second row elements, called sigma (σ) and pi (π). Hydrogen forms only one σ bond.
4. Molecular orbitals can be bonding or antibonding.

3

The Hydrogen Molecule

The 1s orbitals of two hydrogen atoms can combine in two ways to give molecular orbitals. One of these is a bonding σ orbital; the other is an antibonding, σ^* orbital. Bonding molecular orbitals have lower energy (are more stable) than the original atomic orbitals. Antibonding molecular orbitals have higher energy (are less stable) than the original atomic orbitals. The bonding σ orbital holds two electrons, and the antibonding σ^* orbital is empty.

Bonding in Carbon Compounds

The strongest bonds between carbon atoms and other atoms are σ bonds that result from overlap of atomic orbitals along the internuclear axis. Side-by-side overlap of p orbitals leads to a less stable π bond.

Atomic orbitals are combined (mixed) to give hybridized atomic orbitals. These orbitals account for the geometry and properties of molecules, and they follow the rules for VSEPR theory.

sp³ Hybridization of Carbon in Methane

Bonding in methane can be regarded as the formation of covalent bonds between an sp³-hybridized carbon atom and 1s orbital of hydrogen atoms. An sp³-hybrid orbital is constructed from mixing the 2s orbital of an excited state carbon atom, which contains one electron, with three 2p orbitals, each of which also contains one electron. The resulting sp³-hybrid orbitals point at the corners of a tetrahedron. Each of them forms a σ bond with the 1s orbital of a hydrogen atom.

The term % s character is used to describe the contribution of the atomic orbitals to a hybridized orbital. Thus an sp³-hybrid orbital has 25% s character.

sp³ Hybridization of Carbon in Ethane

Ethane and other organic compounds containing four single bonds to carbon atoms consist of sigma bonds to sp³-hybridized carbon atoms arranged at tetrahedral angles to one another. In ethane, two sp³ hybrid orbitals overlap to give a σ bond. The other three sp³ hybrid orbitals on each carbon make σ bonds to hydrogen atoms.

Groups of atoms can rotate about a sigma bond without breaking the bond. The resulting conformations are different temporary arrangements of atoms that still maintain their bonding arrangement.

sp² Hybridization of Carbon in Ethene

The sp² hybrid orbitals of carbon occur in compounds such as ethene that contain a double bond. The overlap of these orbitals with one another or with other orbitals such as an s orbital of hydrogen gives a sigma (σ) bond. The three sp² hybrid orbitals are coplanar and lie 120° to one another. They have 33% s character because they are formed from one 2s orbital and two 2p orbitals. An sp² hybridized carbon also has a 2p orbital that can form a π bond with a neighboring carbon atom in ethene or to a carbon atom in methanal. The σ bond in ethene and other alkenes is stronger than the π bond because there is less orbital overlap in the π bond.

sp Hybridization of Carbon in Ethyne

The sp hybrid orbitals of carbon occur in compounds such as ethyne that contain a triple bond. The overlap of these orbitals with one another or with other orbitals such as an s orbital of hydrogen gives a sigma bond. The sp hybrid orbitals are at 180° to one another. They have 50% s character because they are formed from one 2s orbital and one 2p orbital. Each time there are two sp hybrid orbitals about a carbon atom, there are also two remaining p orbitals that form two π bonds with a neighboring atom, as in the case of another carbon atom in ethyne or a nitrogen atom in cyano compounds.

Effect of Hybridization on Bond Length and Bond Strength

With increasing % s character, the electrons within a hybrid orbital are held closer to the nucleus of the atom. As a consequence, the bond lengths decrease as the % s character increases. And, the strength of the bond increases as % s character increases.

1. C—H bond strengths: ethane (sp³) < ethene (sp²) < ethyne (sp).
2. C—H bonds lengths: ethyne < ethene < ethane.

Hybridization of Nitrogen

Hybridization is not a phenomenon restricted to carbon. It applies to other atoms as well. The only difference is in the number of electrons that are distributed in the orbitals. Nitrogen, a Group VA element, has five valence electrons.

An sp^3-hybridized nitrogen has three half-filled orbitals that can form σ bonds and one filled sp^3 orbital that is a nonbonding electron pair. The orbital containing the nonbonding electron pair and the three half-filled orbitals the bonding are directed to the corners of a tetrahedron. However, the geometry of such molecules is pyramidal, like ammonia, because the position of the atoms, not the electron pairs, defines the molecular geometry.

An sp^2-hybridized nitrogen atom can form three σ bonds and one π bond. The geometry of sp^2 hybridized nitrogen is trigonal planar, and the bond angles around the nitrogen are 120°.

An sp-hybridized nitrogen atom can form two σ bonds with sp orbitals and two π bonds with its half-filled 2p orbitals.

Hybridization of Oxygen

The difference between the hybridization of oxygen compared to nitrogen and carbon is in the number of electrons that are distributed in the orbitals. Oxygen, a Group VIA element, has six valence electrons.

An sp^3-hybridized oxygen atom has two electrons in each of two sp^3 orbitals and one electron in each of the remaining two sp^3 orbitals. The bonded and nonbonded electron pairs are directed to the corners of a tetrahedron. However, the shape of molecules like water is angular.

An sp^2-hybridized oxygen atom has two electrons in two filled sp^2 orbitals and one half-filled sp^2-orbital. The sixth electron is in a 2p orbital, which can form a π bond. Note that the bond angle for σ bonds to sp^2-hybridized orbitals is 120°.

SOLUTIONS TO END-OF-CHAPTER EXERCISES

Atomic Properties

1.1 How many valence shell electrons are in each of the following elements?
(a) N   (b) F   (c) C   (d) O
(e) Cl   (f) Br   (g) S   (h) P

Answers: (a) 5   (b) 7   (c) 4   (d) 6   (e) 7   (f) 7   (g) 6   (h) 5

1.2 Which of the following atoms has the higher electronegativity? Which has the larger atomic radius?
(a) Cl or Br   (b) O or S   (c) C or N   (d) N or O   (e) C or O

Answers: electronegativity: (a) Cl > Br   (b) O > S   (c) N > C   (d) O > N   (e) O > C
Answers: atomic radius: (a) Br > Cl   (b) S > O   (c) C > N   (d) N > O   (e) C > O

Ions and Ionic Compounds

1.3 Write a Lewis structure for each of the following ions.
(a) OH^- (b) CN^- (c) H_3O^+ (d) NO_3^-

Answers:

1.4 Write a Lewis structure for each of the following ions.
(a) NO_2^- (b) SO_3^- (c) NH_2^- (d) CO_3^-

Answers:

Lewis Structures of Covalent Compounds

1.5 Write a Lewis structure for each of the following compounds.
(a) NH_2OH (b) CH_3CH_3 (c) CH_3OH (d) CH_3NH_2 (e) CH_3Cl (f) CH_3SH

Answers:

6

1.6 Write a Lewis structure for each of the following compounds.
    (a) HCN   (b) HNNH   (c) CH_2NH (d) CH_3NO (e) CH_2NOH (f) CH_2NNH_2

Answers:

(a)   H—C≡N: (b)   H—N̈=N̈—H   (c)   H—C(—H)=N̈—H

(d)   H—C(—H)(—H)—N̈=Ö: (e) H—C(—H)=N—Ö—H   (f) H—C(—H)=N̈—N(—H)—H

1.7 Add any required unshared pairs of electrons that are missing from the following formulas.

(a) CH_3—C(=O)—OH (b) CH_3—C(=O)—OCH_3 (c) H—C(=O)—$NHCH_3$

(d) CH_3—S—CH=CH_2 (e) CH_3—C(=N—H)—CH_3 (f) N≡C—CH_2—C≡N

Answers:

(a) CH_3—C(=Ö:)—ÖH   (b) CH_3—C(=Ö:)—ÖCH_3 (c) H—C(=:Ö)—N̈HCH_3

(d) CH_3—S̈—CH=CH_2 (e) CH_3—C(=:N—H)—CH_3 (f) :N≡C—CH_2—C≡N:

1.8 Add any required unshared pairs of electrons that are missing from the following formulas.

(a) CH_3—C(=O)—Cl (b) CH_3—O—CH=CH_2 (c) CH_3—C(=O)—SH

(d) CH_3—CH(—O—CH_3)—O—CH_3 (e) NH_2—C(=O)—O—CH_3 (f) CH_3—O—CH_2—O—CH_3

Answers:

(a) CH_3—C(=:O:)—C̈l: (b) CH_3—Ö—CH=CH_2 (c) CH_3—C(=:O:)—S̈H

(d) CH_3—CH(—Ö—CH_3)—Ö—CH_3 (e) NH_2—C(=:O:)—Ö—CH_3 (f) CH_3—Ö—CH_2—Ö—CH_3

1.9 Using the number of valence electrons in the constituent atoms and the given arrangement of atoms in the compound, write the Lewis structure for each of the following molecules.

(a) CH_2\—N(—CH_2—H)   (b) Cl—C(=O)—Cl

(c) H—N(—H)—C(=O)—N(—H)—H   (d) H—C(—H)(—H)—S(=O)—O—H

Answers:

(a) $CH_2=\overset{..}{N}-CH_3$ (b) $:\overset{..}{\underset{..}{Cl}}-\overset{\overset{\overset{:O:}{\|}}{}}{C}-\overset{..}{\underset{..}{Cl}}:$ (c) $\overset{.}{N}H_2-\overset{\overset{\overset{:O:}{\|}}{}}{C}-\overset{.}{N}H_2$ (d) $CH_3-\overset{\overset{\overset{:S:}{\|}}{}}{C}-\overset{..}{\underset{..}{O}}-H$

1.10 Using the number of valence electrons in the constituent atoms and the given arrangement of atoms in the compound, write the Lewis structure for each of the following molecules.

(a)
```
      H          H
      |          |
 H——C——S——S——C——H
      |          |
      H          H
```

(b)
```
      H    O
      |    ‖
 H——C——C——S——H
      |
      H
```

(c)
```
      H       H
      |       |
 H——C——O——C——Cl
      |       |
      H       H
```

(d)
```
      H       O
      |       ‖
 H——C——O——C——N——H
      |          |
      H          H
```

Answers:

(a) $CH_3-\overset{..}{\underset{..}{S}}-\overset{..}{\underset{..}{S}}-CH_3$ (b) $CH_3-\overset{\overset{\overset{:\overset{..}{O}}{\|}}{}}{C}-\overset{..}{\underset{..}{S}}-H$

(c) $CH_3-\overset{..}{\underset{..}{O}}-\overset{\overset{\overset{H}{|}}{}}{\underset{\underset{H}{|}}{C}}-\overset{..}{\underset{..}{Cl}}:$ (d) $CH_3-\overset{..}{\underset{..}{O}}-\overset{\overset{\overset{:O:}{\|}}{}}{C}-\overset{\overset{\overset{}{}}{}}{\underset{\underset{H}{|}}{N}}-H$

1.11 Two compounds used as dry cleaning agents have the molecular formulas C_2Cl_4 and C_2HCl_3. Write the Lewis structures for each compound.

Answers:

(a)
```
 :Cl         :Cl:
    \       /
     C ═══ C
    /       \
 :Cl:        :Cl:
```

(b)
```
 :Cl         :Cl:
    \       /
     C ═══ C
    /       \
 :Cl:        H
```

1.12 Acrylonitrile, a compound used to produce fibers for rugs, is represented by the formula CH_2CHCN. Write the Lewis structure for the compound.

Answer:
```
 H         H
  \       /
   C ═══ C
  /       \
 H         C≡N:
```

Formal Charge

1.13   Assign the formal charges for the atoms other than carbon and hydrogen in each of the following species.

(a) $H\!-\!\ddot{O}\!-\!C\!\equiv\!N\!:$

(b) $H\!-\!\ddot{O}\!-\!N\!\equiv\!C\!:$

(c) $CH_3\!-\!\overset{\displaystyle CH_3}{\underset{\displaystyle CH_3}{N}}\!-\!\ddot{O}\!:$

(d) $CH_3\!-\!\ddot{N}\!=\!N\!=\!\ddot{N}\!:$

Answers:

(a) $H\!-\!\ddot{O}\!-\!C\!\equiv\!N\!:$

(b) $H\!-\!\ddot{O}\!-\!\overset{+}{\ddot{N}}\!\equiv\!\overset{-}{C}\!:$

(c) $CH_3\!-\!\overset{\displaystyle CH_3}{\underset{\displaystyle CH_3}{\overset{+}{N}}}\!-\!\overset{-}{\ddot{O}}\!:$

(d) $CH_3\!-\!\ddot{N}\!=\!\overset{+}{N}\!=\!\overset{-}{\ddot{N}}\!:$

(a) none of the atoms has a formal charge
(b) nitrogen is +1; carbon is −1
(c) nitrogen is +1; oxygen is −1
(d) nitrogen atoms from left to right have 0, +1, and −1 formal charges

1.14   Assign the formal charges for the atoms other than carbon and hydrogen in each of the following species.

Answers:
(a) oxygen is +1; boron is −1
(b) nitrogen is +1; aluminum is −1
(c) nitrogen is +1; singly bonded oxygen atom is −1
(d) phosphorus is +1; oxygen atom on the right is −1

(a) $CH_3\!-\!\overset{\displaystyle}{\underset{\displaystyle CH_3}{\ddot{O}}}\!-\!BF_3$

(b) $CH_3\!-\!\overset{\displaystyle CH_3}{\underset{\displaystyle CH_3}{N}}\!-\!AlCl_3$

(c) $CH_3\!-\!N\overset{\displaystyle \ddot{O}:}{\underset{\displaystyle \ddot{O}:}{}}$

(d) $CH_3\!-\!\ddot{O}\!-\!\overset{\displaystyle :\ddot{O}CH_3}{\underset{\displaystyle :\ddot{O}CH_3}{P}}\!-\!\ddot{O}:$

1.15   All of the following species are isoelectronic, that is, they have the same number of electrons bonding the same number of atoms. Determine which atoms have a formal charge. Calculate the net charge for each species.

Answers:
(a) carbon is −1; oxygen is +1; total charge is 0
(b) nitrogen is zero; oxygen is +1; total charge is +1
(c) carbon is −1; nitrogen is 0; total charge is −1
(d) both carbon atoms are −1; total charge is −2

(a) $:C\!\equiv\!O\!:$ (b) $:N\!\equiv\!O\!:$ (c) $:C\!\equiv\!N\!:$ (d) $:C\!\equiv\!C\!:$

1.16   All of the following species are isoelectronic, that is, they have the same number of electrons bonding the same number of atoms. Determine which atoms have a formal charge. Calculate the net charge for each species.

(a) $:\ddot{N}\!=\!N\!=\!\ddot{N}\!:$ (b) $:\ddot{O}\!=\!N\!=\!\ddot{O}\!:$

Answers:
(a) central nitrogen atom is +1; the other nitrogen atoms are each −1; the total charge is −1
(b) nitrogen atom is +1; both oxygen atoms are 0; the total charge is +1

1.17   The following species are isoelectronic. Determine which atoms have a formal charge. Calculate the net charge for each species.

        Answers:
        (a) S is +1   (b) N is +1
        (c) O is +1   (d) S is −1
        (e) Br is +1   (f) C is −1

1.18   The following species are isoelectronic. Determine which atoms have a formal charge. Calculate the net charge for each species.

        Answers:
        (a) O is −1   (b) Br is 0
        (c) C is +1   (d) S is 0
        (e) N is 0   (f) N is −1

1.19   Acetylcholine, a compound involved in the transfer of nerve impulses, has the following structure. What is the formal charge on the nitrogen atom? What is the net charge of acetylcholine?

Answer:
The nitrogen atom has a charge of +1; the total charge is +1.

$$CH_3-N(CH_3)(CH_3)-CH_2CH_2-O-C(=O)-CH_3$$

acetylcholine

1.20   Sarin, a nerve gas, has the following structure. What is the formal charge of the phosphorus atom?

Answer:
The phosphorus atom has a charge of +1.

$$CH_3-C(CH_3)(CH_3)-O-P(:O:)(:F:)-CH_3$$

Sarin

Resonance

1.21   The small amounts of cyanide ion contained in the seeds of some fruits are eliminated from the body as SCN⁻. Draw two possible resonance forms for the ion. Which atom has the formal negative charge in each form?

$$^-:\ddot{S}-C\equiv N: \longleftrightarrow :S=C=\ddot{N}:^-$$

Answer:
Sulfur has a −1 charge on the left; nitrogen has a −1 charge on the right.

1.22   Are the following pairs contributing resonance forms of a single species? Formal charges are not shown and have to be added.

        (a) :N̈=N=N̈:   and   :N̈—N≡N: (b) H—C≡N—Ö: and   H—C̈=N=Ö:

Answers: (a) $^-:\ddot{N}=\overset{+}{N}=\ddot{N}:^- \longleftrightarrow :N\equiv \overset{+}{N}-\ddot{N}:^{2-}$

        (b) H—C≡N̈⁺—Ö:⁻ ⟷ H—C̈=N⁺=Ö:

1.23   Write the resonance structure that results when electrons are moved in the direction indicated by the curved arrows for the following amide. Calculate any formal charges that result.

Answer:

1.24   Write the resonance structure that results when electrons are moved in the direction indicated by the curved arrows for acetate. Calculate any formal charges that result.

Answer:

1.25   Write the resonance structure that results when electrons are moved in the direction indicated by the curved arrow for the following electron-deficient ion. To what extent do each of the two resonance forms contribute to the structure of the ion?

Answer:
The alternate resonance form is structurally equivalent to the given resonance form and both contribute equally.

1.26   Write the resonance structure that results when electrons are moved in the direction indicated by the curved arrows for the diazomethane. Do each of the two resonance forms contribute equally to the structure of the ion?

diazomethane

Answer:

Answer:
The alternate resonance form has a negative charge on the carbon atom rather than the nitrogen atom. Because nitrogen is more electronegative than carbon, the original resonance form contributes to a larger extent.

Molecular Shapes

1.27 Based on VSEPR theory, what is the expected value of the indicated bond angle in each of the following compounds?

(a) C—C—N in CH_3—C≡N

(b) C—O—C in CH_3—O—CH_3

(c) C—N—C in CH_3—NH—CH_3

(d) C—C—C in CH_3—C≡C—H

Answers: (a) 180° (b) 109° (c) 109° (d) 180° (e) 180°

1.28 Based on VSEPR theory, what is the expected value of the indicated bond angle in each of the following ions?

(a) C—O—H in CH_3—OH_2^+

(b) C—N—H in CH_3—NH_3^+

(c) O—C—O in $CH_3CO_2^-$

(d) C—O—C in $(CH_3)_2OH^+$

Answers: (a) 109° (b) 109° (c) 109° (d) 109°

1.29 Based on VSEPR theory, what is the expected value of the indicated bond angle in each of the following compounds?

Answers:
(a) 120°
(b) 120°
(c) 120°

(a) C—O—C in CH_3—C(=O)—OCH_3

(b) O—C—N in H—C(=O)—NH_2

(c) O—C—O in CH_3—C(=O)—OCH_3

1.30 Based on VSEPR theory, what is the expected value of the indicated bond angle in each of the following compounds?

Answers:
(a) 109°
(b) 120°
(c) 120°

(a) C—O—C in CH_3—C(=O)—OCH_3

(b) O—C—N in H—C(=O)—NH_2

(c) O—C—O in CH_3—C(=O)—OCH_3

Dipole Moments

1.31 Fluorine is more electronegative than chlorine, but the dipole moment for a C—F bond (1.4 D) is less than the dipole moment for a C—Cl bond (1.5 D). Explain why this is so.

Answer: The carbon—fluorine bond is much shorter than the carbon—chlorine bond.

1.32 Arrange the following bond moments in order of decreasing polarity: H—N, H—O, H—S. Explain the trend that you predict.

Answer:
H — S > H — N > H — O; the difference in electronegativity of the atoms in the O—H bond is larger than that of the atoms in the N—H bond. There is a substantially smaller difference in electronegativity of the atoms in the S—H bond and the bond has a small polarity.

1.33 The dipole moments of both CO, and CS, are zero. However, SCO has a dipole moment. Explain why. Draw the structure of SCO and then an arrow indicating the direction of the dipole moment.

Answer:
The C=O and C=S bond moments are not equal, so they don't cancel each other. The net dipole moment is toward the more electronegative oxygen atom.

1.34 Which compound has the larger dipole moment, acetone or phosgene? Explain why.

acetone

phosgene

Answer:

Acetone has the larger dipole moment because the bond moments of the two C—Cl bonds oppose the C=O dipole moment in phosgene.

1.35   Which compound has the larger dipole moment, *cis*- or *trans*-1,2-dibromoethene? Explain why.

cis-1,2-dibromoethene   trans-1,2-dibromoethene

Answer:

The net resultant of the dipole moments of the carbon—bromine bonds in the *cis* isomer is toward the side of the molecule containing the two carbon—bromine bonds. The bond moments of the carbon—bromine bonds in the *trans* isomer are opposed and therefore cancel one another.

1.36   The dipole moment of chlorobenzene (C_6H_5Cl) is 1.56 D and that of nitrobenzene ($C_6H_5NO_2$) is 3.97 D. The dipole moment of *para*-chloronitrobenzene is 2.57 D. What does this value indicate about the direction of the moments of the two groups with respect to the benzene ring?

chlorobenzene   nitrobenzene   *p*-chloronitrobenzene

Answer:

The two dipole moments must oppose one another to give a resultant that is less than the large dipole bond moment value. The dipole moment of the carbon—chlorine bond is toward the chlorine atom. Thus, the dipole moment of the carbon—nitrogen bond must be toward nitrogen.

Hybridization

1.37   What is the hybridization of each carbon atom in each of the following compounds?

Answers:

(a) from left to right: sp^3, sp^2

(b) from left to right: sp^3, sp^2, sp^2

(c) from left to right: sp^3, sp^2

(d) from left to right: sp^3, sp^2, sp^3

(a) CH_3—C—H (b) CH_3O—CH=CH_2 (c) CH_3—C—OH

(d) CH_3—C—OCH_3

1.38   What is the hybridization of each carbon atom in each of the following compounds?

Answers:

(a) from left to right: sp^2, sp^3

(b) from left to right: sp^3, sp^2, sp^2

(c) from left to right: sp^3, sp^2, sp^3

(d) from left to right: sp, sp^3, sp

(a)   H—C—NH—CH_3 (b) CH_3NH—CH=CH_2

(c) CH_3—C—CH_3 (d) N≡C—CH_2—C≡N

1.39   What is the hybridization of the oxygen atom in each compound in Exercise 1.37?

Answers:

(a) sp^2

(b) sp^3

(c) double-bonded oxygen is sp^2; single-bonded is sp^3

(d) double-bonded oxygen is sp^2; single-bonded is sp^3

1.40   What is the hybridization of the oxygen atom in each compound in Exercise 1.38?

Answers:

(a) sp^3

(b) sp^3

(c) sp^2

(d) sp

1.41   Carbocations and carbanions are unstable organic species with a positive and a negative charge, respectively, on the carbon atom. What is the hybridization of the carbon atom in each ion? What are the H—C—H bond angles?

Answer:
The carbocation is sp^2 hybridized, and the bond angles are 120°. The carbanion is sp^3 hybridized, and the bond angles are 109°.

1.42   Assuming that all of the valence electrons are paired and located in hybrid orbitals, what is the H—C—H bond angle in the reactive species CH_2?

Answer:
The three pairs of electrons, one nonbonded and two bonded, are in a common plane at 120° to one another.

1.43   Write the Lewis structure of CO_2. What is the hybridization of the carbon atom? What is the hybridization of the oxygen atoms?

$$:\ddot{O}=C=\ddot{O}:$$

Answer:
The carbon atom is sp hybridized. The oxygen atoms are sp^2 hybridized.

1.44   Write the Lewis structure of NO_2^+, the nitronium ion. What is the hybridization of the nitrogen atom? What is the hybridization of the oxygen atoms?

$$:\ddot{O}=\overset{+}{N}=\ddot{O}:$$

Answer: The nitrogen atom is sp hybridized. The oxygen atoms are sp^2 hybridized.

1.45   Phosgene ($COCl_2$) is a poisonous gas. Write its Lewis structure and determine the hybridization of the carbon atom.

Answer:
The carbon atom is sp^2 hybridized.

1.46   Carbamic acid is an unstable substance that decomposes to form carbon dioxide and ammonia. Based on the following Lewis structure, what are the hybridizations of the carbon atom and the two oxygen atoms?

Answer:
The carbon atom is sp^2 hybridized. The double-bonded oxygen atom is sp^2; single-bonded oxygen atom is sp^3 hybridized.

Bond Lengths

1.47   The oxygen—hydrogen bond length in both hydrogen peroxide (HO—OH) and hydroxylamine (NH_2—OH) are the same; 96 pm. Explain why.

Answer:
Bond lengths between common sets of atoms tend to be the same and do not depend markedly on the other atoms of the structure.

1.48   The C=N bond length of methyleneimine (CH_2=NH) is 127 pm. Compare this value to the C=C bond length of ethene (133 pm) and suggest a reason for the difference.

Answer:
The C=N bond is shorter than the C=C bond because the atomic radius of nitrogen is smaller than the atomic radius of carbon.

1.49   The nitrogen—oxygen bond lengths of hydroxylamine (NH_2—OH)) and the nitronium ion ($NO_2)^+$ are 145 and 115 pm, respectively. Write their Lewis structures and explain why the bond lengths differ.

1.50 The C—F bond length of CF_4 is 138 pm. The estimated bond length of CF_3^+ is 127 pm. Suggest a reason for the difference between these two values.

Answer:
Each of the three contributing resonance forms in CF_3^+ has a C=F bond which contributes to the overall shortening of the carbon–fluorine bonds in the resonance hybrid structure.

1.51 The carbon–carbon single bond lengths of propane and propene are 154 and 151 pm, respectively. Why do these values differ?

$$CH_3\text{——}CH_2\text{——}CH_3 \qquad CH_3\text{——}CH\text{===}CH_2$$
$$\text{propane} \qquad\qquad \text{propene}$$

Answer:
The bonds are sp^3–sp^3 and sp^3–sp^2 hybridized, respectively. An sp^2-hybridized atom holds the bonding pair of electrons closer to the nucleus, and this leads to a shortening of the bond.

1.52 The carbon—oxygen bond length of dimethyl ether is 142 pm. Predict the lengths of each of the two carbon—oxygen bonds in methyl vinyl ether.

$$CH_3\text{—}O\text{—}CH_3 \qquad CH_3\text{—}O\text{—}CH_2\text{===}CH_2$$
$$\text{dimethyl ether} \qquad\qquad \text{methyl vinyl ether}$$

Answer:
The bond to the CH_3 group should also be 142 pm. The bond of oxygen to the CH group should be shorter than 142 pm because an sp^2-hybridized atom holds the bonding pair of electrons closer to the nucleus, and this leads to a shortening of the bond.

Bond Angles

1.53 What is the C—N—H bond angle in each of the following species?

Answer: (a) 109° (b) 120°

1.54 What is the C—O—H bond angle of protonated methanal?

Answer: 120°

1.55 Diimide (HNNH) is a reactive reducing agent. Draw its Lewis structure. Compare its Lewis structure with that of ethene. Compare the hybridization of the two compounds. What is the H—N—N bond angle in diimide?

Answer:
The hybridization of both the carbon atoms in ethene and the nitrogen atoms in diimide is sp^2. The H—N—N bond angle is 120°.

1.56   What is the H—C—H bond angle in allene ($CH_2=C=CH_2$)? What is the C—C—C bond angle? What is the hybridization of each atom?

Answer:
The H—C—H bond angle is 120°. The C—C—C bond angle is 180°. The hybridization of both terminal atoms is sp^2; the hybridization of the central carbon atom is sp.

1.57   What is the Cl—C—Cl bond angle of the CCl_3^- ion, an intermediate formed by treating CCl_3H with base?

Answer: 109°

1.58   What is the O—N—O bond angle of the nitronium ion $(NO_2)^+$, a reactive intermediate in reactions with benzene compounds?

Answer: 180°

_____

2 PART I: FUNCTIONAL GROUPS AND THEIR PROPERTIES

KEYS TO PART I OF THE CHAPTER

2.1 Functional Groups

Functional groups are structural features of organic compounds other than carbon–carbon single bonds and carbon–hydrogen single bonds. Multiple bonds between carbon atoms and bonds from carbon to atoms such as oxygen, nitrogen, sulfur, and the halogens are components of functional groups. When learning the features of the functional groups, we must pay attention both to their composition and to their structure and bonding.

As we proceed with our study of organic chemistry, we will find that functional groups behave chemically in ways that we can predict based on the number and type of bonds to carbon in each functional group. The chemistry of organic molecules depends on the functional groups that they contain. The only functional groups that do not contain atoms other than carbon and hydrogen contain carbon–carbon multiple bonds, as in ethene (ethylene) and ethyne (acetylene). Benzene, which also contains multiple carbon–carbon bonds, belongs to a separate class of compounds called aromatic hydrocarbons.

2.2 Functional Groups Containing Oxygen

Several types of functional groups contain oxygen. Compounds with a carbon–oxygen and an oxygen–hydrogen bond are **alcohols**. Compounds with two carbon–oxygen bonds are **ethers**.
The oxygen atom forms double bonds to carbon in a **carbonyl group** in several functional groups. If the remaining two single bonds are to other carbon atoms, the compound is a **ketone**. If there is one single bond to a carbon atom and one to a hydrogen atom, the compound is an **aldehyde**.
Compounds with a single bond from an oxygen atom to a carbonyl group are found in **carboxylic acids** and **esters**. In carboxylic acids, the second bond to that oxygen atom is to a hydrogen atom; in esters it is to another carbon atom. Note that a carboxylic acid is not an aldehyde, a ketone, or an alcohol. Both the carbonyl group and the hydroxyl group *together* are considered as a single functional group when they share a common carbon atom.

2.3 Functional Groups Containing Nitrogen

Nitrogen can form functional groups that contain single bonds in amines, double bonds in imines, and triple bonds in nitriles. A nitrogen atom bonded to a carbonyl group is an amide. The amide nitrogen atom may be bonded to any combination of hydrogen atoms or carbon atoms.

2.4 Functional Groups Containing Sulfur or Halogens

Sulfur occurs in functional groups that parallel those of alcohols and ethers. These sulfur-containing compounds are **thiols** and **thioethers**. Halogens can be bonded to sp^3-hybridized carbon atoms or to the sp^2-hybridized carbon atom of a carbonyl group.

2.5 Structural Formulas

Molecular formulas identify the total number of atoms of each element in a molecule. They tell us nothing about the structure of the molecule. **Structural formulas** show how the atoms in the molecule are arranged and which atoms are bonded to each other. A complete structural formula shows every bond. A condensed formula abbreviates the structure by omitting some or all of the bonds and indicating the number of atoms bonded to each carbon atom with subscripts.

Several conventions are used to represent structures in varying degrees of detail and in shorthand form. In general, make sure that each atom has the appropriate number of bonds. **Condensed structural formulas** leave out some bonds, and the bonded atoms are written close to each other. In general, atoms bonded to a carbon atom are usually written right after the carbon atom.

Organic Chemistry Study Guide: Key Concepts, Problems, and Solutions http://dx.doi.org/10.1016/B978-0-12-801889-7.00002-9

2.6 Bond-Line Structures

This section introduces a "shorthand" skill that helps us show the details of chemical structure. Remember that there is a carbon atom at every intersection of two or more lines and at the end of every line. Also remember that there are four bonds to every carbon atom. The bonds from one carbon atom to other carbon atoms and to atoms of other elements are easy to identify; the bonds to hydrogen atoms are not visible in the bond-line structure, and we must carefully account for them. Bond-line structures are a better and faster way to record structural formulas than writing both atoms and bonds.

Remember that the chemistry occurs at the functional groups. Consider, for example, the structure of diphepanol, which is used as a cough suppressant. Can you identify the functional groups? Can you write its molecular formula?

diphepanol

The oxygen atom in diphepanol is part of a hydroxyl group, so the functional group is an alcohol. The nitrogen atom is bonded only to carbon atoms, so it is an amine. The molecular formula is $C_{20}H_{25}NO$.

Here's another example. What are the oxygen-containing functional groups in the herbicide with the commercial name 2,4-D? Its structure is shown below.

2,4-D

One of the oxygen atoms is present as a carbonyl group and a second as a hydroxyl group. They both are bonded to the same carbon atom, so this part of 2,4-D is a carboxylic acid. The third oxygen atom is bonded to two carbon atoms; it is part of an ether.

2.7 Isomers

The composition of a compound does not uniquely establish its structure. For all but the simplest molecules, a group of atoms can usually be bonded in several ways to give different structures called **constitutional isomers**. Distinguishing between structures that are isomers and those that are merely different representations of the same molecule requires practice.

There are many ways to write the structural formula of an organic compound. Two structural formulas with the same molecular formula may look so different that they appear at first glance to represent isomers. To determine if two structures represent isomers, carefully check the bonding sequence in each formula. If the sequence of bonded atoms is the same, the structural formulas represent two views of the same compound. If the sequence of bonded atoms is different, the two structural formulas represent isomers.

PART II: IDENTIFICATION OF FUNCTIONAL GROUPS BY INFRARED SPECTROSCOPY

KEYS TO PART II OF THE CHAPTER

2.8 Spectroscopy

The energy of light is directly proportional to its frequency; $E = h\nu$. Wavelength and frequency are inversely proportional and are related by $\lambda = c/\nu$, where c is the speed of light. As the wavelength of the electromagnetic radiation increases, the corresponding frequency decreases. Spectroscopy is used to probe the physical changes in a molecule as the result of absorption of energy. In infrared spectroscopy, the energy absorbed can change the extent to which a bond stretches or bends.

2.9 Infrared Spectroscopy

Infrared spectroscopy is extremely valuable because it allows us to confirm the presence (and sometimes more importantly the absence) of functional groups. The infrared spectrum is displayed so that absorptions of energy are related to wavelength or wavenumber. The energy of the absorption is indicated by an inverted "peak" pointed down from a baseline.

Infrared absorptions correspond to the stretching of a bond or the bending of a bond angle. The strength of the bond is given by a force constant. Multiple bonds have higher force constants, and their absorbances occur at higher energy. The energy required to stretch a bond is also related to the atomic mass of the bonded atoms. Bonds to hydrogen such as C—H, O—H, and N—H require higher energy than bonds such as C—C, O—C, and N—C.

The amount of energy required to stretch a specific bond in an organic molecule depends on the nature of the bonded atoms and the type of bond between them. The full interpretation of the IR spectrum of a molecule is difficult, but certain functional groups have characteristic absorptions which can be used to propose a structure for an unknown compound.

The spectrum of an unknown compound can be established by comparison to the spectrum of a known compound. If the spectrum of the unknown compound has *all* of the same absorption peaks as a compound of known structure, then the two samples are identical. If the "unknown" has one or more peaks that differ from the spectrum of a known, then the two compounds are not identical, or some impurity in the unknown sample is causing the extra absorptions. If the unknown lacks even one absorption peak that is present in the known structure, then the "unknown" has a different structure than the known one.

2.10 Identifying Hydrocarbons

The energy for the absorbance for a C—H bond depends on the % s character of the bond. With increased % s character, the electrons are more tightly held by an atom, so a bond to that atom requires higher energy to stretch. This difference is used to detect alkene and alkynes providing they have C—H bonds as well as C=C or C≡C bonds.

2.11 Identifying Oxygen-Containing Compounds

The presence of a carbon–oxygen double bond is easily detected by its characteristic strong absorption near the middle of an IR spectrum at a wavenumber of about $1700 \ \text{cm}^{-1}$. The exact location is controlled by the extent to which a dipolar resonance form contributes to the structure. If the dipolar resonance form is stabilized by atoms bonded to the carbonyl group, then the C—O bond has more single bond character, and the energy required to stretch the bond is smaller. Alcohols and ethers both contain C—O bonds that are difficult to confirm unambiguously in IR spectra. However, the presence of an O—H bond in an alcohol is easily detected by a strong absorption on the left of the spectrum in the energy range 3400–$3600 \ \text{cm}^{-1}$.

2.12 Identifying Nitrogen-Containing Compounds

The presence of an N—H bond in an amine is easily detected. Primary amines have two N—H absorbances that occur over a range from 3250 to 3550 cm^{-1}. Secondary amines have a single N—H absorbance that occurs over a range from 3250 to 3550 cm^{-1}. C—N bond stretching occurs in the 1000–1250 cm^{-1} region. C—N peaks are weak. In contrast, the C≡N absorbance, which occurs at around 2250 cm^{-1} is very strong.

2.13 Bending Deformations

The fingerprint region of the presence of the IR spectrum contains many kinds of bending modes. Some of these are readily identified. They provide clues about the substitution pattern on benzene rings.

SOLUTIONS TO END-OF-CHAPTER EXERCISES

Functional Groups

2.1   Identify the functional groups contained in each of the following structures.
        (a) caprolactam, a compound used to produce a type of nylon

Answer: amide

caprolactam

        (b) civetone, a compound in the scent gland of the civet cat

Answer: ketone and double bond

civetone

        (c) DEET, the active ingredient in some insect repellents

Answer: amide and benzene ring

DEET

2.2   Identify the oxygen-containing functional groups in each of the following compounds.
        (a) isopimpinellin, a carcinogen found in diseased celery

Answer: three ethers, ester, and benzene ring

isopimpinellin

        (b) aflatoxin B_1, a carcinogen found in moldy foods

Answer: three ethers, ketone, ester, two double bonds, and benzene ring

aflatoxin B_1

(c) penicillin G, an antibiotic first isolated from a mold.

Answer: two amides, thioether, carboxylic acid, and benzene ring

penicillin G

Molecular Formulas

2.3   Write the molecular formula for each of the following.
      (a) CH_3—CH_2—CH_2—CH_2—CH_3 (b) CH_3—CH_2—CH_2—CH_3 (c) CH_2=CH—CH_2—CH_3
      (d) CH_3—CH_2—$C{\equiv}C$—H (e) CH_3—CH_2—CH_2—CH=CH_2 (f) CH_3—CH_2—$C{\equiv}C$—CH_3

Answers: (a) C_5H_{12} (b) C_4H_{10} (c) C_4H_8
           (d) C_5H_8 (e) C_5H_{10} (f) C_5H_8

2.4   Write the molecular formula for each of the following.
      (a) $CH_3CH_2CH_2CH_2CH_2CH_2CH_2CH_2CH_3$ (b) $CH_3CH_2CH_2CH_2CH_2CH_2CH_2CH_3$ (c) $CH_3CH_2C{\equiv}CH$
      (d) $CH_3CH_2C{\equiv}CCH_3$ (e) $CH_3CH_2CH_2CH$=$CHCH_3$ (f) CH_2=$CHCH_2CH_3$

Answers: (a) C_9H_{20} (b) C_8H_{18} (c) C_4H_6 (d) C_5H_8 (e) C_6H_{12} (f) C_5H_8

2.5   Write the molecular formula for each of the following.
      (a) CH_3—CH_2—$CHCl_2$ (b) CH_3—CCl_2—CH_3 (c) Br—CH_2—CH_2—Br
      (d) CH_3—$CHBr$—$CHBr_2$ (e) CH_3—CF_2—CH_2F (f) F—CH_2—CHF—CH_2—F

Answers: (a) $C_3H_6Cl_2$ (b) $C_3H_6Cl_2$ (c) $C_2H_4Br_2$ (d) $C_3H_5Br_3$ (e) $C_3H_5F_3$ (f) $C_3H_5F_3$

2.6   Write the molecular formula for each of the following.
      (a) CH_3—CH_2—CH_2—OH (b) CH_3—CH_2—O—CH_2—CH_3 (c) CH_3—CH_2—SH
      (d) CH_3—CH_2—S—CH_3 (e) CH_3—CH_2—CH_2—NH_2 (f) CH_3—CH_2—NH—CH_3

Answers: (a) C_3H_8O (b) $C_4H_{10}O$ (c) C_2H_6S (d) C_3H_8S (e) C_3H_9N (f) C_3H_9N

Structural Formulas

2.7   For each of the following, write a condensed structural formula in which only the bonds to hydrogen are not shown.

Answers: (a) Br—CH_2—CH_2—Br
        (b) CH_3—CH_2—CH_2—CH_2—CH_3
        (c) CH_3—CH_2—CH_2—SH
        (d) CH_3—CH_2—CH_2—CH_2—NH_2

2.8   For each of the following, write a condensed structural formula in which only the bonds to hydrogen are not shown.

Answers:
(a) $CH_3-CH_2-CH_2-NH-CH_3$

(b) $CH_3-CH_2-CH_2-O-CH_3$

(c) $CH_3-CH_2-CH_2-CH_2-CCl_3$

(d) $CH_3-CH_2-NH-CH_2-CH_3$

2.9   Write a condensed structural formula in which no bonds are shown for each of the structures in problem 2.7.

Answers: (a) $BrCH_2CH_2Br$
(b) $CH_3CH_2CH_2CH_2CH_3$
(c) $CH_3CH_2CH_2SH$
(d) $CH_3CH_2CH_2CH_2NH_2$

2.10   Write a condensed structural formula in which no bonds are shown for each of the structures in problem 2.8.

Answers: (a) $CH_3CH_2CH_2NHCH_3$
(b) $CH_3CH_2CH_2OCH_3$
(c) $CH_3CH_2CH_2CH_2CCl_3$
(d) $CH_3CH_2NHCH_2CH_3$

2.11   Write a complete structural formula, showing all bonds, for each of the following condensed formulas.
(a) $CH_3CH_2CH_2CH_3$ (b) $CH_3CH_2CH_2Cl$ (c) $CH_3CHClCH_2CH_3$
(d) $CH_3CH_2CHBrCH_3$ (e) $CH_3CH_2CHBr_2$ (f) $CH_3CBr_2CH_2CH_2CH_3$

Answers:

2.12   Write a complete structural formula, showing all bonds, for each of the following condensed formulas.
        (a) CH₃CH₂CH₃   (b) CH₃CH₂CHCl₂   (c) CH₃CH₂CH₂CH₂SH

        (d) CH₃CH₂C≡CCH₃   (e) CH₃CH₂OCH₂CH₃   (f) CH₃CH₂CH₂C≡CH

Answers:

(a) H—C—C—C—H   (with H atoms above and below each carbon)

(b) H—C—C—C—Cl   (with H, H, Cl above and H, H, H below)

(c) H—C—C—C—C—S—H

(d) H—C—C—C≡C—C—H

(e) H—C—C—O—C—C—C—H

(f) H—C—C—C—C≡C—H

Bond-Line Structures

2.13   What is the molecular formula for each of the following bond-line representations?

Answers:
(a) C₉H₁₆O   (b) C₇H₁₆O
(c) C₈H₁₇Br   (d) C₁₁H₁₈

(a)

(b)

(c)

(d)

2.14   What is the molecular formula for each of the following bond-line structures?

Answers:
(a) C₁₀H₁₂O   (b) C₉H₁₁N
(c) C₁₀H₁₆   (d) C₄H₆O₂

(a)

(b)

(c)

(d)

2.15   What is the molecular formula for each of the following bond-line structures?
        (a) a scent marker of the red fox

Answer:
(a) $C_6H_{12}S$

        (b) a compound responsible for the odor of the iris

Answer:
(b) $C_{12}H_{20}O$

        (c) a defense pheromone of some ants

Answer:
(c) $C_{15}H_{22}O$

2.16   What is the molecular formula for each of the following bond-line structures?
        (a) a compound found in clover and grasses

Answer:
(a) $C_9H_6O_2$

        (b) an oil found in citrus fruits

Answer:
(b) $C_{10}H_{18}O$

        (c) a male sex hormone

Answer:
(c) $C_{19}H_{28}O_2$

Isomerism

2.17   Indicate whether the following pairs of structures are isomers or different representations of the same compound.

Answers:
(a) different representations for the same structure
(b) different representations for the same structure
(c) isomers

(a) $Br-\overset{\overset{\displaystyle H}{|}}{\underset{\underset{\displaystyle H}{|}}{C}}-\overset{\overset{\displaystyle H}{|}}{\underset{\underset{\displaystyle H}{|}}{C}}-Br$ and $H-\overset{\overset{\displaystyle Br}{|}}{\underset{\underset{\displaystyle H}{|}}{C}}-\overset{\overset{\displaystyle H}{|}}{\underset{\underset{\displaystyle H}{|}}{C}}-Br$

(b) $CH_3-\overset{\displaystyle CH_2}{\underset{\displaystyle |}{}}$ and $CH_3-CH_2-CH_2-Cl$
$\qquad\qquad CH_2-Cl$

(c) $CH_3-\overset{\displaystyle CH}{\underset{\displaystyle |}{}}-Cl$ and $CH_3-CH_2-CH_2-Cl$
$\qquad\qquad\quad CH_3$

2.18   Indicate whether the following pairs of structures are isomers or different representations of the same compound.

Answers:
(a) isomers
(b) isomers
(c) isomers

(a) $H-\overset{\overset{\displaystyle H}{|}}{\underset{\underset{\displaystyle H}{|}}{C}}-\overset{\overset{\displaystyle Cl}{|}}{\underset{\underset{\displaystyle H}{|}}{C}}-Br$ and $H-\overset{\overset{\displaystyle Cl}{|}}{\underset{\underset{\displaystyle H}{|}}{C}}-\overset{\overset{\displaystyle H}{|}}{\underset{\underset{\displaystyle H}{|}}{C}}-Br$

(b) CH_3-CH_2 and $CH_3-CH-CH_3$
$\qquad\quad CH_2-Cl$ $\qquad\qquad\qquad Cl$

(c) $CH_3-CH-CH_2-Cl$ and $CH_3-CH_2-CH_2-CH_2-Cl$
$\qquad\quad CH_3$

2.19   There are two isomers for each of the following molecular formulas. Draw their structural formulas.
       (a) $C_2H_2Br_2$ (b) C_2H_6O (c) C_2H_4BrCl

       (d) C_2H_7Cl (e) C_2H_7N (f) $C_2H_3Br_3$

Answers:

(a)
```
    H   H                    H   Br
    |   |                    |   |
Br—C—C—Br     and     H—C—C—Br
    |   |                    |   |
    H   H                    H   H
```

(b)
```
    H   H                    H       H
    |   |                    |       |
H—C—C—O     and     H—C—O—C—H
    |   |   |                |       |
    H   H   H                H       H
```

(c)
```
    H   Br                   H   H
    |   |                    |   |
H—C—C—Cl     and     Cl—C—C—Br
    |   |                    |   |
    H   H                    H   H
```

(d)
```
    H   Cl  H                H   H   H
    |   |   |                |   |   |
H—C—C—C—H     and     H—C—C—C—Cl
    |   |   |                |   |   |
    H   H   H                H   H   H
```

(e)
```
    H   H   H                H   H   H
    |   |   |                |   |   |
H—C—N—C—H     and     H—C—C—N—H
    |       |                |   |
    H       H                H   H
```

(f)
```
    H   Br                   H   Br
    |   |                    |   |
H—C—C—Br     and     Br—C—C—Br
    |   |                    |   |
    H   Br                   H   H
```

2.20   There are three isomers for each of the following molecular formulas. Draw their structural formulas.
       (a) $C_2H_3Br_2Cl$ (b) C_3H_8O (c) C_3H_8S

Answers:

(a)
```
    H   Cl                H   H                 Br  H
    |   |                 |   |                 |   |
H—C—C—Br     Br—C—C—Br     Br—C—C—Cl
    |   |                 |   |                 |   |
    H   Br                H   Cl                H   H
```

(b)
```
    H   H       H              H   OH  H           H   H   H
    |   |       |              |   |   |           |   |   |
H—C—C—O—C—H   H—C—C—C—H   H—C—C—C—OH
    |   |       |              |   |   |           |   |   |
    H   H       H              H   H   H           H   H   H
```

(c)
```
    H   H       H              H   SH  H           H   H   H
    |   |       |              |   |   |           |   |   |
H—C—C—S—C—H   H—C—C—C—H   H—C—C—C—SH
    |   |       |              |   |   |           |   |   |
    H   H       H              H   H   H           H   H   H
```

27

Infrared Spectroscopy

2.21   How can infrared spectroscopy be used to distinguish between propanone and 2-propen-l-ol?

propanone

2-propene-1-ol

Answer: The carbonyl group of propanone (acetone) has a strong absorption at 1749 cm^{-1}. 2-Propene-1-ol (allyl alcohol) has an absorption for the carbon–carbon double bond at 1645 cm^{-1} and an absorption for the oxygen-hydrogen bond at 3400 cm^{-1}.

2.22   How can infrared spectroscopy be used to distinguish between 1-pentyne and 2-pentyne?

Answer: 1-Pentyne is a terminal alkyne, so its sp-hybridized C—H bond has an absorption in the 3450 cm^{-1}, and another strong C≡C absorption at 2120 cm^{-1}. 2-Pentyne, which is an internal alkyne, does not have a C—H absorption at 3450 cm^{-1}. Also, the C≡C absorption is so weak that it is barely visible.

2.23   The carbonyl stretching vibration of ketones is at a longer wavelength than the carbonyl stretching vibration of aldehydes. Suggest a reason for this observation.

Answer: The longer wavelength absorption (smaller wavenumber) corresponds to a lower energy vibration. The dipolar resonance form of a ketone is more stable than that of an aldehyde because the extra alkyl group donates electron density. The increased contribution of the resonance form with a carbon–oxygen single bond means that the ketone carbonyl bond absorption requires less energy.

2.24   The carbonyl stretching vibrations of esters and amides occur at 1735 and 1670 cm^{-1}, respectively. Suggest a reason for this difference.

Answer: Both oxygen and nitrogen are inductively electron withdrawing, and they destabilize the dipolar resonance form of the carbonyl group. Since oxygen is more electronegative than nitrogen, this effect is larger for oxygen, so the dipolar resonance form of an ester is less stable that of an amide. The relative ability of the two atoms to donate electrons by resonance is also important. Because nitrogen donates electrons by resonance more effectively than oxygen, there is an increased contribution of a dipolar resonance form for the amide.

2.25   An infrared spectrum of a compound with molecular formula $C_4H_8O_2$ has an intense, broad band between 3500 and 3000 cm^{-1} and an intense peak at 1710 cm^{-1}. Which of the following compounds best fits these data?

I: $CH_3CH_2CO_2CH_3$ II: $CH_3CO_2CH_2CH_3$ III: $CH_3CH_2CH_2CO_2H$

Answer: The absorptions correspond to an O—H and a carbonyl group, respectively. Only the carboxylic acid group of III has both structural features. The other two compounds are esters that would have an absorption corresponding to a carbonyl group but, because esters do not have an O—H group, would have no absorption in the 3500–3000 cm^{-1} region.

2.26   Explain why the carbonyl stretching vibrations of the following two esters differ.

$$CH_2\!\!=\!\!CH\!\!-\!\!CH_2\!\!-\!\!\overset{\displaystyle O}{\overset{\displaystyle \|}{C}}\!\!-\!\!O\!\!-\!\!CH_3 \qquad\qquad CH_3\!\!-\!\!CH\!\!=\!\!CH\!\!-\!\!\overset{\displaystyle O}{\overset{\displaystyle \|}{C}}\!\!-\!\!O\!\!-\!\!CH_3$$
$$1735\ cm^{-1} \qquad\qquad\qquad\qquad 1720\ cm^{-1}$$

Answer: The carbonyl group of the second compound is conjugated with a double bond. As a result, there is some contribution of a resonance form in which the carbon–oxygen bond has single bond character. The increased contribution of the resonance form with a carbon–oxygen single bond means that the carbonyl bond absorption requires less energy.

2.27   Explain how the two isomeric nitration products of isopropylbenzene can be distinguished using infrared spectroscopy.

Answer: The ortho nitro isomer has four adjacent C—H bonds, and the out-of plane bending of these bonds occurs at $748\ cm^{-1}$. The para nitro isomer has two sets of two adjacent C—H bonds, and the out-of plane bending occurs at $866\ cm^{-1}$.

3 INTRODUCTION TO ORGANIC REACTION MECHANISMS

KEYS TO THE CHAPTER

3.1 Acid–Base Reactions

The properties of acids are characterized by K_a and pK_a values. Stronger acids have large K_a values and small pK_a values. For example, alcohols and carboxylic acids have pK_a values in the 16 and 5 range, respectively. The properties of bases are characterized by K_b and pK_b values. Stronger bases have large K_b values and small pK_b values.

Acid–base reactions proceed to favor the weaker of the two possible acids (or the weaker of the two bases). The equilibrium constant for the overall reaction is given by a quotient of two equilibrium constants. Thus, we need to determine the number of powers of 10 by which the equilibrium constants differ. If that difference is 5 powers of 10, for example, then the equilibrium constant is either 10^{-5} or 10^5 depending on your analysis of whether the reaction is favorable or unfavorable.

A curved arrow convention considers the movement of electrons from the tail of the arrow to a point indicated by the arrowhead. Many organic reactions can be described as "have pair-will share." In organic reactions, one species with a nonbonded (or bonded) pair of electrons "donates" an electron pair to an electron-deficient species by forming a covalent bond between the two species.

3.2 Chemical Equilibrium and Equilibrium Constants

Much of the discussion of organic chemical reactions centers on the "driving force" that refers to the magnitude of the equilibrium constant and the change in free energy for the reaction. In the case of reactions with small equilibrium constants, the reaction conditions are usually adjusted to shift the position of equilibrium by taking advantage of Le Chatelier's principle. For example, if an equilibrium constant is small, the equilibrium position can be shifted to the right by removing the products.

3.3 pH and pK Values

The properties of acids are characterized by K_a and pK_a values. Stronger acids have large K_a values and small pK_a values. The properties of bases are characterized by K_b and pK_b values. Stronger bases have large K_b values and small pK_b values.

The equilibrium constant for an acid–base reaction lies on the side of the weaker of the two possible acids (or the weaker of the two bases). The equilibrium constant for the overall reaction is given by the ratio of the two equilibrium constants.

$$HA \ + \ B^- \ \rightleftharpoons \ A^- + HB$$

$$K_{eq} = \frac{K_{HA}}{K_{HB}}$$

3.4 Effect of Structure on Acidity

Four factors—periodic trends, resonance effects, inductive effects, and hybridization effects—influence acidity. The strength of an acid, HA, depends in part upon the strength of the H—A bond. The bond strength decreases as we move down a column of the periodic table. Because bond strength is inversely related to the acidity, the acidity of the halogen acids increases in the order HF < HCl < HBr < HI. For the same reasons, H_2O is a weaker acid than H_2S.

Acidity increases from left to right in a given row of the periodic table. The order of increasing acidity is CH_4 < NH_3 < H_2O < HF. This trend reflects the stabilization of the negative charge, which varies directly with the electronegativity of the atom of the conjugate base. That is, the order of increasing strength of conjugate bases is F^- < OH^- < NH_2^- < CH_3^-. Stabilizing the negative charge in the conjugate base increases K_a. One way the conjugate base is stabilized is by delocalization of the negative charge over two or more atoms. This effect is called *resonance stabilization*. When the conjugate base of an acid is resonance stabilized, acid strength increases substantially. For example, both methanol and ethanoic acid ionize to form conjugate bases with a negative charge on oxygen. However, ethanoic acid is **ten billion** (10^{10}) times more acidic than methanol.

Any atom or group of atoms in an organic molecule that withdraws electron density from the bond between hydrogen and another atom—such as carbon, oxygen, or nitrogen—increases its acidity by an inductive effect.

The acidity of hydrocarbons is related to the hybridization of the carbon atom of the C—H bond. The K_a of a carbon acid increases in the order sp^3 < sp^2 < sp. The order of acidities parallels the contribution of the lower energy of the 2s orbital to the hybrid orbitals in the σ bond.

Organic Chemistry Study Guide: Key Concepts, Problems, and Solutions http://dx.doi.org/10.1016/B978-0-12-801889-7.00003-0

3.5 Standard Free Energy Changes in Chemical Reactions

The standard Gibbs free energy change ($\Delta G°$) is the energy change that occurs in going from the reactants to the products.

$$\Delta G°_{rxn} = \Delta G°_f \text{ (products)} - \Delta G°_f \text{ (reactants)}$$

When the products are more stable than the reactants, $\Delta G°_{rxn}$ is negative, and the reaction is *exergonic*. If the reactants less stable than the products, $\Delta G°_{rxn}$ is positive, and the reaction is *endergonic*.

The following equation describires the relation between the standard free energy change, $\Delta G°_{rxn}$ and the equilibrium constant.

$$\Delta G°_{rxn} = -2.303 RT \log K_{eq}$$
$$R = 8.314 \text{ kj kelvin}^{-1} \text{ mole}^{-1} \ (1.987 \text{ cal kelvin}^{-1} \text{ mole}^{-1})$$
$$T = \text{absolute temperature (kelvin)}$$

3.6 Standard Enthalpy Changes in Chemical Reactions

The heat released or absorbed in a reaction at constant pressure is the **enthalpy change**, $\Delta H°_{rxn}$. If heat flows out of the reaction into the surroundings, the reaction is *exothermic*. For an exothermic reaction, $\Delta H°_{rxn} < 0$. If heat flows into the reaction from the surroundings, the reaction is *endothermic*. For an endothermic reaction, $\Delta H°_{rxn} > 0$.

When a bond forms, energy is released; the process is exothermic. Conversely, breaking a bond requires energy; the process is endothermic. Therefore, the energy change for a chemical reaction reflects the differences in the energies of the bonds that are broken and formed. If the products of a reaction contain less stored energy than the reactants, the net difference is released as heat, $\Delta H°_{rxn}$. The magnitude of the standard enthalpy change for a reaction depends only on the difference in enthalpy between the products and reactants.

3.7 Bond Dissociation Energies

The bond dissociation energy is the energy required—an endothermic process—to break a bond and form two atomic or molecular fragments, each with one electron of the original shared pair. Thus, a very stable bond has a large bond dissociation energy—more energy must be added to cleave the bond. A high bond dissociation energy means that the bond (and molecule) is of low energy and stable. Bond energies depend on the number of bonds between atoms. Even though π bonds are weaker than σ bonds, a double bond, which consists of a σ and π is bond, is stronger than a single bond because there are two bonds.

3.8 Introduction to Reaction Mechanisms

A mechanism is the series of steps that occur as a reactant is converted to a product. The complexity of mechanisms covers a wide range from one-step, concerted mechanisms to complex multistep mechanisms in which a series of intermediates form on the pathway from reactants to products.

Classifying the type of bond cleavage in a particular mechanism requires us to look carefully at the reactant and product, to identify which bond breaks, and how it breaks. Bond cleavage is homolytic if the two resulting fragments each retain one electron from the bond; fragments containing an unpaired electron are called radicals and are highly reactive. For example, *tert*-butylbromide can break into a homolytic process to give a *tert*-butyl **radical** and a bromine atom, each of which has a single, unpaired electron.

Bond cleavage is heterolytic if both electrons in the bond stay with one fragment; the fragment retaining the electron pair acquires a −1 charge, and the other fragment a +1 charge. When a carbon atom is bonded to an electronegative element such as Br, heterolytic cleavage gives the electron pair to the bromine and leaves the carbon atom to which it was bound with a positive charge. The carbon fragment is a **carbocation**, as the following equation shows.

When a carbon atom is bonded to an electropositive element such as H, heterolytic cleavage releases the electropositive species as a cation and leaves the electron with the carbon fragment, which becomes a **carbanion.**

3.9 Structures and Stabilities of Reactive Carbon Intermediates

The stability of a carbon intermediate depends on the number of electrons about the carbon atom and the identity of the attached groups. Both carbocations and radicals, which are electron deficient, are stabilized by larger numbers of alkyl groups. Carbanions already have a sufficient number of electrons, and the supply of additional electron density by attached carbon groups is counterproductive. Thus, the order of stability of carbanions is opposite to that of carbocations. Carbocation and radical stability decrease in the order tertiary > secondary > primary >> methyl.

3.10 Reaction Rate Theory

Reactions occur via one or more transition states in which the bonding patterns correspond to neither the reactants nor the products. The transition state occurs at a maximum point on the minimum energy pathway. This point is at the top of a two-dimensional reaction coordinate diagram. Reactions with a high activation energy occur at a slower rate than those with a lower activation energy. Increasing the temperature increases reaction rates because a larger fraction of molecules possess an energy equal to or greater than the activation energy and can achieve the transition state structure as the temperature increases.

Multistep reactions have more than one transition state, and the lower energy species that forms between transition states is an intermediate. Catalysts provide for a mechanism that occurs via a transition state with a lower activation energy.

The Hammond postulate states that strongly exothermic reactions occur via transition states that more closely resemble the reactant structure. Endothermic reactions occur via transition states that more closely resemble the product. Transition state structures cannot be determined experimentally. The structure of reactants and products is known, and it is often possible to elucidate the structure of intermediates. The structure of the transition state is estimated using the Hammond postulate.

3.11 Stability and Reactivity

The term "stability" of a compound is related to standard free energy change for making a compound from its elements, that is $\Delta G^{\circ}_{\text{formation}}$. If we compare two closely related structural isomers, the one with the more negative $\Delta G^{\circ}_{\text{formation}}$ is more stable. The term stability is used to describe reactants, products, and even intermediates.

The term *reactivity* refers to the *rate* at which a compound reacts. Therefore, reactivity refers to the activation energy required for that substance to form a particular transition state. We must refer to a specific reaction to discuss reactivity. Two compounds can have opposite reactivities depending upon the specific kind of reaction they are undergoing.

SOLUTIONS TO END-OF-CHAPTER EXERCISES

Acids and Bases

3.1   Write the structure of the conjugate acid of each of the following species.

(a) H—O—O—H   (b) NH_2—NH_2 (c) CH_3—S—CH_3

(d) CH_3—O—CH_3 (e) CH_3—NH_2 (f) CH_3—OH

Answers:

(a) H—Ö—Ö⁺—H (b) structure (c) structure

(d) structure (e) structure (f) structure

3.2   Write the structure of the conjugate base of each of the following species.

(a) CH_3—SH   (b) CH_3—NH_2 (c) CH_3—O—SO_3H

(d) CH_2=CH_2 (e) HC≡CH   (f) CH_3—CN

Answers:

(a) structure (b) structure (c) structure

(d) structure (e) H—C≡C:⁻ (f) structure

3.3   Write the structure of the conjugate acid of each of the following species.

(a) CH_2=O   (b) CH_3—NH—CH_3 (c) CH_2=NH   (d) CH_3—$\overset{\text{O}}{\overset{\|}{\text{C}}}$—OH

Answers:

(a) structure (b) structure

(c) structure (d) structure

34

3.4   Write the structure of the conjugate acid of each of the following species.

(a)   (b)   (c)   (d)

Answers:

(a)   (b)   (c)   (d)

3.5   Identify the Lewis acid and Lewis base in each of the following reactions.

(a) CH_3—CH_2—Cl + $AlCl_3$ ⟶ CH_3—CH_2+ + $AlCl_4^-$

(b) CH_3—CH_2—SH + CH_3O^- ⟶ CH_3—CH_2—S^- + CH_3OH

(c) CH_3—CH_2—OH + NH_2^- ⟶ CH_3—CH_2—OH^- + NH_3

(d) $(CH_3)_2N^-$ + CH_3OH ⟶ $(CH_3)_2NH$ + CH_3O^-

Answers:

(a) CH_3—CH_2—Cl is the Lewis base; $AlCl_3$ is the Lewis acid.
(b) CH_3—CH_2—SH is the Lewis acid; CH_3—O^- is the Lewis base
(c) CH_3—CH_2—OH is the Lewis acid; NH_2^- is the Lewis base
(d) $(CH_3)_2N^-$ is the Lewis base; CH_3—OH is the Lewis acid

3.6   Identify the Lewis acid and Lewis base in each of the following reactions.

(a) $(CH_3)_2O$ + HI ⟶ $(CH_3)_2OH+$ + I^-

(b) CH_3—CH_2+ + H_2O ⟶ CH_3—CH_2—OH_2+

(c) CH_3—CH=CH_2 + HBr ⟶ $(CH_3)_2CH+$ + Br^-

(d) CH_3—C≡CH + CH_3NH^- ⟶ CH_3—C≡C^- + CH_3NH_2

Answers:

(a) $(CH_3)_2O$ is the Lewis base; HI is the Lewis acid.
(b) CH_3—CH_2^+ is the Lewis acid; H_2O is the Lewis base.
(c) CH_3—CH=CH_2 is the Lewis base; HBr is the Lewis acid.
(d) CH_3—C≡CH is the Lewis acid; CH_3N^- is the Lewis base.

Equilibrium Constant Expressions

3.7   Write the equilibrium constant expression for the reaction of ethanal and methanol to give an acetal.

$$CH_3CHO + 2\ CH_3OH \rightleftharpoons CH_3CH(OCH_3)_2 + H_2O$$

Answer: $$K_{eq} = \frac{[CH_3CH(OCH_3)_2][H_2O]}{[CH_3CHO][CH_3OH]^2}$$

3.8   Write the equilibrium constant expression for the reaction of acetylene (C_2H_2) to give cyclooctatetraene (C_8H_8).

Answer:

$$K_{eq} = \frac{[C_8H_8]}{[C_2H_2]^4}$$

3.9   How do the equilibrium constant expressions differ for the hydrolysis reaction of ethyl ethanoate (written right to left) and the esterification reaction of ethanol and ethanoic acid (written left to right)? What is the equilibrium constant for the hydrolysis reaction?

Answer:

$$K_{equilibrium} = \frac{[CH_3CO_2CH_2CH_3]\,[H_2O]}{[CH_3CO_2H]\,[CH_3CH_2OH]} = 4.0$$

The hydrolysis reaction written above is the reverse of the esterification reaction. Thus, the equilibrium constant expression for hydrolysis is the reciprocal of the equilibrium constant expression for esterification. The value of the equilibrium constant for hydrolysis is 0.25.

3.10   At equilibrium, the yield of the condensation product of acetone is about 5%. Calculate the equilibrium constant for the reaction.

Answer:

For an initial concentration of acetone equal to x mole liter^{-1}, the theoretical concentration of product for a complete reaction would be $0.5\,x$ mole liter^{-1}. For a 5% yield, the actual concentration is $0.025\,x$ mole liter^{-1}. The equilibrium concentration of reactant is $0.95\,x$ mole liter^{-1} because two moles of reactant are required to give one mole of product. The equilibrium constant is approximately $0.028\,x^{-1}$ mole^{-1} liter.

$$K_{eq} = \frac{[0.025x]}{[0.95x]^2}$$

pH and pK Values

3.11   Without reference to tables of pK_a values, predict the position of the following equilibrium.

Answer: CH_3—CH_2—SH + CH_3—O^- ⇌ CH_3—CH_2—S^- + CH_3—OH

The acid dissociation constants of organic compounds containing atoms within a common group of the periodic table bonded to hydrogen increase down the column. Thus, thiols are more acidic than alcohols. The equilibrium position lies on the side of the equation containing the weaker acid. Therefore, the position of the above equilibrium is to the right, where CH_3—OH is located in the above equilibrium, and $K_{eq} > 1$.

3.12   Without reference to tables of pK_a values, predict the position of the following equilibrium.

$$CH_3\text{—}CH_2\text{—}CO_2H + CH_3\text{—}O^- \xrightleftharpoons{K_{eq}} CH_3\text{—}CH_2\text{—}CO_2^- + CH_3\text{—}OH$$

Answer:

The equilibrium position lies on the side of the equation containing the weaker acid. The acids are methanol located on the right, and acetic acid, which is located on the left of the above equation. Acetic acid is the stronger acid because its conjugate base is resonance stabilized. Thus, the position of the equilibrium is on the right, where the weaker acid, methanol, is found.

3.13   The approximate pK_a values of CH_4 and CH_3OH are 49 and 16, respectively. Which is the stronger acid? Will the equilibrium position of the following reaction lie to the left or to the right?

$$CH_4 \quad + \quad CH_3{-}O^- \quad \overset{K_{eq}}{\rightleftharpoons} \quad CH_3^- \quad + \quad CH_3{-}OH$$

Answer:

The equilibrium position lies on the side of the equation containing the weaker acid. The acids are methanol located on the right, and acetic acid, which is located on the left of the above equation. Acetic acid is the stronger acid because its conjugate base is resonance stabilized. Thus, the position of the equilibrium is on the right, where the weaker acid, methanol, is found.

3.14   The approximate pK_a values of NH_3 and CH_3OH are 36 and 16, respectively. Which is the stronger acid? Will the equilibrium position of the following reaction lie to the left or to the right?

$$CH_3{-}OH + NH_2^- \quad \overset{K_{eq}}{\rightleftharpoons} \quad NH_3 \quad + \quad CH_3{-}O^-$$

Answer:

Methanol is the stronger acid by a factor of 10^{33} in K_a. The equilibrium position lies on the left side of the equation, which contains the weaker acid, methane. The equilibrium constant is 10^{-33}.

Structure and Acid Strength

3.15 Write the structures of the two conjugate acids of hydroxylamine ($NH_2{-}OH$). Which is the more acidic?

stronger acid

Answer:

The acidity of hydrogen atoms bonded to atoms contained in similarly structured compounds increases from left to right within a period of the periodic table. For example, H_2O is a stronger acid than NH_3. The conjugate acid with a proton located on the oxygen atom of hydroxylamine must be a stronger acid than the conjugate acid with a proton located on the nitrogen atom.

3.16 Write the structures of the two conjugate bases of hydroxylamine ($NH_2{-}OH$). Which is the more basic?

stronger base

Answer:

The basicity of atoms contained in similarly structured compounds decreases from left to right within a period of the periodic table. For example, NH_2^- is a stronger base than OH^-. The conjugate base with the charge located on the nitrogen atom of hydroxylamine must be a stronger base than the conjugate base with a charge on the oxygen atom.

3.17 Which is the stronger acid, chloroethanoic acid ($ClCH_2CO_2H$) or bromoethanoic acid ($BrCH_2CO_2H$)? Explain your answer.

Answer:

The inductive electron withdrawal by chlorine is larger than that of bromine because chlorine is more electronegative than bromine. Therefore, electron density is pulled away from the O—H group, and the acidity of chloroacetic acid is greater than the acidity of bromoacetic acid.

3.18 Which acid has the larger pK_a, chloroethanoic acid ($ClCH_2CO_2H$) or dichloroethanoic acid (Cl_2CHCO_2H)? Explain your answer.

Answer:
Dichloroacetic acid is a stronger acid than chloroacetic acid because the two chlorine atoms inductively withdraw more electron density from the O—H group than a single chlorine atom. The pK_a of dichloroacetic acid is therefore smaller than the pK_a of chloroacetic acid.

3.19 Based on the pK_a values of substituted butanoic acids (Section 3.4), predict the pK_a of 4-chlorobutanoic acid.

Answer:
The pK_a of the substituted chlorobutanoic acids increases with increasing distance separating the chlorine atom and the acidic site. Thus, the pK_a of the 4-chloro compound is greater than 4.02, the pK_a of the 3-chloro compound. It is also less than the pK_a of butanoic acid, which is 4.82.

3.20 Explain the trends in the pK_a values of the following ammonium ions.

$$CH_3CH_2CH_2CH_2\overset{+}{N}H_3 \qquad\qquad CH_3—O—CH_2CH_2CH_2CH_2\overset{+}{N}H_3$$

$$pK_a = 10.6 \qquad\qquad\qquad pK_a = 9.9$$

$$N{\equiv}C—CH_2CH_2CH_2CH_2\overset{+}{N}H_3$$

$$pK_a = 7.8$$

Answer:
The order of decreasing pK_a values indicates that the groups bonded to the nitrogen atom of the ammonium ions increase in ability to inductively withdraw electron density. Although oxygen is more electronegative than nitrogen, the nitrile has a triple bond and is a much more polar group.

3.21 Explain why the hydrogen of the CH_3 of propene is more acidic than hydrogen of the CH_3 of propane.

Answer:
The conjugate base of propane has its negative charge localized on a single carbon atom. The conjugate base of propene has its negative charge delocalized over two carbon atoms, as shown by two contributing resonance structures.

3.22 Ethanonitrile (CH_3CN) is a stronger acid than ethane. Explain why.

Answer:
The conjugate base of ethane has its negative charge localized on a single carbon atom. The conjugate base of ethanonitrile has its negative charge delocalized with some of the charge located on the more electronegative nitrogen atom as shown in one of the two contributing resonance structures.

3.23 The pK_a of acetic acid (CH_3CO_2H) is 4.8. Explain why the carboxylic acid group of amoxicillin (pK_a = 2.4), a synthetic penicillin, is more acidic than acetic acid, whereas the carboxylic acid group of indomethacin (pK_a = 4.5), an anti-inflammatory analgesic used to treat rheumatoid arthritis, is of comparable acidity.

amoxicillin

Indomethaxan

Answer:

In amoxicillin, the acidic —CO_2H group is bonded to a carbon atom that is also bonded to a nitrogen atom that inductively withdraws electron density and increases the acidity of the O—H group. In indomethacin, the —CO_2H group is bonded to a carbon atom that is not directly bonded to any electronegative groups. The nitrogen atom in indomethacin is one atom farther removed than that in amoxicillin.

3.24 The pK_a of the OH group of phenobarbital is 7.5, whereas the pK_a of CH_3OH is 16. Explain why phenobarbital is significantly more acidic.

phenobarbital

Answer:

The greatly increased acidity of the O—H group in phenobarbital reflects the resonance stabilization of the conjugate base, in which the charge is delocalized over two oxygen atoms.

3.25 The N—H bond of ammonia is not very acidic (pK_a = 33). However, the pK_a for the N—H bond of sulfanilamide, a sulfa drug, is 10.4. Suggest a reason for the higher acidity of sulfanilamide.

sulfanilamide

Answer:

The nitrogen atom is bonded to a sulfur atom that has two oxygen atoms, which are electronegative, and hence withdraw electron density from the N—H bond.

3.26 The pK_a of sulfadiazine, a sulfa drug, is 6.5. Why is this compound more acidic than sulfanilamide?

sulfadiazine

Answer:

The nitrogen atom is bonded to a carbon atom of a ring that has two nitrogen atoms, which are electronegative, and hence withdraw electron density from the N—H bond.

Equilibrium Constant and Free Energy

3.27 A reaction has K_{eq} = 1 × 10^{-5}. Are the products more or less stable than the reactants? Is the reaction exergonic or endergonic?

Answer: For an equilibrium constant less than 1, the products are less stable than the reactants. Such a reaction is endergonic.

3.28 Which reaction would be exergonic, one with K_{eq} = 100 or one with K_{eq} = 0.01?

Answer: The reaction with K = 100 proceeds further to completion and is more exergonic than a reaction with K = 0.01.

3.29 Can a reaction have K_{eq} = 1? What relationship would exist between the free energies of the reactants and products?

Answer: Yes, a reaction can have K_{eq} =1 and have products and reactants of equal stability—that is, $\Delta G°_{rxn}$ = 0.

3.30 Which reaction has an equilibrium constant greater than 1, one with $\Delta G°_{rxn}$ = + 15 kJ mole^{-1} or one with $\Delta G°_{rxn}$ = −15 kJ mole^{-1}?

Answer: Spontaneous reactions are exergonic and have $\Delta G°_{rxn}$ < 0. The reaction with $\Delta G°_{rxn}$ = −15 kJ mole^{-1} is exergonic.

3.31 The $\Delta G°_{rxn}$ for the following reaction is +2 kJ mole^{-1}. What is K_{eq} at 25 °C?

Answer: Using the relationship $\Delta G°_{rxn}$ = −2.303 RT log K, the equilibrium constant is 0.4. Remember that 25 °C is 298 K and that $\Delta G°_{rxn}$ must be expressed in kJ mole^{-1}.

$$CH_3SH \ + \ HBr \ \rightleftharpoons \ CH_3Br \ + \ H_2S$$

3.32 The equilibrium constant for the isomerization of butane to 2-methylpropane is 4.9. What is $\Delta G°_{rxn}$?

Answer: Using the relationship $\Delta G°_{rxn}$ = −2.303 RT log K, the $\Delta G°_{rxn}$ = −3.9 J mole^{-1}.

Bond Cleavage and Reaction Intermediates

3.33 Write the structure of the radical formed by abstraction of a hydrogen atom by a chlorine atom for each of the following compounds.

(a) CH_3CH_3 (b) CH_3Cl (c) CH_2Cl_2

Answers:

(a)
```
      H   H
      |   |
  H—C—C·
      |   |
      H   H
```

(b)
```
      H
      |
  H—C·
      |
      Cl
```

(c)
```
      Cl
      |
  H—C·
      |
      Cl
```

3.34 Write the structures of all possible radicals formed by abstraction of a hydrogen atom by a chlorine atom for each of the following compounds.

Answers:

(a)
```
      H   H   H
      |   |   |
  H—C—C—C·      and
      |   |   |
      H   H   H
```
n-propyl radical

```
      H       H
      |       |
  H—C—C·—C—H
      |   |   |
      H   H   H
```
iso-propyl radical

(b)
```
      H   H   H   H
      |   |   |   |
  H—C—C—C—C·     and
      |   |   |   |
      H   H   H   H
```
n-butyl radical

```
      H   H       H
      |   |       |
  H—C—C—C·—C—H
      |   |   |   |
      H   H   H   H
```
sec-butyl radical

(c)
```
      H       H
      |       |
  H—C—C·—C—H      and
      |   |   |
      H  CH₃  H
```
tert-butyl radical

```
      H   H   H
      |   |   |
  H—C—C—C·
      |   |   |
      H  CH₃  H
```
iso-butyl radical

3.35 The oxygen–chlorine bond of methyl hypochlorite (CH_3—O—Cl) can cleave heterolytically. Based on the electronegativity values of chlorine and oxygen, predict the charges on the cleavage products.

Answer: Oxygen is more electronegative than chlorine. Thus, the electrons of the O—Cl bond will remain with oxygen and the products should be CH_3O^- and Cl^+.

3.36 2-Chloropropane reacts with the Lewis acid $AlCl_3$ to give $AlCl_4^-$ and a carbon intermediate. What is the intermediate?

Answer: $AlCl_3$ combines with Cl^- to give $AlCl_4^-$. Thus, the C—Cl bond cleaves heterolytically. The intermediate is a carbocation $(CH_3)_2CH^+$.

3.37 Hydrogen peroxide (H—O—O—H) reacts with a proton to give a conjugate acid, which undergoes heterolytic, oxygen–oxygen bond cleavage to yield water. What is the second product?

Answer: Heterolytic cleavage that places the O—O bonding electrons on the oxygen atom of water leaves a cation with the positive charge on the oxygen atom of the HO group.

3.38 Benzoyl peroxide is used in creams to control acne. It is an irritant that causes proliferation of epithelial cells. It undergoes a homolytic cleavage of the oxygen–oxygen bond. Write the structure of the product, indicating all of the electrons present on all of the oxygen atoms.

Answer: Homolytic cleavage leaves one electron of the pair of electrons in the O—O bond with each of the two equivalent radical fragments, giving the structures shown below.

benzoyl peroxide

Stability of Reactive Intermediates

3.39 Arrange the following intermediates in order of increasing stability.

I   II   III

Answer: The order of increasing stability is II < I < Ill, which corresponds to primary < secondary < tertiary.

3.40 Arrange the following intermediates in order of increasing stability.

I   II   III

Answer: The order of increasing stability is I < III < II, which corresponds to primary < secondary < tertiary.

3.41 Explain why more energy is required for heterolytic bond cleavage of the carbon–bromine bond of 1-bromopropane than is needed to cleave the carbon–bromine bond of 2-bromopropane.

CH$_3$CH$_2$CH$_2$Br
1-bromopropane

CH$_3$CHBrCH$_3$
2-bromopropane

Answer: Heterolytic cleavage of a C—Br bond produces a carbocation and a bromide ion. The secondary carbocation derived from 2-bromopropane is more stable than the primary carbocation derived from 1-bromopropane, so formation of the primary carbocation requires more energy.

3.42 Explain why less energy is required to cleave the carbon–chlorine bond of 3-chloropropene than that needed to cleave the carbon–chlorine bond of 1-chloropropane.

CH₂=CHCH₂Cl
3-cloropropene

CH₃CH₂CH₂Cl
1-chloropropane

Answer: Heterolytic cleavage of a C—Cl bond produces a carbocation and a chloride ion. A resonance-stabilized primary carbocation is derived from 3-chloropropene and is thus more stable than the primary carbocation derived from 1-chloropropane, in which the charge is localized.

resonance-stabilized allyl carbocation

3.43 Chloroform (CHCl₃) reacts with a strong base in an unusual elimination reaction to give dichlorocarbene (CCl₂). Write the Lewis structure for this species. What features of the chlorine atoms might stabilize this carbene compared to CH₂?

Answer: The dichlorocarbene is electron deficient; there are only four bonding electrons and a lone pair of electrons about the carbon atom. However, either of the chlorine atoms can share one of its lone pairs of electrons in contributing resonance forms. The delocalization of electrons makes CCl₂ more stable than CH₂.

3.44 Draw the Lewis structure of OH⁺. (a) How does it differ from OH⁻? Is OH⁺ a nucleophile or an electrophile?

Answers:
(a) OH⁺ has one less unshared electron pair than hydroxide.
(b) OH⁺ is an electron-deficient species, so it is an electrophile.

Activation Energy and Rates of Reaction

3.45 Given the following information about two reactions, which one will occur at the faster rate at a common temperature?

Reaction	$\Delta H°_{rxn}$	E_a
A ⟶ X	−120 kJ mole⁻¹	+100 kJ mole⁻¹
B ⟶ Y	−100 kJ mole⁻¹	+120 kJ mole⁻¹

Answer: The reaction converting A to X has the lower activation energy (E_a), so it proceeds at the faster rate.

3.46 Given the information given in Exercise 3.44, which one is more exothermic?

Answer: The reaction converting A to X has the more negative $\Delta H°_{rxn}$, so it is more exothermic.

3.47 Given the activation energies for the following free radical reactions, which one occurs at the faster rate?

CH₄ + F ⟶ CH₃· + HF E_a = 5 kJ mole⁻¹

CH₄ + Cl· ⟶ CH₃· + HCl E_a = 16 kJ mole⁻¹

Answer: The reaction of methane with a fluorine atom has the lower activation energy (E_a), so it proceeds at the faster rate.

3.48 Consider the activation energies for the following nucleophilic substitution reactions. Which reaction occurs at the faster rate?

$$CH_3-I + Cl^- \longrightarrow CH_3-Cl + I^- \qquad E_a = 104 \text{ kJ mole}^{-1}$$

$$CH_3-I + Br^- \longrightarrow CH_3-Br + I^- \qquad E_a = 96 \text{ kJ mole}^{-1}$$

Answer: The reaction of bromide ion with iodomethane has the lower activation energy (E_a), so it proceeds at the faster rate.

Kinetic Order of Reaction

3.49 Sodium cyanide reacts with chloroethane by the following equation. When the concentration of cyanide ion is tripled, the reaction rate triples. When the concentration of chloroethane doubles, the reaction rate doubles. What is the overall kinetic order of the reaction? Write the rate equation for the reaction.

$$N{\equiv}C{:}^- + CH_3-CH_2-Cl \longrightarrow CH_3-CH_2-C{\equiv}N{:} + Cl^-$$

Answer: The rate of reaction is first order in both chloroethane and cyanide ion, so it is second order overall. The rate equation is:

$$\text{rate} = k\,[CH_3CH_2Cl][CN^-].$$

3.50 Reaction of *tert*-butyl alcohol with concentrated HBr gives *tert*-butyl bromide. When the concentration of the alcohol is doubled, the reaction rate doubles. When the concentration of acid is tripled, the reaction rate triples. If more bromide ion in the form of sodium bromide is added, the rate is unaffected. What is the kinetic order with respect to each reactant? What is the overall kinetic order of the reaction?

Answer: The rate of reaction is first order in both *tert*-butyl alcohol and hydrogen ion and is zero order in bromide ion, so it is second order overall. The rate law is:

$$\text{rate} = k\,[(CH_3)3OH][H^+].$$

Reaction Mechanisms

3.51 Identify the processes of bond cleavage and bond formation for each of the following reactions.

Answers:
(a) Homolytic cleavage of a C—H bond and homogenic formation of a H—Br bond.
(b) Homolytic cleavage of a Br—Br bond and homogenic formation of a C—Br bond.

3.52 Identify the processes of bond cleavage and bond formation for each of the following reactions.

(a)
$$HO^- + CH_3-\underset{\underset{CH_3}{|}}{\overset{\overset{CH_3}{|}}{C}}+ \longrightarrow CH_3-\underset{\underset{CH_3}{|}}{\overset{\overset{CH_3}{|}}{C}}-OH$$

(b)
$$CH_3-\underset{\underset{CH_3}{|}}{\overset{\overset{CH_3}{|}}{C}}-Cl \longrightarrow CH_3-\underset{\underset{CH_3}{|}}{\overset{\overset{CH_3}{|}}{C}}+ \quad + \quad Cl^-$$

Answers:
(a) Heterogenic formation of a C—O bond.
(b) Heterolytic cleavage of a C—Cl bond.

3.53 In the presence of a strong acid, *tert*-butyl alcohol acts as a base. The resulting conjugate acid produces water and an intermediate. Write the structure of the intermediate. What type of bond cleavage occurs?

$$H_3C-\underset{\underset{CH_3}{|}}{\overset{\overset{CH_3}{|}}{C}}-\overset{..}{\underset{..}{O}}-H \xrightarrow{H^+} H_3C-\underset{\underset{CH_3}{|}}{\overset{\overset{CH_3}{|}}{C}}-\overset{+}{\underset{H}{\overset{..}{O}}}-H \longrightarrow H_3C-\underset{\underset{CH_3}{|}}{\overset{\overset{CH_3}{|}}{C}}+ \quad + \quad \overset{..}{\underset{H}{O}}-H$$

Answer:
Heterogenic cleavage of a C—O bond.

3.54 Dimethyl ether (CH_3—O—CH_3) can be prepared by adding a strong base such as NaH to methanol (CH_3OH) and then adding iodomethane (CH_3I) to the reaction mixture. Write plausible steps for this reaction.

Answer:

step 1
$$H-\underset{\underset{H}{|}}{\overset{\overset{H}{|}}{C}}-\overset{..}{\underset{..}{O}}-H + :H^- \longrightarrow H-\underset{\underset{H}{|}}{\overset{\overset{H}{|}}{C}}-\overset{..}{\underset{..}{O}}:^- + H-H$$

step 2
$$H-\underset{\underset{H}{|}}{\overset{\overset{H}{|}}{C}}-\overset{..}{\underset{..}{O}}:^- + H-\underset{\underset{H}{|}}{\overset{\overset{H}{|}}{C}}-\overset{..}{\underset{..}{I}}: \longrightarrow H-\underset{\underset{H}{|}}{\overset{\overset{H}{|}}{C}}-\overset{..}{\underset{..}{O}}-\underset{\underset{H}{|}}{\overset{\overset{H}{|}}{C}}-H + :\overset{..}{\underset{..}{I}}:^-$$

Reaction Coordinate Diagrams

3.55 What are the differences between a reaction intermediate and a transition state.

Answer: An intermediate, although short lived, can be detected experimentally. A transition state is a transient species whose structure can only be postulated.

3.56 A reaction occurs in three steps. How many transition states are there? How many intermediates form?

Answer: There are three transitions states—one for each step. There are two intermediates. One is formed from step 1 and reacts in step 2. The second intermediate is formed from step 2 and reacts in step 3.

3.57 Draw a reaction coordinate diagram for a two-step exothermic reaction in which the second step is a rate-determining.

Answer:

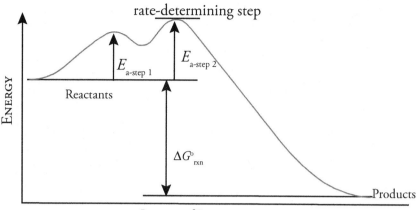

Hammond Postulate

3.58 The ΔH°_{rxn} for abstracting each of the possible hydrogen atoms of propane by a bromine atom is indicated below. Based on the data and the fact that the starting materials are the same, what might be surmised about the relative energies of activation for the two reactions? Do the transition states more closely resemble the reactants or the products?

CH_3—CH_2—CH_3 + Br CH_3—CH_2—$CH_2^\bullet$ + HBr $\Delta H^\circ = +42$ kJ mole^{-1}

CH_3—CH_2—CH_3 + Br CH_3—$\overset{\bullet}{CH}$—CH_3 + HBr $\Delta H^\circ = +29$ kJ mole^{-1}

Answer: The activation energy must be larger for the first reaction listed because it is the more endothermic reaction. The transition state for this reaction must more closely resemble the product than for the second reaction.

3.59 The ΔH°_{rxn} for abstracting each of the possible hydrogen atoms of propane by a bromine atom is indicated below. Based on the data and the fact that the starting materials are the same, what might be surmised about the relative energies of activation for the two reactions? Do the transition states more closely resemble the reactants or the products?

CH_3—CH_2—CH_3 + Cl$\bullet$ CH_3—CH_2—$CH_2^\bullet$ + HCl $\Delta H^\circ = -20$ kJ mole^{-1}

CH_3—CH_2—CH_3 + Cl$\bullet$ CH_3—$\overset{\bullet}{CH}$—CH_3 + HCl $\Delta H^\circ = -33$ kJ mole^{-1}

Answer: The activation energy must be smaller for the second reaction because it is the more exothermic reaction. The transition state for this reaction must more closely resemble the reactant than for the first reaction.

4

ALKANES AND CYCLOALKANES STRUCTURES AND REACTIONS

KEYS TO THE CHAPTER

4.1 Classes of Hydrocarbons

In this section, we encountered new terms that we will continue to use throughout the text. **Hydrocarbons** contain only carbon and hydrogen. **Saturated hydrocarbons** contain only carbon–carbon single bonds; **unsaturated hydrocarbons** contain carbon–carbon multiple bonds. **Alkanes** have only carbon atoms bonded in chains of atoms. **Cycloalkanes** have only carbon atoms bonded in a ring of atoms. Compounds without rings are acyclic; compound with rings are cyclic. Other atoms may be found in some rings. Atoms other than carbon within rings are **heteroatoms**, and the compounds are **heterocyclic**.

4.2 Alkanes

Normal alkanes consist of a continuous chain of carbon atoms; **branched alkanes** have some carbon atoms bonded to more than two other carbon atoms. The general formula for an alkane is C_nH_{2n+2}, whether it is normal or branched. If the number of carbon atoms are known, the number of hydrogen atoms and the molecular formula are known. Thus, inspecting the molecular formula and comparing it to the reference molecular formula expected for an alkane provides a clue about the identity of other structural features. The C_nH_{2n+2} formula is the reference.

A carbon atom is classified as primary (1°), secondary (2°), or tertiary (3°) when it has 1, 2, or 3 alkyl groups, respectively, bonded to it. A carbon atom is quaternary (4°) when it has 4 alkyl groups bonded to it.

4.3 Nomenclature of Alkanes

The nomenclature rules in this section form the foundation on which we will base all other nomenclature. Here is a brief summary.

1. Locate the longest carbon chain, called the parent chain.
2. Identify the groups that are substituents attached to the parent chain.
3. Number the parent chain to give the branching carbon atoms and other substituents the lowest possible numbers.
4. Use a prefix to the name of the parent chain to identify the name and location of all branches and other substituents.
5. Each substituent must be assigned a number to indicate its position. Thus, if two methyl groups are bonded to C-2 in a chain of carbon atoms, the name "2-dimethyl" as part of the prefix is incorrect; two methyl groups bonded to C-2 must be designated as 2,2-dimethyl. To determine the numbering of the substituents to use in the prefix, choose the point of first difference.
6. List the names of substituents alphabetically. Note that the prefixes di, tri, etc., do not affect the alphabetic method of listing alkyl groups. For example, ethyl is listed before dimethyl because it is the "e" of ethyl that takes precedence over the "m" of methyl.
7. The most common alkyl groups are methyl, ethyl, propyl, isopropyl, butyl, *sec*-butyl, isobutyl, and *tert*-butyl.

4.4 Conformations of Alkanes

The study of the chemical and physical properties of different conformations of compounds, called conformational analysis, formed a basis for understanding the relationships between structure and properties. The energy difference between the conformation of a molecule in its most stable conformation and that required for the molecule in the transition state affects the rates of reactions.

Conformations of Ethane

The conformation of a molecule refers to different arrangements of atoms in a molecule that result from rotation about carbon–carbon sigma bonds. The conformations (conformers) of ethane have low energy forms that are "staggered" and high energy conformations that are "eclipsed." The staggered conformation is the most stable; the eclipsed conformation is the least stable. In general, the bonding electron pairs of the carbon–hydrogen bonds of neighboring carbon atoms tend to stay as far apart as possible.

Newman projection formulas give us a method for conveying three-dimensional information in two dimensions. The energy difference between one staggered conformation and another is equal to that required to get past an eclipsed conformation. The energy difference is called **torsional strain**. For each carbon–hydrogen bond, the contribution to the torsional strain is 4.2 kJ mole⁻¹.

Organic Chemistry Study Guide: Key Concepts, Problems, and Solutions http://dx.doi.org/10.1016/B978-0-12-801889-7.00004-2

Conformations of Propane

Rotation about a carbon–carbon bond of propane is similar to that of ethane. However, the eclipsed conformation now has a hydrogen–methyl interaction in addition to two hydrogen–hydrogen interactions, so the barrier to rotation is larger. The resulting increase in energy is attributed to van der Waals repulsion between atoms. Although small in this case, the **steric hindrance** between atoms is larger when atoms are larger or brought closer together.

Conformations of Butane

With butane, the comparison of conformations becomes more interesting. In fact, the concepts introduced here are important to the understanding of many other phenomena such as the stability of cycloalkanes and many reactions that we will encounter later. There are two nonequivalent staggered conformations—the *anti* conformation and the *gauche* conformation, which have a **dihedral angle**, or **torsion angle** of 180° and 60°, respectively. The *anti* conformation is the more stable because there is van der Waals repulsion between the two methyl groups in the gauche conformation.

anti butane   gauche butane

4.5 Cycloalkanes

Cycloalkanes, as their name tells us, contain rings. Some cyclic compounds have atoms that are shared between two or more rings. These are **spirocyclic, bridged-ring,** and **fused-ring compounds**.

The general formula for an alkane is C_nH_{2n+2}. Each ring in a compound reduces the number of hydrogen atoms by 2 relative to an alkane because a ring contains an extra carbon–carbon bond and, therefore, two fewer carbon–hydrogen bonds. Thus, the general formula for cycloalkanes with one ring is C_nH_{2n}, the formula for compounds with two rings is C_nH_{2n-2}, and so on.

Geometric isomers can result when two or more substituents are attached to the ring at different carbon atoms. If the substituents are on the same side of the ring, the compound is the *cis* isomer. When substituents are on opposite sides of the ring, the compound is the *trans* isomer. Geometric isomers are one type of **stereoisomer**.

Cycloalkanes are named by prefixing the term *cyclo-* to a name giving the number of carbon atoms in the ring. The number 1 carbon atom is selected based on the importance of a functional group or alkyl group attached to the ring. The direction of numbering is selected to give the lowest combination of numbers to the remaining substituents at the point of first difference. Geometric isomers are identified with the appropriate *cis-* or *trans-* prefix.

Small ring compounds are unstable due to **ring strain**, which is the result of the small bond angles required to maintain the structure. The most severely strained compounds are cyclopropane and cyclobutane.

4.6 Conformations of Cycloalkanes

The small ring compounds cyclopropane and cyclobutane are strained rings. Their total strain energy is a combination of bond angle strain and torsional strain. In cylcopropane, the carbon–carbon bonds are highly strained, and there is also steric strain because the carbon–hydrogen bonds of adjacent carbon atoms are eclipsed. In cyclobutane, there is considerable bond angle strain and some eclipsing strain. A little twisting decreases the torsional strain, but the bonds giving rise to the torsional strain are still close. In cyclopentane, the bond angle strain is small because the bond angles are nearly tetrahedral. However, torsional interactions still occur in the molecule. Twisting cyclopentane into an envelop conformation alleviates but does not completely eliminate torsion angle strain.

The most stable conformation of cyclohexane is a "chair" in which there are three axial bonds pointed up and three pointed down. There are also six equatorial bonds, which point out around the ring; three are pointed slightly upward and three slightly downward.

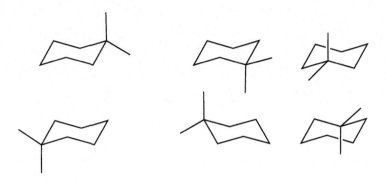

4.7 Conformation Mobility of Cyclohexane

The chair conformation of cyclohexane can change by a chair–chair interconversion, or "flip," and this process changes the orientations of all bonds. The equatorial bonds become axial and vice versa. Chair–chair interconversion passes through a "boat" conformation. Boat conformations are unstable and exist in vanishingly small amounts.

4.8 Conformations of Monosubstituted Cyclohexanes

Substituents bonded to the cyclohexane ring have a conformational preference for the equatorial position. Substituents in the axial position are sterically hindered to some degree because they are within the van der Waals radii of the axial hydrogen atoms at the C-3 and C-5 positions. This interaction is called a 1,3-diaxial interaction.

The energy differences between the axial and equatorial conformations of monosubstituted cyclohexanes are listed in Table 4.5. These values represent the magnitude of the two 1,3-diaxial interactions, and they depend on the size of the atom, the length of the bond, the polarizability of the atom, and the number of atoms bonded to the atom directly bonded to the cyclohexane ring.

4.9 Conformations of Disubstituted Cyclohexanes

In disubstituted cyclohexanes, not only the stability of the two possible conformations but also the relative stability of the geometric isomers depend on two factors. One is the inherent conformational preference of each substituent, and the other is any possible steric interaction between the two groups themselves. The *trans*-1,2-, the *cis*-1,3-, and the *trans*-1,4 dimethyl compounds are most stable in diequatorial conformations.

trans-1,2-dimethyl-
cyclohexane

cis-1,3-dimethyl-
cyclohexane

trans-1,4-dimethyl-
cyclohexane

These isomers are more stable than their respective equatorial/axial geometric isomers, because the axial methyl group has an unfavorable 1,3-diaxial interaction.

The most stable isomers for compounds with two different substituents are also the *trans*-1,2, the *cis*-1,3, and the *trans*-1,4 compounds. The difference in energy between either the isomers or the alternate conformations of each compound can be calculated by considering the conformational preferences of each group.

4.10 Polycyclic Molecules

The isomeric decalins provide the models for the rings that occur in fused ring compounds such as steroids. In *trans*-decalin, the hydrogens at the ring junction are both axial; in the *cis* isomer, one is equatorial and one is axial.

trans-decalin   *cis*-decalin

4.11 Physical Properties of Alkanes

Both alkanes and cycloalkanes have nonpolar covalent bonds. Thus, only van der Waals forces control the intermolecular interactions between neighboring molecules and those between solute and solvent molecules. Boiling points increase with increasing molecular weight and decrease with branching.

SOLUTIONS TO END-OF-CHAPTER EXERCISES

Molecular Formulas

4.1   Does each of the following molecular formulas for an acyclic hydrocarbon represent a saturated compound?

Answers: Acyclic saturated compounds have the general formula C_nH_{2n+2}. Only (b) and (d) meet this requirement.

(a) C_6H_{12} (b) C_5H_{12} (c) $C8H_{16}$ (d) $C10H_{22}$.

4.2   Can each of the following formulas correspond to an actual acyclic or cyclic molecule?
(a) C_6H_{14} Answer: This formula is possible because it has $2n + 2$ hydrogen atoms.

(b) $C_{10}H_{23}$ Answer: This formula is impossible because it has more than $2n + 2$ hydrogen atoms and also has an odd number of hydrogen atoms.

(c) C_7H_{14} Answer: This formula is possible because it has $2n$ hydrogen atoms, which can result from either a cyclic structure or unsaturation.

(d) C_5H_{14} Answer: This formula is impossible because it has more than $2n + 2$ hydrogen atoms.

4.3   Beeswax contains approximately 10% hentriacontane, a normal alkane with 31 carbon atoms. What is the molecular formula of hentriacontane? Write a completely condensed formula of hentriacontane.

Answer: The number of carbon atoms represented by n in the general formula for alkanes is 31. The number of hydrogen atoms must be $2n + 2$ or 64. The completely condensed formula for hentriacontane is $CH_3(CH_2)_{29}CH_3$.

4.4   Hectane is a normal alkane with 100 carbon atoms. What is the molecular formula of hectane? Write a completely condensed formula of hectane.
Answer: For $n = 100$, the value of $2n + 2$ is 202. The molecular formula is $C_{188}H_{202}$. The completely condensed formula for this normal alkane is $CH_3(CH_2)_{98}CH_3$.

Structural Formulas

4.5   Redraw each of the following so that the longest continuous chain is written horizontally.

(a) CH_3—CH_2
                    |
                    CH_2—CH_3

(b) CH_2—CH_2—CH—CH_2—CH_3
             | |
             CH_3 CH_2—CH_3

(c) CH_3—CH—CH_2—CH_3
                    |
                    CH_2—CH_3

(d) CH_3—CH—CH—CH_3
                    | |
                    CH_3 CH_2—CH_3

Answers:

(a) CH_3—CH_2—CH_2—CH_3

(b) CH_3—CH_2—CH_2—CH—CH_2—CH_3
                                      |
                                      CH_2—CH_3

(c) CH_3—CH_2—CH—CH_2—CH_3
                         |
                         CH_3

(d) CH_3—CH—CH—CH_2—CH_3
                | |
                CH_3 CH_3

4.6   Redraw each of the following so that the longest continuous chain is written horizontally.

(a)
$$CH_3-CH-CH_2$$
with CH_3 and CH_3 below the CH and CH_2

(b)
CH_3 above
$$CH_3-CH-CH_2$$
with CH_2-CH_3 below the CH

(c)
$$CH_3-CH-CH_2-CH_3$$
$$CH_3-CH-CH_2-CH_3$$

(d)
CH_2-CH_3 (upper)
$$CH_3-CH-CH_2$$
$$CH_2-CH_3$$ (lower)

Answers:

(a) $$CH_3-CH-CH_2-CH_3$$ with CH_3 below the CH

(b) $$CH_3-CH_2-CH-CH_2-CH_2-CH_3$$ with CH_3 below the CH

(c) $$CH_3-CH_2-CH-CH-CH_2-CH_3$$ with CH_3 and CH_3 below the two central carbons

(d) $$CH_3-CH_2-CH-CH-CH_2-CH_3$$ with CH_3 and CH_3 below the two central carbons

4.7   Which of the following structures represent the same compound?

I
$$CH_3-CH-CH-CH_2-CH_3$$
with CH_3 and $CH-CH_2-CH_3$ below

II
$$CH_3-CH-CH-CH_2-CH_2$$
with CH_3, CH_2-CH_3, CH_3 below

III
$$CH_3-CH-CH_2-CH_3$$
$$CH_3-CH-CH_2-CH_3$$

IV
$CH_2-CH_2-CH_3$ (upper)
$$CH_3-CH-CH-CH_3$$
$$CH_2-CH_3$$ (lower)

Answer: Both I and II have a chain of six carbon atoms with a methyl group at C-2 and an ethyl group at C-3. Both III and IV have a chain of seven carbon atoms with methyl groups at C-3 and C-4.

4.8   Which of the following structures represent the same compound?

I
$$CH_3-CH-CH-CH_2-CH_3$$
with CH_3 and CH_3 below

II
$$CH_3-CH-CH_2-CH_2$$
with CH_2-CH_3 and CH_3 below

III
$$CH_3-CH-CH_2-CH_3$$
$$CH_3-CH-CH_3$$

IV
CH_2-CH_3 (upper)
$$CH_3-CH-CH_2$$
$$CH_2-CH_3$$ (lower)

Answer: Both I and III have a chain of five carbon atoms with methyl groups at C-2 and C-3. Both III and IV have a chain of six carbon atoms with a methyl group at C-3.

52

Alkyl Groups

4.9 What is the common name for each of the following alkyl groups?

(a) CH_3—

(b) CH_3—CH_2
　　　　|
　　　CH_2—

(c) CH_3—CH—CH_2—CH_3
　　　　　|

(d) CH_3—CH—CH_2—
　　　　　|
　　　　CH_3

Answers: (a) methyl　　(b) propyl　　(c) *sec*-butyl　　(d) isobutyl

4.10 What is the common name for each of the following alkyl groups?

(a) CH_3—CH_2—

(b) CH_3—CH—
　　　　　|
　　　　CH_3

(c) CH_3—CH_2—CH_2—CH_2—

(d)
　　　　CH_3
　　　　　|
　CH_3—C—
　　　　　|
　　　　CH_3

Answers: (a) ethyl　　(b) isopropyl　　(c) butyl　　(d) *tert*-butyl

4.11 What is the common name for each of the following alkyl groups?

(a) CH_3—CH_2
　　　　|
　　　CH_2—CH_2—

(b) CH_3—CH—CH_3
　　　　　|
　　　　CH_2—

(c) CH_3—CH—CH_2—CH_3
　　　　　|
　　　　CH_2—

(d) CH_3—CH—CH_2—CH_2—
　　　　　|
　　　　CH_3

Answers: (a) butyl　　(b) 2-methylpropyl　　(c) 2-methylbutyl　　(d) 3-methylbutyl

4.12 What is the IUPAC name for each of the following alkyl groups?

(a) CH_3—CH—CH_2—
　　　　　|
　　　　CH_3

(b) CH_3—CH—CH_2—
　　　　　|
　　　　CH_2—CH_3

(c) CH_3—CH—CH_2—CH_2—
　　　　　|
　　　　CH_2—CH_3

(d)
　　　　CH_2—CH_3
　　　　　|
　CH_3—C—CH_2—CH_3
　　　　　|
　　　　CH_3

Answers: (a) 2-methylpropyl　　(b) 2-methylbutyl　　(c) 3-methylpentyl　　(d) 1,1-dimethylpropyl

4.13 The spermicide octoxynol-9 is used in diverse contraceptive products. Name the alkyl group to the left of the benzene ring.

$$CH_3\text{—}\underset{\underset{CH_3}{|}}{\overset{\overset{CH_3}{|}}{C}}\text{—}CH_2\text{—}\underset{\underset{CH_3}{|}}{\overset{\overset{CH_3}{|}}{C}}\text{—}\boxed{\bigcirc}\text{—}O\text{—}(CH_2)_8O\text{—}CH_2CH_2OH$$

octoxynol-9

Answer: 1,1,3,3-tetramethylbutyl

4.14 The name vitamin E actually refers to a series of closely related compounds called tocopherols. Name the complex alkyl group present in α-tocopherol.

$(CH_2CH_2CH_2CH)_3\text{—}CH_3$

Answer: 4,8,12-trimethyltridecyl

Nomenclature of Alkanes

4.15 Give the IUPAC name for each of the following compounds.

(a) $CH_3\text{—}\underset{\underset{CH_2\text{—}CH_3}{|}}{CH}\text{—}CH_3$

(b) $CH_2\text{—}CH_2\text{—}CH\text{—}CH_2\text{—}CH_3$ with CH_3 and CH_3 substituents

(c) $CH_3\text{—}CH\text{—}CH_2\text{—}CH_2$ with CH_3 and CH_3 substituents

(d) $CH_3\text{—}CH_2\text{—}\underset{\underset{CH_3}{|}}{CH}\text{—}CH_2\text{—}CH_3$

(e) $CH_3\text{—}CH\text{—}CH_2$ with CH_3 and $CH_2\text{—}CH_3$ substituents

(f) $CH_2\text{—}CH_2\text{—}CH\text{—}CH_3$ ... CH_3

Answers: (a) 2-methylbutane (b) 3-methylhexane (c) 2-methylpentane (d) 3-methylpentane (e) 2-methylpentane (f) 5-methylnonane

4.16 Give the IUPAC name for each of the following compounds.

(a) $CH_3\text{—}CH\text{—}CH$ with CH_3 and $CH_2\text{—}CH_2\text{—}CH_3$ substituents

(b) $CH_3\text{—}CH\text{—}CH_2$ CH_3 with CH_3 and $CH_2\text{—}CH\text{—}CH_3$ substituents

(c) $CH_3\text{—}CH\text{—}CH_2\text{—}CH_3$ / $CH_3\text{—}CH\text{—}CH_2\text{—}CH_2\text{—}CH_3$

(d) $CH_3\text{—}CH\text{—}CH_2\text{—}CH\text{—}CH_3$ with $CH_2\text{—}CH_3$ and $CH_2\text{—}CH_3$ substituents

(e) $CH_3\text{—}CH\text{—}CH_2\text{—}CH_2\text{—}CH_3$ with $CH_2\text{—}CH_3$ substituent (f) $CH_3\text{—}CH_2\text{—}CH\text{—}CH_2$ with $CH_3\text{—}CH_2$ and CH_3 substituents

Answers: (a) 2-methylhexane (b) 2,5-dimethylhexane (c) 3,4-dimethylheptane (d) 3,5-dimethylheptane (e) 3-methylhexane (f) 3-ethylpentane

54

4.17   Give the IUPAC name for the following compound.

$$CH_3—CH_2—CH_2—CH_2—CH—CH_2—CH_2—CH_2—CH_2—CH_3$$
$$CH_3—CH_2—CH—CH_2—CH_3$$

Answer: 5-(1-ethylpropyl)decane

4.18   Give the IUPAC name for the following compound.

$$CH_3—CH_2—CH_2—CH_2—CH—CH_2—CH_2—CH_2—CH_3$$
$$CH_3—C—CH_3$$
$$CH_2—CH_3$$

Answer: 5-(1,1-dimethylpropyl)nonane

4.19   Write the structural formula for each of the following compounds.
(a) 3-methylpentane   (b) 3,4-dimethylhexane   (c) 2,2,3-trimethylpentane
(d) 4-ethylheptane   (e) 2,3,4,5-tetramethylhexane

Answers: (a) $CH_3—CH_2—CH—CH_2—CH_3$
$$CH_3$$

(b) $CH_3—CH_2—CH—CH—CH_2—CH_3$
$$CH_3 \quad CH_3$$

(c) $$CH_3$$
$$CH_3—C—CH—CH_2—CH_3$$
$$CH_3 \quad CH_3$$

(d) $CH_3—CH_2—CH_2—CH—CH_2—CH_2—CH_3$
$$CH_2CH_3$$

(e) $CH_3—CH—CH—CH—CH_2—CH_3$
$$CH_3 \quad CH_3 \quad CH_3$$

55

4.20   Write the structural formula for each of the following compounds.
     (a) 2-methylpentane   (b) 3-ethylhexane (c) 2,2,4-trimethylhexane
     (d) 2,4-dimethylheptane   (e) 2,2,3,3-tetramethylpentane

Answers: (a) CH_3—CH—CH_2—CH_2—CH_3
                 |
                CH_3

(b) CH_3—CH_2—CH—CH_2—CH_2—CH_3
                 |
              CH_2CH_3

(c) CH_3—CH—CH—CH—CH_2—CH_3
           | | |
          CH_3 CH_3 CH_3

(d) CH_3—CH—CH—CH_2—CH_2—CH_2—CH_3
           | |
         CH_3 CH_3

            CH_3 CH_3
             | |
(e) CH_3—C—C—CH_2—CH_3
           | |
         CH_3 CH_3

4.21   Write the structural formula for each of the following compounds.
     (a) 4-(1-methylethyl)heptane   (b) 5-(1,1-dimethylethyl)nonane   (c) 5-(1-methylpropyl)decane

Answers: (a) CH_3—CH_2—CH_2—CH—CH_2—CH_2—CH_3
                         |
                CH_3—CH—CH_3

(b) CH_3—CH_2—CH_2—CH_2—CH—CH_2—CH_2—CH_2—CH_3
                             |
                  CH_3—C—CH_3
                           |
                     CH_3

(c) CH_3—CH_2—CH_2—CH_2—CH—CH_2—CH_2—CH_2—CH_2—CH_3
                             |
                  CH_3—CH—CH_2—CH_3

4.22   Write the structural formula for each of the following compounds.
        (a) 5-(2-methylpropyl)nonane   (b) 4-butyl nonane   (c) 5-(2,2-dimethylpropyl)decane

Answers: (a)

$$CH_3—CH_2—CH_2—CH_2—CH—CH_2—CH_2—CH_2—CH_3$$
$$|$$
$$CH_2—CH—CH_3$$
$$|$$
$$CH_3$$

(b)

$$CH_3—CH_2—CH_2—CH_2—CH—CH_2—CH_2—CH_2—CH_3$$
$$|$$
$$CH_2—CH_2—CH_2—CH_3$$

(c)

$$CH_3—CH_2—CH_2—CH_2—CH—CH_2—CH_2—CH_2—CH_2—CH_3$$
$$|$$
$$CH_3—C—CH_2—CH_3$$
$$|$$
$$CH_3$$

Isomers

4.23   There are nine isomeric C_7H_{16} compounds. Name the isomers that have a single methyl group as a branch.

Answers: 2-methylhexane and 3-methylhexane

4.24   There are nine isomeric C_7H_{16} compounds. Name the isomers that have two methyl groups as branches and are named as dimethyl-substituted pentanes.

Answers: 2,2-dimethylpentane, 3,3-dimethylpentane, 2,3-dimethylpentane, and 2,4-dimethylpentane

Classification of Carbon Atoms

4.25   Classify each carbon atom in the following compounds as primary, secondary, or tertiary.

(a) $CH_3—CH_2—CH_2—CH_2—CH_3$ (b) $CH_3—CH_2—CH—CH_2—CH_3$
$$|$$
$$CH_3$$

(c)
$$CH_3$$
$$|$$
$$CH_3—C—CH_2—CH_3$$
$$|$$
$$CH_3$$

(d) $CH_3—CH—CH—CH_3$
$$|\quad|$$
$$CH_3\ CH_3$$

Answers:

(a)
$$1° \quad 2° \quad 2° \quad 2° \quad 1°$$
$$CH_3—CH_2—CH_2—CH_2—CH_3$$

(b)
$$1° \quad 2° \quad 3° \quad 2° \quad 1°$$
$$CH_3—CH_2—CH—CH_2—CH_3$$
$$|$$
$$CH_3$$
$$1°$$

(c)
$$1° \quad 3° \quad 3° \quad 1°$$
$$CH_3—CH—CH—CH_3$$
$$|\qquad|$$
$$CH_3 \quad CH_3$$
$$1° \qquad 1°$$

57

4.26   Classify each carbon atom in the following compounds as primary, secondary, tertiary, or quaternary.

(a) CH_3—CH—CH_2—CH—CH_3 (b) CH_3—CH_2—CH_2—CH_3
           | |
           CH_3 CH_3

(c) CH_3—CH—CH_3 (d) CH_3—C—C—CH_3
              | | |
      CH_3—CH—CH_3 CH_3 CH_3

with (d) top: CH_3 CH_3

Answers:

         $1°$ $3°$ $2°$ $3°$ $1°$ $1°$ $2°$ $2°$ $1°$
(a) CH_3——CH——CH_2——CH——CH_3 (b) CH_3——CH_2——CH_2——CH_3
                 | |
                 CH_3 CH_3
                 $1°$ $1°$

         $1°$ $3°$ $1°$ $1°$ $1°$
(c) CH_3——CH——CH_3 CH_3 CH_3
                 | $1°$ | | $1°$
         CH_3——CH——CH_3 (d) CH_3——C——C——CH_3
         $1°$ $3°$ $1°$ | |
                                           $4°$ CH_3 CH_3 $4°$
                                                    $1°$ $1°$

4.27   Draw the structure of a compound with molecular formula C_5H_{12} that has one quaternary and four primary carbon atoms.

Answer: CH_3
                    |
        CH_3——C——CH_3
                    |
                   CH_3

4.28   Draw the structure of a compound with molecular formula C_6H_{14} that has two tertiary and four primary carbon atoms.

Answer: $1°$ $3°$ $1°$
          CH_3——CH——CH_3
                     |
          CH_3——CH——CH_3
          $1°$ $3°$ $1°$

          2,3-dimethylbutane

58

4.29   Determine the number of primary, secondary, tertiary, and quaternary carbon atoms in each of the following compounds.

Answers:
(a) four primary and one quaternary
(b) three primary, one secondary, and one tertiary
(c) three primary, two secondary, and one tertiary
(d) four primary and two tertiary

(a)
$$CH_3-\overset{\underset{\displaystyle \underset{1^\circ}{CH_3}}{\vert}}{\underset{4^\circ}{\overset{\overset{\displaystyle \overset{1^\circ}{CH_3}}{\vert}}{C}}}-CH_3$$

(b) $CH_3-\underset{\underset{1^\circ}{\overset{\vert}{CH_3}}}{\overset{3^\circ}{CH}}-\overset{2^\circ}{CH_2}-\overset{1^\circ}{CH_3}$

(c) $\overset{1^\circ}{CH_3}-\underset{\underset{1^\circ}{\overset{\vert}{CH_3}}}{\overset{3^\circ}{CH}}-\overset{2^\circ}{CH_2}-\overset{2^\circ}{CH_2}-\overset{1^\circ}{CH_3}$

(d) $\overset{1^\circ}{CH_3}-\underset{\underset{1^\circ}{\overset{\vert}{CH_3}}-\underset{1^\circ}{\overset{3^\circ}{CH}}-\overset{1^\circ}{CH_3}}{\overset{3^\circ}{CH}}-\overset{1^\circ}{CH_3}$

4.30   Determine the number of primary, secondary, tertiary, and quaternary carbon atoms in each of the following compounds.

Answers:
(a) six primary, one secondary, and two quaternary
(b) four primary, one secondary, and two tertiary
(c) three primary, two secondary, and one tertiary
(d) five primary and three tertiary

(a) $CH_3-\underset{\underset{CH_3}{\overset{\vert}{\vert}}}{\overset{\overset{CH_3}{\overset{\vert}{\vert}}}{C}}-CH_2-\underset{\underset{CH_3}{\overset{\vert}{\vert}}}{\overset{\overset{CH_3}{\overset{\vert}{\vert}}}{C}}-CH_3$

(b) $\overset{1^\circ}{CH_3}-\overset{2^\circ}{CH_2}-\underset{\underset{1^\circ}{\overset{\vert}{CH_3}}}{\overset{3^\circ}{CH}}-\overset{2^\circ}{CH_2}-\overset{1^\circ}{CH_3}$

(c) $CH_3-CH_2-\underset{\overset{\vert}{CH_3}}{CH}-CH_2-CH_3$

(d) $CH_3-\underset{\overset{\vert}{CH_3}}{CH}-\underset{\overset{\vert}{CH_3}}{CH}-\underset{\overset{\vert}{CH_3}}{CH}-CH_2-CH_3$

59

Cycloalkanes

4.31   Write condensed planar formulas for each of the following compounds.
  (a) chlorocyclopropane   (b) 1,1-dimethylcyclobutane   (c) cyclooctane

Answers:

(a)   (b)   (c)

4.32   Write condensed planar formulas for each of the following compounds.
  (a) bromocyclobutane   (b) 1,1-dichlorocyclopropane   (c) cyclopentane

Answers:

(a)   (b)   (c)

4.33   Name each of the following compounds.

(a)   (b)   (c)   (d)

Answers: (a) 1,1-dimethylcycloheptane   (b) cyclodecane   (c) *trans*-1,2-dichlorocyclohexane   (d) 1,1-dichlorocyclohexane

4.34   Name each of the following compounds.

Answers: (a) butylcyclooctane
  (b) isopropylcyclohexane
  (c) *cis*-1,3-dimethylcyclohexane
  (d) 1,1,3-trichlorocyclohexane   (a)   (b)

(c)   (d)

4.35   A saturated refrigerant has the molecular formula C_4F_8. Draw structural formulas for two possible isomers of this compound.

Answers:

(a)   (b)

60

4.36   How many isomeric saturated hydrocarbons have the molecular formula C_5H_{10}?

Answers: Five compounds have this formula: (i) cyclopentane, (ii) methylcyclobutane, (iii) 1,1-dimethylcyclopropane,
          (iv) *cis*-1,2-dimethylcyclopropane, and (v) *trans*-1,2-dimethylcyclopropane

Bicyclic compounds
4.37   What is the molecular formula of each of the following compounds?

Answers: (a) $C_{10}H_{16}$ (b) $C_{10}H_{12}O$ (c) $C_7H_{13}N$ (d) C_6H_6

4.38   What is the molecular formula of each of the following compounds?

Answers: (a) $C_{10}H_{18}O$ (b) $C_{10}H_{15}$ (c) $C_{10}H_{14}$ (d) C_8H_{14}

Polycyclic compounds
4.39   How many rings are present in each of the following polycyclic compounds?

Answers: (a) tricyclic   (b) bicyclic   (c) tricyclic

4.40   How many rings are present in each of the following polycyclic compounds?

Answers: (a) pentacyclic   (b) tricyclic   (c) bicyclic

Properties of Hydrocarbons

4.41 Which of the isomeric C_8H_{18} compounds has the highest boiling point? Which has the lowest boiling point?

Answer: Octane has the highest boiling point. All of its isomers have branches and should have lower boiling points. 2,2,3,3-Tetra-methylbutane has the largest number of branches and should have the lowest boiling point.

4.42 The boiling point of methylcyclopentane is lower than the boiling point of cyclohexane. Suggest a reason why.

Answer: Methylcyclopentane compared to cyclohexane has a more compact structure and the difference results from the same phenomena observed for branched alkanes.

Newman Projection Formulas of alkanes

4.43 Draw the Newman projection of the staggered conformation of 2,2-dimethylpropane around the C-1 to C-2 bond.

Answer:

4.44 Draw the Newman projections of the two possible staggered conformations of 2,3-dimethylbutane around the C-2 to C-3 bond.

Answer:

4.45 Draw the Newman projections of the two possible staggered conformations of 2-methylbutane around the C-2 to C-3 bond. Which is the more stable?

Answer:

more stable

4.46 Draw the Newman projections of the two possible staggered conformations of 2,2-dimethylpentane around the C-3 to C-4 bond. Which is the more stable?

Answer:

more stable

Stabilities of Acyclic Conformations

4.47   Do you expect the barrier to rotation around the central bond for CH_3—CH_2—SiH_2—CH_3 to be smaller or larger than the barrier to rotation for butane? Why?

   Answer: The silicon–carbon bond length is longer than a carbon–carbon bond. Thus, the distance between the two methyl groups is larger in the silicon compound than in butane. The barrier to rotation should be lower in the silicon compound because all of the sets of eclipsing bonded pairs of electrons are separated by a larger distance.

4.48   Draw a potential energy diagram for rotation around the C-2 to C-3 bond of 2,2-dimethylbutane.

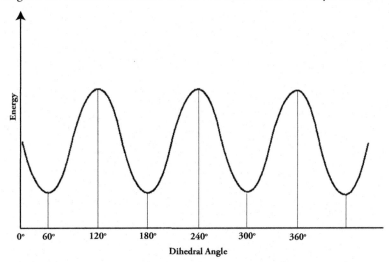

4.49   Draw a potential energy diagram for rotation around the C-2 to C-3 bond of 2-methylbutane.

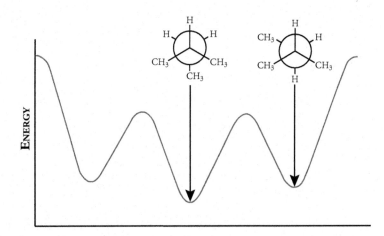

4.50   2-Chloroethanol (ClCH$_2$CH$_2$OH) is most stable in the gauche conformation. Suggest a reason for this fact.

Answer: There is an attractive interaction between chlorine and the hydroxyl group. The interaction is a dipole–dipole attraction that resembles a hydrogen bond.

4.51   1-Chloropropane is most stable in the gauche conformation. What does this fact indicate about the interaction of chlorine and a methyl group in this compound?

Answer: There is an attractive van der Waals interaction between chlorine and the methyl group that results from the polarizability of the chlorine atom.

4.52   Draw the two staggered conformations of 1,2-dichloroethane. Which of the conformations has a dipole moment? The dipole moment of 1,2-dichloroethane is 1.1 D. Does this fact provide any information about the composition of the mixture of conformations?

Answer: The *anti* conformation does not have a dipole moment because the bond moments of the C—Cl bonds cancel. The gauche conformation has a dipole moment. The observed dipole moment indicates that some of the compound must exist in the gauche conformation.

4.53   Ethylene glycol (HOCH$_2$CH$_2$OH) forms intramolecular hydrogen bonds. Does this fact provide any information about the composition of the mixture of conformations?

Answer: The *anti* conformation cannot form intramolecular hydrogen bonds because the hydroxyl groups are widely separated. The hydroxyl groups are sufficiently close in the gauche conformation to form an intramolecular hydrogen bond. The structure has five atoms in a hydrogen-bonded "ring" counting the atoms starting from the hydroxyl hydrogen atom of one hydroxyl group to the oxygen atom of the other hydroxyl group.

Conformations of Cyclohexanes

4.54   Draw the most stable conformation of the equatorial form of methylcyclohexane showing the relationship of the methyl hydrogen atoms to the hydrogen atom at C-1.

Answer:

4.55   Draw the conformation of the axial form of methylcyclohexane showing the relationship of the methyl hydrogen atoms to C-3 and C-5.

Answer:

4.56   Draw the most stable chair conformation of each of the following compounds.
       (a) *trans*-l-fluoro-3-methylcyclohexane   (b) *trans*-l-*tert*-butyl-3-methylcyclohexane   (c) *trans*-1,2-dimethylcyclohexane

Answers:

4.57   Draw the most stable chair conformation of each of the following compounds.
       (a) *cis*-1,1,4-trimethylcyclohexane   (b) *trans*-1,1,3-trimethylcyclohexane   (c) *cis*-1-fluoro-4-ethylcyclohexane

Answers:

65

4.58   Why is the steric strain caused by the *tert*-butyl group so different from those of methyl, ethyl, and isopropyl groups?

Answer: In any conformation of the *tert*-butyl compound, there is a methyl group located over the cyclohexane ring and there is severe steric repulsion between it and the axial hydrogen atoms at C-3 and C-5.

4.59   Within experimental error, the steric strain caused by a bromine atom is the same as that of a chlorine atom. Taking into account the "size" of the atoms and the length of the carbon–halogen bond, explain these data.

Answer: The C—Br bond is longer than the C—Cl bond, so the distance separating the bromine atom from the C-3 and C-5 hydrogen atoms is larger. In addition, the bromine atom is more polarizable than the chlorine atom, and its electrons may be more easily distorted away from the steric congestion in the axial conformation.

4.60   *cis*-1,3-Cyclohexanediol is most stable in a diaxial conformation. Suggest a reason for this "unusual" stability.

Answer: The hydroxyl groups are close enough in the diaxial conformation to form an intramolecular hydrogen bond. The hydrogen-bonded structure has 6 atoms in a hydrogen-bonded "ring," counting the atoms starting from the hydroxyl hydrogen atom of one hydroxyl group to the oxygen atom of the other hydroxyl group.

4.61   *trans*-1,3-Di-*tert*-butylcyclohexane exists in a twist boat conformation, rather than a chair conformation. Why?

Answer: An axial *tert*-butyl group exists in the *trans* isomer, which has a steric repulsion of 22 kJ mole^{-1} (Table 4.6). This repulsion is eliminated in the twist boat conformation, even though it is 22 kJ mole^{-1} less stable than the chair conformation because both *tert*-butyl groups can occupy equatorial-like positions in this conformation.

4.62   The diaxial conformation of *cis*-1,3-dimethylcyclohexane is 23 kJ mole^{-1} less stable than the diequatorial conformation. Why is this value larger than twice the steric strain of a methyl group?

Answer: The steric strain results from repulsion between an axial group and the axial hydrogen atoms at C-3 and C-5. Repulsion between an axial group and larger atoms at the C-3 and C-5 positions must be substantially larger. In the diaxial conformation of *cis*-1,3-dimethylcyclohexane, that repulsion is between two methyl groups.

4.63   The diaxial conformation of *cis*-l-chloro-3-methylcyclohexane is 16 kJ mole^{-1} less stable than the diequatorial conformation. Why is this value larger than the sum of the steric strains of a chlorine atom and a methyl group?

Answer: The steric strains all due to repulsion between an axial group and the axial hydrogen atoms at C-3 and C-5. Repulsion between an axial methyl group and a larger atom at C-1, such as chlorine, is substantially larger.

Bicyclic compounds
4.64   An isomerization equilibrium between *cis*-decalin and *trans*-decalin can be established by heating the mixture to about 300 °C in the presence of a palladium catalyst. The *trans* isomer predominates. Why is the *trans* isomer more stable than the *cis* isomer?

Answer: The ring junction in *cis*-decalin is equatorial axial, so there are always unfavorable 1,3-diaxial interactions; but, in *trans*-decalin, the ring junction is diequatorial, and there are fewer unfavorable steric interactions.

Pd / C catalyst

$K_{eq} > 1$

cis ring junction   *trans* ring junction

Steroids

4.65   Examine the structure of an A/B (*trans*) steroid skeleton and determine whether each of the following is in an equatorial or axial location.
(a) a 2α hydroxyl group   (b) a 3α chlorine atom   (c) a 6α amino (—NH₂) group
(d) an 11β bromine atom   (e) a 12β cyano group

Answers: (a), (c), and (e) have their groups in equatorial positions.

(a)

2α hydroxyl group

(b)

3α chlorine atom

(c)

6α amino group

(d)

11β bromine atom

(e)

12β cyano group

4.66   Examine the structure of an A/B (*cis*) steroid skeleton and determine whether each of the following is in an equatorial or axial location.
(a) a 1β hydroxyl group   (b) a 3α chlorine atom   (c) a 6α amino (—NH₂) group
(d) an 11α bromine atom   (e) a 12α cyano group

Answers: (b) and (d) have their groups in equatorial positions.

(a)

1β hydroxyl group

(b)

3α chlorine atom

(c)

6β amino group

(d)

11α bromine atom

(e)

12α cyano group

5

ALKENES
STRUCTURES AND PROPERTIES

KEYS TO THE CHAPTER

5.1 Alkenes

A double bond in an alkene consists of one sigma bond and one pi bond. A triple bond in an alkyne consists of one sigma bond and two pi bonds.

On the average, the electrons in π bonds are more reactive than those in σ bonds—the subject of Chapter 6. Virtually all of the reactions of alkenes and alkynes (Chapter 7) involve the reactivity of π bonds. Two or more double bonds separated by one carbon–carbon single bond constitute conjugated double bonds. A compound may contain one set or a whole series of conjugated double bonds. Double bonds separated by more than one carbon–carbon single bond are not conjugated. The chemistry of conjugated double bonds is a more specialized subject than the reactions of alkenes, and it is discussed in Chapter 11.

5.2 Structure and Bonding of Alkenes

Functional groups are structural features of organic compounds other than carbon–carbon single bonds. Alkenes consist of a sigma bonded framework, which for the most part we can ignore, and two electrons in a π bond that results from the side-by-side overlap of two 2p orbitals. As a consequence of the π bond, the two carbon atoms and the four atoms bonded to these atoms are coplanar.

The concept of the % s character of the hybrid orbitals of an alkene is the first of many times throughout the study of organic chemistry that we will use this terminology. The % s character affects the bond energy of a σ bond. In order to apply the concept, we have to be careful to understand what is required in a chemical reaction—for example, does the process occur by homolytic or heterolytic cleavage of the bond? In the case of a bond dissociation, homolytic cleavage occurs, and each atom retains one electron in the bond. If the electrons are held more tightly by an atom as a consequence of its larger % s character, then it will take more energy to break the bond. That is why the bond dissociation energy increases with increased % s character.

Bond distances also depend on % s character. If electrons are held more tightly by the carbon atom as a result of its hybridization, then the bond length must be shorter.

Time and time again, we will find that there are advantages not only to classifying compounds by their functional groups but also to fine tuning the classification by subclasses. Fortunately, the terminology of **degree of substitution** is straightforward. Simply count the number of alkyl groups attached to the double-bonded carbon atoms.

5.3 Unsaturation Number

Before we can determine what functional groups might be part of a structure, we have to limit the possibilities based on the molecular formula. When we look at molecular formulas, we have to recognize likely possibilities for functional groups.

The presence of double or triple bonds in a structure is indicated by the **degree of unsaturation**. Each multiple bond diminishes the maximum number of hydrogen atoms by two. Each pair of "missing" hydrogen atoms can correspond to a pi bond. However, it may also signal the presence of a ring. Based on this one criterion, we can't say which structural feature is present. But we have limited the possibilities. The effect of other atoms on the unsaturation number is summed up as follows:

1. Oxygen and sulfur have no effect.
2. Count halogen atoms as hydrogen atoms.
3. Add a hydrogen.

5.4 Geometric Isomerism

As a consequence of the restricted rotation about a carbon–carbon double bond, alkenes can exist as geometric isomers if two different groups are bonded to each carbon atom of the double bond. If either carbon atom has two identical groups, then only one compound is possible.

Cycloalkenes usually have a *cis* configuration. Bridging a *trans* configuration in a cycloalkene requires at least an eight-membered ring due to geometric constraints.

Organic Chemistry Study Guide: Key Concepts, Problems, and Solutions http://dx.doi.org/10.1016/B978-0-12-801889-7.00005-4

5.5 The *E,Z* Designation of Geometric Isomers

To decide whether a geometric isomer is *E* or *Z* we apply a set of **sequence rules**. The most important criterion for deciding on the priority of the groups is atomic number. Furthermore, just one atom of high atomic number has a higher priority than any number of atoms with lower atomic numbers. For example, —CH_2Br has a higher priority than —CCl_3 because bromine has a higher atomic number than chlorine. Another criterion is the "point of first difference", which means keep going until something important is found in a chain of atoms that makes it more important than a second group of atoms. Thus, the group —$CH_2CH_2CHFCH_2CH_3$ has a higher priority than —$CH_2CH_2CH_2CHBrCH_3$ because the fluorine atom is found sooner on the chain (at C-3) than the bromine atom (which is at C-4).

5.6 Nomenclature of Alkenes

We name alkenes by a set of rules that parallel those used for alkanes, with the added complication of the double bond and possibility of geometric isomers. Some of the rules are the same, and others are closely related to those used for alkanes. The double bond takes priority in numbering the longest chain, which must contain the double bond. Only the first number of the two carbon atoms in the double bond is used to name a compound. In cycloalkenes, one of the two carbon atoms of the double bond is given the number 1 and the numbering of the carbon atoms then proceeds through the second carbon atom of the double bond. The name does not include a number locating the double bond, but numbers are required for the location of substituents such as alkyl groups.

5.7 Physical Properties of Alkenes

As we have observed for alkanes, trends in physical properties are related to the size and structure of the compounds studied. Boiling points for a homologous series of alkenes increase with molecular weight, reflecting the increase in strength of London forces. The identity and geometry of the groups attached to the double-bonded carbon atoms determine whether an alkene is polar or nonpolar. If one isomer is polar and the other is nonpolar, the polar compound has a higher boiling point.

5.8 Oxidation of Alkenes

The heats of combustion of alkenes allow us to compare the relative stabilities of isomeric compounds. For isomers, the same number of moles of carbon dioxide and water is formed. Thus, a comparison of the heats of combustion indicates the difference in the enthalpy content of the isomers. Three generalizations can be made based on the data depicted in Figure 5.4 in the text. These are:

1.   Branched isomers are more stable than unbranched ones, so they have smaller heats of combustion.
2.   More highly substituted alkenes are more stable, so they have smaller heats of combustion.
3.   Alkenes with the *E* configuration are more stable than alkenes with the *Z* configuration, so they have smaller heats of combustion.

The increased stability of alkenes with increased substitution results from the release of electron density from the sp^3-hybridized alkyl groups toward the sp^2-hybridized atoms of the carbon–carbon double bond. The electron donating capacity of alkyl groups toward sp^2-hybridized centers is a common feature that explains many chemical reactions that we will encounter in later chapters.

5.9 Reduction of Alkenes

The reaction of an alkene with hydrogen to give an alkane, called hydrogenation, is a reduction reaction. The order of reactivity of alkenes decreases with increased substitution of the double bond. Thus, the reaction shows some regioselectivity, which means that one double bond in a compound with two or more double bonds can often be reduced in preference over another double bond. This regioselectivity, which is the tendency of a reaction to generate one isomer preferentially over another, is another concept that we will encounter many times as we proceed. Transition metal catalysts such as platinum, palladium, Adams catalyst (PtO_2), and a special form of nickel called Raney nickel can be used. Although the hydrogenation reaction is usually carried out under heterogeneous conditions, one catalyst known as the Wilkinson catalyst is used for hydrogenation of alkenes under homogeneous conditions. Neither heterogeneous nor homogeneous catalysts reduce functional groups such as the carbonyl group of ketones, carboxylic acids, or esters under the relatively mild conditions required to hydrogenate a carbon–carbon double bond.

5.10 Mechanism of Catalytic Hydrogenation

Heterogeneous catalytic hydrogenation occurs on the surface of the metal and transfers the two hydrogen atoms to the carbon atoms by a *syn* addition process. For many alkenes, there is no difference in the two faces of the double bond, so there is no stereoselectivity in the reaction. However, the environment near the double bond can sometimes make one face less accessible to the transfer of hydrogen. When hydrogen adds to one face, the atoms bonded to the carbon atoms of the double bond are "pushed" to the opposite side. Because hydrogen adds from the sterically less hindered side, the groups are forced into a more sterically hindered environment.

5.11 Heats of Hydrogenation

The measurement of the heats of hydrogenation ($\Delta H° < 0$) of isomeric alkenes is used to determine their relative stabilities. The values of $\Delta H°_{hydrogenation}$ are all negative, and the reference point is the saturated hydrocarbon. The size of the term indicates how stable the alkene is relative to its isomers. Heats of hydrogenation can be used to compare the relative stabilities of isomeric alkenes only if the same alkane results from both compounds. For example, isomers with different degrees of branching cannot be directly compared because they don't produce the same saturated hydrocarbon when hydrogenated. We can still reach some basic conclusions about alkene stability if the isomers are reasonably similar.

1.   Branched isomers are more stable than unbranched isomers.
2.   Alkene stability increases with increasing substitution.
3.   An alkene with the *E* configuration is more stable than its *Z* isomer.

SUMMARY OF REACTIONS   1. Heterogeneous Catalytic Hydrogenation

(a)   H$_2$ / Pd/C

(b)   2H$_2$ / Raney Ni

(c)   H$_2$ / Pd/C   1 atm

(d)   H$_2$ / PtO$_2$

2. Homogeneous Catalytic Hydrogenation

H$_2$ [C$_6$H$_5$)$_3$P]$_3$RhClH

SOLUTIONS TO EXERCISES

Molecular Formulas

5.1   What is the molecular formula for a compound with each of the following structural features?
    (a)   six carbon atoms and one double bond
    (b)   five carbon atoms and two double bonds
    (c)   seven carbon atoms, a ring, and one double bond

Answers: (a) C_6H_{12} (b) C_5H_8 (c) C_7H_{12}

5.2   What is the molecular formula for a compound with each of the following structural features?
    (a)   four carbon atoms and two double bonds
    (b)   ten carbon atoms and two rings
    (c) ten carbon atoms, two rings, and five double bonds

Answers: (a) C_4H_6 (b) $C_{10}H_{18}$ (c) $C_{10}H_8$

5.3   Write the molecular formula for each of the following compounds.

Answers: (a) C_8H_{16} (b) C_6H_{12}
         (c) $C_{10}H_{16}$ (d) C_8H_{14}

5.4   Write the molecular formula for each of the following compounds.

Answers: (a) C_8H_{12} (b) $C_{10}H_{14}$
         (c) C_7H_{10} (d) C_7H_{14}

Classification of Alkenes

5.5   Classify each double bond in the alkenes in Exercise 5.3 by its substitution pattern.
Answers: (a) trisubstituted (b) trisubstituted (c) disubstituted   (d) trisubstituted

5.6   Classify each double bond in the alkenes in Exercise 5.4 by its substitution pattern.
Answers: (a) di- and trisubstituted   (b) di- and trisubstituted   (c) disubstituted   (d) trisubstituted

5.7   Indicate the degree of substitution of the double bond in each of the following compounds.

Answer: (a) trisubstituted   (a)   Cholesterol, a steroid

(c)   Saffrole, a carcinogen found in sassafras root

(d)   Tamoxifen, a drug used in the treatment of breast cancer

$OCH_2CH_2N(CH_3)_2$

Answer: (c) monosubstituted

Answer: (d) tetrasubstituted

5.8   Indicate the degree of substitution of all double bonds in each of the following compounds, polyenes found in natural oils.

Answers: (a) di-, tri, and trisubstituted (a) zingiberene   (b) β-selenene   (c) limonene
         (b) both disubstituted
         (c) di- and trisubstituted

Unsaturation Number

5.9   Calculate the unsaturation number for each of the following compounds.
       (a) camphor, $C_{10}H_{16}O$ (b) nicotine, $C_{10}H_{14}N_2$
       (c) vitamin B6, $C_8H_9NO_2$ (d) hexachlorophene, $C_{13}H_6O_2Cl_6$

Answers: (a) 3   (b) 5   (c) 5   (d) 8

5.10   Calculate the unsaturation number for each of the following compounds.
       (a) β-carotene, $C_{40}H_{56}$ (b) amphetamine, $C_9H_{13}N$ (c) DDT, $C_{11}H_9Cl_5$ (d) aspirin, $C_9H_8O_4$

Answers: (a) 13   (b) 5   (c) 5   (d) 6

5.11   Calculate the unsaturation number for each of the following compounds.
       (a) vitamin A, $C_{20}H_{30}O$ (b) sucrose, $C_{12}H_{22}O_{11}$ (c) vitamin B2, $C_{17}H_{20}N_4O_6$ (d) saccharin, $C_7H_5NO_3S$

Answers: (a) 2   (b) 2   (c) 10   (d) 6

5.12   Calculate the unsaturation number for each of the following compounds.
       (a) L-dopa, $C_9H_{11}NO_4$ (b) prontosil, $C_{12}H_{13}N_5O_2S$
       (c) testosterone, $C_{19}H_{28}O_2$ (d) phenobarbital, $C_{12}H_{12}N_2O_3$

Answers: (a) 5   (b) 9   (c) 6   (d) 8

Geometric Isomers

5.13   Which of the following molecules can exist as *cis* and *trans* isomers?
  (a) $CH_3CH=CHBr$ (b) $CH_2=CHCH_2Br$ (c) $CH_3CH=CHCH_2Cl$ (d) $(CH_3)_2C=CHCH_3$

Answer: only (a) and (c)

5.14   Which of the following molecules can exist as *cis* and *trans* isomers?
  (a) $CH_3CH=CBr_2$ (b) $CH_2=CHCHBr_2$ (c) $CH_3CH=CHCHCl_2$ (d) $CH_3CH_2CH=C(CH_3)_2$

Answer: only (c)

5.15   Which of the following molecules can exist as *cis* and *trans* isomers?
  (a) 1-hexene   (b) 3-heptene   (c) 4-methyl-2-pentene   (d) 2-methyl-2-butene

Answer: only (b) and (c)

5.16   Which of the following molecules can exist as *cis* and *trans* isomers?
  (a) 3-methyl-l-hexene   (b) 3-ethyl-3-heptene   (c) 2-methyl-2-pentene   (d) 3-methyl-2-pentene

Answer: only (d)

E,Z System of Nomenclature

5.17   Select the group with the highest priority in each of the following sets.
  (a) $—CH(CH_3)_2$ $—CHClCH_3$, $—CH_2CH_2Br$

Answer: $—CH(CH_3)_2$

  (b) $—CH_2CH=CH_2$, $—CH_2CH(CH_3)_2$, $—CH_2C≡CH$

Answer: $—CH_2C≡CH$

  (c) $—OCH_3$, $—N(CH_3)_2$, $—C(CH_3)_3$
Answer: $—OCH_3$

5.18   Select the group with the highest priority in each of the following sets.

Answers:

74

5.19   Assign the *E* or *Z* configuration to each of the following antihistamines.

Answer: (a) *Z* (a) pyrrobutamine

Answer: (b) *Z* (b) triprolidine

Answer: (c) *Z* (c) chloroprothixene

5.20   Assign the *E* or *Z* configuration to each of the following hormone antagonists used to control cancer.

(a)   Chlomiphene

$OCH_2CH_2N(CH_2CH_3)_2$

Answer: (a) *E*

(b)   Tamoxifen

$OCH_2CH_2N(CH_3)_2$

Answer: (b) *Z*

(c)   Nitromifene

OCH_2CH_2

H_3CO NO_2

Answer: (c) *E*

5.21   Draw the structural formula for each of the following pheromones with the indicated configuration.
(a) sex pheromone of Mediterranean fruit fly, *E* isomer

$CH_3CH_2CH=CH(CH_2)_4CH_2OH$

Answer: (a) *E*

$$\begin{array}{c} H \qquad\qquad CH_2(CH_2)_3CH_2OH \\ C=C \\ CH_3CH_2 \qquad\quad H \end{array}$$

(b) sex pheromone of honey bee, *E* isomer

$CH_3CO(CH_2)_4CH_2CH=CHCO_2H$

Answer: *E*

(b)

$$\begin{array}{c} H \qquad\quad CH_2(CH_2)_4-\overset{\overset{\displaystyle O}{\|}}{C}-CH_3 \\ C=C \\ HO_2C \qquad\quad H \end{array}$$

75

(c) defense pheromone of termite, *E* isomer

$$CH_3(CH_2)_{12}CH=CHNO_2$$

Answer: (c) *E*

5.22   Assign the configuration at all double bonds where geometrical isomerism is possible in each of the following sex pheromones.

Answer: (a) left to right: *Z, E*   (a) European vine moth

Answer: (b) left to right: *E, Z*   (b) pink bollworm moth

Answer: (c) *Z*   (c) Japanese beetle

Nomenclature of Alkenes

5.23   Name each of the following compounds.

Answers: (a) 2-methyl-1-propene   (b) 2,3-dimethyl-2-butene
(c) 2-methyl-2-butene   (d) (*E*)-2,3-dichloro-2-pentene

5.24   Name each of the following compounds.

Answers: (a) (*E*)-1,2-dibromo-1-propene   (b) 1-bromo-3-methyl-2-butene
(c) 3-chloro-2-methyl-2-pentene   (d) (*Z*)-3,4-difluoro-3-hexene

5.25   Name each of the following compounds.

Answers: (a) 1,2-dimethylcyclohexene   (b) 1,3-dimethylcylclohexene
         (c) cyclodecene   (d) 6-ethyl-1-methylcyclohexene

5.26   Name each of the following compounds.

Answers: (a) 1-ethylcyclohexene   (b) 1,5-dimethylcyclopentene
         (c) cyclooctene   (d) 1,6-diethylcyclohexene

5.27   Draw a structural formula for each of the following compounds.
        (a) 2-methyl-2-pentene   (b) 1-hexene
        (c) (Z)-2-methyl-3-hexene (d) (E)-5-methyl-2-hexene

Answers:

(a) 2-methyl-3-pentene

(b) 1-hexene

(c) Z-2-methyl-3-hexene

(d) E-5-methyl-2-hexene

5.28   Draw a structural formula for each of the following compounds.
        (a) (E)-1-chloropropene   (b) (Z)-2,3-dichloro-2-butene
        (c) 3-chloropropene   (d) 4-chloro-2,4-dimethyl-2-hexene

Answers:

(a) E-1-chloropropene

(b) Z-2,3-dichloro-2-butene

(c) 3-chloropropene

(d) 4-chloro-2,4-dimethyl-2-hexene

5.29   Draw a structural formula for each of the following compounds.
        (a) cyclohexene   (b) 1-methylcyclopentene
        (c) 1,2-dibromocyclohexene   (d) 4,4-dimethylcyclohexene

Answers:

(a) cyclohexene

(b) 1-methylcyclopentene

(c) 1,2-dibromocyclohexene

(d) 4,4-dimethylcyclohexene

5.30   Draw a structural formula for each of the following compounds.
        (a) cyclopentene   (b) 3-methylcyclohexene
        (c) 1,3-dibromocyclopentene   (d) 3,3-dichlorocyclopentene

Answers:

(a) cyclopentene

(b) 3-methylcyclohexene

(c) 1,3-dibromocyclopentene

(d) 3,3-dichlorocyclopentene

Physical Properties

5.31   The dipole moment of hexane is 0.09 D, but the dipole moment of 1-hexene is 0.4 D. Explain the reason for the difference.
Answer: The carbon–carbon double bond of 1-hexene is not symmetrically substituted, and the single alkyl group is electron donating to the sp^2-hybridized carbon atom.

5.32   Which isomer of 2-butene has the larger dipole moment?
Answer: The bond moments of the methyl groups bonded to the sp^2-hybridized carbon atoms of *cis*-2-butene reinforce one another and there is a net dipole moment. In the *trans* isomer, the two bond moments are opposed; therefore, they cancel, and there is no dipole moment.

5.33   The dipole moment of 2-methylpropene is 0.5 D, but the dipole moment of 1-butene is 0.3 D. Explain why these values differ.
Answer: The two bond moments of the methyl groups bonded to the sp^2-hybridized carbon atom are additive in 2-methylpropene and are larger than the bond moment of the single ethyl group of 1-butene.

5.34   The dipole moment of chloroethene is 1.4 D. Predict the dipole moment of *cis*-1,2-dichloroethene.
Answer: The dipole moment should be larger than 1.4 D but less than 2.8 D because there will be some partial cancellation of the component of the dipole moment along the carbon–carbon bond axis. There will be reinforcement of the component of the dipole moment perpendicular to the bond axis.

5.35   *cis*-1-Bromopropene has a higher boiling point than *cis*-1-chloropropene but has the smaller dipole moment. Explain why.

Answer: The bromine atom is more polarizable and the resulting London forces are stronger than those resulting from the chlorine atom.

5.36   The boiling points of 1-hexene and 2,3-dimethyl-2-butene are 63.5 and 73 °C, respectively. Suggest a reason for this difference.

Answer: 2,3-Dimethyl-2-butene has a small, compact shape which allows close approach of molecules and results in increased London forces. The conformational flexibility of 1-hexene results in a larger effective volume and, therefore, a higher boiling point.

Heats of Combustion of Alkenes

5.37   The difference between the heats of combustion of *cis-* and *trans-*2-butenes is about 4.2 kJ mole^{-1}, but the difference between those of *cis-* and *trans-*4,4-dimethyl-2-pentenes is about 16 kJ mole^{-1}. Explain why these two values differ significantly.

Answer: There is substantial van der Waals repulsion of a methyl group and a *tert-*butyl group in *cis-*4,4-dimethyl-2-pentene which makes this isomer much less stable than the *trans* isomer. The van der Waals repulsion between the two methyl groups in *cis-*2-butene is significantly smaller.

5.38   The difference between the heats of combustion of *cis-* and *trans-*2,2,5,5-tetramethyl-3-hexenes is about 40 kJ mole^{-1}. Explain this very large difference.

Answer: There is a very large van der Waals repulsion of two *tert-*butyl groups in the *cis* isomer.

5.39   Which of the following two isomers should have the larger heat of combustion? Explain why.

Answer: Compound II has the larger angle strain resulting from the carbon–carbon double bond in the four-membered ring. Compound II is less stable and releases more energy in a combustion reaction.

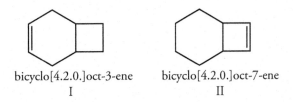

bicyclo[4.2.0.]oct-3-ene   bicyclo[4.2.0.]oct-7-ene
               I   II

5.40   Although 1-methylcyclopropene is a trisubstituted alkene and methenecyclopropane is a disubstituted alkene, the heat of combustion of 1-methylcyclopropene is larger by about 42 kJ mole^{-1}. Explain why.

Answer: Both sp^2-hybridized carbon atoms of 1-methylcyclopropene are part of the three-membered ring, giving rise to larger angle strain than the single sp^2-hybridized carbon atom of methenecyclopropane.

methylenecyclopropane   1-methylcyclopropene

5.41   Arrange the following compounds in order of increasing heats of combustion: 3-methyl-1-butene, 2-methyl-1-butene, 2-methyl-2-butene.

Answer: 2-methyl-2-butene < 2-methyl-1-butene < 3-methyl-1-butene, which is the order of decreasing degree of alkyl substitution of the double bond.

5.42   Arrange the following compounds in order of increasing heats of combustion.

     I   II   III

Answer: I < Ill < II, which is the order of decreasing degree of alkyl substitution of the double bond.

Hydrogenation of Alkenes

5.43   How many moles of hydrogen gas will react at atmospheric pressure with each of the following compounds?
(a) 1,4-cyclooctadiene   (b) 4-vinylcyclohexene
(c) 2,4-dimethyl-1,4-pentadiene   (d) 2-methyl-1,3-cyclohexadiene

Answers: (a) 2   (b) 2   (c) 2   (d) 2

5.44   How many moles of hydrogen gas will react at atmospheric pressure with each of the following compounds?

Answer: (a) 6   (a) Vitamin A, contained in freshwater fish

Answer: (b) 3   (b) zingiberene, found in oil of ginger

(c) ergosterol, a form of vitamin D

Answer: (c) 3

5.45   Oil of marjoram contains α-terpinene, whose molecular formula is $C_{10}H_{16}$. Hydrogenation using the Adams catalyst yields $C_{10}H_{20}$. How many double bonds and how many rings do the α-terpinene contain?

Answer: There are two double bonds because two moles of H_2 are added to the molecular formula. There is one ring because the unsaturation number of the product is 1.

5.46   The wax found on apples contains α-farnesene, whose molecular formula is $C_{15}H_{26}$. Hydrogenation using palladium on charcoal yields $C_{15}H_{32}$. How many double bonds and how many rings do the α-farnesene contain?

Answer: There are three double bonds because three moles of H_2 are added to the molecular formula. There are no rings because the unsaturation number of the product is 0.

Hydrogenation of Alkenes

5.47 Consider each compound of the following pairs of isomeric hydrocarbons and determine whether or not there should be a substantial difference in their heats of hydrogenation. Explain why. Indicate the compound with the higher heat of hydrogenation where possible.

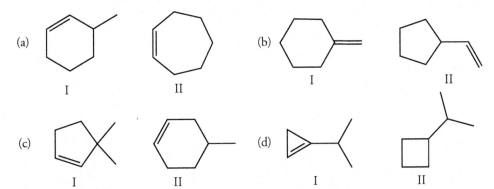

Answers: (a) There is no difference in the heats of hydrogenation because the degree of substitution of the double bonds is identical.
(b) The heat of hydrogenation of methenecyclohexane should be smaller because it is disubstituted whereas the isomeric vinyl cyclopentane is monosubstituted.
(c) There is no difference in the heats of hydrogenation because the degree of substitution of the double bonds is identical.
(d) The cyclopropene compound should have the larger heat of hydrogenation because the strain of the double bond in the smaller ring makes the compound less stable.

5.48 There are three isomeric methylcyclopentenes. Which compound has the smallest heat of hydrogenation?

Answer: 1-Methylcyclopentene has the smallest heat of hydrogenation because it has a trisubstituted double bond. The 3-methyl and 4-methyl compounds have disubstituted double bonds.

5.49 There are three isomeric methylcyclopentenes. Which compound has the smallest heat of hydrogenation?

Answer: I < III < II, which is the order of decreasing degree of substitution of the double bond.

5.50 The standard heat of hydrogenation of of bicyclo[4.2.0]oct-7-ene is larger than that of the isomeric bicyclo[4.2.0]oct-3-ene. Based on this information, which compound is more stable? What feature of the structures of the two compounds is responsible for this difference in stability?

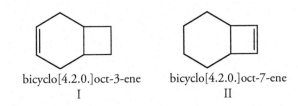

bicyclo[4.2.0.]oct-3-ene   bicyclo[4.2.0.]oct-7-ene
I   II

Answer: The compound with a double bond in a six-membered ring is more stable. The compound with a double bond in the four-membered ring is more strained and of higher energy than the isomer, so hydrogenation of the cyclobutene will release a larger amount of energy.

5.51   Although ethylidenecyclohexane and 1-ethylcyclohexene are both trisubstituted alkenes, the latter compound predominates in an equilibrium reaction. Based on this information, which compound has the larger heat of hydrogenation?

1-ethylidenecyclohexane   1-ethylcyclohexene

Answer: Both compounds yield the same saturated compound when hydrogenated. Since 1-ethylcyclohexene predominates in an equilibrium mixture, it is more stable. Thus, ethylidenecyclohexane has a larger heat of hydrogenation.

5.52   The data given in Exercise 5.51 support the observation that isomers with double bonds within rings (endocyclic) are more stable than isomers with double bonds between a carbon in the ring and a carbon outside the ring (exocyclic). Explain why the heats of hydrogenation of 1-methylcyclohexane and methenecyclohexane, which are −107 and −116 kJ mole⁻¹, respectively, cannot be used to support this generalization.

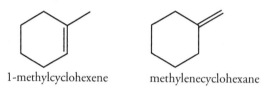

1-methylcyclohexene   methylenecyclohexane

Answer: The degree of substitution is not the same, so they cannot be directly compared.

5.53   The heat of hydrogenation of the bicyclic hydrocarbon shown below is approximately 270 kJ mole⁻¹. Why is this value so much larger than those listed in Table 5.3 for alkenes?

Answer: The double bond is present in "two" strained cyclobutene rings, which increases the energy of the compound and makes it un stable.

5.54   Explain why the $\Delta H°_{reaction}$ for the following two isomerization reactions are negative. Why is the $\Delta H°_{reaction}$ for the second reaction more negative than for the first reaction?

$$CH_2 \longrightarrow CH_3 \qquad \Delta H°_{rxn} = -3.7 \text{ kJ mole}^{-1}$$

$$CH_2 \longrightarrow CH_3 \qquad \Delta H°_{rxn} = -16.2 \text{ kJ mole}^{-1}$$

Answer: There is an increase in the degree of substitution from reactant to product which makes both reactions favorable. However, the four-membered ring compound has more strain in the product as the result of having two sp²-hybridized carbon atoms in the ring. Thus, the reaction is less favorable.

5.55   The difference between the heats of hydrogenation of (E)- and (Z)-4,4-dimethyl-2-pentenes is approximately 14.6 kJ mole⁻¹. Compare this difference with the difference between the heats of hydrogenation of (E)- and (Z)-2-butenes. Why do the two values differ?

Answer: The van der Waals repulsion between the methyl and tert-butyl groups in (Z)-4,4-dimethyl-2-pentene is much greater than between two methyl groups is (Z)-2-butene, so the difference between the (E) and (Z) isomers is greater also.

5.56   The heats of hydrogenation of the geometric isomers of 2,2,5,5-tetramethyl-3-hexene differ by 39 kJ mole⁻¹. Explain why this difference is so large compared to other geometric isomers.

Answer: There is a very large van der Waals repulsion of two tert-butyl groups in the cis isomer which makes the compound far less stable.

5.57   Write the structure of the product obtained by hydrogenating the following diester using PtO_2 and hydrogen gas at atmospheric pressure.

Answer:

5.58   Write the product obtained by the catalytic hydrogenation of the sex pheromone of the European vine moth at atmospheric pressure using PtO_2 and hydrogen gas.

Answer:

$$CH_3(CH_2)_{10}CH_2—O—\overset{\overset{\displaystyle O}{\|}}{C}—CH_3$$

5.59   Deuterium gas can be used to deuterate compounds using the Adams catalyst. The reaction proceeds by the same mechanism as for hydrogenation. Write the product of the reaction of 1-ethyl-2-methylcyclohexene with D_2.

Answer:

5.60   Which of the two isomeric caranes is the major product of the hydrogenation of 3-carene using the Adams catalyst?

3-carene

Answer: The second product, with the methyl group and the cyclopropane ring on the same side of the six-membered ring, is the major product because hydrogen is added from the sterically less hindered side, which is the face opposite the cyclopropane ring.

5.61   Which of the double bonds of limonene is hydrogenated at the faster rate? Comment on the likelihood that selective hydrogenation may occur.

limonene

Answer: The disubstituted double bond will be hydrogenated somewhat faster than the trisubstituted double bond, but selective hydrogenation is not likely.

5.62   Explain why the hydrogenation of compound I occurs at a faster rate than the hydrogenation of compound II.

I   II

Answer: Approach of hydrogen from the "top" of compound I occurs much more easily than for compound II, where the *tert*-butyl group sterically hinders this face of the double bond.

5.63   Evaluate the degree of substitution of the double bonds of bisabolene and determine whether stereoselective reduction of a double bond is possible.

bisabolene

Answer: There are two trisubstituted double bonds and a tetrasubstituted double bond. All should react slowly in a hydrogenation reaction, and there should be little selectivity.

5.64   Write the structure of the product obtained by catalytic reduction of each of the following compounds using the Wilkinson catalyst and one molar equivalent of hydrogen gas.

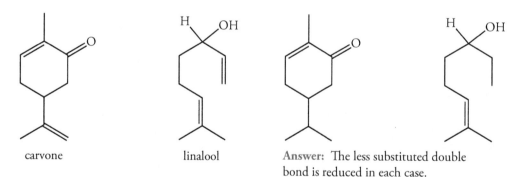

carvone   linalool   **Answer:** The less substituted double bond is reduced in each case.

6 ALKENES ADDITION REACTIONS

KEYS TO THE CHAPTER

6.1 Characteristics of Addition Reactions

The reactivity of alkenes depends on the properties of the π bond, whose electrons that are larger distances from carbon atoms than the electron in σ bonds. The π electrons are more open to attack by reagents than the σ bonds, which are located in a more protected environment between atoms. The chemistry of the π bond is most easily understood using Lewis acid–base conventions. The bonding electron pair in the pi bond can act as a Lewis base and is susceptible to attack by Lewis acids. Lewis acids are electrophiles. Electrophiles react with double bonds in addition reactions. The electrophile reacts with π electrons to form a bond to a carbon atom, and as a result, the π bond is broken.

Addition reactions of alkenes result in the incorporation of two atoms or groups of atoms on adjacent carbon atoms which share the double bond. Reagents such as HBr and H_2O add a hydrogen atom and a second species, which in these cases are bromine and the hydroxyl group, respectively. Reagents such as Br_2 can provide two equivalent groups, in which case one bromine atom adds to each of the two carbon atoms in the original π bond.

The addition reaction of a reagent X—Y to an alkene results in the destruction of one σ bond and one π bond, between two carbon atoms and between X and Y. However, two σ bonds form: one carbon atom bonds to X and the other carbon atom bonds to Y. Carbon–carbon π bonds are weaker than σ bonds. Thus, the addition reaction is exothermic because the net result is the replacement of a weak π bond by a stronger σ bond. All the common reagents we discussed in this chapter give addition products in exothermic reactions.

Different reactants add to the π bond in two different ways known as ***anti* addition** and ***syn* addition**. Thus, the geometry of the added groups in the final products gives information about the reaction pathway. This type of study is most easily done with cycloalkenes because the products are geometric isomers.

Although several types of reagents are discussed in this chapter, the most common theme is that of attack by electrophiles. As a consequence of the attack of an electrophile on the π electrons to form a bond to the carbon atom, the other carbon atom becomes electron deficient and has a positive charge. The subsequent chemistry of this carbocation is one of the common threads of addition reactions initiated by electrophiles.

6.2 Addition of Hydrogen Halides

Hydrogen halides react with alkenes to give alkyl halides. This process is called electrophilic addition because the addition of an electrophilic proton initiates the reaction. Addition of hydrogen halides is **regiospecific**. That is, not only does the hydrogen halide specifically react with the π bond, but also it gives only one of the two possible products. This result is summarized by **Markovnikov's rule**. *HX compounds add to double bonds so that the hydrogen atom bonds to the carbon atom of the double bond containing the largest number of directly bonded hydrogen atoms.* Therefore, we can easily predict the product of a reaction of hydrogen halides to a double bond.

6.3 Mechanistic Basis of Markovnikov's Rule

The proposed mechanism of electrophilic addition explains Markovnikov's rule. A proton (or other electrophlie species) adds to a π bond to give a carbocation. Thus, the direction of the addition—that is, its regiospecificity—depends on the stability of the two possible carbocations that could form. The hydrogen atom adds to the carbon atom of the double bond containing the largest number of directly bonded hydrogen atoms. This is the least substituted carbon atom. As a consequence, the carbocation formed must have the positive charge on the more substituted carbon atom. The reaction *rates* for the two possible ways that a hydrogen ion can add are controlled by the energy of the two possible transition states, and *not* by the stability of the intermediate. The first step in the addition reaction is "uphill," it is endothermic. The structure of the possible intermediate carbocations resembles the two possible transition states. Thus, the stability of a carbocations reflects the stability of the transition state leading to it.

Organic Chemistry Study Guide: Key Concepts, Problems, and Solutions http://dx.doi.org/10.1016/B978-0-12-801889-7.00006-6

6. 4 Carbocation Rearrangements

Organic chemistry would be a simple subject if all of the atoms of a reactant stayed in place in the conversion to product in reactions such as an electrophilic addition reaction. However, reactive intermediates such as carbocations can undergo rearrangement reactions. Either a hydride ion or an alkyl group with a negative charge can move from a center adjacent to the carbocation center to form a new bond and generate a positive charge at that adjacent center. The driving force for such reactions is the formation of a more stable carbocation. Hydride ion shifts occur when a secondary or primary carbocation is generated adjacent to a tertiary center. Shifts of alkyl groups most commonly occur when a carbocation is generated at a carbon atom adjacent to a quaternary center. Any of the alkyl groups bonded to that quaternary carbon can migrate. Thus, a mixture of products can result. Even ring bonds of cycloalkanes can migrate, resulting in rings of different size than the reactant.

6.5 Hydration of Alkenes

The mechanism of the addition of water to a double bond is similar to that of the addition of hydrogen halides. A hydrogen ion first attacks the π bond, followed by capture of the carbocation by the nucleophilic oxygen atom of water. Note that it is water that reacts, *not* a hydroxide ion. Remember that the reaction occurs under acidic conditions, in which case the most available nucleophilic material is water. Subsequent loss of a proton from the oxonium ion gives a product which appears to have been the result of addition of a hydrogen ion and a hydroxide ion.

The reaction is readily reversible; the reverse reaction is dehydration. The dehydration reaction is much like viewing a film in reverse. The same things happen but in reverse sequence. That is, the mechanism for the forward is the same as the mechanism of the reverse reaction. This the **principle of microscopic reversibility**.

6.6 Addition of Halogens

The mechanism for addition of halogens such as chlorine or bromine to a double bond differs from the mechanism for the addition of hydrogen halides. The reaction rate is sensitive to the same features of the alkenes. As in the addition of hydrogen halides or water, compounds that give more stable carbocations are also more reactive in addition of bromine or chlorine. Thus, the carbon atom of the double bond bears some positive charge in the transition state. However, rearrangement products are not observed, and the stereochemistry of the product of the reaction is the result of a net *anti* addition.

Halogenation occurs by way of a cyclic bromonium or chloronium ion in which the positive charge is largely on the halogen atom but is distributed to some extent to the two carbon atoms. Of the two carbon atoms, the more substituted atom has the greater positive charge. The charge is insufficient to induce rearrangement reactions but sufficient to distinguish between the two sites for subsequent attack by a nucleophile. This difference in charge distribution is shown the formation of halohydrins, where water attacks the more substituted carbon atom because it has the higher positive charge.

6.7 Addition of Carbenes

Carbenes, which contain divalent carbon atoms, are electron-deficient species and behave as electrophiles toward the π bond of alkenes. However, they are electrically neutral, and there is no associated nucleophile. The reaction is a simultaneous formation of two σ bonds to give a cyclopropane ring. The Simmons–Smith reagent is used as a convenient way to form cyclopropane rings. The stereochemistry of the groups of the alkene are unaltered in the product cyclopropane.

6.8 Epoxidation of Alkenes

The addition of an oxygen atom to a double bond results in a cyclic ether known as an **epoxide.**
A number of peroxide reagents are available for the formation of epoxides. In every case, the reaction proceeds by a cyclic-concerted process in which the stereochemistry of the groups bonded to the original carbon atoms of the double bond are unchanged in the product.

6.9 Dihydroxylation of Alkenes

Dihydroxylation reactions convert alkenes to vicinal diols (glycols). The reaction of alkenes with osmium tetroxide occurs by concerted cyclic mechanism that simultaneously places one oxygen atom on each carbon atom of the original double bond. Subsequent reaction of the cyclic intermediate generates the two hydroxyl groups. The stereochemistry of the addition reaction is *syn*.

6.10 Ozonolysis of Alkenes

Alkenes react with ozone to give intermediates that subsequently react under work-up conditions to give two products that result from cleaving both bonds of the original carbon–carbon double bond. Those two carbon atoms are identified in the product by the location of oxygen atoms. Under reductive work-up conditions, either aldehydes or ketones can result. Under oxidative work-up conditions, either carboxylic acids or ketones can form.

SUMMARY OF REACTIONS

1. Electrophilic Addition of Hydrogen Halides

(methylenecyclohexane) + HBr ⟶ (1-bromo-1-methylcyclohexane)

2. Addition of Water (Hydration)

(2-methyl-2-pentene) + H_2O —$\xrightarrow{H_2SO_4}$→ $H-\overset{H}{\underset{H}{C}}-\overset{OH}{\underset{CH_2CH_3}{C}}-CH_3$

3. Addition of Halogens

(cyclohexene) + Br_2 ⟶ (trans-1,2-dibromocyclohexane)

(2-methyl-2-pentene) + Cl_2 ⟶ $H-\overset{H}{\underset{Cl}{C}}-\overset{Cl}{\underset{CH_2CH_3}{C}}-CH_3$

(1-methylcyclohexene) + Br_2 / H_2O ⟶ (1-methyl-2-bromocyclohexanol)

4. Addition of Carbenes

(cyclopentene) —$\xrightarrow[(CH_3)_3O^-]{CHCl_3}$→ (6,6-dichlorobicyclo[3.1.0]hexane)

(cis-2-pentene) —$\xrightarrow[Zn(Cu)]{CH_2I_2}$→ (cyclopropane product)

5. Epoxidation of Alkenes

(cis-2-pentene) —$\xrightarrow[CH_2Cl_2]{MCPBA}$→ (epoxide product)

6. Dihydroxylation of Alkenes

(cyclohexene) —$\xrightarrow[NMO]{OsO_4}$→ (cis-1,2-cyclohexanediol)

$$\underset{H}{\overset{H}{}}C=C\underset{CH_2CH_3}{\overset{H}{}} + KMnO_4 \xrightarrow{pH\ 7} H-\underset{\underset{H}{|}}{\overset{\overset{OH}{|}}{C}}-\underset{\underset{H}{|}}{\overset{\overset{OH}{|}}{C}}-CH_2CH_3 + MnO_2$$

7. Ozonolysis of Alkenes

$$\xrightarrow[Zn\ /\ H^+]{O_3}$$

$+\ CH_2{=}O$

$$\xrightarrow[H_2O_2]{O_3}$$

$+\ HCO_2H$

8. The Grubbs Reaction

$$\xrightarrow{\text{Grubbs Catalyst}}$$

$+\ CH_2{=}CH_2$

(E)-2-pentene

SOLUTIONS TO EXERCISES

Syn and *Anti* Addition

6.1   The indirect hydration of an alkene using a procedure called hydroboration–oxidation transforms 1-methylcyclohexene into *trans*-2-methylcyclohexanol. Describe the stereochemistry of the net addition reaction.

Answer: The net result of the reaction is *syn* addition; it is *anti*-Markovnikov.

trans-2-methyl-cyclohexanol

6.2   Reaction of 1,2-dimethylcyclopentene with potassium permanganate yields the following compound. What is the stereochemistry of the net addition reaction?

Answer: The hydroxyl groups are *cis* because it is *syn* addition.

Electrophiles and Markovnikov Addition

6.3   Predict the structure of the addition product of IN_3 and 1-pentene. The mechanism occurs by electrophilic attack followed by capture of the carbocation by a nucleophile.

Reaction:

Explanation: The heterolytic cleavage of the I—N bond of the IN_3 compound gives I^+ and N_3^-. Addition of the electrophile I^+ at C-1 of 1-pentene yields a secondary carbocation at C-2 which is captured by the nucleophile N_3^-. The structure of the product is shown above.

6.4   Based on the information given in the following equation, outline the mechanism of the reaction of the reagent INCO.

Mechanism:

Explanation: Heterolytic cleavage of the I—N bond of the INCO compound gives I^+ and NCO^-. Addition of the electrophile I^+ at the double bond from face of the ring opposite the axial methyl group gives a cyclic iodonium ion. To achieve net *trans* addition, the nucleophilic NCO^- ion must attack at the indicated atom to open the ring of the cyclic iodonium ion via a *trans* diaxial arrangement. Attack at the other carbon atom would give the diequatorial isomer.

6.5 Write the product of the reaction of HBr with each of the following compounds.

(a) 2-methyl-1-butene   (b) 2-methyl-2-butene   (c) (Z)-2-hexene   (d) (E)-3-methyl-2-pentene

Answers:

(a)

(b)

(c)

(d)

6.6 Write the product of the reaction of HBr with each of the following compounds.

(a)   (b)   (c)   (d)

Answers:

(a)   (b)   (c)   (d)

6.7 Reaction of 1,6-dimethylcyclohexene with HBr by an electrophilic addition mechanism yields two products. What are the two compounds?

Answer: The reaction proceeds through a planar carbocation intermediate that can be attacked by bromide from either side to give two geometric isomers.

91

6.8 Reaction of 1,2-dimethylcyclohexene with HCl by an electrophilic addition mechanism yields two products. What are the two compounds?

Reaction:

Explanation: The alkyl selenium ion, RSe⁺, is an electrophile that attacks the double bond to give a cyclic selenium ion which is then opened by nucleophilic attack of the oxygen atom of the carboxyl group. Note that the oxygen and selenium atoms are *trans* in the product.

6.9 The electrophilic addition of HCl to 3,3,3-trifluoropropene gives 1-chloro-3,3,3-trifluoropropane, an anti-Markovnikov addition product. Consider the structure of the intermediate carbocations possible for the two modes of addition and suggest a reason for the observed regioselectivity.

Reaction:

Explanation: Two carbocations are possible. Since the trifluoromethyl group is strongly electron withdrawing, the carbocation with a positive charge at C-1 is more stable than the carbocation with a positive charge at C-2.

6.10 The electrophilic addition of HCl to chloroethene yields 1,1-dichloroethane. Based on resonance structures, account for the observed regioselectivity.

Reaction:

Explanation: Although chlorine is electron withdrawing, it can stabilize a carbocation by a resonance effect, which accounts for the observed regiospecificity

6.11 Reaction of 3,3-dimethy1-1-butene with HI gives a mixture of unrearranged product and rearranged product in the ratio 90:10. Account for the difference in this ratio compared to that for addition of HCl (Section 7.4).

Reaction:

2-iodo-3-methylbutane
90%
(expected product)

2-iodo-2-methylbutane
10%
(rearranged product)

Answer: The amount of rearranged product is smaller with HI. Thus the iodide ion must capture the carbocation prior to rearrangement more efficiently than does chloride ion.

6.12   Reaction of 3,3-dimethyl-l-butene with HBr gives a mixture of two addition products in the ratio 70:30. Based on Exercise 6.11, predict the structures of the two products.

2-bromo-3-methylbutane
70%
(expected product)

2-bromo-2-methylbutane
30%
(rearranged product)

Explanation: The major product is 2-bromo-3,3-dimethylbutane, which is expected from a normal addition reaction. The bromide ion can capture the carbocation prior to its rearrangement more efficiently than does the chloride ion, but less efficiently than the iodide ion. The minor product is 2-bromo-2,3-dimethylbutane, which results from a 1,2 hydride shift of a methyl group.

Hydration of Alkenes

6.13   Write the product of hydration of each of the following compounds assuming that no rearrangement occurs.

(a) 2-methyl-l-butene   (b) 2-methyl-2-butene   (c) (Z)-2-hexene   (d) (E)-3-methyl-2-pentene

Answers:

(a)

(b)

(c)

(d)

6.14 Write the product of hydration of each of the following compounds assuming that no rearrangement occurs.

(a) (b) (c) (d)

Answers:

(a) [structure] + H_2O → [structure with OH and CH$_3$, CH$_2$]

(b) [structure] + H_2O → [structure with CH$_3$, OH]

(c) [structure] + H_2O → [bicyclic structure with OH]

(d) [structure] + H_2O → [bicyclic structure with HO]

6.15 Hydration of either 2-methyl-l-butene or 2-methyl-2-butene yields the same alcohol. What is its structure? Explain why the same compound forms from both alkenes.

Answer: The same tertiary carbocation is formed from either of the two alkenes. The product is a tertiary alcohol.

2-methyl-1-butene

2-methyl-2-butene

$$CH_3-\overset{\overset{\displaystyle H}{|}}{\underset{\underset{\displaystyle H}{|}}{C}}-\overset{\overset{\displaystyle OH}{|}}{\underset{\underset{\displaystyle CH_3}{|}}{C}}-CH_3$$

6.16 Hydration of 2,3-dimethyl-2-butene is a slower reaction than the hydration of 2,3-dimethyl-l-butene under the same reaction conditions. Suggest a possible explanation.

Answer: The structure of the tertiary carbocation is the same for both alkenes. However, the rate of the reaction depends on the difference in energy between the reactant and the transition state. The 2,3-dimethyl-2-butene is the more stable isomer because it has the more highly substituted double bond. Thus, the energy required to achieve the transition state is larger for this compound.

Addition of Bromine to Alkenes

6.17 Write the product of the reaction of Br_2 with each of the following compounds.

    (a) 2-methyl-l-butene   (b) 2-methyl-2-butene   (c) (Z)-2-hexene (d) (E)-3-methyl-2-pentene

Answers:

6.18 Write the product of the reaction of Br_2 with each of the following compounds.

Answers:

6.19   Reaction of 3-methylcyclohexene with bromine in CCl$_4$ gives a mixture of two products. Explain why two products result.

Answer: The bromide ion may attack either of two carbon atoms of the bromonium ion.

6.20   Reaction of 3-bromocyclohexene with HBr gives an "unusual" product, *trans*-1,2-dibromocyclohexane. Explain its origin using an appropriate mechanism and intermediate.

Answer: A proton adds at C-1, giving a secondary carbocation at the C-2 atom, which is then temporarily captured by the bromine atom at C-3. The intermediate is thus a bromonium ion that subsequently reacts with bromide ion.

6.21   The reaction of cyclohexene with bromine in water as the solvent yields the alcohol *trans*-2-bromocyclohexanol. Explain why.

Answer: The reaction of the bromonium ion with water is similar to the reaction of the chloronium ion with water to give a chlorohydrin. The nucleophilic water attacks on the "face" of the original alkene that is opposite the bromine atom of the bromonium ion.

6.22   Reaction of cyclohexene with an aqueous bromine solution saturated with sodium chloride gives a mixture of *trans*-2-bromocyclohexanol and a compound with the molecular formula C$_6$H$_{10}$BrCl. What is the structure of the latter compound?

Answer: The reaction of the bromonium ion with chloride ion is similar to that of its reaction with bromide ion in the reaction of bromine. The nucleophilic chloride ion attacks on the "face" of the original alkene that is opposite the bromine atom of the bromonium ion to give *trans*-1-bromo-2-chlorocyclohexane.

6.23 Bromination of 3,3-dimethyl-1-butene in methanol (CH_3OH) gives a mixture of the expected dibromo compound and a bromoether. Explain the origin of the two products.

Answer: The bromonium ion, which has a bromine atom bridging C-1 and C-2, may react with methanol as well as with Br_2, thus producing a mixture of the dibromo compound and a bromoether.

$(CH_3)_3CCHCH_2Br$
|
Br

$(CH_3)_3CCHCH_2Br$
|
OCH_3

6.24 Based on the information given in Exercise 6.21, predict the structure of the chloroalcohol formed in the reaction of methylenecyclohexane with an aqueous chlorine solution.

Answer: Attack of chlorine forms a chloronium ion which has primary carbocation character at the original methylene carbon atom and tertiary carbocation character at the ring atom. Methanol is expected to attack the tertiary center to give the product shown below.

6.25 Based on the information given in Exercise 6.21, predict the structure of the chloroalcohol formed in the reaction of methylenecyclohexane with an aqueous chlorine solution.

Answer: Attack of chlorine forms a chloronium ion which has primary carbocation character at the original methylene carbon atom and tertiary carbocation character at the ring atom. Methanol is expected to attack the tertiary center to give the product shown below.

6.26 Reaction of 4-penten-l-ol with aqueous bromine gives the indicated cyclic bromoether. Write a mechanism for its formation.

Answer: The bromonium ion undergoes intramolecular attack by the oxygen atom of the hydroxyl group followed by loss of a proton.

Addition of Carbenes to Alkenes

6.27 Chlorocarbene (CHCl) can be produced from dichloromethane using butyl lithium ($CH_3CH_2CH_2CH_2^-$ Li^+), but cannot be produced using potassium *tert*-butoxide. Suggest a reason why not.

Answer: The *tert*-butoxide ion is not as strong a base as butyl lithium and cannot remove the proton from dichloromethane, which is a weaker acid than trichloromethane.

6.28   Addition of dichlorocarbene to *cis*-2-butene gives a mixture of two isomeric compounds. Explain why.

Answer: The chlorocarbene can add to place the chlorine group either *cis* or *trans* to the two methyl groups.

6.29   Write the products of the following reaction.

Answer:

6.30   Write the products of the following reaction.

Answer:

6.31   Based on the stated electrophilicity of dichlorocarbene, predict the relative reactivities of 1-butene and *trans*-2-butene with dichlorocarbene.

Answer: The methyl groups of the disubstituted *trans*-2-butene should supply electrons to the double bond and increase the availability of electrons to the electrophilic dichlorocarbene. The monosubstituted double bond of 1-butene will be less reactive.

6.32   Predict the relative electrophilicities of dichlorocarbene and chlorocarbene.

Answer: Based on inductive electron withdrawal of the electronegative chlorine atoms, dichlorocarbene should be more electrophilic.

6.33   Reaction of 1,1-dichloroethane with butyllithium does not give a carbene. Why?

Answer: The base can remove a proton from the methyl group and cause an elimination reaction to give chloroethene.

6.34   Dichlorocarbene can be formed by heating sodium trichloroacetate. Propose a mechanism for this reaction.

Answer:

trichloroacetate

Epoxidation of Alkenes

6.35   Write the structure of the epoxide obtained from the reaction of *trans*-9-octadecen-l-ol with MCPBA.

Answer: The *trans* stereochemistry of the groups about the double bond is retained in the epoxide product.

6.36   The following epoxide is an intermediate in the synthesis of disparlure, the sex attractant of the gypsy moth. Write the structure of the unsaturated alcohol used to produce the epoxide.

Answer: The *cis* stereochemistry of the epoxide must exist about the double bond of the unsaturated alcohol.

6.37   Predict which of the two isomeric epoxides will be produced from the following bicyclic unsaturated compound.

Answer: The first compound forms because the oxygen of the epoxide is delivered to the double bond from the face that is not sterically hindered by the axial methyl group.

99

6.38   Oxidation of bicyclo[2.2.1]hept-2-ene gives the indicated epoxide. Write the structure of an alternative epoxide product and explain why this compound is not produced.

Answer: The first compound, which is *exo*, forms because the oxygen of the epoxide is delivered to the double bond from the face that is less sterically hindered.

6.39   Write the structure of the epoxide expected from the reaction of the following diene with one molar equivalent of MCPBA.

Answer: The more highly substituted double bond should react with MCPBA. The oxygen atom should be placed on the side of the ring opposite the vinyl group.

6.40   Arrange the following compounds in order of increasing rate of reaction with MCPBA.
        I: 5-methyl-1-hexene   II: 3-methyl-2-hexene   III: 4-methyl-2-hexene   IV: 2,3-dimethyl-2-pentene

Answer: The order of increasing rate of reactivity is the same as the order of increasing degree of substitution, which is I < III < II < IV.

Dihydroxylation of Alkenes

6.41   Describe the visual appearance of the reaction that occurs when *cis*-2-pentene reacts with potassium permanganate. How could this reagent be used to distinguish between the isomers *cis*-2-pentene and cyclopentane?

Answer: The purple color of the permanganate ion disappears, and a brown precipitate of manganese dioxide results. These results are not obtained with cyclopentane.

6.42   Write the product of the reaction of vinylcyclohexane with potassium permanganate.

Answer: A diol forms when hydroxyl groups are added to the vinyl group.

6.43 The *exo* face of bicyclo[2.2.1]hept-2-ene (norbornene) is less sterically hindered than the *endo* face. Based on this information, write the product of reaction of norbornene with $KMnO_4$.

Answer:

exo

6.44 Write the structure of the diol that forms when OsO4 reacts with the following alkenes.

Answers:

(a)

(b)

(c)

Ozonolysis of Alkenes

6.45 Write the product(s) of ozonolysis of each of the following compounds under reductive workup conditions.

Answers:

(a)

(b)

(c)

(d)

101

6.46   Write the product(s) of ozonolysis of each of the following compounds under oxidative workup conditions.

Answers:

(a)

(b)

(c)

(d)

6.47   How can you distinguish between 1,3-cyclohexadiene and 1,4-cyclohexadiene based on their ozonolysis products?

Answer: 1,4-Cyclohexadiene gives only one ozonolysis product, but 1,3-cyclohexadiene gives two.

1,4-cyclohexadiene

1,3-cyclohexadiene

6.48   How can you distinguish between 1,3-cyclohexadiene and 1,4-cyclohexadiene based on their ozonolysis products?

Answer: 1-Methylcyclohexene gives a dicarbonyl compound that is an aldehyde and a ketone. Both 3-methylcyclohexene and 4-methylcyclohexene give dialdehydes.

6.49　Write the products of ozonolysis using reductive workup conditions for each of the three isomeric methylcyclohexenes and classify the carbonyl group present in each product.

Answer:　1-Methylcyclohexene gives a dicarbonyl compound that is an aldehyde and a ketone. Both 3-methylcyclohexene and 4-methylcyclohexene give dialdehydes.

6.50　A hydrocarbon of molecular formula C_9H_{14} is found in sandalwood oil. Ozonolysis of the hydrocarbon followed by oxidative workup gives the following diketone. Draw the structure of the hydrocarbon.

6.51　A hydrocarbon component of a pheromone of a species of moth reacts with ozone followed by reductive workup to give the following compounds. Draw a structure of the hydrocarbon. How many geometric isomers are possible with this structure?

Answer:　There are two double bonds in the pheromone and E-Z isomers are possible about each bond, giving rise to four isomers. The 6(E), 9(E) isomer is shown.

6.52 Two isomeric unsaturated carboxylic acids, oleic acid and elaidic acid, melt at 13 and 45 °C, respectively. Ozonolysis of either compound under oxidative conditions yields the following two compounds. What are possible structures of the two compounds? Why do they give the same products?

Answer: The two compounds are geometric isomers. When the double bond is cleaved, that structural distinction no longer exists and the resulting fragments are identical. Oleic acid has configuration (Z), and elaidic acid has configuration (E).

$$CH_3(CH_2)_7\overset{\overset{\displaystyle O}{\|}}{C}-OH \qquad HO-\overset{\overset{\displaystyle O}{\|}}{C}(CH_2)_7\overset{\overset{\displaystyle O}{\|}}{C}-OH$$

oleic acid (Z) isomer

elaidic acid (E) isomer

6.53 An unsaturated fatty acid isolated from brain tissue has the molecular formula $C_{24}H_{40}O_2$. Hydrogenation yields an unbranched carboxylic acid with the molecular formula $C_{24}H_{46}O_2$. Ozonolysis of the fatty acid under reductive conditions yields two equivalents of 1,3-propanedial and one equivalent each of hexanal and an aldehydic acid with the formula $C_{12}H_{22}O_3$. Write the structure of the most stable isomer that is most consistent with these data. How many other isomeric compounds are also consistent with the data?

Answer: There are three double bonds in the compound, and E-Z isomers are possible about each double bond, so eight geometric isomers are possible. The 12(E), 15(E), and 18(E) isomer is shown.

$$CH_3(CH_2)_4\overset{\overset{\displaystyle O}{\|}}{C}H \qquad H\overset{\overset{\displaystyle O}{\|}}{C}CH_2\overset{\overset{\displaystyle O}{\|}}{C}H$$

hexanal

propanedial

$CH_3(CH_2)_3CH_2$ ⌇ $CH_2(CH_2)_9CO_2H$

7

ALKYNES

In many respects, the chemistry of alkynes closely resembles that of alkenes. Both classes of compounds have π bonds that dominate their chemical reactivity, which is largely addition of electrophiles. Moreover, at least some of the synthetic methods used to produce alkynes are the same as those used for the synthesis of alkenes, namely, elimination reactions. Terminal alkynes have one important new feature. The C—H bond of terminal alkynes is sufficiently acidic for the proton to be removed by strong bases. As a consequence, the conjugate base formed (a carbanion) is a nucleophile.

7.1 Occurrence and Uses of Alkynes

Alkynes are less common than alkenes in naturally occurring materials. The few examples cited that are of interest have multiple conjugated triple bonds. Carbon–carbon triple bonds are contained in a few drugs, including oral contraceptives.

7.2 Structure and Properties of Alkynes

A triple bond in an alkyne consists of one sigma bond and two pi bonds. As a result of the geometry of the sp hybrid orbitals, the two carbon atoms of the triple bond and the two atoms directly attached are collinear. There are two classes of alkynes—**monosubstituted** (terminal) and **disubstituted** (internal).

The greater % s character of the sp-hybridized carbon atom of alkynes strongly affects the properties of the bond of that carbon atom. The electrons in the bond are held more tightly by the carbon atom, and as a consequence, the homolytic cleavage of the C—H bond requires a greater amount of energy. The length of the C—H bond as well as bonds to other atoms is shorter than for sp^2 and sp^3 bonds of the same type.

The bond energy of the carbon–carbon triple bond reflects the less effective bonding of π electrons. The bond energy of the carbon–carbon triple bond is substantially less than three times the bond energy of a carbon–carbon single bond.

The heats of formation of alkynes containing 10 or fewer carbon atoms are positive because they contain a triple bond that is less stable than carbon–hydrogen and carbon–carbon single bonds. The heats of formation of disubstituted alkynes are less positive than the heats of formation of isomeric monosubstituted alkynes.

Alkynes are relatively nonpolar molecules, and their boiling points are controlled by London forces. Terminal alkynes have small dipole moments that are slightly larger than the dipole moments of terminal alkenes. Internal alkynes have no dipole moment. The chemical properties of alkynes are similar to the properties of alkenes. The only difference is that there are twice as many π bonds to react.

7.3 Nomenclature

Alkynes are named by selecting the longest continuous carbon chain that contains the triple bond. The chain is numbered to assign the lowest number to the first carbon atom of the triple bond. Alkyl groups and halogens are disregarded in selecting the direction of numbering unless the same number for the triple bond is obtained from either end of the chain. For compounds containing both double and triple bonds, the chain is numbered from the end nearer the first multiple bond. However, in equivalently placed multiple bonds, the double bond takes precedence over triple bonds in the direction of numbering. Compounds with both double and triple bonds are called enynes, not ynenes.

7.4 Acidity of Terminal Alkynes

Although weakly acidic, terminal alkynes can be converted to their conjugate bases called alkynide ions. The pK_a is approximately 25. Thus, a base whose conjugate acid has a pK_a greater than 25 must be used to abstract the hydrogen ion. Hydroxide ion is not sufficiently basic, but the amide ion is. Because the pK_a value of ammonia is approximately 36, the equilibrium constant for the reaction of an alkyne with amide ion is 10^7. This reaction is used to produce alkynide ions for use as nucleophiles in displacement of a halide ion from a haloalkane to synthesize alkynes.

Organic Chemistry Study Guide: Key Concepts, Problems, and Solutions http://dx.doi.org/10.1016/B978-0-12-801889-7.00007-8

7.5 Hydrogenation of Alkynes

When the hydrogenation of alkynes is catalyzed with finely divided platinum, palladium, or nickel, hydrogenation of alkynes is complete and produces alkanes. Hydrogenation requires one mole of hydrogen gas for each π bond in a compound, so triple bonds require two moles of hydrogen gas.

It is possible to stop the hydrogenation of alkynes after adding one molar equivalent of hydrogen, giving alkenes as the product. This is accomplished by using a specially prepared catalyst. Hydrogenation of alkynes with Lindlar catalyst produces *cis*-alkenes by *syn* addition, whereas hydrogenation using lithium in liquid ammonia produces *trans*-alkenes by *anti* addition.

7.6 Electrophilic Addition Reactions

An unsymmetrical reagent such as HBr adds to a triple bond in a characteristic way given by Markovnikov's rule. The hydrogen atom adds to the less substituted carbon atom of the triple bond. The bromine atom adds to the more substituted carbon atom of the triple bond. The addition product has the two added atoms *trans* in the resulting alkene, although the stereoselectivity may be low.

Hydrogen bromide adds more slowly to triple bonds than to double bonds. However, after one mole of hydrogen bromide has added, the resulting double bond is less reactive as a result of the electron withdrawing bromine atom. As a consequence, it is possible to obtain the product formed from the addition of one mole of HBr. When the second mole of HBr adds, the product has two hydrogen atoms added to the carbon atom that was less substituted originally. Two bromine atoms are located on the other carbon atom.

Two moles of bromine will add to compounds with triple bonds. The initial addition product has two *trans* bromine atoms. Continued addition yields a tetrabromoalkane.

Hydration of alkynes results in the Markovnikov addition of one mole of water to give an enol that rearranges to give a ketone.

7.7 Synthesis of Alkynes

Alkynes can be prepared from vicinal or geminal dihalides by a double dehydrohalogenation using a strong base such as $NaNH_2$ in liquid ammonia. Vicinal dihalides are obtained from the addition of a halogen to an alkene. The number of moles of base required for the reaction depends on the type of alkyne produced. A terminal alkyne produced in the synthesis is deprotonated by the reacting base, and thus a total of three moles of amide ion is required. Upon work up with water, the terminal alkyne forms.

SUMMARY OF REACTIONS

1. Hydrogenation of Alkynes
A. **Hydrogenation with Palladium on Carbon Catalysts**

$$R-C\equiv C-R' \xrightarrow[\text{Pd/C}]{2\,H_2} R-\underset{\underset{H}{|}}{\overset{\overset{H}{|}}{C}}-\underset{\underset{H}{|}}{\overset{\overset{H}{|}}{C}}-R'$$

alkyne   alkane

B. *Syn* **hydrogenation with Lindlar Catalyst**

$$R-C\equiv C-R'+H_2 \xrightarrow{\text{Lindlar catalyst}} \underset{R}{\overset{H}{}}C=C\underset{R'}{\overset{H}{}}$$

alkyne   *cis*-alkene

C. *Anti* **Hydrogenation with Sodium in Liquid Ammonia**

$$R-C\equiv C-R' \xrightarrow{Na\,/\,NH_3(l)} \underset{H}{\overset{R}{}}C=C\underset{R'}{\overset{H}{}}$$

alkyne   *trans*-alkene

2. Electrophilic Addition Reactions of Alkynes
A. **Addition of Hydrogen Halides**

$$R-C\equiv C-R' + HBr \longrightarrow \underset{H}{\overset{R}{}}C=C\underset{R'}{\overset{Br}{}}$$

$$\underset{H}{\overset{R}{}}C=C\underset{R'}{\overset{Br}{}} + HBr \longrightarrow R-\underset{\underset{H}{|}}{\overset{\overset{H}{|}}{C}}-\underset{\underset{Br}{|}}{\overset{\overset{Br}{|}}{C}}-R'$$

B. **Addition of Halogens**

$$R-C\equiv C-R' + Br_2 \longrightarrow \underset{Br}{\overset{R}{}}C=C\underset{R'}{\overset{Br}{}}$$

$$\underset{Br}{\overset{R}{}}C=C\underset{R'}{\overset{Br}{}} + Br_2 \longrightarrow R-\underset{\underset{Br}{|}}{\overset{\overset{Br}{|}}{C}}-\underset{\underset{Br}{|}}{\overset{\overset{Br}{|}}{C}}-R$$

C. Hydration of Alkynes

Step 1. Formation of enol

$$R—C≡C—H \xrightarrow[\text{H}_2\text{SO}_4 / \text{HgSO}_4]{\text{H}_2\text{O}}$$

enol

Step 2. Spontaneous conversion of enol to ketone

$$\xrightarrow[\text{concerted}]{\text{1,3-proton shift}}$$

enol   ketone

$$—C≡C—H \xrightarrow[\text{HgSO}_4]{\text{H}_2\text{O}/\text{H}_2\text{SO}_4}$$

3. Synthesis of Alkynes
A. Synthesis of Alkynes by Dehydrohalogenation

$$R—\underset{\underset{Br}{|}}{\overset{\overset{Br}{|}}{C}}—\underset{\underset{H}{|}}{\overset{\overset{H}{|}}{C}}—R' \; + \; 2\,\text{NaNH}_2 \xrightarrow{\text{NH}_3\,(l)} R—C≡C—R' \; + \; 2\,\text{NH}_3 \; + \; 2\,\text{NaBr}$$

geminal dibromide

$$R—\underset{\underset{H}{|}}{\overset{\overset{Br}{|}}{C}}—\underset{\underset{Br}{|}}{\overset{\overset{H}{|}}{C}}—R' \; + \; 2\,\text{NaNH}_2 \xrightarrow{\text{NH}_3\,(l)} R—C≡C—R' \; + \; 2\,\text{NH}_3 \; + \; 2\,\text{NaBr}$$

vicinal dibromide

$$\text{CH}_3(\text{CH}_2)_3—\underset{\underset{H}{|}}{\overset{\overset{Cl}{|}}{C}}—\underset{\underset{Cl}{|}}{\overset{\overset{H}{|}}{C}}—H \xrightarrow[\text{2. H}_2\text{O}]{\text{1. 3NaNH}_2 / \text{NH}_3\,(l)} \text{CH}_3(\text{CH}_2)_3 —C≡C—H$$

1,2-dichlorohexane   1-hexyne

B. Synthesis of Alkynes by Alkylation

$$R—C≡C—H \xrightarrow{\text{NaNH}_2 / \text{NH}_3\,(l)} R—C≡C:^- \; \text{Na}^+ + \text{NH}_3$$

terminal alkyne   alkynide

$$R—C≡C:^- \; + \text{R}'\text{CH}_2\text{Br} \longrightarrow R—C≡C—\text{CH}_2\text{R}'$$

primary alkyl halide

SOLUTIONS TO END-OF-CHAPTER EXERCISES

Structures of Alkynes

7.1 What is the molecular formula of each of the following compounds that contain carbon–carbon triple bonds?
(a) mycomycin, an antibiotic

Answer: $C_{13}H_{10}O$ $H—C≡C—C≡C—CH=C=CH—CH=CH—CH_2CO_2H$

mycomycin

(b) capillin, a skin fungicide

Answer: $C_{12}H_8O$

capillin

(c)) ichthyothereol, a convulsant

Answer: $C_{15}H_{18}O$

ichthyothereol

7.2 Classify the triple bond in each of the following drugs. MDL 18962 is a drug used in breast cancer therapy. RU 486 is a drug used to induce abortion and may be useful in cancer therapy.

Answers:
(a) MDL 1280 monosubstituted

(b) RU 486 disubstituted

MDL 18962

RU 486

7.3 Write the molecular formula for the compounds with each of the following structural features.
(a) six carbon atoms and one double bond   (b) five carbon atoms and two double bonds
(c) seven carbon atoms, a ring, and one double bond   (d) four carbon atoms and one triple bond

Answers: (a) C_6H_{12} (b) C_5H_8 (c) C_7H_{12} (d) C_4H_6

7.4 What is the molecular formula for the compounds with each of the following structural features?
(a) four carbon atoms and two triple bonds   (b) four carbon atoms, a double bond, and a triple bond
(c) ten carbon atoms and two rings   (d) ten carbon atoms, two rings, and five double bonds

Answers: (a) C_4H_2 (b) C_4H_4 (c) $C_{10}H_{18}$ (d) $C_{10}H_8$

Properties of Alkynes

7.5 The heats of formation of 1-pentyne and 2-pentyne are 144 and 128.6 kJ mole^{-1}, respectively. Which compound is more stable? Based on this information, which compound has the larger heat of combustion?

Answer: One sp-hybridized carbon atom decreases the carbon–carbon bond length relative to the sp^3–sp^3 bond of propane from 154 to 146 pm. Changing a second sp^3-hybridized carbon atom to sp should decrease the bond length to 138 pm.

7.6 The heats of formation of 1-pentyne and 1,4-pentadiene are 144 and 106 kJ mole^{-1}, respectively. What does this information indicate about the relative stability of a triple bond compared to two double bonds?

Answer: The heats of formation are positive, and both compounds are unstable with respect to the elements. Thus, the compound with the lower positive heat of formation is more stable assuming that the entropies of formation of the two compounds are approximately equal. In this case, 2-pentyne, which has the more substituted triple bond, is the more stable. The heat of combustion measures the heat energy released ($\Delta H° < 0$) when carbon dioxide and water are formed. The less stable (highest energy) isomer releases more energy in the combustion reaction. In this case, 1-pentyne has the more negative heat of combustion.

7.7 The heats of formation of 1-propyne and 1,2-propadiene (allene) are 185 and 190 kJ mole^{-1}, respectively. Assuming that an equilibrium can be established, which compound would be present in the larger amount?

Answer: The heats of formation are positive for both compounds and both are unstable with respect to the elements. Thus, the compound with the lower positive heat of formation is more stable assuming that the entropies of formation of the two compounds are approximately equal. In this case, 1,4-pentadiene is the more stable isomer, so two double bonds are more stable than one triple bond.

7.8 Predict the direction of the dipole moment of 1-propyne. Why is its dipole moment larger than that of 1-propene?

Answer: The positive end of the dipole is the methyl group because the sp-hybridized carbon atom in the middle has greater s character and attracts electron density from the sp^3-hybridized carbon of the methyl group. The dipole moment of propene is smaller than that of propyne because the electrons are less strongly attracted to the sp^3-hybridized carbon atom of propene.

7.9 The boiling points of 1-alkynes are higher than those that of the 1-alkenes with the same number of carbon atoms. Suggest reasons for this fact.

Answer: The 1-alkynes are slightly more polar than 1-alkenes. The electrons in the two π bonds of an alkyne are more polarizable than the electrons in the single π bond of an alkene.

7.10 The boiling points of 3,3-dimethyl-l-butyne and 1-hexyne are 39.5 and 71.3 °C, respectively. Explain why the values are so different for these two isomers.

Answer: 3,3-Dimethyl-1-butyne has a more compact and somewhat spherical structure. The London forces for such compounds are smaller than for cylindrical structures such as 1-hexyne.

7.11 The boiling points of terminal alkynes are lower than the boiling points of isomeric internal alkynes. Is this fact consistent with the dipole moments of the compounds? If not, what other structural factors might contribute to the difference in the boiling points?

Answer: Terminal alkynes have a larger dipole moment than internal alkynes. Thus terminal alkynes should have higher boiling points than isomeric internal alkynes if polarity were the only structural feature determining this physical property. There is decreased freedom of motion of more carbon atoms of internal alkynes compared to terminal alkynes. Thus, the shape of the internal alkyne allow stronger London forces between the more linear chains.

Nomenclature

7.12 Name each of the following compounds.

(a) $CH_3CH_2CH_2C\equiv CH$ (b) $(CH_3)_3CC\equiv CCH_2CH_3$ (c) $CH_3\!-\!\!-\!C\equiv C\!-\!CH\!-\!CH_3$
$$\overset{|}{CH_2CH_3}$$

Answers: (a) 1-pentyne   (b) 2,2-dimethyl-3-hexyne   (c) 4-methyl-2-hexyne

7.13   Name each of the following compounds.

Answers: (a) 4,5-dibromo-2-hexyne (a) $CH_3CHBrCHBr$—$C\equiv C$—$CH_2(CH_2)_2CH_3$

(b) 1-chloro-3-octyne

(b) $ClCH_2$—CH_2—$C\equiv C$—$\underset{\underset{Cl}{|}}{CH}$—$CH_3$

(c) 2-chloro-6-methyl-3-oc-
tyne

(c) CH_3—$\underset{\underset{CH_2CH_3}{|}}{\overset{\overset{H}{|}}{C}}$—$CH_2$—$C\equiv C$—$\underset{\underset{Cl}{|}}{CH}$—$CH_3$

7.14   Write the structural formula for each of the following compounds.
        (a) 2-hexyne   (b) 3-methyl-l-pentyne   (c) 5-ethyl-3-octyne

Answers: (a) $CH_3CH_2CH_2$—$C\equiv C$—CH_3 (b) CH_3—CH_2—$\underset{\underset{CH_3}{|}}{CH}$—$C\equiv C$—H

2-hexyne

3-methyl-l-pentyne

(c) $CH_3CH_2CH_2\underset{\underset{CH_2CH_3}{|}}{CH}$—$C\equiv C$—$CH_2CH_3$

5-ethyl-3-octyne

7.15   Write the structural formula for each of the following compounds.
        (a) 3-heptyne   (b) 4-methyl-l-pentyne   (c) 5-methyl-3-heptyne

Answers: (a) $CH_3CH_2CH_2$—$C\equiv C$—CH_2CH_3 (b) CH_3—$\underset{\underset{CH_3}{|}}{CH}$—$CH_2$—$C\equiv C$—H

3-heptyne

4-methyl-l-pentyne

(c) CH_3—CH_2—$\underset{\underset{CH_3}{|}}{CH}$—$C\equiv C$—$CH_2CH_3$

5-methyl-3-heptyne

7.16   Write the structural formula for 4-ethynyl-l,5-nonadien-7-yne.

Answer: CH_3—$C\equiv C$—$CH=CH$—$\underset{\underset{\underset{\underset{H}{|}}{\overset{\overset{C}{|||}}{C}}}{|}}{CH}$—$CH_2$—$CH=CH_2$

4-ethynyl-l,5-nonadien-7-yne

7.17   Write the structural formula for 1-ethyl-3-(2-propynyl)cyclopentene.

Answer: CH_2—$C\equiv C$—H

1-ethyl-3-(2-propynyl)cyclopentene

111

7.18   What is the IUPAC name for the group —C≡C—CH$_3$?

Answer: Propargyl

7.19   Which of the drugs listed in Exercise 7.2 contains a propargyl group?

Answer: (a) MDL 1280 contains a propargyl group, —CH$_2$—C≡C—H.

Acidity of Terminal Alkynes

7.20   Diisopropylamide ion [(CH$_3$)$_2$CH]$_2$N$^-$ is a strong base commonly used in organic reactions. Is it expected to be a stronger or weaker base than the amide ion?

Answer: The two isopropyl groups are inductively electron donating relative to hydrogen, and they increase the electron density on the negatively charged nitrogen atom. Therefore, diisopropylamide ion is a stronger base than the amide anion.

7.21   Suggest an experimental procedure to prepare 1-deuterio-l-propyne from propene.
Answer: Prepare the conjugate base of 1-propyne using a strong base such as the amide ion. Then add D$_2$O to the reaction mixture. Since the conjugate base of propyne is a stronger base than DO$^-$, it reacts with D$_2$O to give 1-deutero-1propyne.

1-deutero-1-propyne

Hydrogenation Reactions

7.22   How many moles of hydrogen gas will react with each of the following compounds?
        (a) CH$_3$=CH=CH—C≡CH   (b) HC≡C—C≡C—H

        (c) CH$_2$=CH—C≡C—CH=CH$_2$ (d) HC≡C—C≡C—C≡CH

Answers: (a) three (b) four (c) four (d) six

7.23   How many moles of hydrogen gas will react with each of the compounds listed in Exercise 7.1?

Answers: (a) eight (b) four (c) seven

7.24   Which compound should have the larger heat of hydrogenation for the addition of two moles of hydrogen gas, 1-pentyne or 1,4-pentadiene? Why?

Answer: The heat of formation of 1-pentyne is more positive than the heat of formation of 1,4-pentadiene. In the hydrogenation reaction, pentane is the common product. Because heats of hydrogenation are negative, the energy difference between 1-pentyne and pentane is larger than the energy difference between 1,4-pentadiene and pentane.

7.25   Stearolic acid is converted to oleic acid by hydrogenation using the Lindlar catalyst. Elaidic acid is the product obtained by sodium/ammonia reduction of stearolic acid. Write the structures of oleic and elaidic acids.

$$CH_3(CH_2)_7C≡C(CH_2)_7CO_2H$$
stearolic acid

Answer:

oleic acid (Z isomer)   elaidic acid (Z isomer)

112

7.26   Disparlure, the pheromone of the gypsy moth, can be prepared by reduction of an alkyne followed by epoxidation of the alkene. What alkyne is required? What is the configuration of the alkene? What reagents are required for reduction of the alkyne?

$$CH_3(CH_2)_8CH_2\text{''''} \overset{\overset{\displaystyle O}{\diagup\diagdown}}{\underset{\underset{\displaystyle H}{|}}{C}-\underset{\underset{\displaystyle H}{|}}{C}} \text{''''}(CH_2)_4CH(CH_3)_2$$

disparlure

Answer: The required alkene is (*Z*)-2-methyl-7-octadecene, which can be prepared by catalytic hydrogenation of 2-methyl-7-octadecyne using the Lindlar catalyst.

$$CH_3(CH_2)_8CH_2\text{——}C\equiv C\text{—}CH_2(CH_2)_8CH_3 \xrightarrow{\text{H}_2 \text{ / Lindlar catalyst}}$$

disparlure

7.27   The pheromone of the grape berry moth is indicated below. How could this compound be prepared from a related alkyne. Would the ester functional group be affected by the reaction conditions?

Answer: The (*Z*) configuration can be achieved by reduction of a structurally related alkyne using the sodium in liquid ammonia. The reaction conditions would result in abstraction of a proton from the alcohol, but it would be replaced in the workup of the reaction mixture. Its structure is shown below.

$$CH_3CH_2\text{—}C\equiv C\text{—}CH_2(CH_2)_6CH_2\text{—}O\text{—}\overset{\overset{\displaystyle O}{\|}}{C}\text{—}CH_3$$

7.28   (*E*)-11-Tetradecen-l-ol is one of the intermediate compounds required to synthesize the sex pheromone of the spruce budworm. How could this compound be prepared from an appropriate alkyne? Would the reaction conditions affect the hydroxyl group?

Answer: The (*E*) configuration can be achieved by reduction of a structurally related alkyne using the sodium in liquid ammonia. The reaction conditions would result in abstraction of a proton from the alcohol, but it would be replaced in the workup of the reaction mixture.

$$CH_3CH_2C\equiv CCH_2(CH_2)_8CH_2OH \xrightarrow{\text{Na /NH}_3 \text{ (l)}}$$

(*E*)-11-Tetradecen-l-ol

7.29   Draw the structure of the product of the reaction of the following compound with hydrogen using the Lindlar catalyst.

$$HO-CH_2CH_2-CH=\overset{\overset{\displaystyle CH_3}{|}}{C}-C\equiv C-H \xrightarrow{\text{H}_2 \text{ / Lindlar Catalyst}} HO-CH_2CH_2-CH=\overset{\overset{\displaystyle CH_3}{|}}{C}-CH=CH_2$$

7.30   Draw the structure of the product of the reaction of the following compound with hydrogen using the Lindlar catalyst.

$$CH_3-C\equiv C-(CH_2)_3CH_2-\overset{\overset{\displaystyle O}{\|}}{C}-OH \xrightarrow{\text{H}_2 \text{ / Lindlar catalyst}}$$

Z-isomer

Electrophilic Addition Reactions

7.31   Addition of one mole of HCl to 2-hexyne gives a mixture of two products in approximately equal amounts. Draw their structures.

$$CH_3-C\equiv C-CH_2CH_2CH_3 \xrightarrow[\text{1 equiv.}]{\text{HCl}}$$

2-hexyne

+

7.32   Draw the structure of the addition of one mole of DBr to 1-propyne.

Answer: *Trans* addition of DBr gives the following (*E*) isomer.

$$H-C\equiv C-CH_3 \xrightarrow{\text{DBr (1 mole)}}$$

1-propyne

7.33   Predict the product of the addition of one mole of Br_2 to 1-penten-4-yne.

Answer: The bromine adds *anti* to the triple bond to give the following product, which has no dipole moment.

$$\xrightarrow{\text{Br}_2}$$

1-pentene-4-yne

7.34   Draw the structure of the compound resulting from the addition of one molar equivalent of bromine to acetylene dicarboxylic acid. What is the dipole moment of the product?

$$HO_2C-C{\equiv}C-CO_2H$$
acetylene dicarboxylic acid

Answer: The bromine adds *anti* to the triple bond to give the following product, which has no dipole moment.

7.35   Hydration of one of the following two compounds yields a single ketone product. The other compound yields a mixture of ketones. Which one yields only the single ketone product? Why?

$$CH_3CH_2C-C{\equiv}C-CH_2CH_3$$
I

$$CH_3C-C{\equiv}C-CH_2CH_2CH_3$$
II

Answer: Compound I is a symmetrical alkyne and C-3 and C-4 are structurally equivalent. Hydration of this alkyne gives a single product with a carbonyl carbon atom located at C-3. Compound II forms two products. One has a carbonyl carbon atom located at C-3 and the other at C-2.

7.36   Hydration of 4-methyl-2-pentyne gives the following compounds in the indicated amounts. Suggest a reason for the observed product ratio.

Answer: The C-2 and C-3 carbon atoms each contain an alkyl group. Thus, based on inductive effects alone, there should be no regioselectivity, and either of them could react to give an intermediate enol in a hydration reaction. However, there is a small difference in the steric environments of each carbon atom. The slight regioselectivity in the reaction may be the result of this difference. The major product has the carbonyl oxygen atom at the least sterically hindered C-2 position.

Synthesis of Alkynes

7.37   Write the structure of all compounds that could yield the following alkyne upon dehydrohalogenation.

7.38 Which isomer, 2,2-dibromopentane or 3,3-dibromopentane, would give the better yield of 2-pentyne using sodium amide as the base?

Answer: 3,3-Dibromopentane can yield only 2-pentyne because C-2 and C-4 are equivalent in this symmetrical molecule. 2,2-Dibromopentane can yield 1-pentyne by elimination of hydrogen atoms at C-1 and 2-pentyne by elimination of hydrogen atoms at C-3.

7.39 Would the following reaction provide a good yield of the indicated product? Explain.

$$CH_3CH_2CH_2CBr_2CH_3 \xrightarrow{\text{NaNH}_2} CH_3CH_2—C\equiv C—CH_3$$

Answer: No, because elimination can result by abstraction of hydrogen atoms at either C-1 or C-3 to give a mixture of 1-pentyne and 2-pentyne.

7.40 Write the product of the reaction of 1,6-dibromohexane with excess sodium acetylide.

$$BrCH_2(CH_2)_4CH_2Br \xrightarrow[\text{(excess)}]{H—C\equiv C:^-} H—C\equiv C—CH_2(CH_2)_2CH_2—C\equiv C—H$$

7.41 Predict the product of the reaction of one equivalent of the alkynide of 1-propyne and 1-bromo-5-fluoropentane.

Answer: The carbon–fluorine bond is much stronger than the carbon–bromine bond, so bromide ion will be displaced to give 8-fluoro-2-octyne.

$$BrCH_2(CH_2)_3CH_2F \ + \ CH_3—C\equiv C:^- \longrightarrow CH_3—C\equiv C—(CH_2)_3CH_2F$$
$$\text{8-fluoro-2-octyne.}$$

7.42 Draw the structure of the final product of the following series of reactions.

$$H—C\equiv C—H \xrightarrow[\text{2. CH}_3\text{CH}_2\text{CH}_2\text{I}]{\text{1. NaNH}_2} \xrightarrow[\text{2. CH}_3\text{CH}_2\text{I}]{\text{1. NaNH}_2} \ ? \qquad CH_3CH_2—C\equiv C—(CH_2)_2CH_3$$
$$\text{3-heptyne}$$

7.43 Outline the steps of a synthesis of 2,2-dimethyl-3-octyne using reactants having no more than six carbon atoms.

Answer: Prepare the acetylide salt of 3,3-dimethyl-1-butyne and react it with 1-bromobutane. Note that reaction of the acetylide salt of 1-hexyne with 2-bromo-2-methylpropane would give only an elimination product because the alkyl halide is tertiary.

8 STEREOCHEMISTRY

KEYS TO THE CHAPTER

8.1 Configuration of Molecules

Stereoisomers have different configurations. The term "configuration" refers to the arrangement of atoms in space. Geometric isomers, which we previously studied in cycloalkanes and alkenes, are also stereoisomers.

8.2 Mirror Images and Chirality

Some molecules have mirror images that are not **superimposable**. Such molecules are **chiral**. Molecules that have a **plane of symmetry are achiral**; they are superimposable on their mirror image.

A **stereogenic center** in an organic molecule is a carbon atom bonded to four different atoms or groups of atoms. It is also called a **chiral center**. By inspecting the atoms or groups of atoms bonded to each carbon atom in a molecule, we can easily identify any chiral centers. If a carbon atom is bonded to two or more identical atoms or groups, such as two hydrogen atoms or two methyl groups, it is not a chiral center. *If a carbon atom is bonded to four different atoms or groups, it is a chiral center, and the molecule has a nonsuperimposable mirror image.* The two possible isomers having different configurations at a chiral center are **enantiomers.**

Another way to identify a molecule as chiral or achiral is to look for a plane of symmetry. A plane of symmetry can bisect atoms, groups of atoms, and bonds between atoms. In a molecule with a plane of symmetry, one side of the molecule is the mirror image of the other side. Thus, a molecule with a plane of symmetry is achiral. If a molecule contains two or more chiral centers and does not have a plane of symmetry, it is chiral. If is has a plane of symmetry, it is an achiral *meso* compound.

Pairs of enantiomers have the same physical properties but behave differently in a chiral environment such as a chiral binding site in an enzyme. Most of the molecules isolated from living organisms are chiral. They generate a chiral environment that allows distinctions to be made between enantiomers.

8.3 Optical Activity

Each member of a pair of enantiomers rotates the plane of polarized light in an instrument called a **polarimeter.** This phenomenon is called **optical activity**. The rotation observed for one enantiomer is equal in magnitude but opposite in direction for the other enantiomer. A chiral substance that rotates plane-polarized light clockwise is **dextrorotatory**; a chiral substance that rotates plane-polarized light counterclockwise is **levorotatory.** The amount of rotation under defined standard experimental conditions is the specific rotation. **Optical purity** is a measure of the excess of one enantiomer over another in a mixture.

8.4 Fischer Projection Formulas

Enantiomers in a Fischer projection are drawn according to the following conventions:
1. Arrange the carbon chain vertically with the most oxidized group (—CHO in glyceraldehyde) at the "top."
2. Place the carbon atom at the chiral center in the plane of the paper. It is C-2 in glyceraldehyde.
3. C-2 is bonded to four groups, the CHO group and the CH_2OH group extend behind the plane of the page, and the hydrogen atom and the hydroxyl group extend up and out of the plane.
4. Project these four groups onto a plane. The carbon atom at the chiral center usually not shown in this convention. It is located at the point where the bond lines cross. The vertical lines project away from the viewer. The horizontal lines project toward the viewer.

$$
\begin{array}{c}
\text{CHO} \\
\text{H} \!\!-\!\!\!\!-\!\!\!\!-\!\! \text{OH} \\
\text{CH}_2\text{OH}
\end{array}
$$

Organic Chemistry Study Guide: Key Concepts, Problems, and Solutions http://dx.doi.org/10.1016/B978-0-12-801889-7.00008-X

8.5 Absolute Configuration

The Kahn–Ingold–Prelog configurational nomenclature system, which is the same for both *E,Z* geometric isomers and chiral molecules, gives an unambiguous description of the absolute configuration of a molecule.

Priority is assigned to atoms based on the atomic numbers of directly bonded atoms. Atoms farther down the chain are ignored even though they may have still higher atomic numbers. Thus, a fluorine atom has a higher priority than a carbon atom even if that carbon atom is bonded to three chlorine atoms; for example, F— > —CCl₃. The chlorine atoms in this case are irrelevant because the comparison is between the atomic numbers of fluorine and carbon.

Once the priority order of the atoms or groups of atoms bonded to the chiral carbon atom has been determined, the molecule is viewed through the bond to the lowest priority group. The other three groups then lie on a circle. If the movement from priorities 1→ 2→ 3 is clockwise, the molecule is *R*; if the motion is counterclockwise, the configuration is *S*.

The assignment of *R* or *S* configuration to a compound does not identify its optical rotation as being either (+) or (−). The direction of optical rotation is experimentally determined with a polarimeter. The absolute configuration is experimentally determined by X-ray crystallography.

8.6 Molecules with Two or More Stereogenic Centers

Some molecules have two or more stereogenic centers. The resulting stereochemistry depends on whether those centers are equivalent or nonequivalent. **Equivalent** sterogenic centers have identical sets of substituents. For *n* nonequivalent centers, there are 2^n stereoisomers. Some of those isomers are pairs of enantiomers. These stereoisomers have opposite configurations at every center and are thus mirror images. All other stereoisomers are termed **diastereomers.**

The configuration of each stereogenic center is determined independently. Then, the configuration of each center is written as *R* or *S*. For example, the enantiomer of a molecule with a stereogenic center 2*S*,3*R* is 2*R*,3*S*. Any other combination—2*S*,3*S* or 2*R*,3*R*— is a diastereomer.

Compounds with two or more equivalent stereogenic centers have fewer stereoisomers than predicted by the 2^n formula. Some of the stereoisomers have a plane of symmetry and are not optically active; they are **meso compounds**. For two chiral centers, the configurations are *R,S*, which is the same as *S,R* because of the plane of symmetry. The isomers *R,R* and *S,S* are optically active and are enantiomers.

8.7 Cyclic Compounds with Stereogenic Centers

Cyclic compounds can have stereogenic centers. We apply the same rules to assign configuration to cyclic compounds and acyclic compounds. The only difference is that we eventually return to the stereogenic center as we move around the ring. However, in a chiral compound, the point of first difference is reached before that time.

Cyclic compounds having two nonequivalent stereogenic centers can exist in four stereoisomeric forms. An interesting feature of these molecules is seen when there are equivalent stereogenic centers. In those cases, there is at least one plane of symmetry. That plane, in some cases, bisects bonds, and in other cases bisects the atoms of the ring. In this latter case, it also bisects the atoms bonded to the stereogenic centers.

8.8 Separation of Enantiomers

Enantiomers have the same physical properties and, therefore, cannot be separated by physical methods. However, diastereomers have different physical properties and can be separated. Figure 8.18 illustrates the conversion of a mixture of enantiomers into a mixture of diastereomers. The diastereomers are separated, after which they are broken down to obtain one enantiomer from one diastereomer and the other enantiomer from the second diastereomer. Chiral chromatography provides a way to separate enantiomers based upon their diastereomeric interactions with a chiral column support.

8.9 Reactions at Stereogenic Centers

If a reaction at a stereogenic center does not change the bonds to the stereogenic center, then the configuration at that center is unchanged.

Reactions at the stereogenic center affect the configuration of the molecule. If the product has a configuration opposite that of the reactant, we postulate a transition state in which the nucleophile attacks opposite the bond to the leaving group and inverts the configuration as the reaction occurs.

Radical reactions proceed through a planar intermediate, which is achiral. Thus, subsequent reaction with another radical can occur with equal probability from either side of the plane of the molecule. The result is a racemic mixture.

8.10 Formation of Compounds with Stereogenic Centers

Formation of compounds with one stereogenic center from achiral compounds using achiral reagents cannot yield a single stereoisomer. However, in an enzyme catalyzed process, the reaction of an achiral compound generates a single stereoisomer. Such reactions are **stereospecific**.

In some reactions where two stereogenic centers are generated from an achiral substrate, some mechanistic information is obtained based on the diastereomers formed. For example, the formation of two equivalent centers might give a mixture of the R,R compound and the S,S compound. Although not optically active, that result is different than a process that gives the R,S (*meso*) compound.

8.11 Reactions that Form Diastereomers

If a new stereogenic center is generated in a reaction of a substrate that already has a stereogenic center, then a mixture of diastereomers results. The amounts of these isomers are not equal because the new center is generated in a chiral environment. The excess of one diastereomer over another is called the **stereoselectivity** of the reaction.

8.12 Prochiral Centers

In a chiral environment, two apparently equivalent groups can be distinguished, and the resulting product of a reaction involving those groups is chiral. The atomic center at which optical activity may result is prochiral. The equivalent groups bonded to the prochiral center are enantiotopic and are designated **pro-*R*** or **pro-*S*** to indicate the potential configuration, R or S, if the group is replaced.

Groups at a prochiral center in a molecule that contains a chiral center are **diastereotopic.** The "faces" of a planar site or functional group that contains a center that can be converted into a stereogenic center are designated as **re** or **si** depending on the priority ranking of the three groups and their arrangement using the R,S rules.

SOLUTIONS TO IN-CHAPTER PROBLEMS

Chirality

8.1 Which of the following isomeric methylheptanes has a chiral center?

(a) 2-methylheptane   (b) 3-methylheptane   (c) 4-methylheptane

Answers: (a) none (b) one (c) none

$$CH_3-CH_2-\underset{\underset{H}{|}}{\overset{\overset{CH_3}{|}}{C}}-CH_2-CH_2-CH_2-CH_3$$

3-methylheptane

8.2 Which of the following isomeric bromohexanes has a chiral center?

(a) l-bromohexane   (b) 2-bromohexane   (c) 3-bromohexane

Answers: (a) none (b) one (c) one

(b) $$CH_3-\underset{\underset{H}{|}}{\overset{\overset{Br}{|}}{C}}-CH_2-CH_2-CH_2-CH_3$$

2-bromohexane

(c) $$CH_3-CH_2-\underset{\underset{H}{|}}{\overset{\overset{Br}{|}}{C}}-CH_2-CH_2-CH_3$$

3-bromohexane

8.3 Which of the compounds with molecular formula $C_5H_{11}Cl$ has a chiral center?

Answer: 2-chloropentane and 2-chloro-3-methylbutane each have one chiral center.

$$CH_3-\underset{\underset{H}{|}}{\overset{\overset{Cl}{|}}{C}}-CH_2-CH_2-CH_3$$

2-chloropentane

$$CH_3-\underset{\underset{H}{|}}{\overset{\overset{Cl}{|}}{C}}-\underset{\underset{H}{|}}{\overset{\overset{CH_3}{|}}{C}}-CH_3$$

2-chloro-3-methylbutane

8.4 Which of the compounds with molecular formula $C_3H_5Cl_2$ has a chiral center?

Answer: 1,2-dichloropentane has one chiral center.

$$CH_3-\underset{\underset{H}{|}}{\overset{\overset{Cl}{|}}{C}}-\underset{\underset{H}{|}}{\overset{\overset{H}{|}}{C}}-Cl$$

1,2-dichloropropane

8.5 Which of the following isomeric methylheptanes has a chiral center?

Answers: (a) none (b) two (c) none (d) one

(a)

(b)

(c)

(d)

8.6 How many chiral centers does each of the following cyclic compounds have?

Answers: (a) two (b) one (c) none (d) none

(a) OH / C—CH$_2$CH$_3$ / C (cyclohexane)

(b) cyclohexane—CHOHCH$_3$

(c) cyclohexane—CH$_2$CH$_2$OH

(d) cyclohexane with OH and CH$_2$CH$_3$

8.7 How many chiral centers does each of the following barbiturates have?

Answers: (a) one (b) two (c) one (d) one

(a) phenobarbital

(b) secobarbital

(c) hexobarbital

(d) amobarbital

121

8.8 How many chiral centers does each of the following drugs have?

Answers: (a) none (b) one   (c) none
(d) two

(a)   phenylbutazone, used to treat gout

(b)   ibuprofen, an analgesic

S enantiomer, most effective

(c)   chlorphentermine, a nervous system
        stimulent

(d)   chloramphenicol, an antibiotic

8.9 How many chiral carbon atoms are in each of the following synthetic anabolic steroids?

Answers: (a) six   (b) seven

(a)   Dianabol

(b)   stanozolol

8.10 Determine the number of chiral centers in the male sex hormone testosterone and in the female sex hormone estradiol.

(a)   testosterone

(b)   estradiol

Answers: (a) six   (b) five

122

Plane of Symmetry

8.11 Determine whether each of the following compounds has a plane of symmetry.

Answer: Only (b) and (c) have a plane of symmetry.

(a)

(b) ——————— plane of symmetry

(c)

(d)

plane of symmetry

8.12 Determine whether each of the following compounds has a plane of symmetry.

Answer: Only (a) and (d) have a plane of symmetry.

(a) ———Br plane of symmetry

(b) H_3C ... CH_2 ... CH_3

(c) H_3C ... CH_3

(d) H_3C ... CH_3 plane of symmetry

Optical Activity

8.13 Lactic acid in the blood has a specific rotation of +2.6°. A sample of lactic acid obtained from sour milk has a specific rotation of −2.6°. How do these compounds differ?

Answer: The compounds are enantiomers.

8.14 Optically pure (S)-(+)-citronellol from citronella oil has a specific rotation of +5.3°. An enantiomer of optically pure (S)-(+)-citronellol is obtained from geranium oil. What is its specific rotation?

Answer: The specific rotation of the enantiomer is −5.3°.

8.15 The configuration of naturally occurring MSG, which has a specific rotation of +24° is S. Is the assignment of configuration based upon the sign of the optical rotation correct?

Answer: No, it is incorrect. An assignment based upon the optical rotation of a compound is not related to its configuration. An S isomer can have a positive sign of rotation.

8.16 Carvone obtained from spearmint oil is the (R)-(−)-enantiomer. Explain the meaning of both terms within parentheses.

Answer: The R refers to the configuration at the chiral center. The (−) refers to the sign of the optical rotation of the compound.

8.17 A solution of 3 g of menthol in 50 mL of ethanol is prepared and a sample is placed in a 10-cm tube. The optical rotation is +3.0°. What is the specific rotation of menthol?

Answer: The concentration is 0.06 g/mL. The 10-cm tube is 1 dm long. Using the observed rotation and substituting into the equation to calculate specific rotation in Section 8.3, the specific rotation is 50°.

8.18 The specific rotation of (R)-2-bromobutane in ethanol is −23.1°. A solution of the compound in a 1-dm tube has $[\alpha]_D = 55°$. What is the concentration of the compound in grams per 100 mL?

Answer: Using the equation to calculate specific rotation, the concentration is calculated as 2.4 g/mL.

8.19 The specific rotation of (+)-2-butanol as a pure liquid is +13.9°. A synthetic sample of 2-butanol has an optical rotation of −4.5°. What is the composition of the sample?

Answer: The synthetic sample has a majority of the enantiomer of the opposite configuration of the reference (+) isomer. The optical purity of the synthetic sample is calculated as (4.5/13.9) × 100% = 32%, with the majority being the (−) isomer. If there is x% of the (+) isomer, there is (100 − x) % of the − isomer. Solving the equation 32% = (100−x) − x, there is 66% of the (−) isomer and 34% of the (+) isomer.

8.20 The specific rotation of the S enantiomer of MSG, a flavor enhancer, is +24°. What is the optical purity of a synthetic sample whose a is +6°? What are the percentages of the two enantiomers in the sample?

Answer: The synthetic sample has a majority of the reference (S) isomer. The optical purity of the synthetic sample is calculated as (6/24) × 100% = 25%, with the majority being the (S) isomer. If there is x% of the (S) isomer, there is (100 − x) % of the (R) isomer. Solving the equation 25% = x − (100−x), there is 62% of the (S) isomer and 38% of the (R) isomer.

Fischer Projection Formulas

8.21 Draw the Fischer projection formula of the following enantiomer of naturally occurring threonine obtained from proteins. Draw all diastereomers as well.

Answer: The structure must first be rotated around the C-2 to C-3 bond to obtain an eclipsed conformation of the CH_3 and CO_2H groups and turned to place the most oxidized group at the top. The Fischer projection of the enantiomer given is the first of the series of four stereoisomers depicted below.

8.22 What stereochemical relationship exists between any and all pairs of the following structures of carbohydrates?

Answer: Compounds I and II are enantiomers, as are compounds III and IV. All other pairs of stereoisomers are diastereomers.

Priority Rules

8.23 Arrange the groups in each of the following sets in order of increasing priority.

(a) —OH, —SH, —SCH$_3$, —OCH$_3$
Answer: —OH < —OCH$_3$ < —SH < —SCH$_3$

(b) —CH$_2$Br, —CH$_2$Cl, —Cl, —Br
Answer: —CH$_2$Cl < —CH$_2$Br < —Cl < —Br

(c) —CH$_2$—CH=CH$_2$, —CH$_2$—O—CH$_3$, — CH$_2$—C≡CH, —C≡C—CH$_3$
Answer: —CH$_2$—CH=CH$_2$ < — CH$_2$—C≡CH < —C≡C—CH$_3$ < —CH$_2$—O—CH$_3$

(d) —CH$_2$CH$_3$, —CH$_2$OH, —CH$_2$CH$_2$Cl, —OCH$_3$
Answer: —CH$_2$CH$_3$ < —CH$_2$CH$_2$Cl < —CH$_2$OH < —OCH$_3$

8.24 Arrange the groups in each of the following sets in order of increasing priority.

Answers:

(a)

(b)

(c)

(d) —C≡H   —C≡N   —N≡C

(a)

(b)

(c)

(d) —C≡H   <   —C≡N   <   —N≡C

8.25 Examine the chiral carbon atom in each of the following drugs and arrange the groups from low to high priority.

(a) ethchlorvynol, a sedative-hypnotic

Answer: CH$_2$CH$_3$ <   —CH=CHCl < —C≡CH < —OH

(b) chlorphenesin carbamate, a muscle relaxant

Answer: —H < —CH$_2$O—Cl < —CH$_2$O—C—NH$_2$ < —OH

(c) mexiletine, an antiarrhythmic

Answer: ——H < ——CH₃ < ——CH₂O—[2,6-dimethylphenyl] < ——NH₂

8.26 Examine the chiral carbon atom in each of the following drugs and arrange the groups from low to high priority.

(a)   brompheniramine, an antihistamine

Answer: ——H < ——CH₂N(CH₃)₂ < ——[4-bromophenyl] < ——[2-pyridyl]

(b)   fluoxetine, an antidepressant

Answer: ——H < ——CH₂CH₂NHCH₃ < ——[phenyl] < ——O—[4-(trifluoromethyl)phenyl]

126

(c)   baclophen, an antispastic

Answer: —H < —CH$_2$CO$_2$H < —Cl < —CH$_2$NH$_2$

R,S Configuration

8.27 Draw the structure of each of the following compounds.

(a) (R)-2-chloropentane   (b) (R)-3-chloro-1-pentene   (c) (S)-3-chloro-2-methylpentane

Answers:

8.28 Draw the structure of each of the following compounds.

(a) (S)-2-bromo-2-phenylbutane (b) (S)-3-bromo-1-hexyne   (c) (R)-2-bromo-2-chlorobutane

Answers:

8.29 Assign the configuration of each of the following compounds.

Answers: (a) S (b) S (c) R (d) S

8.30 Assign the configuration of each of the following compounds.

Answers: (a) S (b) S (c) S (d) S

8.31 Assign the configuration of terbutaline, a drug used to treat bronchial asthma.

Answer:
(*R*)

terbutaline

8.32 Assign the configuration of the following hydroxylated metabolite of diazepam, a sedative.

Answer: *S*. View from below the plane of the page!

Diastereomers

8.33 Assign the configuration of each of the following compounds.

Answers: (a) *R,R* (b) *S,S* (c) *R,S*
(d) *S,S*

8.34 Assign the configuration of each of the following compounds.

Answers: (a) *R,R* (b) *R,S* (c) *R,R*
(d) *R,S*

8.35 Assign the configuration of each of the following compounds.

(a) [structure with D, H, CH₃, H, H₃C, D] (b) [structure with D, CH₃, H₃C, H, H, H] (c) [structure with H, H, H, H₃C, CH₃, D]

Answers: (a) *R,R* and not *meso* (b) *R,S* and is *meso*
(c) *S,S* and not *meso*

8.36 Assign the configuration of each stereogenic center in the following structures. Based on the assignment, determine if the structure is *meso*.

Answers: (a) *R,R* and not *meso* (b) *R,S* and is *meso*
(c) *S,S* and not *meso*

(a) [structure with HO, H, H, CH₃, H₃C, OH] (b) [structure with HO, OH, H₃C, H, H, CH₃] (c) [structure with D, H, HO, CH₃, H₃C, OH]

8.37 Ribose is optically active, but ribitol, its reduction product, is optically inactive. Why?

Answer: Ribitol has a plane of symmetry bisecting the Fischer projection through the C-3 atom and its attached C—H and C—OH bonds. Ribose does not have a plane of symmetry at that point because the "top" and "bottom" groups are not equivalent.

```
      CHO                    CH2OH
  H——————OH            H——————OH
  H——————OH     ———————H——————OH——————— plane of symmetry
  H——————OH            H——————OH
      CH2OH                  CH2OH
      ribose                 ribitol
```

8.38 Which of the following carbohydrate derivatives are *meso* compounds?

Answer: (a) and (b) are meso compounds. Compound (a) has a plane of symmetry bisecting the Fischer projection through C-3 and its attached carbonyl oxygen atom. Compound (b) has a plane of symmetry bisecting the Fischer projection between C-3 and C-4.

```
(a)      CH2OH          (b)      CH2OH           (c)      CO2H
     H——————OH              H——————OH                H——————OH
         ═══O ———————   HO——————H ——— plane       H——————OH
     H——————OH          HO——————H     of symmetry      CH2OH
         CH2OH               H——————OH
                                CH2OH

     meso compounds
```

8.39 5-Hydroxylysine is an amino acid isolated from collagen. Determine the number of stereoisomers possible.

Answer: There are two nonequivalent stereogenic centers. One center, at C-2, bears an amino group. The second center, located at C-4, bears an hydroxyl group. There are four possible stereoisomers.

$$NH_2CH_2\overset{\overset{\displaystyle OH}{|}}{C}HCH_2CH_2\overset{\overset{\displaystyle NH_2}{|}}{C}H\overset{\underset{\displaystyle O}{\|}}{C}OH$$

5-hydroxylysine

8.40 Consider the structure of pantothenic acid (vitamin B_3) and determine the number of stereoisomers possible.

$$HOCH_2—\overset{\overset{\displaystyle CH_3}{|}}{\underset{\underset{\displaystyle CH_3}{|}}{C}}—\overset{\overset{\displaystyle O}{\|}}{C}H\overset{\overset{}{|}}{\underset{\underset{\displaystyle OH}{}}{}}CNHCH_2CH_2CO_2H$$

pantothenic acid

Answer: There is only one stereogenic center. There are two enantiomers.

$$HOCH_2—\overset{\overset{\displaystyle CH_3}{|}}{\underset{\underset{\displaystyle CH_3}{|}}{C}}—CHCNHCH_2CH_2CO_2H$$

pantothenic acid

8.41 There are four isomeric 2,3-dichloropentanes, but only three isomeric 2,4-dichloropentanes. Explain why.

Answer: The two stereogenic centers in 2,3-dichloropentane are nonequivalent, so there are four possible stereoisomers. The two stereogenic centers in 2,4-dichloropentane are equivalent. A plane of symmetry can be placed through the C-3 atom. Thus, there are two enantiomers and one *meso* compound possible for this isomer.

stereoisomers of 2,3-dichloropentane

meso compound   enantiomers

2,4-dichloropentane

8.42 Which of the following structures are *meso* compounds?

Answer: (a) and (c) are *meso* compounds. Compound (a) has a plane of symmetry bisecting the Fischer projection through C-3 and its attached C—H and C—Cl bonds. Compound (c) has a plane of symmetry bisecting the Fischer projection between C-2 and C-3. The symmetry planes are shown in red.

(a)

CH₂Cl
Br——H
H——Cl
Br——H
CH₂Cl

meso compound

(b)

CH₃
H——OH
H——OH
HO——H
HO——H
CH₂OH

(c)

CH₂OH
H——OCH₃
H——OCH₃
CH₂OH

meso compound

Cyclic Compounds

8.43 Which of the following compounds has a plane of symmetry?

(a) *cis*-1,2-dibromocyclobutane   (b) *trans*-1,2-dibromocyclobutane

(c) *cis*-1,3-dibromocyclobutane   (d) *trans*-1,3-dibromocyclobutane

Answer: (a) and (c) have a plane of symmetry and are *meso* compounds. (b) and (d) do not have a plane of symmetry.

8.44 Which of the following structures has a plane of symmetry?
Answer: (a) and (b) have a plane of symmetry (shown in red) and are *meso* compounds. (c) does not have a plane of symmetry.

8.45 Assign the configuration of each stereogenic center in the following structures.

8.46 Assign the configuration of each stereogenic center in the following structures.

Resolution of Enantiomers

8.47 Reaction of a racemic mixture of A_R,A_S with a resolving agent X_R yields diastereomers. The A_R—X_R isomer is less soluble than A_R—A_S. Consequently, the A_S isomer is obtained optically pure. Describe the experimental results if X_S were available as a resolving agent.

Answer: The X_S compound would give a diastereomeric mixture of A_R—X_S and A_S—X_S compounds. The A_R—X_S compound, which is the enantiomer of the A_S—X_R compound, will be the less soluble. Consequently the A_R compound can be obtained optically pure.

8.48 Resolution of a racemic mixture yields one enantiomer with $[\alpha]_D = +44°$ and another enantiomer with $[\alpha]_D = -33°$. One enantiomer is optically pure. Which one? What is the optical purity of the other enantiomer?

Answer: The enantiomer with the larger specific rotation is the pure compound. Thus, the enantiomer with the +44 rotation is pure. The other sample is 75% optically pure.

Reactions of Chiral Compounds

8.49 (R)-(–)-Lactic acid is converted into a methyl ester when it reacts with methanol. What is the configuration of the ester? Can you predict its sign of rotation?

Answer: The methyl ester also has the R configuration. The bonds to the stereogenic center are *not* changed, nor is the priority series of the groups bonded to that center. The —CO_2H group has the second highest priority in the acid, as does the —CO_2CH_3 group of the methyl ester. The sign of rotation of the ester cannot be predicted.

(R)-(–)-Lactic acid

8.50 Free radical chlorination of (S)-2-bromobutane gives a mixture of compounds resulting from attack at any of the four nonequivalent carbon–hydrogen bonds. The products of reaction at C-1 and C-4 are both optically active. Explain why.

Answer: The products have a chlorine atom bonded to carbon atoms that are not stereogenic centers, but the arrangement of the groups bonded to C-2 is unchanged by the reaction, so the products are still optically active.

(S)-2-bromobutane   (S)-1-chloro-3-bromobutane   (S)-1-chloro-2-bromobutane

8.51 Free radical chlorination of (S)-2-fluorobutane gives a 31% yield of 2-chloro-2-fluorobutane. What is the expected stereochemistry of the product?

Answer: The reaction occurs via a free radical that results from abstraction of hydrogen from C-2 atom, which is the stereogenic center. The radical is achiral and can react with chlorine from either face of this plane. Thus, a racemic mixture results.

(S)-2-fluorobutane   (R)-2-chloro-2-fluorobutane   (S)-2-chloro-2-fluorobutane

8.52 Free radical chlorination of (S)-2-bromobutane at the C-2 atom gives an optically inactive product, but reaction at C-3 gives an optically active product. Explain why.

Answer: The reaction occurs via a free radical that results from abstraction of hydrogen. C-2 is the stereogenic center, and loss of hydrogen from this center yields an achiral intermediate that can react with chlorine from either face of its plane. Thus, a racemic mixture results. The C-3 atom is not a stereogenic center but becomes one once substituted by chlorine. A mixture of 2S,3S and 2S,3R product results. It has a net optical activity because the optical rotations of the diastereomers do not cancel.

9

HALOALKANES AND ALCOHOLS
INTRODUCTION TO NUCLEOPHILIC SUBSTITUTION AND ELIMINATION REACTIONS

KEYS TO THE CHAPTER

9.1 Functionalized Hydrocarbons

In this chapter, we discuss the chemistry of haloalkanes and alcohols in which the halogen or hydroxyl group is bonded to an sp^3-hybridized carbon atom. Molecules with sp^2-hybridized carbon atoms bonded to a halogen or a hydroxyl group have different chemistry. We will discuss these compounds in later chapters. Haloalkanes and alcohols are classified as 1°, 2°, and 3° by the same method used to classify carbon atoms in alkanes.

9.2 Nomenclature of Haloalkanes

The rules for naming haloalkanes are very similar to the rules for naming alkanes (Section 4.3) and alkenes (Section 5.6). Halogen atoms and branching alkyl groups have equal priorities in terms of their positions in the parent chain. If no other functional groups are present, the chain is numbered from the end closest to the first substituent, whether it is a halogen or an alkyl group. However, the double bond of an alkene or the triple bond of an alkyne takes precedence in numbering a carbon chain, regardless of where the halogen is located or how many halogens there may be. The concept of the priority of one functional group over another is expanded with each new functional group we will study in later chapters. It is explicitly part of *R,S* configurational nomenclature.

9.3 Nomenclature of Alcohols

The common names of simple alcohols are based on the name of the alkyl group bonded to the hydroxyl group, as in methyl alcohol. The IUPAC method of naming alcohols is based on the longest continuous carbon chain that contains a hydroxyl group. The chain is numbered to give the carbon atom bonded to the hydroxyl group the lowest possible number. The suffix *-ol* is added to the stem of the name of the parent alkane. Note that in contrast to halogens, which have lower priority than multiple bonds in assigning names, the hydroxyl group has a higher priority than multiple bonds. Compounds with the hydroxyl group bonded to a ring are numbered from the carbon atom containing the hydroxyl group, and subsequent numbers are assigned in the direction to give the lowest possible numbers for any other structural features.

9.4 Structure and Properties of Haloalkanes

The electronegativities of the halogens decrease, and their polarizabilities increase from top to bottom in Group VII of the periodic table. The boiling points of homologous haloalkanes compounds increase in the same order, so the intermolecular attractive forces also increase from top to bottom. This indicates that polarizability is more important than bond polarity in determining the physical properties of haloalkanes. Other factors such as molecular shape and the extent of branching also influence the intermolecular forces and physical properties of haloalkanes.

9.5 Structure and Properties of Alcohols

Intermolecular hydrogen bonds between alcohols dominate their physical properties. Alcohols have higher boiling points than alkanes of similar molecular weight as a result of intermolecular hydrogen bonding. Hydrogen bonding between alcohol molecules and water also accounts for the solubility of alcohols. Alcohols serve as solvents for polar compounds, especially those that can also form hydrogen bonds with the solvent.

Organic Chemistry Study Guide: Key Concepts, Problems, and Solutions http://dx.doi.org/10.1016/B978-0-12-801889-7.00009-1

9.6 Organometallic Compounds

The Grignard reagent is a highly reactive organomagnesium compound formed by reacting a haloalkane with magnesium in an ether solvent. When a Grignard reagent is exposed to water, an alkane is produced. This reaction is useful in the synthesis of deuterium-substituted compounds. The carbon atom of a Grignard reagent has a partial negative charge and behaves as a nucleophile.

Organolithium compounds can also act as carbon nucleophiles. Organolithium compounds are very strong bases, and they are employed in many synthetic procedures. A Gilman reagent is used to "couple" two carbon groups by forming a new sigma bond. One group is provided by an organo-halogen compound; the second is provided by the Gilman reagent. We will also discuss the Gilman reaction again in Chapter 17.

9.7 Reactions of Haloalkanes

We begin to explore the chemistry of haloalkanes in this section. We will continually encounter nucleophilic substitution and the competing elimination reaction. Nucleophiles can displace a halide ion as a leaving group in a substitution reaction. However, nucleophiles are often sufficiently basic to remove a proton from the carbon atom adjacent to the carbon atom bearing the halogen. The overall result is an elimination reaction.

9.8 Nucleophilic Substitution Reactions of Haloalkanes

Nucleophilicity refers to the ability of a nucleophile to displace a leaving group in a substitution reaction. We will describe trends in nucleophilicity in Chapter 10. Most common nucleophiles have a negative charge. However, it is the nonbonding electron pair that is important. For example, water, alcohols, ammonia, and amines are nucleophiles even though they are electrically neutral.

We must keep track of pairs of electrons in nucleophilic substitution reactions. In all cases, in this section, a nonbonded electron pair of a nucleophile forms a new bond to carbon, and the leaving group departs with a pair of electrons. If the nucleophile has a negative charge, and it reacts with a neutral substrate, then the leaving group also has a negative charge.

Although all of the examples of nucleophiles we discussed allow the synthesis of new compounds, two are especially interesting. Both cyanide ion and alkynide ions react with primary haloalkanes to give a product with a new carbon–carbon bond. The products, a nitrile or a terminal alkyne, can be converted to many other functional groups.

9.9 Mechanisms of Nucleophilic Substitution Reactions

In Chapter 6, we discussed electrophilic addition to unsaturated compounds. The essential features of S_N1 and S_N2 mechanisms are presented in this section. We will expand upon these reactions in Chapter 10.

The S_N2 reaction mechanism is based in part on kinetic experiments. The key points are

1. Both the substrate and the nucleophile are present in the transition state.
2. The nucleophile attacks along the same axis that the leaving group departs. (We will discuss the stereochemical consequences of this process in Chapter 10.)

The rate of the S_N2 reaction depends on the structure of the substrate. The order of reactivity is methyl > primary > secondary >> tertiary. Steric hindrance caused by the groups bonded to the reacting carbon center severely hinder approach of the nucleophile in tertiary compounds, and tertiary compounds do *not* react by an S_N2 mechanism. The ability of the nucleophile to displace the leaving group is improved in secondary compounds, and still further in primary
compounds, so both can react with nucleophiles by an S_N2 mechanism. In chapter 10, we will discuss the stereochemical consequences of S_N2 reactions of chiral substrates.

A different mechanism accounts for substitution reactions at sterically hindered sites. Again, the mechanism is based on kinetic data. The rate determining step is the ionization of the haloalkane to give a carbocation that is subsequently captured by the nucleophile in a faster second step. Since only one molecule is present in the transition state for the reaction, it is unimolecular and is designated S_N1. Thus, the rate of the reaction depends only on the substrate concentration. The order of reactivity is tertiary > secondary > primary. In fact, it is unlikely that a primary compound would react by this mechanism when it has the S_N2 process as an available option. Why do tertiary compounds react so much faster by the S_N1 mechanism compared to secondary compounds, and why don't primary compounds react by this mechanism? The answer turns on carbocation stability: a highly unstable primary carbocation would form in an S_N1 reaction, so the reaction proceeds by an S_N2 mechanism instead. In chapter 10, we will explore the stereochemical consequences of S_N1 reactions of chiral substrates.

Substrates that react via the S_N1 mechanism may give rearranged products. This process is exactly like the rearrangements described in Section 6.4. The only difference is in the reaction that leads to the carbocation. In the case of alkenes, the carbocation resulted from addition of a proton to a π bond. We now find that the same types of carbocations form in an S_N1 reaction.

A carbocation can rearrange by a 1,2-hydride shift or 1,2-methide shift. The driving force for such rearrangements is the formation of a more stable carbocation. Hydride ion shifts occur when a secondary or primary carbocation is generated adjacent to a tertiary center. Shifts of alkyl groups most commonly occur when a carbocation is generated at a carbon atom adjacent to a quaternary center. Any of the alkyl groups bonded to that quaterary carbon can migrate. Thus, a mixture of products can result. As shown in Exercise 9.14, even ring bonds of cycloalkanes can migrate, resulting in rings of different size than the reactant.

9.10 Reactions of Alcohols

The reactions of alcohols can occur in several ways that differ in the number and type of bonds cleaved. These are:

1. Cleavage of the oxygen–hydrogen bond in an acid–base reaction.

2. Cleavage of the carbon–oxygen bond in a nucleophilic substitution reaction.

3. Cleavage of the carbon–oxygen bond as well as the carbon–hydrogen bond at the carbon atom adjacent to the carbon atom bearing the hydroxyl group in an elimination reaction.

4. Cleavage of the oxygen–hydrogen bond as well as the carbon–hydrogen bond at the carbon atom bearing the hydroxyl group in an oxidation reaction.

9.11 Acid–Base Reactions of Alcohols

In general, alcohols are somewhat weaker acids than water. Ethanol, 2-propanol, and 2-methyl-2-propanol have pK_a values of 15.9, 18.0, and 19.0, respectively. These values indicate a decrease in acidity from 1° to 2° to 3° alcohols. This agrees with the general principle that alkyl substituents are inductively electron-donating; this effect supplies more electron density to the oxygen atom, so it strengthens the O—H bond.

Also, electronegative substituents near the carbon atom bearing the hydroxyl group increase its acidity. Halogen substituents inductively withdraw electron density from the oxygen atom and weaken the O—H bond. The halogens also stabilize the negative charge of the conjugate base—an alkoxide ion.

Alcohols, like water, can be protonated. The product is a conjugate acid known as an alkyloxonium ion.

9.12 Substitution Reactions of Alcohols

The hydroxyl group of alcohols react with hydrogen halides such as HBr to give haloalkanes. The order of reactivity is tertiary > secondary > primary. Hydrogen bromide suffices to form bromoalkanes, but zinc chloride is required as a catalyst for the reaction with hydrogen chloride.
The substitution reactions of alcohols parallel that of haloalkanes. However, hydroxide ion is not the leaving group. Protonation of the hydroxyl group must occur to allow water to become the leaving group. In general, a weaker base is a better leaving group than a stronger base. Since hydroxide ion is a stronger base than water, it is a poor leaving group in both S_N1 and S_N2 reactions.

The order of reactivity of alcohols in S_N1 reactions is tertiary > secondary > primary. This order parallels the stability of the carbocation intermediates that form in the reaction. This order of reactivity is reversed for S_N2 reactions; however, tertiary alcohols do *not* react by an S_N2 mechanism.

9.13 Alternate Methods for the Synthesis of Alkyl Halides

Since undesirable rearrangements occur in acid-catalyzed reactions of alcohols, other methods have been developed to synthesize alcohols that do not require acid. In this section, we discussed two additional reagents that convert alcohols to haloalkanes. They are used for secondary and primary alcohols which react slowly with hydrogen halides.

Thionyl chloride is used to convert alcohols to chloroalkanes. The by-products are sulfur dioxide and hydrogen chloride, both of which escape from the solution as gases. Phosphorus tribromide is used to convert alcohols to bromoalkanes. The by-product, phosphorous acid, is soluble in water.

9.14 Elimination Reactions

The elimination reactions we consider result in loss of atoms from adjacent carbon atoms and is called a 1,2-elimination or a (β-elimination). β-Elimination reactions occur by either E1 or E2 mechanisms. An E1 mechanism is similar to an S_N1 mechanism in one key respect: it is a unimolecular reaction in which a carbocation intermediate forms in the rate determining step. And, as in S_N1 reactions, the carbocation can, and often does, rearrange to give several products. E1 reactions typically occur in the dehydration of tertiary alcohols. E1 reactions compete with S_N1 reactions. E2 reactions, like S_N2 reactions, are bimolecular processes in which there are two species in the transition state. E2 reactions are observed for primary and secondary alkyl halides and alcohols. The reactions of alcohols in both E1 and E2 reactions are acid catalyzed.

9.15 Regioselectivity in Dehydrohalogenation

Dehydrohalogenaton occurs to give a predominance of the most substituted alkene. This regioselectivity results in the so-called Zaitsev product. This product is the most stable alkene. The alkene with the most alkyl groups bonded to the carbon atoms of the double bond is favored. If geometric isomers are possible, the more stable *trans* isomer predominates.

9.16 Mechanisms of Dehydrohalogenation

There are two mechanisms for dehydrohalogenation. The E2 reaction occurs for primary haloalkanes and usually for secondary haloalkanes. The E1 process occurs with tertiary haloalkanes. In either case, the rate depends on the leaving groups. The reactivity order is iodo- > bromo- > chloroalkane.

The E2 mechanism is a concerted process in which a base abstracts a proton while a carbon–carbon double bond forms and simultaneously a halide ion leaves. Because a double bond starts to develop in the transition state, the energy of the transition state is lower for more highly substituted compounds. The stereochemistry of the E2 process involves an *anti* periplanar arrangement of the carbon–hydrogen bond and the bond from a carbon atom to the leaving group. This arrangement provides the necessary geometric alignment for the emerging orbitals required to form a π bond.

SUMMARY OF REACTIONS

1. Formation and Reactivity of Grignard Reagents

2. Coupling Reaction of Gilman Reagent and Halogen Compounds

$$(CH_3)_2Cu^- \ Li^+ \ + \ CH_3(CH_2)_6CH_2I \ \longrightarrow \ CH_3(CH_2)_6CH_3$$

1-iodooctane   nonane

trans-1-iodononene   *trans*-6-tetradecene

3. Nucleophilic Substitution of Haloalkanes

$$CH_3CH_2CH_2CH_2CH_2Cl \ + \ CH_3O^- \ \longrightarrow \ CH_3CH_2CH_2CH_2CH_2OCH_3$$

4. Synthesis of Haloalkanes from Alcohols

$$CH_3CH_2CH_2CH_2CH_2OH \ + \ HCl \ \longrightarrow \ CH_3CH_2CH_2CH_2CH_2Cl$$

139

5. Dehydrohalogenation of Haloalkanes

$$\xrightarrow[\text{CH}_3\text{CH}_2\text{OH}]{\text{CH}_3\text{CH}_2\text{O}^-}$$

$$\xrightarrow[\text{(CH}_3)_3\text{COH}]{\text{(CH}_3)_3\text{COK}}$$

6. Dehydration of Alcohols

$$\xrightarrow{\text{H}_2\text{SO}_4}$$

7. Synthesis of Haloalkanes from Alcohols

$$\text{CH}_3\text{CH}_2\text{CH}_2\text{CH}_2\text{CH}_2\text{OH} + \text{HCl} \longrightarrow \text{CH}_3\text{CH}_2\text{CH}_2\text{CH}_2\text{CH}_2\text{Cl}$$

+ HBr ⟶

+ SOCl₂ ⟶

3 —CH₂CH₂OH + PBr₃ ⟶ —CH₂CH₂Br

8. Dehydrohalogenation of Haloalkanes

$$\xrightarrow[\text{CH}_3\text{CH}_2\text{OH}]{\text{CH}_3\text{CH}_2\text{O}^-}$$

$$\xrightarrow[\text{(CH}_3)_3\text{COH}]{\text{(CH}_3)_3\text{COK}}$$

9. Dehydration of Alcohols

SUMMARY OF REACTIONS

ANSWERS TO END-OF-CHAPTER EXERCISES

9.1 Classify each of the following haloalkanes.

Answers:
(a) secondary
(b) tertiary
(c) secondary
(d) tertiary

(a) $CH_3-CH_2-\underset{\underset{Br}{|}}{CH}-CH_2-CH_3$

(b) $CH_3-\underset{\underset{Cl}{|}}{\overset{\overset{CH_3}{|}}{C}}-CH_2-CH_3$

(c) $CH_3-CH_2-\underset{\underset{F}{|}}{CH}-CH_3$

(d) $CH_3-\underset{\underset{CH_3}{|}}{\overset{\overset{CH_3}{|}}{C}}-I$

9.2 Classify each of the following alcohols.

Answers:
(a) primary
(b) secondary
(c) secondary
(d) primary

(a) $CH_3-\underset{\underset{CH_3}{|}}{\overset{\overset{CH_3}{|}}{C}}-CH_2-OH$

(b) $CH_3-\underset{\underset{OH}{|}}{CH}-CH_2-\underset{\underset{CH_3}{|}}{CH}-CH_2-CH_3$

(c) $CH_3-\underset{\underset{OH}{|}}{CH}-CH_2-CH_2-CH_3$

(d) $CH_3-\underset{\underset{CH_3}{|}}{\overset{\overset{CH_3}{|}}{C}}-CH_2-CH_2-OH$

9.3 Classify each of the hydroxyl groups in the following vitamins.

Answers:
(a) phenol and primary
(b) primary
(c) primary at top, all others secondary

(a) pyridoxal (vitamin B_6)

(b) thiamine (vitamin B_1)

(c) riboflavin (vitamin B_2)

9.4 Classify each of the hydroxyl groups in the following steroids.

(a)   digitoxigenin, a cardiac glycoside

(b) hydrocortisone, an antiinflammatory drug

(c) norethindrone, an oral contraceptive

Nomenclature of Haloalkanes

9.5 What is the IUPAC name for each of the following compounds?

(a) vinyl fluoride   (b) allyl chloride   (c) benzyl bromide

Answers: (a) fluoroethene   (b) 3-chloro-l-propene   (c) (bromomethyl)benzene

9.6 What is the IUPAC name for each of the following compounds?

(a) $(CH_3)_3CCH_2Cl$ (neopentyl chloride)   (b) $(CH_3)_2CHCH_2CH_2Br$ (isoamyl bromide)

(c) $C_6H_5CH_2CH_2F$ (phenethyl fluoride)

Answers: (a) 1-chloro-2,2-dimethylchloropropane   (b) 1-bromo-3-methylbutane   (c) 1-fluoro-2-phenylethane

9.7 Draw the structure of each of the following compounds.

(a) *cis*-l-bromo-2-methylcyclopentane   (b) 3-chlorocyclobutene

(c) (*E*)-1-fluoro-2-butene   (d) (*Z*)-l-bromo-1-propene

Answers:

9.8 What is the IUPAC name for each of the following compounds?

(a)

trans-1-ethyl-4-iodocycloheptane

(b)

1-chloro-l-propylcyclohexane

(c)

trans-1-bromo-2-cyclobutylcyclodecane

(d)

2-bromo-3-cyclohexyl-1-propanol

Nomenclature of Alcohols

9.9 Write the structural formula of each of the following compounds

(a) 2-methyl-2-pentanol (b) 2-methyl-1-butanol (c) 2,3-dimethyl-1-butanol (d) cyclopentanol (e) *trans*-2-methylcyclohexanol

(f) 1,3-propandiol (g) 1,2,4-butanetriol

Answers:

$$CH_3-\underset{\underset{OH}{|}}{\overset{\overset{CH_3}{|}}{C}}-\underset{\underset{H}{|}}{\overset{\overset{H}{|}}{C}}-CH_2-CH_3$$

(a) 2-methyl-2-pentanol

$$CH_3-CH_2-\underset{\underset{H}{|}}{\overset{\overset{CH_3}{|}}{C}}-CH_2-OH$$

(b) 2-methyl-1-butanol

$$CH_3-CH_2-\underset{\underset{CH_3}{|}}{\overset{\overset{CH_3}{|}}{C}}-CH_2-OH$$

(c) 2,3-dimethyl-1-butanol

—OH

(d) cyclopentanol

$$HO-CH_2-CH_2-CH_2-OH$$

(f) 1,3-propanediol

(e) *trans*-2-methylcyclohexanol

$$HO-CH_2-\underset{\underset{OH}{|}}{\overset{\overset{CH_3}{|}}{C}}-CH_2-OH$$

(g) 1,2,4-butanetriol

9.10 What is the IUPAC name for each of the following compounds?

(a) 2-methyl-3-pentanol   (b) 3-ethyl-3-pentanol   (c) 4-methyl-2-pentanol   (d) l-ethylcyclohexanol   (e) *cis*-3-ethylcyclopentanol

(f) 1,2-hexanediol   (g) 1,2,3,4,5,6-hexanehexol

Answers:

(a) 2-methyl-3-pentanol

(b) 3-ethyl-3-pentanol

(c) 4-methyl-2-pentanol

(d) 1-ethylcyclohexanol

(e) *cis*-3-ethylcyclopentanol

(f) 1,2-hexanediol

(g) 1,2,3,4,5,6-hexanehexol

9.11 What is the IUPAC name for each of the following compounds?

Answers:
(a) 3-ethyl-2-hexanol
(b) 4,7-dimethyl-5-decanol
(c) 4-methyl-3-hexanol
(d) 5,6-dimethyl-3-heptanol

145

9.12 What is the IUPAC name for each of the following compounds?

Answers:
(a) *trans*-4-methylcyclohexanol
(b) 4-cyclopentyl-1-butanol
(c) *trans*-2-bromocyclooctanol
(d) 2-bromo-3-cyclohexyl-1-propanol

(a)

(b)

(c)

(d)

9.13 Name the sex attractant of the Mediterranean fruit fly.

Answer:
(*E*)-6-nonen-1-ol

9.14 Name the following compound, which used as a mosquito repellent.

Answer:
2-ethyl-1,3-hexanediol

$$CH_3 \!-\! CH_2 \!-\! CH_2 \!-\! CH \!-\! CH \!-\! CH_2 \!-\! CH_3$$
$$\qquad\qquad\qquad\quad OH \quad CH_2OH$$

Properties of Haloalkanes

9.15 Which compound is more polar, methylene chloride (CH_2Cl_2) or carbon tetrachloride (CCl_4)?
Answer: Methylene chloride is more polar because it has a dipole moment. Carbon tetrachloride has no dipole moment.

9.16 Tribromomethane is more polar than tetrabromomethane, but their boiling points are 150 and 189 °C, respectively. Explain why the more polar compound has the lower boiling point.
Answer: The polar tribromomethane has the lower molecular weight and is less polarizable than the higher-molecular-weight tetrabromomethane.

9.17 The dipole moment of (*Z*)-1,2-dichloroethene is 1.90 D. Predict the dipole moment of the *E* isomer.
Answer: The (*E*) isomer has no dipole moment because the bond moments of the two C—Cl groups cancel.

9.18 The dipole moment of 1,2-dichloroethane is 1.19 D. What does this value indicate about the conformational equilibrium of this compound?
Answer: The *anti* conformation has no dipole moment because the bond moments of the two C—Cl groups cancel one another. Therefore, there must be a substantial amount of the gauche conformation in the conformational equilibrium mixture to give a dipole moment of 1.19 D.

Physical Properties of Alcohols

9.19 1,2-Hexanediol is very soluble in water but 1-heptanol is not. Explain why these two compounds with similar molecular weights have different solubilities.
Answer: The two hydroxyl groups in 1,2-hexanediol can form more hydrogen bonds with water, thus greatly increasing its solubility.

9.20 Ethylene glycol and 1-propanol boil at 198 and 97 °C, respectively. Explain why these two compounds with similar molecular weights have different boiling points.
Answer: Ethylene glycol has two hydroxyl groups and can form more intermolecular hydrogen bonds. Therefore, its boiling point is higher than that of 1-propanol.

9.21 Explain why 1-butanol is less soluble than 1-propanol in water.
Answer: The nonpolar hydrocarbon portion of the molecule is larger in 1-butanol than in 1-propanol.

9.22 Suggest a reason why 2-methyl-1-propanol is much more soluble than 1-butanol in water.

Answer: The 2-methyl-1-propanol molecule has a more spherical shape, so it interferes less with the network of hydrogen-bonded water molecules.

Organometallic Reagents

9.23 Devise a synthesis of 1-deutero-1-methylcyclohexane starting from 1-methylcyclohexene.

Answer:

9.24 Devise a synthesis of 1,4-dideuterobutane starting from any organic compound that does not contain deuterium.

Answer:

$$BrCH_2CH_2CH_2CH_2Br \xrightarrow[\text{ether}]{Mg} BrMgCH_2CH_2CH_2CH_2BrMg$$

$$D\text{---}CH_2CH_2CH_2CH_2\text{---}D$$

9.25 Devise two syntheses to prepare 2-methyloctane using reagents containing alkyl groups with five or fewer carbon atoms.

Answer: Using a Gilman reagent and an alkyl halide, two possible combinations are (1) the Gilman reagent lithium di(3-methyl-1-iodobutyl) cuprate and 1-iodobutane and (2) the Gilman reagent lithium dibutyl cuprate and 3-methyl-l-iodobutane. The reaction with lithium dibutyl cuprate is shown below.

9.26 Write the products of the following reactions for Gilman reagents that contain primary alkyl groups.

Answers:

Nomenclature of Alcohols

9.27 Write the structure of the product obtained for each of the following combinations of reactants.

(a) 1-chloropentane and sodium iodide (b) 1,3-dibromopropane and excess sodium cyanide (c) benzyl chloride and sodium acetylide

(d) 2-bromobutane and sodium hydrosulfide (NaSH)

Answers:

(a) $CH_3CH_2CH_2CH_2CH_2Cl \xrightarrow{\text{NaI}} CH_3CH_2CH_2CH_2CH_2\text{——}I$

(b) $BrCH_2CH_2CH_2Br \xrightarrow[\text{excess}]{\text{NaCN}} N\equiv C\text{——}CH_2CH_2CH_2\text{——}C\equiv N$

(c)

$\xrightarrow{HC\equiv C^- Na^+}$

(d) $CH_3CH_2\overset{\overset{\displaystyle Br}{|}}{C}HCH_3 \xrightarrow{\text{NaSH}} CH_3CH_2\overset{\overset{\displaystyle SH}{|}}{C}HCH_3$

9.28 What haloalkane and nucleophile are required to produce each of the following compounds?

(a) $CH_3CH_2CH_2C\equiv CH$ (b) $(CH_3)_2CHCH_2CN$ (c) $CH_3CH_2SCH_2CH_3$ (d) $C_6H_5CH_2OH$

Answers:

(a) $CH_3CH_2CH_2Br \xrightarrow{HC\equiv C^- Na^+} CH_3CH_2CH_2\text{——}C\equiv CH$

(b) $CH_3\overset{\overset{\displaystyle CH_3}{|}}{C}HCH_2Br \xrightarrow{\text{NaCN}} CH_3\overset{\overset{\displaystyle CH_3}{|}}{C}HCH_2\text{——}C\equiv N$

(c) $CH_3CH_2Br \xrightarrow{\text{NaSCH}_2CH_3} CH_3CH_2\text{——}S\text{——}CH_2CH_3$

(d)

$\xrightarrow{\text{NaOH}}$

Mechanism of Nucleophilic Substitution Reactions

9.29 Which compound in each of the following pairs reacts at the faster rate with sodium iodide in an S_N2 process to yield an alkyl iodide?

(a) 1-chlorohexane or 2-chlorohexane (b) bromocyclohexane or 1-bromo-1-methylcyclohexane

(c) 2-bromo-4-methylpentane or 2-bromo-2-methylpentane

Answers: (a) 1-Chlorohexane reacts fastest because it is a primary haloalkane; 2-chlorohexane is a secondary haloalkane.
(b) Bromocyclohexane reacts fastest because it is a secondary haloalkane; 1-bromo-1-methylcyclohexane is a tertiary haloalkane.
(c) 2-Bromo-4-methylpentane reacts fastest because it is a secondary haloalkane; 2-bromo-2-methylpentane is a tertiary haloalkane.

9.30 Rank the following compounds in order of increasing S_N2 reactivity with a common nucleophile.

I: 1-bromohexane   II: 1-bromo-2-methylpentane   III: l-bromo-3-methylpentane

Answer: All are primary halogen compounds. The differences in rates are due to steric hindrance of the methyl groups in II and III. The order of reactivity is II < III < I based on the distance of the methyl group from the site of the reaction.

9.31 Which compound in each of the following pairs reacts at the faster rate in an S_N1 process under the same reaction conditions?

(a) bromocyclohexane or 1-bromo-1-methylcyclohexane (b) 2-bromobutane or l-bromo-2-methylpropane

(c) 2-bromobutane or 2-methyl-2-bromobutane

Answers: (a) 1-Bromo-l-methylcyclohexane is a tertiary haloalkane and is more reactive in an S_N1 reaction than the secondary haloalkane.
(b) 2-Bromobutane is a secondary haloalkane and is more reactive under S_N1 conditions than l-bromo-2-methylpropane, which is a primary haloalkane.
(c) 2-Bromo-2-methylbutane is a tertiary haloalkane and is more reactive under S_N1 conditions; 2-bromobutane is a secondary haloalkane.

9.32 Which compound in each of the following pairs reacts at the faster rate in an S_N1 process under the same reaction conditions?

I: 2-bromohexane   II: 2-bromo-2-methylpentane   III: 1-bromo-2-methylpentane

Answer: The order of reactivity is III < I < II, which is in the order primary < secondary < tertiary.

9.33 Predict the product of the reaction of one molar equivalent of sodium iodide with 1,3-dichlorohexane.

Answer: The product is 3-chloro-l-iodohexane, which results from S_N2 displacement of the primary halogen by iodide ion rather than displacement of the secondary halogen at C-3.

1,3-dichlorohexane

9.34 Treatment of the following compound with sodium sulfide yields C_4H_8S. What is the structure of the product? How is it formed?

$$Br—CH_2—CH_2—CH_2—CH_2—CH_2—Br + NaS \rightarrow C_4H_8S$$

Answer: Displacement of one chloride ion by sulfide gives an intermediate that can displace a second halide ion in an intramolecular S_N2 reaction.

9.35 Reaction of the following compound with water under S_N1 conditions yields a mixture of two alcohols. Explain why.

Answer: The compound has a tertiary C—Cl bond which reacts under S_N1 conditions to give a tertiary carbocation. Capture of the carbocation can occur from either face of the planar carbocation, resulting in two geometric isomers.

9.36 Reaction of either 3-bromo-1-butene or (*Z*)-1-bromo-2-butene with water under S_N1 conditions yields the same product. Explain why.

Answer: The same resonance-stabilized allyl carbocation results from either of the two compounds under S_N1 conditions. Therefore, only one product forms.

9.37 The rate of reaction of *cis*-1-bromo-4-*tert*-butylcyclohexane with methylthiolate (CH_3S^-) is faster than for the *trans* isomer. Suggest a reason for this difference.

Answer: The reaction of the *trans* isomer requires attack of the nucleophile by a path over the cyclohexane ring to displace the equatorial halogen. This process is more sterically hindered than attack of the nucleophile to displace the axial halogen in the *cis* isomer in a path in the equatorial plane of the ring.

cis-1-bromo-4-*tert*-butylcyclohexane
(nucleophile not sterically hindered)

trans-1-bromo-4-*tert*-butylcyclohexane
(nucleophile sterically hindered)

9.38 Which of the following two compounds reacts at the faster rate with sodium cyanide?

Answer: Compound I is less sterically hindered and reacts at the faster rate.

I

II

Acid–Base Properties of Alcohols

9.39 1,1,1- Trichloro-2-methyl-2-propanol is used as a bacteriostatic agent. Compare its pK_a to that of 2-methylpropanol.

Answer: The trichloro compound is more acidic as a result of inductive electron withdrawal by the three chlorine atoms. Its pK_a is 12.87, much less the pK_a value of 2-methyl-2-propanol, which is 19.

9.40 Which base is the stronger, methoxide ion or *tert*-butoxide ion? Explain your reasoning.

Answer: The *tert*-butoxide ion is a stronger base than methoxide ion, because *tert*-butyl alcohol is a weaker acid than methanol. The electron-releasing alkyl groups of *tert*-butyl alcohol destabilize the anion.

Formation of Alkyl Halides from Alcohols

9.41 Rank the following compounds according to their rates of reaction with HBr.

I

II

III

Answer: The reaction will occur by an S_N1 mechanism, so the order of reactivity parallels the order of carbocation stability: III < I < II.

9.42 Rank the following compounds according to their rates of reaction with HCl and $ZnCl_2$.

$$CH_3-CH_2-\underset{\underset{CH_3}{|}}{\overset{\overset{CH_3}{|}}{C}}-OH \qquad CH_3-CH_2-\underset{\underset{CH_3}{|}}{CH}-CH_2OH \qquad CH_3-\underset{\underset{OH}{|}}{CH}-\underset{\underset{CH_3}{|}}{\overset{\overset{H}{|}}{C}}-CH_3$$

$$\text{I} \qquad\qquad\qquad \text{II} \qquad\qquad\qquad \text{III}$$

Answer: The reaction will occur by an S_N1 mechanism, so the order of reactivity parallels the order of carbocation stability: II < III < I.

9.43 Write the structure of the product of reaction for each of the following compounds with PBr_3.

Answers:

$$CH_3-\underset{\underset{CH_3}{|}}{\overset{\overset{CH_3}{|}}{C}}-CH_2-CH_2-OH \quad\xrightarrow{PBr_3}\quad CH_3-\underset{\underset{CH_3}{|}}{\overset{\overset{CH_3}{|}}{C}}-CH_2-CH_2-Br$$

$$\text{I}$$

$$CH_3-\underset{\underset{CH_2CH_3}{|}}{\overset{\overset{OH}{|}}{C}}-CH_2-CH_3 \quad\xrightarrow{PBr_3}\quad CH_3-\underset{\underset{CH_2CH_3}{|}}{\overset{\overset{Br}{|}}{C}}-CH_2-CH_3$$

$$\text{II}$$

$$CH_3-\underset{\underset{CH_3}{|}}{\overset{\overset{H}{|}}{C}}-CH_2-\underset{\underset{OH}{|}}{CH}-CH_3 \quad\xrightarrow{PBr_3}\quad CH_3-\underset{\underset{CH_3}{|}}{\overset{\overset{H}{|}}{C}}-CH_2-\underset{\underset{Br}{|}}{CH}-CH_3$$

$$\text{III}$$

9.44 Write the structure of the product of reaction for each of the following compounds with $SOCl_2$.

Answers:

(a) $\xrightarrow{SOCl_2}$

(b) $\xrightarrow{SOCl_2}$

(c) $\xrightarrow{SOCl_2}$

9.45 Reaction of 3-buten-1-ol with HBr yields a mixture of two products: 3-bromo-l-butene and 1-bromo-2-butene. Explain why. (Hint: The reaction of this allyl alcohol occurs via an S_N1 process.)

1-bromo-2-butene   3-bromo-l-butene

Answer: A resonance-stabilized allyl carbocation forms as an intermediate under these S_N1 conditions, and it can react with Br⁻ to give two different products.

9.46 The rate of reaction of the following unsaturated alcohol with HBr is faster than the rate of reaction of the saturated alcohol. Explain why.

resonance-stabilized carbocation

tertiary carbocation

Answer: Both compounds are tertiary alcohols. However, the first compound yields a tertiary allylic carbocation in an S_N1 reaction. As a result, the transition state for formation of this intermediate has a lower activation energy, and the reaction occurs at a faster rate.

9.47 Which of the compounds in Exercises 9.41 and 9.42 may yield rearranged products?

Answer: Compound I in Exercise 9.41 can rearrange from a secondary carbocation intermediate to give a tertiary carbocation. Compound II in Exercise 9.42 can rearrange from a primary carbocation intermediate to give a tertiary carbocation. Compound III in Exercise 9.42 can rearrange from a secondary carbocation intermediate to give a tertiary carbocation.

I Question 9.41 $2°$ carbocation $3°$ carbocation

CH_3—CH_2—CH—CH_2OH $\longrightarrow$ CH_3—CH_2—C—CH_2^+

II Question 9.42 $1°$ carbocation

1,2-hydride shift

CH_3—CH_2—C—CH_3

$3°$ carbocation

CH_3—CH—C—CH_3 $\longrightarrow$ CH_3—C—C—CH_3 $\longrightarrow$ CH_3—C—C—CH_3

III Question 9.42 $1°$ carbocation $3°$ carbocation

9.48 The reaction of 2-octanol with HBr gives 2-bromooctane and 3-bromooctane in a 13:1 ratio. Explain how 3-bromooctane forms in this reaction.

CH_3—CH—CH_2—$CH_2(CH_2)_3CH_3$ $\longrightarrow$ CH_3—CH—CH_2—$CH_2(CH_2)_3CH_3$

OH

1,2-hydride shift

CH_3—CH_2—CH—$CH_2(CH_2)_3CH_3$

Answer: The carbocation with positive charge at C-2 can rearrange by a hydride shift of hydrogen at C-3 to give a carbocation with positive charge at C-3. Each carbocation can capture Br^-.

153

Regioselectivity in Dehydrohalogenation

9.49 Consider each of the following isomeric compounds with the molecular formula $C_6H_{13}Br$. Which ones will give only a terminal monosubstituted alkene when they undergo dehydrobromination by an E2 process?

I: $CH_3CH_2CH_2CH_2CH_2CH_2Br$ II: $CH_3CHBrCH_2CH_2CH_2CH_3$

III: $CH_3CH_2CHBrCH_2CH_2CH_3$ IV: $(CH_3)_2CHCH_2CH_2CH_2Br$

V: $(CH_3)_2CHCH_2CHBrCH_3$ VI: $(CH_3)_2CHCHBrCH_2CH_3$

$$CH_3$$
VII: $CH_3CH_2\overset{|}{C}HCH_2CH_2Br$ VIII: $CH_3CH_2\overset{\overset{\displaystyle CH_3}{|}}{C}HCHBrCH_3$

$$CH_3$$
IX: $CH_3CH_2\overset{|}{C}BrCH_2CH_3$ X: $BrCH_2\overset{\overset{\displaystyle CH_3}{|}}{\underset{\underset{\displaystyle CH_3}{|}}{C}}CH_2CH_3$

XI: $(CH_3)_3CCH_2CH_2Br$ XII: $(CH_3)_3CCHBrCH_3$

$$CH_3$$
XIII: $(CH_3)_2CH\overset{|}{C}HCH_2Br$ XIV: $(CH_3)_2CH\overset{\overset{\displaystyle CH_3}{|}}{C}BrCH_3$

Answer: Only compounds I, IV, VII, and XI can lose HBr in an E2 reaction to give a single product. Each is a primary bromoalkane.

9.50 Consider each of the compounds in Exercise 9.49. Which ones can undergo dehydrobromination by an E2 process to give only a terminal disubstituted alkene?
Answer: Only compound XIII can yield a terminal disubstituted alkene.

XIII

9.51 Consider each of the compounds in Exercise 9.49. Which ones can undergo dehydrobromination by an E1 process?

Answer: Only compounds IX and XIV are tertiary haloalkanes, and they will undergo dehydrobromination by an E1 process.
9.52 Consider each of the compounds in Exercise 9.49. Which ones cannot undergo dehydrobromination?

Answer: Compound XIII cannot undergo dehydrobromination because the β carbon does not have a hydrogen.

9.53 Consider each of the compounds in Exercise 9.49. Which ones can undergo dehydrobromination to give at least one set of E, Z stereoisomers among the products?

Answer: Compounds II, III, V, VI, VIII, and IX can undergo dehydrobromination to give at least one set of E, Z stereoisomers among the products?

9.54 Consider each of the compounds in Exercise 9.49. Which ones can undergo dehydrobromination to give a trisubstituted alkene among the products? Which ones can undergo dehydrobromination to give a tetrasubstituted alkene among the products?

Answer: Compounds VI, VIII, and IX give trisubstituted products. XIV can give a tetrasubstituted alkene.

154

9.55 How many alkenes can form from each of the following compounds via an E2 process? Write the structure of each alkene.

(a) l-bromopentane (b) 2-chlorohexane (c) 3-iodoheptane (d) S-bromononane

Answers:

9.56 How many alkenes can form from each of the following compounds via an E2 process? Write the structure of each alkene.

(a) 3-bromo-2-methylhexane (b) 2-chloro-3-methylhexane (c) 3-iodo-4-ethylhexane (d) 4-bromo-4-methylheptane

Answers:

9.57 What bromocycloalkane can give each of the following unsaturated compounds in the best yield by an E2 process?

9.58 Which of the following unsaturated compounds can be obtained in good yield by an E2 process from a bromocycloalkane?

Answer: Only (a) and (c) can be converted to a single product. No bromocycloalkane for (b) or (d) can be made that leads to a single product.

(a)

(b)

(c)

(d)

Stereoelectronic Effects in Dehydrohalogenation

9.59 The following isomer undergoes an E2 reaction about 1000 times slower than any of the other stereoisomers of 1,2,3,4,5,6-hexachlorocyclohexane. Why?

Answer: The required *trans*-diaxial arrangement of C—H and C—Cl cannot be achieved even by a ring flip. In any chair conformation, all chlorine atoms are in axial positions and all hydrogen atoms are in equatorial positions with a dihedral angle of 60°.

lindane

9.60 One of the following two isomeric bicyclic compounds undergoes an E2 elimination much faster than the other. Identify the compound that reacts at the faster rate and explain why.

Answer: The fused rings of *trans*-decalin are conformationally rigid. Compound I has an axial C—Br bond that is in a position to undergo *trans*-diaxial elimination. Compound II does not have the required *trans*-diaxial arrangement, so the reaction is much slower.

9.61 What is the configuration of the alkene formed by the elimination of one molar equivalent of HBr from the following compound?

Answer: The *anti* periplanar arrangement of the C—H and C—Br bonds in the conformation shown gives the (*E*) isomer shown to the right.

9.62 An E2 elimination of 1-chloro-1,2-diphenylethane can yield a mixture of (*E*)- and (*Z*)-1,2-diphenylethene. How would the *E/Z* ratio of isomers for this reaction compare to the *E/Z* ratio for the E2 elimination of 2-bromopentane?

(*E*)-1,2-diphenylethene
(major product)

(*Z*)-1,2-diphenylethene
(minor product)

(*E*)-2-pentene
(major product)

(*Z*)-2-pentene
(minor product)

Answer: There would be a larger *E/Z* ratio of isomers for 1,2-diphenyiethene because the two phenyl groups present a larger steric hindrance in the formation of the (*Z*) isomer than the methyl and ethyl groups do.

9.63 When menthyl chloride reacts with sodium ethoxide in ethanol, the only alkene product is 2-menthene. Explain why.

menthyl chloride

2-menthene

menthyl chloride

$\xrightarrow{E_2}$ 2-menthene

Answer: In the conformation obtained by a ring flip, there is only one axial C—H bond situated to undergo an *anti* periplanar E2 elimination. It gives 2-menthene.

9.64 When neomenthyl chloride reacts with sodium ethoxide in ethanol, the only alkene product is 3-menthene. Explain why.

neomenthyl chloride

3-menthene

Answer: There are two C—H bonds situated to undergo *anti* periplanar E2 eliminations. The ratio of products is determined by the degree of substitution of the double bonds. 3-Menthene is the more stable compound because it has a trisubstituted double bond.

9.65 Draw the structure of the alkene formed in an E2 elimination of the following compound.

Answer: The *anti* periplanar arrangement of the C—H and C—Cl bonds gives the alkene shown to the right. It is the more stable, tetrasubstituted double bond.

9.66 Which of the following compounds reacts at the faster rate in an E2 elimination reaction?

I

II

Answer: Compound I has axial C—H and C—Cl bonds as shown in the structure and can undergo an *anti* periplanar E2 elimination reaction. Compound II has an equatorial C—Cl bond, so it would react at a slower rate.

9.67 The following compound cannot undergo dehydrobromination under either El or E2 conditions. Explain why.

Answer: There is no *anti* periplanar C—H bond as required for an E2 elimination reaction. The bridgehead carbon cannot form the required planar arrangement for the carbocation that would have to be generated in an E1 reaction.

1-bromobicyclo[2.2.2]octane

9.68 Explain why 1-*tert*-butylcyclopropene is difficult to synthesize by a dehydrohalogenation reaction.

—C(CH₃)₃

1-*tert*-butylcyclopropene

Answer: The reaction of either the required 1-halo or *trans*-2-halo compound could undergo an E2 elimination because there is a *syn* periplanar arrangement of the required bonds. However, the resulting compound is very strained. Therefore, a high energy transition state would be required for elimination.

9.69 Reaction of 1-bromo-2-deutero-2-phenylethane with *tert*-butoxide in *tert*-butyl alcohol gives a 7:1 ratio of deuterated and nondeuterated phenylethenes. Write the structures of the products. What does the data suggest about the ease of abstraction of deuterium versus hydrogen?

—CHDCH₂Br

major product

minor product

Answer: The energy required to break a C—D bond is larger than for a C—H bond. This energy difference is reflected in the transition state for the E2 reaction where these bonds are partially broken.

9.70 Dehydrobromination of each of the following compounds gives a single product. One compound yields a cycloalkene containing deuterium, and the other yields a cycloalkene that does not contain deuterium. Which compound is which?

Answer: Compound II has axial C—H and C—Br bonds and can undergo an *anti* periplanar E2 elimination reaction in which the deuterium is retained. Compound I cannot easily undergo an E2 elimination reaction. However, it can eliminate the deuterium atom situated *trans* to the bromine atom in a twisted conformation, which improves the relationship between the two bonds required for elimination of DBr.

9.71 Although the following reaction of the deuterated bicyclic compound with a strong base occurs at a somewhat slow rate, it gives the indicated product by an E2 mechanism. Explain why the product forms.

Answer: The E2 reaction occurs from a *syn* periplanar arrangement of the C—D and C—Br bonds, even though breaking a C—D bond requires more energy than breaking a C—H bond. The only C—H bond available for an elimination reaction is at a 120° dihedral angle to the C—Br bond, which is unfavorable for an E2 reaction.

9.72 One of the following 2,3-dichlorobicyclo[2.2.1]heptanes undergoes an E2 elimination using potassium *tert*-butoxide in *tert*-butyl alcohol about 100 times as fast as the other. Which compound reacts at the faster rate? The same product, 2-chlorobicyclo[2.2.1]hept-l-ene, forms in both reactions.

Answer: The E2 reaction of the compound on the right occurs from a *syn* periplanar arrangement of the C—H and C—Cl bonds in the *exo* positions. The only C—H bond available for an elimination reaction in the first compound is at a 120° dihedral angle to the C—Cl bond, which is unfavorable for an E2 reaction.

Dehydration of Alcohols

9.73 Draw the structure of the dehydration product(s) when each of the following compounds reacts with sulfuric acid. If more than one product forms, predict the major isomer assuming that no rearrangement reactions occur.

(a) CH₃—C(CH₃)(CH₂CH₃)—OH →(H₂SO₄) [two products, second labeled major]

(b) CH₃CH₂—C(OH)(CH₃)—CH₂CH₃ →(H₂SO₄) [three products, third labeled major]

(c) CH₃CH₂CH(OH)—CH₂CH₂CH₃ →(H₂SO₄) [four products]
These two isomers have comparable stability.

(d) CH₃CH₂—C(OH)(CH₂CH₃)—CH₂CH₂CH₃ →(H₂SO₄) [three products]
These three isomers have comparable stability.

9.74 Draw the structure of the dehydration product(s) when each of the following compounds reacts with sulfuric acid. If more than one product forms, predict the major isomer assuming that no rearrangement reactions occur.

(a) [decalin structure with CH₃ and OH] →(H₂SO₄) [product]

(b) [cyclohexanol with CH₂CH₃] →(H₂SO₄) [two products]
These two isomers have comparable stability

(c) [cyclobutane with CH₂CH₃ and OH] →(H₂SO₄) [two products, second labeled major]

162

9.75 Write the expected product of the acid-catalyzed dehydration of l-phenyl-2-propanol. The reaction is more rapid than the dehydration of 2-propanol. Explain why.

1-phenyl-2-propanol

H₂SO₄

1,2-hydride shift

resonance-stabilized benzyl carbocation

-H⁺

(E)-1-phenyl-1-propene

Answer: The major product is the more highly substituted alkene. The reaction proceeds through a resonance-stabilized, benzyl carbocation intermediate, and the product has a double bond that can interact with the benzene ring giving a resonance-stabilized structure. Therefore, the dehydration is more rapid than for 2-propanol.

9.76 1,2-Diphenylethanol dehydrates extremely easily. Explain why.

1,2-Diphenylethanol

H₂SO₄

(E)-1,2-Diphenylethene

Answer: The alkene has a double bond that can interact with the two benzene rings giving a resonance-stabilized structure.

9.77 Dehydration of *cis*-2-methylcyclohexanol yields two products in a 5:1 ratio. What are the structures of the two products?

H₂SO₄

1-methylcyclohexene
major

+

3-methylcyclohexene

Answer: 1-Methylcyclohexene, which has a trisubstituted double bond, predominates over 3-methylcyclohexene, which has a disubstituted double bond.

9.78 Dehydration of cyclododecanol yields two isomeric products in approximately equal amounts. Catalytic hydrogenation of either compound yields cyclododecane. What are the structures of the two products?

Answer: The two compounds formed in the dehydration reaction are geometric isomers (E)-cyclodecene and (Z)-cyclodecene.

Carbocation Rearrangement in S_N1 and E1 Reactions

9.79 The following isomerization reactions occur in some industrial processes. Write a mechanism that accounts for each step. Indicate whether each reaction is energetically favorable or unfavorable.

Answer: The first step, a 1,2-methide shift, converts a 2° carbocation into a 1° carbocation. Therefore, it is unfavorable. However, the second step, a 1,2-hydride shift, converts a 1° carbocation into a 3° carbocation. Thus, the pathway from the secondary to the tertiary carbocation is favorable overall.

9.80 Write a mechanism that accounts for each step of the rearrangement of the carbocation shown below. Indicate whether each reaction is energetically favorable or unfavorable.

Answer: The first step, a 1,2-hydride shift, converts a primary carbocation into a tertiary carbocation and is favorable. The second step, another 1,2-hydride shift, converts a tertiary carbocation into a secondary carbocation and is unfavorable.

9.81 Ethylidenecyclohexane and l-ethylcyclohexene can be equilibrated using an acid catalyst. Write a mechanism that accounts for this conversion.

Answer: The rearrangement occurs by protonation of the double bond to give a tertiary carbocation, which is followed by loss of a proton from C-2 of the cyclohexane ring.

9.82 4-Methylcyclohexene isomerizes to 1-methylcyclohexene over alumina (an acidic substance). Write a mechanism that accounts for this conversion.

Answer: First, protonation of the double bond at C-4 gives a secondary carbocation at C-3. Second, a hydride shift from C-2 to C-3 gives a secondary carbocation at C-2. Third, loss of a proton from C-1 of the cyclohexane ring forms the 1-methylcyclohexene isomer.

Rearrangement in Dehydration Reactions

9.83 Dehydration of 2,2,4-trimethyl-3-pentanol with acid gives a complex mixture of the alkenes in the indicated percentages. Write a mechanism that accounts for each product.

I: 2,3,4-trimethyl-1-pentene   29%   II: 2,4,4-trimethyl-1-pentene   24%

III: 3,3,4-trimethyl-1-pentene 2%   IV: 2,4,4-trimethyl-2-pentene   24%

V: 3,3,4-trimethyl-2-pentene   18%   VI: 2-isopropyl-3-methyl-1-butene   3%

Answer: The initial secondary carbocation at C-3 can rearrange either by a hydride ion shift from C-4 to give a tertiary carbocation (blue arrow) or by a methide ion shift from C-2 to give a different tertiary carbocation (red arrow). These three carbocations account for most of the alkene products. Compounds III and VI are formed from loss of a proton by carbocations that result from further rearrangement of the one of the tertiary carbocations. Compounds III and VI are formed in significantly smaller amounts than the other four.

9.84 Dehydration of 2,2-dimethylcyclohexanol with acid gives the following isomeric alkenes. Write a mechanism that accounts for each product.

1,2-dimethylcyclohexene

isopropylidinecyclopentane

Answer: The initial secondary carbocation can rearrange by either a 1,2-methide shift (red arrow) or a methylene group shift (blue arrow) of the cyclohexane ring. Each process forms a tertiary carbocation. Subsequent loss of a proton from each of these carbocations gives the observed products.

9.85 1-Methylcyclopentene is one of the dehydration products obtained from l-cyclobutyl-l-ethanol. Write a mechanism that accounts for this reaction.

Answer: The initial secondary carbocation can rearrange by migration of a methylene group of the cyclobutane ring to give another secondary carbocation. The driving force of the reaction is the decrease in ring strain. Subsequent loss of a proton from the cyclopentyl carbocation gives 1-methylcyclopentene.

9.86 3,3-Dimethylcyclopentene is one of the dehydration products obtained from 2-cyclobutyl-2-propanol. Write a mechanism that accounts for this reaction.

Answer: The initial tertiary carbocation can rearrange by migration of a methylene group of the cyclobutane ring to give a secondary carbocation. Although the carbocation is less stable, the driving force of the reaction is the decrease in ring strain. Subsequent loss of a proton from the cyclopentyl carbocation gives 3,3-dimethylcyclopentene.

9.87 1-*tert*-Butylcyclohexene is one of several dehydration products obtained from 1,2,2-trimethylcycloheptanol. Two rearrangements are required for this transformation. Write a mechanism accounting for these reactions.

Answer: The initial tertiary carbocation can rearrange by migration of a methylene group at C-3. Another tertiary carbocation results, but there is a small reduction of ring strain since a cycloheptane ring is converted into a more stable cyclohexane ring. A subsequent 1,2-methide shift forms a tertiary butyl group and generates a tertiary carbocation with the charge on the six-membered ring. This carbocation loses a proton to give a trisubstituted double bond.

9.88 Dehydration of 2-methyl-2-spiro[4.4]nonanol gives a mixture containing 1-methyl-6-bicyclo[4.3.0]nonene. Write a mechanism that accounts for formation of this product.

2-methyl-2-spiro[4.4]nonanol

1-methyl-6-bicyclo[4.3.0]nonene

Answer: The initial tertiary carbocation can rearrange by migration of a methylene group at the adjacent carbon atom that is part of the other cyclopentane ring. Another tertiary carbocation results, but there is a reduction of ring strain as a cyclopentane ring is converted into a cyclohexane ring. This carbocation loses a proton to give a trisubstituted double bond.

10 NUCLEOPHILIC SUBSTITUTION AND ELIMINATION REACTIONS

Keys to the Chapter

10.1 Nucleophilicity and Basicity

Nucleophilicity refers to the ability of a nucleophile to displace a leaving group in a substitution reaction. We describe various trends in nucleophilicity in this section. There is a trend within a period, a trend within a group, a trend based on the charge of the nucleophile, and a steric effect of the nucleophile. Nucleophiles are also bases, and they can abstract protons in elimination reactions. However, although nucleophilicity and basicity are related to the availability of the same electron pair, the reactions of a series of nucleophiles do not necessarily parallel those of the same species as bases.

Within a period, nucleophilicity parallels basicity and decreases from left to right in the periodic table for elements in similarly structured species with the same charge. For example, hydroxide ion is a better nucleophile than fluoride ion. Nucleophilicity is decreased by hydrogen bonding of protic solvents such as alcohols.

Within a group, the order of nucleophilicity is opposite to the order of basicity. This order of nucleophilicity is related to the polarizability of the nucleophile. The order $I^- > Br^- > Cl^-$ is one that we encounter many times in the study of reaction mechanisms. Another important relationship is $RS^- > RO^-$.

Charge has a large effect on nucleophilicity. A species with a negative charge is more nucleophilic than a neutral species with a similar structure. For example, alkoxide ion, RO^-, is a better nucleophile than ROH.

A nucleophile must approach a carbon reaction center to form a bond. Therefore, steric hindrance affects the rate of reaction. Sterically hindered nucleophiles react at a slower rate than similarly charged, smaller nucleophiles containing the same nucleophilic element. For example, *tert*-butoxide reacts more slowly than ethoxide in S_N2 reactions.

10.2 Stereochemistry of Substitution Reactions

The stereochemistry of a nucleophilic substitution reaction is also a powerful probe of the reaction mechanism. In an S_N2 reaction, a chiral substrate undergoes inversion of configuration because the nucleophile, the reactive center of the substrate, and the leaving group are colinear. The substrate turns "inside out," like an umbrella in the wind, during the reaction. In the S_N1 mechanism, a planar carbocation forms. It is achiral. Thus, the stereochemical result of an S_N1 reaction is the formation of a racemic product. The degree of racemization depends on the degree to which the leaving group leaves the site of the carbocation prior to its capture by the nucleophile. If both sides of the plane of the carbocation are symmetrically solvated, then complete racemization occurs. If one side is still shielded by the leaving group, then there is some residual net inversion.

10.3 S_N2 Versus S_N1 Reactions

The prediction of which mechanism will prevail for a specific reaction is not always an absolute process. Four factors determine whether an S_N1 or S_N2 reaction prevails.

1. The structure of the substrate
2. The nucleophile
3. The leaving group
4. The solvent

Organic Chemistry Study Guide: Key Concepts, Problems, and Solutions http://dx.doi.org/10.1016/B978-0-12-801889-7.00010-8

1. The Structure of the Substrate

The order of reactivity in S_N1 reactions is tertiary > secondary > primary. The order of reactivity in S_N2 reactions is primary > secondary > tertiary. At the extremes, tertiary substrates usually react by an S_N1 mechanism, and primary compounds usually react by an S_N1 mechanism. The reaction of secondary substrates may proceed by either mechanism depending on the nucleophile, leaving group, and solvent.

The effect of branching at the β carbon atom and conjugation of allyl and benzyl carbocations also have an important effect on S_N1 and S_N2 reactions. Branching at the β carbon atom decreases the rate of an S_N2 reaction due to steric hindrance. Conjugation of the allyl and benzyl carbocations increases their stability. Primary allyl and benzyl carbocations are about as stable as unconjugated secondary carbocations. Secondary allyl and benzyl carbocations are about as stable as unconjugated tertiary carbocations.

2. The Nucleophile

The nucleophile can affect which mechanism prevails in those borderline substrates such as secondary compounds. A charged nucleophile tends to favor the S_N2 mechanism. The structurally related neutral nucleophile tends to favor an S_N1 mechanism.

3. The Leaving Group

The leaving group strongly affects the rate of both S_N1 and S_N2 reactions. Any feature that stabilizes the negative charge of the leaving group increases the rate of the reaction. The same polarizability order of the halide ions as nucleophiles is seen in the order of their reactivity as leaving groups: $I^- > Br^- > Cl^-$. The range of rates is larger for S_N1 reactions than for the S_N2 reactions because the differences in the stabilities of carbocations are larger than the energy differences between different transition states for S_N2 reactions. For leaving groups containing the same element, such as oxygen, the order of leaving group abilities parallels their basicity. Weak bases are better leaving groups. Thus, leaving groups that are derived from strong acids are the best leaving groups.

4. The Solvent

Polar solvents stabilize charged intermediates such as carbocations better than nonpolar solvents. Thus, the rate of an S_N1 reaction increases with increasing solvent polarity. The effect of solvent polarity on S_N2 reactions is much smaller because there is less charge separation in the transition state. Aprotic solvents increase the effective nucleophilicity of the nucleophile because decreased solvation of the nonbonding electron pair makes that pair more available for reaction with a substrate.

10.4 Mechanisms of Elimination Reactions

An E2 reaction is a concerted, *bimolecular* process in which a base extracts a proton from the β-carbon of the substrate and a bond to the leaving group departs simultaneously. Both the substrate and the base are present in the transition state for the reaction. An El reaction occurs in two steps, the first step is rate determining, and does not involve the base. Thus, the formation of a carbocation in an El reaction is the same as the first step in the S_N1 process. The preferred conformation for an E2 reaction is *anti* periplanar. Based on the *anti* periplanar transition state for an E2 mechanism, the stereochemistry of the alkene can be predicted based on the stereochemistry of both stereogenic centers of the substrate. First, write the structure with the correct configuration at the centers containing the hydrogen atom to be eliminated and the leaving group. Then, rotate about the carbon–carbon bond to the staggered conformation with the hydrogen atom and the leaving group *anti* to one another. This conformation has the correct geometry for the bonded groups that will remain in the alkene.

The rate of an E2 reaction is slower when C—D bond breaks than when a C—H bond breaks. This is called a deuterium isotope effect. If a C—H (or C—D) bond is not broken in the rate determining step, as in an El reaction, there is no deuterium isotope effect.

The basicity of the base increases the rate of an E2 reaction. Thus, the amount of E2 product increases even for substrates that can react by an El reaction. The size of the base affects the regiochemistry. More hindered bases abstract protons from less sterically hindered sites and increase the amount of the less substituted alkene.

Many factors control the competition of S_N1, S_N2, E2, and E1 reactions. In general, tertiary haloalkanes react by S_N1 and E1 mechanisms. The ratio of substitution and elimination products depends on the balance between the basicity and the nucleophilicity of the reagent. Primary haloalkanes react by S_N2 and E2 mechanisms. The amount of elimination increases with the basicity of the nucleophile and with its steric size. Secondary haloalkanes are much more sensitive to the conditions of the reaction, and S_N1, S_N2, E2, and E1 processes occur to varying degrees. Highly polarizable nucleophiles favor an S_N2 reaction; neutral nucleophiles are more apt to be seen in an S_N1 reaction. Aprotic solvents favor S_N2 reactions; polar solvents enhance the rate of S_N1 reactions. Effective nucleophiles that are weak bases, such as thiolates, favor substitution over elimination. Strong bases that are less effective nucleophiles, such as *tert*-butoxide, favor elimination.

Solutions to End-of-Chapter Exercises

Nucleophilicity

10.1 Hydroxylamine (NH_2OH) is a nucleophile. Write its Lewis structure. Which atom supplies the electrons in nucleophilic substitution reactions?

hydroxylamine

Answer: The nucleophilicity for the atoms within a period decreases from left to right. Thus, the nitrogen atom supplies the electrons in nucleophilic substitution reactions.

10.2 Thiocyanate ion (SCN^-) reacts with alkyl halides to give thiocyanate products (R—SCN). The cyanate ion (OCN^-) reacts to form isocyanate products (R—NCO). Write the Lewis structures of the ions. Explain the difference in the sites of reactivity for the two ions.

Answer: The nucleophilicity for the atoms within a family increases from top to bottom. Thus, the sulfur atom of the thiocyanate ion is more nucleophilic than the oxygen atom of the cyanate ion. In the case of the cyanate ion, the nitrogen atom is more basic and a better nucleophile than the oxygen atom, so it forms isocyanate products, R—NCO.

10.3 Reaction of methoxide ion with an alkyl halide to give dimethyl ether is about 100 times faster than the reaction of acetate ion with an alkyl halide to give an ester, methyl acetate. Explain this observation

Answer: The negative charge is localized in the methoxide ion, which makes the ion an effective nucleophile. The negative charge is delocalized in the acetate ion, and the decreased charge on the oxygen atom decreases its nucleophilicity.

10.4 Diethylphenylphosphine reacts with iodoethane about 10^3 times faster than the nitrogen analog, diethylaniline. Explain this result.

Answer: The nucleophilicity for the atoms within a family increases from top to bottom as the atoms become more polarizable. Thus, the phosphorus atom of the phosphine is more nucleophilic than the nitrogen atom of the aniline.

10.5 Dimethyl sulfide, $(CH_3)_2S$, reacts with iodomethane to displace iodide ion twice as fast as diethyl sulfide, $(CH_3CH_2)_2S$. Explain why.

Answer: Dimethyl sulfide is a smaller nucleophile than diethyl sulfide. Nucleophilicity decreases with increasing size of the nucleophile because the larger nucleophile experiences steric hindrance in the transition.

10.6 Triethylarsine, $(CH_3CH_2)_3As$, reacts with iodomethane only four times as fast as dimethyl selenide, $(CH_3)_2Se$. Compare this difference in rate with the rate difference of ammonia relative to water, about 3×10^5.

Answer: The nucleophilicity for the atoms within a period decreases from left to right. Thus, if the structures are similar, the nucleophilicity of an arsenic compound will be greater than that of a selenium compound, and the nucleophilicity of a nitrogen compound will be greater than that of an oxygen compound. The sizes of ammonia and water are similar. However, triethylarsine is larger than dimethyl selenide. The larger size of triethylarsine decreases its nucleophilicity and decreases its rate of reaction.

Stereochemistry of Substitution Reactions

10.7 Reaction of (R)-$(-)$-2-butanol with HBr yields a mixture of 87% (S)-$(+)$-2-bromobutane and 13% (R)-$(-)$-2-bromobutane. What is the optical purity of the product? What is the mechanism for this substitution reaction?

Answer: The optical purity is 87% − 13% = 74%. Because a mixture of enantiomers results, the mechanism must be S_N1. However, the leaving group (H_2O) substantially shields one face of the carbocation as it leaves, and the bromide ion reacts with substantial inversion of configuration.

10.8 Reaction of (R)-2-methyl-l-butanol with HBr yields 1-bromo-2-methylbutane. Predict the configuration of the product. What is the mechanism for this substitution reaction?

Answer: The mechanism for a substitution reaction at a primary center is S_N2. The stereogenic center at C-2 does not change in this reaction. In the process, the highest priority group —CH_2OH is changed into the highest priority group —CH_2Br. Thus, the configuration is unchanged and is R.

10.9 The rate of incorporation of radioactive iodide into optically active 2-iodooctane in acetone as solvent leads to racemization at twice the rate of incorporation of radioactive iodine. Explain how these data support an S_N2 mechanism.

$$CH_3(CH_2)_5\overset{\overset{\textstyle I}{|}}{\underset{\underset{\textstyle H}{|}}{C}}{-}CH_3 \ + \ {}^*I^{-} \longrightarrow CH_3(CH_2)_5\overset{\overset{\textstyle I^*}{|}}{\underset{\underset{\textstyle H}{|}}{C}}{-}CH_3 \ + CH_3(CH_2)_5\overset{\overset{\textstyle H}{|}}{\underset{\underset{\textstyle I^*}{|}}{C}}{-}CH_3 \ + \ I^{-}$$

2-iodooctane
(optically active)

racemization rate equals twice
the rate of I* incorporation

Answer: Each time a radioactive iodide ion reacts by an inversion process in the S_N2 mechanism, the resulting molecule of inverted product cancels the optical rotation of one molecule of the reactant. Thus, two molecules of optically active 2-iodobutane are effectively "lost" (in terms of their optical rotation) for every one molecule that reacts with radioactive iodide.

10.10 The reaction of (S)-2-bromooctane with cyanide ion gives a cyano compound with an R configuration. However, reaction of (S)-2-bromooctane with iodide ion followed by reaction of the alkyl iodide with cyanide ion gives a cyano compound with the S configuration. Explain these data.

Answer: In the substitution reaction with cyanide ion, the highest priority bromine group is replaced by a high priority cyanide group. The reaction occurs by inversion, and the net result is a product with the R configuration. In the two step reaction sequence, two inversion steps occur, resulting in net retention of configuration. First, replacing bromide by iodide ion yields the R product, which subsequently is converted into the S product when cyanide replaces iodide in the second step.

10.11 *trans*-l-Chloro-3-methylcyclopentane reacts with sodium iodide in acetone to give *cis*-1-iodo-3-methylcyclopentane. What is the mechanism of this reaction?

Answer: Since the stereochemistry at C-1 is inverted when iodide replaces chloride, the reaction must occur by an S_N2 mechanism.

10.12 Write the product expected from the reaction of *cis*-1-bromo-2-methylcyclopentane with cyanide ion.

Answer: The stereochemistry at C-1 is inverted when cyanide ion replaces bromide ion. Thus, the product has the *trans* configuration.

10.13 The following compound has the *R* configuration. Draw the product expected from the reaction of this compound in ethanol, indicating the stereochemistry.

Answer: The compound is a tertiary halide, which reacts by an S_N1 mechanism, so the reaction gives a racemic mixture of ethyl ethers.

10.14 (*S*)-l-Chloro-l-phenylethane reacts in a 20% water–80% acetone solution to give a 51:49 ratio of (*R*)- and (*S*)-1-phenyl-1-ethanols. Explain why the product is highly racemic even though the reactant is a secondary alkyl halide.

Answer: The benzene ring stabilizes the secondary carbocation, which is a benzylic carbocation, by resonance. The benzylic carbocation can be attacked from either side of the plane to give an almost completely racemized product.

10.15 The reactant in the following reaction has the *S* configuration. Based on the composition of the product mixture, what is the mechanism of the reaction?

Answer: A primary tosylate tends to react by the S_N2 mechanism. However, in this case, the benzene ring can stabilize a primary benzyl carbocation by resonance and allow the reaction to occur by the S_N1 mechanism. However, the leaving tosylate group, which is quite large, substantially shields one face of the carbocation as it leaves, so the acetate ion reacts with substantial inversion of configuration.

10.16 The reactant in the following reaction has the *S* configuration. Based on the composition of the product mixture, what is the mechanism of the reaction?

Answer: A secondary chloroalkane can react by an S_N1 or S_N2 mechanism depending on reaction conditions and other structural features of the compound. In this case, the secondary carbon is benzylic. Thus, the benzene ring can stabilize a secondary carbocation—it is a benzyl carbocation—by resonance, and the reaction then occurs by the S_N1 mechanism. The product is an almost equal mixture of enantiomers. The slight excess of inverted product forms because the leaving chloride shields one face of the carbocation, so trifluoroethoxide ion reacts from the back side.

10.17 What is the configuration of the tosylate prepared from (*R*)-2-butanol? What is the configuration of the iodide obtained by reacting that tosylate with iodide ion in acetone?

Answer: A formation of the tosylate ester occurs by displacement of a chloride ion from the sulfur atom of *p*-toluenesulfonyl chloride (TosylCl) by the oxygen atom of the alcohol. The reaction occurs with retention of configuration because no bonds at the C-2 stereogenic center are affected. The highest priority hydroxyl group in the reactant is transformed into the highest priority tosylate group in the product. Thus, the configuration is still *R*. Subsequent displacement of the tosylate group by iodide occurs with inversion of configuration. The highest priority tosylate group in the reactant is replaced by a high priority iodide group in the product. Thus, the configuration of the inverted product is *S*.

10.18 Write the product expected from the reaction of each of the following compounds in aqueous acetone.

Answer: Both compounds are tertiary bromides that react with nucleophiles such as water in aqueous acetone by an S_N1 mechanism. A mixture of compounds with equatorial and axial hydroxyl groups results. The composition of the mixture is the same for both compounds I and II.

Reactivity in Substitution Reactions

10.19 1-Bromo-1,1-diphenylethane reacts very rapidly in ethanol. Explain why.

Answer: The bromide ion leaves readily because the resulting carbocation is both tertiary and benzylic. The carbocation is resonance stabilized by the two benzene rings.

1-bromo-1,1-diphenylethane   tertiary, benzylic carbocation

10.20 4-Chloro-2,2,4,6,6-pentamethylheptane reacts in aqueous acetone about 500 times faster than *tert*-butyl chloride does. Explain this observation.

Answer: The tertiary carbocation has two *tert*-butyl groups and a methyl group bonded to the positively charged carbon atom compared to three methyl groups in the *tert*-butyl carbocation. Steric crowding around the C-4 atom of the reactant is decreased in the carbocation compared to the reactant as the chloride ion leaves.

4-Chloro-2,2,4,6,6-pentamethylheptane

10.21 3-Bromo-1-butene and (*E*)-1-bromo-2-butene react at the same rate in aqueous acetone. Explain why. An identical mixture of two substitution products is obtained from either compound. What are the structures of the products?

Answer: The same resonance-stabilized carbocation results from both compounds. Attack of water at either C-1 or C-3 gives the same mixture of alcohols.

(*E*)-1-bromo-2-butene

10.22 The following compound reacts in methanol to rapidly replace one of the two bromine atoms by a methoxy group. Which bromine atom is replaced?

benzylic carbocation

Answer: The bromine atom nearer the benzene ring is replaced because the resulting secondary carbocation is benzylic, and therefore it is resonance stabilized.

176

10.23 The following sulfonium ion reacts in 80% ethanol–20% water to give 36% 2-methyl-1-propene. The remaining 64% of the product is a mixture of two substitution products. What are the substitution products? *tert*-Butyl chloride reacts under the same conditions to give the identical mixture of products. Explain this observation.

Answer: Dimethyl sulfide is the leaving group and a *tert*-butyl carbocation forms. The substitution products are an alcohol resulting from nucleophilic attack of water on the tertiary carbocation and an ethyl ether resulting from nucleophilic attack of ethanol. The same products are formed from *tert*-butyl chloride because the same carbocation is formed when the chloride ion leaves.

10.24 The relative rates of substitution of bromide by ethoxide in ethanol for methyl, ethyl, propyl and butyl bromides are 1, 0.057, 0.018, and 0.013, respectively. Explain these data.

Answer: The methyl compound reacts fastest because the three hydrogen atoms do not present any steric hindrance to the nucleophile in the S_N2 reaction. The other three compounds all have one alkyl group and two hydrogen atoms in the path of the nucleophile. Each alkyl group is primary, and their steric sizes are similar. The slight decrease in rate suggests that as the length of the chain increases there is some increase in steric hindrance. Note that the change is larger when comparing a methyl group to an ethyl group. Extending the chain to a propyl group does not decrease the rate of reaction by as large a factor.

10.25 Trifluoromethanesulfonyl chloride reacts with alcohols to form sulfonate esters. Would you expect the "trifylates" to be more or less reactive than the methanesulfonate esters?

Answer: The fluorine atoms withdraw electron density from the sulfonyl group and also stabilize the trifylate ion compared to the methanesulfonate ion. Thus, the trifluoromethanesulfonate ion is a better leaving group, and therefore the trifylate is more reactive.

10.26 A nitro group is electron withdrawing. Would you expect the sulfonate esters of *p*-nitrobenzenesulfonic acid to be more or less reactive than the tosylates?

Answer: The nitro group inductively withdraws electron density from the sulfonyl group, and it also stabilizes the sulfonate ion compared to the tosylate ion. Thus, the nitro-substituted sulfonate ion is a better leaving group, so its esters are more reactive.

p-nitrobenzenesulfonic acid

10.27 Explain why ethanol (CH_3CH_2OH) can solvate both cations and anions. Explain why dimethyl ether, $(CH_3)_2O$, is a poorer solvent for ionic compounds. Discuss the solvation characteristics of both solvents for both cations and anions.

Answer: The lone pair electrons of the oxygen atom of both the alcohol and the ether coordinate with cations. The partially positive hydrogen atom of the hydroxyl group of alcohols can help solvate anions. Because the ether cannot solvate anions, the solubility of ionic compounds in ether is much smaller than in alcohols.

10.28 *trans*-1-Iodo-3-methylcyclopentane reacts with KF in DMF to give a fluoro compound. What is its configuration? The iodo compound reacts with KF in ethanol to give products that do not contain fluorine. Explain these data.

Answer: In an aprotic solvent, the fluoride ion is sufficiently nucleophilic to react with the iodo compound in an S_N2 mechanism that occurs with inversion of configuration. The product of this reaction is *cis*-1-fluoro-3-methylcyclopentane. In ethanol, the fluoride is solvated and is not an effective nucleophile. The substitution product is an ethyl ether.

177

cis-1-fluoro-3-methylcyclopentane

Solvent Effect in Substitution Reactions

10.29 The structure of hexamethylphosphoramide is shown below. Its dielectric constant is 30. How do you expect HMPA to affect the rates of S_N1 and S_N2 reactions?

hexamethylphosphoramide
(HMPA)

Answer: Hexamethylphosphoramide (HMPA) is very polar, as indicated by its dielectric constant. It has no protic sites and is thus aprotic. HMPA can be used for substitution reactions. Because it has a high dielectric constant, it will favor S_N1 processes. Because it is aprotic, it will accelerate S_N2 processes.

10.30 The rate constant for the displacement of iodide from iodomethane by fluoride ion is 10^6 times faster in dimethyl formamide than in methanol. Explain why.

Answer: The fluoride ion is solvated by methanol, and its nucleophilicity is much less than it is in an aprotic solvent such as dimethylformamide.

10.31 Methyl tosylate reacts with halide ions in water. The rate constants for the reaction in water decrease in the order $k_I > k_{Br} > k_{Cl}$. The rate constants for the reactions in acetone stand in the order $k_I < k_{Br} < k_{Cl}$. Explain these data.

Answer: The reaction rates in water reflect the nucleophilicity of the solvated anion. The degree of solvation decreases as ionic size increases. Thus, iodide ion is the best nucleophile of the series. In the aprotic acetone solvent, the order of nucleophilicity is related to the strength of the ions as bases. Chloride ion is the strongest base of the series.

10.32 The equilibrium constant for the reaction of bromomethane with iodide ion is 15 in water and 0.6 in acetone. Explain these data.

$$CH_3Br + I^- \rightleftharpoons CH_3I + Br^-$$

Answer: The position of the equilibrium in water is affected by the degree of solvation of the anion. Bromide ion is more strongly solvated than iodide ion, and the reaction tends to proceed to the right as written. In acetone as solvent, the position of the equilibrium is controlled by factors other than the stability of the anions, neither of which is solvated.

Elimination Reactions

10.33 Attempted displacement of iodide ion by fluoride in acetone usually fails because elimination products result. Explain why elimination is favored over substitution.

Answer: Fluoride ion is the strongest base of the halides because HF is the weakest acid of the hydrogen halides. In an aprotic solvent the fluoride ion is not solvated, and it is an even stronger base than in protic solvents. Elimination is favored in reactions with strong bases.

10.34 The product mixture obtained in the reaction of isobutyl bromide with sodium ethoxide in ethanol contains 62% 2-methyl-1-propene. The reaction using potassium *tert*-butoxide in *tert*-butyl alcohol contains 92% 2-methyl-1-propene. Explain why.

Answer: The *tert*-butoxide ion is not as nucleophilic as ethoxide ion because it is sterically larger. Thus, substitution reactions occur less readily with *tert*-butoxide ion. In addition, the *tert*-butoxide ion is a stronger base than the ethoxide ion, and a larger fraction of elimination product results with a stronger base.

10.35 The product mixture obtained in the reaction of *sec*-butyl bromide with 1 M sodium ethoxide in ethanol contains 78% unsaturated material. What are the products and which of them should predominate? Using 4 M sodium ethoxide in ethanol, the product mixture is 91% unsaturated material. Why?

Answer: The unsaturated products are 1-butene, *cis*-2-butene, and *trans*-2-butene. The 2-butenes are more substituted alkenes and are the major products. The alkenes result in part from an El mechanism derived from loss of a proton from the carbon atom adjacent to the carbocation center, as well as an E2 mechanism in which either ethanol or ethoxide ion abstracts a proton from the reactant. Increasing the concentration of ethoxide ion increases the amount of product in the E2 mechanism.

10.36 The unsaturated compounds obtained in the reaction of 2-bromo-2,3-dimethylbutane with the alkoxide of 3-ethyl-3-pentanol are 92% 2,3-dimethyl-l-butene and 8% 2,3-dimethyl-2-butene. Compare these data with the data for reaction of *tert*-butoxide with the same compound (Section 10.5).

Answer: The alkoxide derived from 3-ethyl-3-pentanol is a more sterically hindered base than *tert*-butoxide ion. Their base strengths are similar. The increased amount of the 1-butene indicates that the more sterically hindered base does not abstract the tertiary hydrogen atom at C-3 as readily and is far more likely to abstract the primary hydrogen atom. Thus, the less substituted alkene predominates.

10.37 E2 reactions of tosylates occur using alkoxide ions as bases in the related alcohol solvent. Determine the stereochemistry of the 2-phenyl-2-butene formed from reaction of the tosylate of (2*R*,3*R*)-3-phenyl-2-butanol.

Answer: Arrange the structure in a conformation so that the hydrogen atom at C-3 and the tosylate group are *anti* periplanar. The structure of the product has the (*E*) configuration.

10.38 The tosylate of *cis*-2-phenylcyclohexanol undergoes an elimination reaction much more rapidly with *tert*-butoxide in *tert*-butyl alcohol than does the *trans* isomer. The product is exclusively 1-phenylcyclohexene. Explain these data.

Answer: In the *cis* isomer, the tosylate group and the hydrogen at the C-2 atom are *anti* periplanar and elimination can readily occur.

Answer: In the *trans* isomer, elimination cannot occur unless the chair undergoes a ring flip to place the tosylate in an axial position. However, in this conformation, the product will not be 1-phenylcyclohexene because the phenyl ring will be in an axial position. Elimination will occur by abstraction of the hydrogen atom at C-6 to give the isomeric 3-phenylcyclohexene.

10.39 The following products are obtained from the E2 reaction of (2S,3R)-2-bromo-3-deuterio-butane using sodium ethoxide in ethanol. Explain why. Predict the products from the E2 reaction of (2S,3S)-2-bromo-3-deuteriobutane.

trans-2-butene   (Z)-2-deutero-2-butene

Answer: Arrange the structure in a conformation so that the hydrogen atom at C-3 and the bromide ion are *anti* periplanar. This conformation gives (Z)-2-deutero-2-butene. An alternate conformation has the deuterium atom at C-3 and the bromide ion in an *anti* periplanar arrangement. This conformation gives *trans*-2-butene.

trans-2-butene

(Z)-2-deutero-2-butene

Answer: Reversing the configuration at C-3 from *R* to S gives a (Z) compound without deuterium and an (E) compound containing deuterium.

(E)-2-deutero-2-butene

cis-2-butene

180

10.40 The E2 reaction of l-bromo-2-deutero-2-phenylethane gives the following compounds. Explain why the indicated percentage of each compound is formed.

87%   13%

Answer: The carbon–deuterium bond is not as easily broken in an E2 reaction as the carbon–hydrogen bond. Thus, the major product contains the deuterium atom. It is formed by abstraction of the hydrogen atom from C-2.

13%   87%

10.41 The E2 reaction of each of the following compounds with sodium methoxide in methanol proceeds regiospecifically to give different compounds. What is the structure of the compounds derived from each stereoisomer?

I   II

Answer: Compound I has a hydrogen atom at the bridgehead position that may be abstracted to give a tetrasubstituted alkene. The elimination reaction of compound II can only occur by abstraction of a proton from the methyl group.

10.42 Predict the E2 product formed in the reaction of each of the following compounds. Which compound reacts at the faster rate?

I

II

Answer: Compound I has an axial bromine atom that is *anti* periplanar to the hydrogen atom. This compound reacts at a faster rate than compound II, which has an equatorial bromine atom. The elimination product of compound II does not contain deuterium.

11

CONJUGATED ALKENES
AND
ALLYLIC SYSTEMS

KEYS TO THE CHAPTER

In this chapter, we have focused on compounds and intermediates that have conjugated double bonds. Conjugated alkenes are resonance stabilized, and this property affects both the structures of the conjugated molecules and the stabilities of reaction intermediates derived from them. Although we can describe these properties with conventional Lewis structures, molecular orbital theory provides far deeper insights into their properties and reactions. We will use the ideas developed in this chapter again in the next chapter when we discuss the properties and reactions of aromatic compounds.

11.1 Classes of Dienes

Conjugated dienes and higher polyenes contain a series of alternating single and double bonds. Isolated dienes have more than one single bond separating the two double bonds and are regarded as two separate alkenes. Cumulated dienes have two double bonds sharing a common carbon atom.

Terpenes are naturally occurring compounds that are synthesized in cells from isoprene units. Isoprene is a conjugated butadiene with a methyl branch. The terpenes are named according to the number of five-carbon units that make up their skeleton.

11.2 Stability of Conjugated Dienes

The effect of the interaction of two double bonds is the first case of the interaction of functional groups that we have discussed. Others such as the interaction of the carbon–oxygen double bond of a carbonyl group with a carbon–carbon double bond will be considered in later chapters. The π electrons in a conjugated system are *delocalized*.

Conjugated dienes are stabilized by resonance interaction of the double bonds. Therefore, these dienes are of lower energy than dienes in which the double bonds do not interact. Thus, the energy released upon hydrogenation is smaller than predicted based on the heats of hydrogenation of the component double bonds. The difference between the experimental value and the predicted value based on isolated double bonds is the resonance energy.

11.3 Molecular Orbital Models of Conjugated Systems

Molecular orbitals are made from **linear combinations of atomic orbitals**. We obtain these orbitals by adding or subtracting the equations that describe the energies of atomic orbitals. Adding orbitals with the proper sign of the wave function corresponds to constructive overlap of atomic orbitals. This combination gives rise to bonding **molecular orbitals**. Subtracting the equations for the wave functions of atomic orbitals with opposite signs gives destructive overlap and produces an **antibonding molecular orbital**.

Molecular orbitals may be **symmetric** or **antisymmetric** based on the sign of the molecular orbital at one point compared to a related point on the other side of a **nodal plane**. The electron density at this nodal plane is zero. The bonding molecular orbital of ethene is symmetric; the antibonding molecular orbital of ethene is antisymmetric with respect to a nodal plane. The energies of the molecular orbitals increase as the number of nodal planes increases.

A group of molecular orbitals for a polyene can be separated into a group of bonding molecular orbitals (corresponding to the number of double bonds) and an equal number of antibonding molecular orbitals. In a neutral polyene, all of the π electrons are in the bonding molecular orbitals. Our picture of the degree of double bond character between adjacent carbon atoms in a polyene is based on whether the various contributing molecular orbitals are bonding or antibonding. For the C-2 to C-3 bond in butadiene, the π_2 provides no electron density, but π_1 does provide electron density, so there is some double bond character in this bond.

Organic Chemistry Study Guide: Key Concepts, Problems, and Solutions http://dx.doi.org/10.1016/B978-0-12-801889-7.00011-X

11.4 Effects of Conjugation on Diene Structure

Conjugation accounts for the partial double bond character of the C-2 to C-3 bond of butadiene. The bond length is shorter than predicted based on the hybridization of the component sp^2 orbitals that form the sigma bond. This shortening is attributed to the double bond character as a result of the contribution of the π_1 molecular orbital.

To maintain the overlap of the 2p orbitals that make up the molecular orbitals, the atoms of the double bonds must be coplanar. Two conformations are observed, consistent with this requirement. The *s-trans* conformation is more stable than the *s-cis* conformation due to steric hindrance in the *s-cis* conformation. Rotation about the C-2 to C-3 bond in butadiene requires more energy than rotation about a single bond without double bond character. The rotational energy barrier is due to both a torsional component in which there is some steric interference and the loss of resonance energy that occurs when the π orbitals are orthogonal in the transition state of the process.

11.5 Allylic Systems

Allylic halides react to give allylic carbocations under S$_N$1 conditions. The positive charge of an allyl carbocation or any substituted allylic carbocation is distributed between two "end" carbon atoms. There is no charge on the "center" carbon atom. However, the charge distribution is equal only in the allyl carbocation itself. Substituted allylic carbocations necessarily have unequal charge distributions based on the identity of the attached groups. This effect is seen in the product distribution of the S$_N$1 reactions of substituted allyl chlorides. The major product corresponds to capture of a nucleophile at the more highly substituted center because the positive charge is better stabilized at that center.

Allylic radicals have an electron deficiency at either end of the radical. (There is no radical character at the "center" carbon atom.) Delocalization in the allyl radical is reflected in the lower bond energy of the C—H bond that gives rise to the radical. *N*-bromosuccinimide (NBS) reacts with allyl systems to give selective halogenation in which a bromine atom replaces an allylic hydrogen.

11.6 Molecular Orbitals of Allylic Systems

The molecular orbitals formed from an odd number of p orbitals in an allylic system are arranged somewhat differently from the arrangement of molecular orbitals made from an even number of p orbitals. First, there is one molecular orbital, called a **nonbonding orbital**, in which the nodal plane contains the central carbon atom. This means that there is no π electron density at the central carbon atom. Second, the energy of the nonbonding orbital is the same as the contributing atomic orbital. Hence, there is no net stabilization as a result. The symmetry of the molecular orbitals is defined in the same way as for polyenes. The lowest energy molecular orbital is symmetric, and the symmetry alternates with each higher energy molecular orbital. And, as the number of nodal planes increases, the energy increases. The nodal plane in an allylic system, or any π system made from an odd number of 2p orbitals, may be at an atom or between atoms.

The distribution of electrons among the molecular orbitals follows Hund's rule. The allyl cation has electrons only one electron in π_1, and the resulting positive charge is felt at the terminal carbon atoms, corresponding to the π_2 orbital. This orbital contains one electron in the allyl radical and two electrons in the allyl anion.

11.7 Electrophilic Conjugate Addition Reactions

Conjugates dienes undergo electrophilic addition. However, rather than forming a localized carbocation in the first step of the reaction, as in alkenes, an allylic carbocation is formed. The nucleophile can then add to either of two carbon atoms that bear the positive charge of the allylic carbocation. Therefore, either 1,2- or 1,4 addition can occur. The amounts of 1,2- and 1,4-addition products depend on the groups bonded to the two "end" carbon atoms of the allylic system.

If the 1,2-addition product of the reaction is stable, and does not have sufficient energy to revert to an allyl cation by ionization of the original nucleophile, then the result is termed **kinetic control**. This means that the product that forms fastest is the major product. However, if the energy of the system is high enough, the 1,2-product ionizes. Repeating this process eventually gives the most thermodynamically stable product, hence the term thermodynamic control.

The product of kinetic control may be influenced by the stability of the intermediate and the partial positive charge at the two "ends" of the allylic carbocation. However, the product of thermodynamic control is influenced only by structural features of the product, such as the stability of the remaining double bond. Products with the more highly substituted double bond are favored.

11.8 The Diels–Alder Reaction

The Dields–Alder reaction is a powerful synthetic method for the formation of compounds that contain six-membered rings. The addition of a diene to a dienophile produces a cyclohexene ring. The reaction is concerted, and the stereochemistry of the products is the same as the stereochemistry of the reactants. In bicylic ring systems, the *endo* isomer predominates.

11.9 Spectroscopy

The energy of light is directly proportional to its frequency. The energy of light is given by the relation $E = h\nu$. The wavelength of light is inversely proportional to the frequency, as given by $\lambda = c/\nu$. Spectroscopy is used to probe the physical changes in a molecule as the result of absorption of energy.

11.10 Ultraviolet–Visible Spectroscopy

The portions of the electromagnetic spectrum known as the ultraviolet and the visible regions are associated with energies sufficient to cause electronic transitions of conjugated systems. The position of the absorption peak is referred to as λ_{max}. Each electronic transition corresponds to promotion of an electron from the highest occupied molecular orbital (HOMO) to the lowest energy unoccupied molecular orbital (LUMO).

The λ_{max} of a compound increases as the number of conjugated π bonds increases. Some conjugated systems absorb light in the visible region of the spectrum.

Summary of Reactions

1. Allylic Bromination

2. Electrophilic Conjugate Addition

1,2-addition   1,4-addition

3. Diels–Alder Reaction

Diene   Dienophile

SOLUTIONS TO END-OF-CHAPTER EXERCISES

Classes of Polyenes

11.1   Which of the following compounds has conjugated double bonds?

(a)   (b)   (c)

(d)

Answer: Conjugated compounds have only one single bond separating the double bonds. Only compound (a) has conjugated double bonds. In (b), the double bonds are separated by two single bonds. In (c), four single bonds separate the two double bonds. In (d), each of the three double bonds is separated by three single bonds.

11.2   Which of the following compounds has conjugated double bonds?

(a)   (b)   (c)

(d)

Answer: The compounds in (a), (b), and (c) all have conjugated double bonds. In (d), the closest double bonds are the one on the left and the vinyl group, shown as a branch. However, there are two intervening single bonds, so the double bonds are not conjugated.

11.3   How many compounds in each of the following sets of isomeric compounds contain conjugated double bonds?

(a)

(b)

Answer: (a) Only the middle compound has conjugated double bonds. (b) The first and second compounds have conjugated double bonds.

11.4   Classify the double bonds in each of the following compounds.
       (a) mycomycin, an antibiotic.

$$H\!-\!C\!\equiv\!C\!-\!C\!\equiv\!C\!-\!CH\!=\!C\!=\!CH\!-\!CH\!=\!CH\!-\!CH\!=\!CH\!-\!CO_2H$$
             cumulated   conjugated

Answer: The double bonds shown in read are cumulated; those shown in blue are conjugated. What about the alkynes? The triple bonds are separated by single bonds. We know that one pair of 2p orbitals makes a π bond that is perpendicular to a second pair of 2p orbitals making the other π bond. Thus, one pair of two conjugated π bonds is perpendicular to a second pair of two conjugated π bonds.

(b) vitamin A₂, contained in freshwater fish

(c) humulene, a compound found in hops

CH₃

CH₃—

OH

CH₃

CH₃

CH₃

Answers: (a) All of the double bonds in vitamin A₂ are conjugated; (b) humulene has no conjugated double bonds.

11.5   Cyanodecapentayne has been identified in intergalactic space by radio astronomers. How many conjugated π bonds are in this compound?

H—C≡C—C≡C—C≡C—C≡C—C≡C—C≡N

Answer: The series of five carbon–carbon triple bonds and the carbon–nitrogen triple bond are all separated by single bonds. One pair of 2p orbitals makes a π bond that is perpendicular to a second pair of 2p orbitals making the other π bond. Thus, there are six conjugated π bonds perpendicular to a second set of six conjugated π bonds.

11.6   How many conjugated π bonds are in lycopene, the red pigment in tomatoes?

Answer: Lycopene has 16 conjugated double bonds.

Terpenes
11.7   Classify each of the following terpenes and divide it into isoprene units.

Answers:
(a) monoter-
pene
(b) sesquiter-
pene
(c) sesquiterpene

11.8   Classify each of the following terpenes and divide it into isoprene units.

Answers:
(a) sesquiter-
pene
(b) monoterpene
(c) sesquiterpene

Stability of Polyenes

11.9 Which of the following octadienes has the least exothermic heat of hydrogenation?

I II III

Answer: Conjugated dienes are more stable than isomeric compounds with isolated double bonds. This energy difference is called the resonance energy. Because they are more stable, conjugated dienes release a smaller amount of energy upon hydrogenation. Compound III is conjugated, and therefore, it has a smaller heat of hydrogenation than compounds I and II.

11.10 Estimate the heat of hydrogenation for converting each of the following isomers to hexane.
(a) (*E*)-1,3-hexadiene   (b) (2*E*,4*E*)-hexadiene   (c) (*E*)-1,4-hexadiene   (d) 1,5-hexadiene

Answer: Use -125 kJ mole^{-1} and -116 kJ mole^{-1} for the heats of hydrogenation of terminal monosubstituted and trans-disubstituted double bonds, respectively.
(a) Add −125 kJ mole^{-1} and −116 kJ mole^{-1} for the terminal monosubstituted and trans-disubstituted double bonds but decrease the quantity by 15 kJ mole^{-1} resonance energy because the bonds are conjugated. The predicted value is −226 kJ mole^{-1}.
(b) Add −116 kJ mole^{-1} and −116 kJ mole^{-1} for the two trans disubstituted double bonds but decrease the quantity by 15 kJ mole^{-1} resonance energy because the double bonds are conjugated. The predicted value is −217 kJ mole^{-1}.
(c) Add −125 and -116 kJ mole^{-1} for the isolated terminal monosubstituted and trans-disubstituted double bonds. The predicted value is −241 kJ mole^{-1}.
(d) Add −125 kJ mole^{-1} and −125 kJ mole^{-1} for the two terminal monosubstituted double bonds, which are isolated. The predicted value is −250 kJ mole^{-1}.

11.11 In acid solution, 1,4-cyclohexadiene isomerizes to a mixture containing 1,3-cyclohexadiene. Write a mechanism to account for this rearrangement. Which of the two isomers should form the major component of the equilibrium mixture?

Answer: Protonate C-1 of the π bond to give a carbocation at C-2 of the original double bond. Deprotonate at the saturated C-3 atom of the original compound to give the conjugated 1,3-cyclohexadiene. This compound is more stable because it is resonance-stabilized, and it is the major component of the equilibrium mixture.

11.12 Compare the relative stabilities of the following two dienes. If no competing reactions occur, indicate how the two compounds could be equilibrated using an acid catalyst.

I II

Answer: The two dienes are similarly substituted, so equal amounts of the two compounds exist in an equilibrium mixture. To achieve equilibrium, protonate the terminal carbon atom of the conjugated diene six-membered ring on the left. Write an alternate resonance form of the allylic carbocation. Deprotonate at a methine carbon atom adjacent to the carbocation center in this resonance form.

Molecular Orbitals of Polyenes

11.13　Which of the bonding molecular orbitals of 1,3,5,7-octatetraene resembles the Lewis structure for this compound?

Answer: The highest occupied bonding molecular orbital, which is π_4. It has three nodal planes separating the system into four "isolated" π bonds, which is the conventional Lewis structure.

Lewis structure

1,3,5,7-octatetraene

π_4 antisymmetric, three nodal planes

π_3 symmetric, two nodal planes

π_2 antisymmetric, one nodal plane

π_1 symmetric, no nodal planes

Energy

11.14　Determine the symmetry of each molecular orbital of 1,3-butadiene and suggest another guideline that could be added to your list of ways to compare the energies of molecular orbitals.

Answer: The symmetry of π_1 through π_4 is symmetric, antisymmetric, symmetric, and antisymmetric. The highest occupied bonding molecular orbital of a conjugated polyene with n double bonds, which is π_n, has $n - 1$ vertical nodal planes separating the system into n "isolated" π bonds. The highest energy, occupied molecular orbital (HOMO) corresponds to the conventional Lewis structure.

Allylic Systems

11.15 Write contributing resonance forms for the radical formed by abstraction of the bold hydrogen atom in each structure.

(a) $CH_3—CH_2—\overset{\cdot}{\underset{\underset{CH_3}{|}}{C}}—CH=CH_2 \longleftrightarrow CH_3—CH_2—\underset{\underset{CH_3}{|}}{C}=CH—\overset{\cdot}{C}H_2$

(b) $\overset{\cdot}{C}H_2—\underset{\overset{|}{CH_3}}{C}=CH—CH_2—CH_3 \longleftrightarrow CH_2=\underset{\overset{|}{CH_3}}{C}—\overset{\cdot}{C}H—CH_2—CH_3$

(c) $CH_3—CH_2—\underset{\underset{CH_3}{|}}{C}=CH—\overset{\cdot}{C}H—CH_3 \longleftrightarrow CH_3—CH_2—\overset{\cdot}{\underset{\underset{CH_3}{|}}{C}}—CH=CH—CH_3$

(d) $CH_3—\underset{\overset{|}{CH_3}}{C}=CH—\overset{\cdot}{C}H—CH_3 \longleftrightarrow CH_3—\underset{\overset{|}{CH_3}}{\overset{\cdot}{C}}—CH=CH—CH_3$

11.16 Write alternate resonance forms for each of the following ions.

(a) $CH_3—CH_2—CH=CH—\overset{+}{C}H_2 \longleftrightarrow CH_3—CH_2—\overset{+}{C}H—CH=CH_2$

(b) $CH_3—\underset{\overset{|}{CH_3}}{C}=CH—\overset{+}{C}H_2 \longleftrightarrow CH_3—\underset{\overset{|}{C}+}{\overset{|}{C}}—CH=CH_2$

with CH_3 above

(c) $CH_3—CH_2—\underset{\underset{CH_3}{|}}{C}=CH—\overset{+}{C}H_2 \longleftrightarrow CH_3—CH_2—\overset{+}{\underset{\underset{CH_3}{|}}{C}}—CH=CH_2$

(d) $CH_2=\underset{\overset{|}{CH_3}}{C}—\overset{+}{C}H—CH_3 \longleftrightarrow \overset{+}{C}H_2—\underset{\overset{|}{CH_3}}{C}=CH—CH_3$

11.17 The rate of reaction of 1-chloro-3-methyl-2-butene in ethanol to give substitution products is about 6×10^3 times faster than the rate for allyl chloride. Explain why.

Answer: The positively charged carbon atoms of the allyl carbocation are both primary.

$$CH_2=CH—CH_2{}^+ \longleftrightarrow {}^+CH_2—CH=CH_2$$

Answer: In contrast, the positively charged carbon atoms of the carbocation derived from 1-chloro-3-methyl-2-butene are primary and tertiary. Hence this carbocation is more stable than the allyl carbocation. The stability of this intermediate is reflected in the transition state that generates it. As a consequence, the activation energy required for the reaction is smaller, and the reaction occurs at a faster rate.

191

11.18   Which of the following two compounds would react faster with HCl to produce an alkyl halide?

11.18   Answer: Allylic carbocations are generated when the protonated oxygen atom of the alcohol leaves as water in an S_N1 reaction. The positively charged carbon atoms of the resonance hybrid of the carbocation derived from compound I are primary and secondary. The positively charged carbon atoms of the resonance hybrid of the carbocation derived from compound II are primary and tertiary. Thus, the carbocation derived from II is more stable than that derived from I and the activation energy required for the reaction of II is smaller and the reaction occurs at a faster rate.

11.19   3-Isopentenyl pyrophosphate and 2-isopentenyl pyrophosphate are intermediates in the biosynthesis of terpenes. The pyrophosphate ion is a good leaving group. One of the two compounds reacts more readily to form a carbocation than the other. Which one and why?

3-isopentenyl pyrophosphate   2-isopentenyl pyrophosphate

Answer: The carbon atom bearing the pyrophosphate group in both compounds is primary. However, that carbon atom in 2-isopentenyl pyrophosphate is allylic. Thus, a carbocation forms more readily for this compound, and the reaction rate is greater than for the reaction of 3-isopentenyl pyrophosphate.

11.20   Write the structure of the major substitution product expected from the reaction of 1-methyl-3-cyclohexen-1-ol with HBr.

11.21   Write the structures of the compounds expected for the allylic bromination of methenecyclohexane using one molar equivalent of NBS.

methylenecyclohexane

11.22 The reaction of NBS with 1-octene gives the following products in the indicated yields. Account for each of these products. Explain why the indicated yields are "expected".

(E)-l-bromo-2-octene 44%   (Z)-l-bromo-2-octene 39%   3 -bromo-1-octene 17%

$$CH_3(CH_2)_4{\rule{0.6cm}{0.4pt}}CH_2{\rule{0.6cm}{0.4pt}}CH{=}CH_2 \xrightarrow{\text{NBS}} CH_3(CH_2)_4{\rule{0.6cm}{0.4pt}}\overset{\cdot}{C}H{\rule{0.6cm}{0.4pt}}CH{=}CH_2$$

1-octene

$$CH_3(CH_2)_4{\rule{0.6cm}{0.4pt}}CH{=}CH{\rule{0.6cm}{0.4pt}}CH_2\cdot$$

Answer: Both the (E) and (Z) isomers are derived from abstraction of bromine by the radical at its primary C-1 atom, shown in one of the two contributing resonance forms. These compounds are the major products because they have the more substituted double bond, and the reaction of the radical at the primary center is less sterically hindered. The isomeric 3-bromo-1-octene is derived from abstraction of bromine by the radical at its secondary C-3 atom.

$$CH_3(CH_2)_4{\rule{0.6cm}{0.4pt}}CH_2{\rule{0.6cm}{0.4pt}}CH{=}CH_2 \xrightarrow{\text{NBS}} \left[\begin{array}{c} CH_3(CH_2)_4{\rule{0.6cm}{0.4pt}}\overset{\cdot}{C}H{\rule{0.6cm}{0.4pt}}CH{=}CH_2 \\ \updownarrow \\ CH_3(CH_2)_4{\rule{0.6cm}{0.4pt}}CH{=}CH{\rule{0.6cm}{0.4pt}}CH_2\cdot \end{array} \right]$$

1-octene

39%   44%

$$+ \quad CH_3(CH_2)_4{\rule{0.6cm}{0.4pt}}\underset{\underset{Br}{|}}{C}H{\rule{0.6cm}{0.4pt}}CH{=}CH_2$$
19%

11.23 Alkenes can be chlorinated at the allylic position by *tert*-butyl hypochlorite, $(CH_3)_3C{\rule{0.3cm}{0.4pt}}O{\rule{0.3cm}{0.4pt}}Cl$, which undergoes homolytic cleavage to give the *tert*-butoxy radical and a chlorine atom. Reaction of (E)-4,4-dimethyl-2-pentene with this compound gives two C_3H_9Cl products in the ratio 93:7. What are the structures of these two products?

Answer: 1-Chloro-4,4-dimethyl-2-pentene is the major isomer because the double bond is disubstituted and the chlorine atom is abstracted from the tert-butyl hypochlorite by a primary radical. The isomeric product 3-chloro-4,4-dimethyl-1-pentene has a monosubstituted double bond, and the chlorine atom is abstracted by a secondary radical.

1-chloro-4,4-dimethyl-2-pentene
93%

3-chloro-4,4-dimethyl-1-pentene
7%

11.24   (a) Which compound should undergo allylic bromination at the faster rate, 1,3-pentadiene or 1,4-pentadiene? (b) How would the composition of the product mixtures compare?

Answer: (a) The same resonance stabilized radical is generated by abstraction of a hydrogen atom at C-5 of 1,3-pentadiene or abstraction of a hydrogen atom at C-3 of 1,4-pentadiene. Thus, the mixture of brominated products is the same for both compounds.

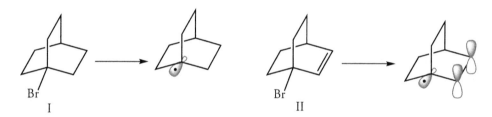

Answer: (b) 1,3-Pentadiene is a conjugated diene, and is more stable than 1,4-pentadiene. Thus, the activation energy for abstraction of its hydrogen atom is *larger* because both isomers yield the same intermediate, and the transition states are of similar energy. Therefore, 1,4-pentadiene would undergo allylic bromination at a faster rate.

11.25   The C—Br bond dissociation energies of the following two compounds are essentially equal. Explain why the allylic compound does not have a lower bond dissociation energy.

Answer: To stabilize an allylic system, the 2p orbitals must be coplanar. In compound II, the orbital of the radical at the bridgehead carbon cannot overlap with the 2p orbitals making up the π bond, so no resonance stabilization occurs.

Molecular Orbitals of Allylic Systems

11.26   The highest energy π electrons of the pentadienyl anion are found in the π_3 molecular orbital. Write a representation of this orbital and predict the location of the negative charge for the ion.

Answer: Each molecular orbital has 2π electrons. The π_3 molecular orbital has two vertical nodal planes. They are located at C-2 and C-4. Hence, there is electron density only at C-1, C-3, and C-5. The negative charge is distributed at these three atoms.

11.27   Reaction of the pentadienyl cation with a nucleophile occurs by interaction with an empty molecular orbital. Which orbital? On the basis of this analysis, predict the carbon atoms at which the nucleophile will bond. Is this prediction consistent with the products predicted by writing conventional Lewis resonance forms for the cation?

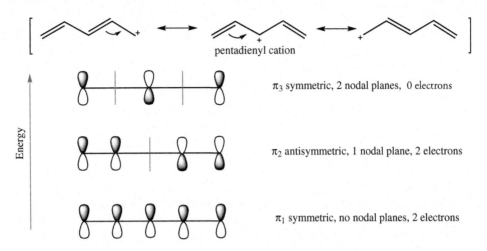

Answer: A nucleophile bonds to the lowest energy, vacant orbital. In the cation, the four electrons are located in π_1 and π_2. Thus, the reaction occurs at the π_3 molecular orbital. The nucleophile can bond only at C-1, C-3, and C-5. These atoms correspond to the atoms having a positive charge in the three contributing resonance forms of the pentadienyl cation.

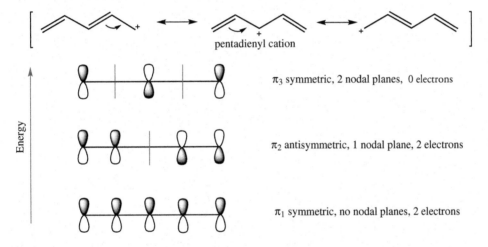

Conjugate Addition Reactions

11.28   Explain why only one product forms in the addition of one molar equivalent of HBr to 1,3-cyclohexadiene.

Answer: Unlike the addition reactions to 1,3-butadiene, where the double bond may be located in two different places, there is only one position for the double bond in the product cyclohexene. In both 1,2 and 1,4 addition, the bromine is located at a C-3 position using the double bond to number the ring. The product is 3-bromocyclohexene.

11.29   Write the structure of the products formed in the addition of one molar equivalent of DBr to 1,3-cyclohexadiene, indicating the location of the deuterium.

Answer: Unlike the similar reaction in Exercise 11.28, the products of 1,2 and 1,4 addition of DBr are different because the deuterium positions can be distinguished.

11.30   Explain why the extent of 1,2- versus 1,4-addition cannot be determined for the reaction of 1,3-pentadiene with one molar equivalent of HCl.

Answer: Adding a proton to the C-1 carbon atom of 1,3-pentadiene gives a resonance stabilized allylic carbocation with equal electron density at C-2 and C-4, which are also structurally equivalent. Capture of the carbocation by chloride ion at C-2 and C-4 gives the same product, 4-chloro-2-pentene.

11.31   Write the structures of the products of the addition of one molar equivalent of DCl to 1,3-pentadiene. Can one determine the amounts of 1,2- and 1,4-addition reactions?

Answer: Adding a deuterium to C-1 of 1,3-pentadiene gives a resonance-stabilized allylic carbocation with equal electron density at C-2 and C-4. However, unlike the reaction with HCl in Exercise 11.31, C-2 and C-4 are not structurally equivalent with respect to the two terminal carbon atoms, one of which contains a deuterium atom. Capture of the carbocation by chloride ion at C-2 of the original compound gives a 1,2-addition product. Adding chloride at C-4, which corresponds to a 1,4-addition, gives a different product. Thus, the amount of each addition process can be determined.

11.32   1,3,5-Hexatriene reacts with one molar equivalent of bromine to give only 1,2- and 1,6-addition products. Write the structures of these products. Why is there no 1,4-addition product?

Answer: Both the 1,2 and 1,6 addition products have conjugated double bonds. These products are more stable than the product of 1,4 addition, which has isolated double bonds.

$$CH_2\!\!=\!\!CH\!-\!CH\!\!=\!\!CH\!-\!CH\!\!=\!\!CH_2 \xrightarrow{Br_2}$$

1,3,5-hexatriene

Br Br
| |
$CH_2\!\!-\!\!CH\!-\!CH\!\!=\!\!CH\!-\!CH\!\!=\!\!CH_2$

1,2-addition

Br   Br
| |
$CH_2\!\!-\!\!CH\!\!=\!\!CH\!-\!CH\!\!=\!\!CH\!-\!CH_2$

1,6-addition

Br   Br
| |
$CH_2\!\!=\!\!CH\!-\!CH\!-\!CH\!\!=\!\!CH\!-\!CH_2$

1,4-addition

11.33 Reaction of 1,3-butadiene with one molar equivalent of bromine at —15 °C gives a 60:40 mixture of two products. At 60 °C, the product ratio is 10:90. Write the structures of the two products. Explain why different product ratios are observed at the two temperatures.

Answer: At 60 °C, the more stable isomer is favored. This compound, which contains a disubstituted double bond, is the result of 1,4-addition, and is the isomer obtained in 90% yield. The 1,2- addition product is less stable because it has a monosubstituted double bond. At –15 °C, the amounts of the two products are kinetically controlled.

$CH_2\!\!=\!\!CH\!-\!CH\!\!=\!\!CH_2$

1,3-butadiene

$\xrightarrow[-15°\ C]{Br_2}$

Br Br
| |
$CH_2\!\!=\!\!CH\!-\!CH\!-\!CH_2$ Product of kinetic control is favored at -15° C.

$\xrightarrow[60°\ C]{Br_2}$

Br   Br
| |
$CH_2\!\!-\!\!CH\!\!=\!\!CH\!-\!CH_2$

Product of thermodynamic control is favored at 60° C.

11.34 Reaction of 2,3-dimethyl-1,3-butadiene with one molar equivalent of HBr gives only one product. Write its structure and explain why this product is favored.

Answer: The product is 1-bromo-2,3-dimethyl-2-butene, which is the result of a 1,4 addition reaction. The double bond is tetrasubstituted, compared to a disubstituted double bond in the 1,2-addition product, which is not observed.

H^+

CH_3 CH_3
| |
$CH_2\!\!=\!\!C\!-\!C\!\!=\!\!CH_2$

2,3-dimethyl-1,3-butadiene

$\xrightarrow[1\ mole]{HBr}$

CH_3 CH_3
| |
$CH_3\!-\!C\!-\!C\!\!=\!\!CH_2 \longleftrightarrow CH_3\!-\!C\!\!=\!\!C\!-\!CH_2^+$

CH_3 CH_3
| |
$CH_3\!-\!C\!-\!C\!\!=\!\!CH_2$
      |
      Br

1,2-addition product,
not formed

CH_3 CH_3
| |
$CH_3\!-\!C\!\!=\!\!C\!-\!CH_2Br$

1,4-addition product,
only product

11.35   Reaction of 1-phenyl-1,3-butadiene with one molar equivalent of Cl_2 gives only one product. Write its structure and explain why this product is favored.

Answer: The initial electrophilic attack is at the terminal position of the diene because the resulting allylic carbocation is resonance stabilized by the benzene ring. Attack at the carbon atom bonded to the benzene ring would give an allylic carbocation lacking stabilization by the benzene ring.

1-phenyl-1,3-butadiene

allylic carbocation conjugated with benzene ring

allylic carbocation not conjugated with benzene ring

Product with double bond conjugated with benzene ring.

Product with double bond that is not conjugated with benzene ring.

11.36   Write the structures of the products from the reaction of 1,3-butadiene with an aqueous bromine solution.

$$CH_2\!\!=\!\!CH\!-\!CH\!\!=\!\!CH_2 \xrightarrow[H_2O]{Br_2}$$

1,3-butadiene

$$\underset{\underset{OH}{|}}{CH_2}\!\!=\!\!CH\!-\!\underset{\underset{}{|}}{CH}\!-\!\underset{\underset{Br}{|}}{CH_2}$$

+

$$\underset{\underset{OH}{|}}{CH_2}\!\!-\!CH\!\!=\!\!CH\!-\!\underset{\underset{Br}{|}}{CH_2}$$

11.37 A single chloro alcohol forms in the reaction of 2-methyl-1,3-butadiene with an aqueous chlorine solution. Write its structure and explain why this product is favored.

Answer: Attack of the electrophilic chlorine cation at the C-4 position would give an allylic carbocation that has positive charge distributed between a secondary and a primary carbon atom. However, the electrophilic chlorine cation adds to C-1, giving a more stable allylic carbocation that has positive charge distributed between a tertiary and a primary carbon atom.

1,2-addition product

Answer: However, the electrophilic chlorine cation adds to C-1, giving a more stable allylic carbocation that has positive charge distributed between a tertiary and a primary carbon atom. This is the preferred path, and 1,4-addition predominates.

1,4-addition product

UV Spectroscopy

11.38 The λ_{max} values of naphthalene, anthracene, and tetracene are 314, 380, and 480 nm, respectively. Suggest a reason for this order of the wavelength of maximum absorption. Are any of the compounds colored?

naphthalene   anthracene   tetracene

Answer: The increased number of aromatic rings gives a more conjugated system, which decreases the energy difference between the highest occupied molecular orbital and the lowest unoccupied molecular orbital. The 380 nm value for anthracene is slightly outside the wavelength of light of the visible spectrum. However, the 480 value for tetracene indicates that light in the blue region is absorbed. As a result, the compound will have an orange color. See Table 11.2.

11.39   How many conjugated double bonds are contained in lycopene? Compare the conjugation in this compound to that of
         β-carotene. Using this information, predict the color of lycopene.

lycopene

β-carotene

Answer: Both compounds have 11 conjugated double bonds. There are two more double bonds in lycopene than in β-carotene, but they
         are not part of the conjugated system. Thus, lycopene has the same color as β-carotene, which is yellow-orange.

11.40   How might 2,4-hexadiyne be distinguished from 2,5-hexadiyne by ultraviolet spectroscopy?

Answer: 2,4-Hexadiyne has two conjugated triple bonds and absorbs light in the ultraviolet region. The triple bonds in 2,5-hexadiyne are
         isolated and do not absorb in the ultraviolet region.

11.41   The λ_{max} values of 2,4,6-octatriyne and 2,4,6,8-decatetrayne are 207 and 234 nm, respectively. Explain why these values differ.

Answer: 2,4-Hexadiyne has two conjugated triple bonds and absorbs light in the ultraviolet region. The triple bonds in 2,5-hexadiyne are
         isolated and do not absorb in the ultraviolet region.

11.42   Each of the following compounds is an indicator. At pH 7, one appears violet and the other blue-green. Assign each color to the
         proper indicator.

Answer: The compound on the left has a dimethylamino group that increases the degree of conjugation and stabilizes the positive charge
on the nitrogen on the upper right across the benzene rings. Therefore, the compound on the left absorbs light at a longer wavelength. The
violet and blue green colors correspond to wavelengths of 510 and 540 nm, respectively. Thus, the compound on the left is blue-green and
the compound on the right is violet.

11.43   The λ_{max} of phenol is 210 nm in ethanol. Explain why adding sodium hydroxide to the solution shifts this absorption to 235 nm.

Answer: Adding hydroxide converts phenol to its conjugate base, a phenolate anion. Extended conjugation in the anion shifts the λ_{max} of phenolate to a higher wavelength.

phenol
(λ_{max} 210 nm)

phenolate
(λ_{max} 235 nm)

11.44   The λ_{max} values of benzene and p-methylaniline (p-toluidine) are 204 and 235 nm, respectively. However, when HCl is added, the λ_{max} of benzene is unchanged, whereas the λ_{max} of methylaniline changes to 207 nm. Explain the difference in the λ_{max} values for the two compounds and the effect of acid on the spectrum.

Answer: Adding HCl converts aniline to its conjugate acid, an anilinium ion. Extended conjugation in aniline is lost when this occurs, and the λ_{max} of p-toluidine shifts to a lower wavelength.

12 ARENES AND AROMATICITY

12.1 Aromatic Compounds

The most common aromatic compounds contain a benzene ring that may have one or more of its hydrogen atoms replaced by substituents of varying complexity. Although many of these compounds are termed "aromatic" because of their odor, that property is a human physiological response and is not the criterion used to classify them.

Several benzene rings can be fused to include two common carbon atoms between rings. The possibilities are endless, as more and more rings can be fused. The common feature in all such compounds is the alternating series of single and double bonds within the rings. The structures of benzene, naphthalene, anthracene, and phenanthrene are the simplest arenes and the ones most often encountered.

12.2 Covalent Structure of Benzene

Benzene consists of a six-membered ring in which all carbon and hydrogen atoms are identical. We now see that the structures that Kekulé proposed for benzene are contributing structures to a resonance hybrid that we can draw with equivalent Lewis structures.

12.3 Aromaticity and the Resonance Stabilization of Benzene

The chemical criterion for aromaticity is the lack of reactivity toward reagents, such as bromine, that normally react readily with carbon–carbon double bonds. Benzene owes its stability to the resonance stabilization of its cyclic, conjugated π bonds.

The resonance energy of benzene is determined by comparing its heat of hydrogenation to that expected by the heat hydration of three isolated double bonds. The resonance energy for benzene is substantially larger than for cyclohexadiene.

12.4 Molecular Orbitals of Benzene

Benzene has six molecular orbitals that form a delocalized π system; three of benzene's molecular orbitals are bonding and each has two electrons; three are antibonding, and they are vacant.

One way to determine the number of molecular orbitals and their relative energies in benzene is to inscribe a hexagon in a circle. Place one vertex at the six o'clock position. The points of contact of the hexagon with the circle give the relative energies of the molecular orbitals. A horizontal line through the center of the circle bisects the polygon and corresponds to an energy level of zero; that is, any orbitals having this energy are nonbonding. Vertices below this level correspond to bonding orbitals; vertices above this level are antibonding.

12.5 Aromaticity and the Hückel Rule

To be aromatic, a molecule must be cyclic, planar, and it must contain only sp²-hybridized atoms. The number of π electrons in the delocalized system must equal $4n + 2$, where n is an integer. (The value of n is not necessarily the number of carbon atoms in the ring, or the number of π molecular orbitals in a ring.)

The "$4n + 2$ rule" is known as the Hückel rule. It predicts that cyclic π systems having 2 ($n = 0$), 6 ($n = 1$), 10 ($n = 2$), and 14 ($n = 3$) electrons will be unusually stable, that is, they will be aromatic. For example, benzene has six π electrons, $4n + 2 = 6$ when $n = 1$.

Organic Chemistry Study Guide: Key Concepts, Problems, and Solutions http://dx.doi.org/10.1016/B978-0-12-801889-7.00012-1

12.6 Heterocyclic Aromatic Compounds

The criteria described above for establishing aromaticity also apply to heterocyclic compounds. Heteroatoms in the ring must be sp^2 hybridized and the ring must contain $4n+2$ π electrons. The most common heteroatoms are nitrogen, oxygen, and sulfur. The nitrogen atom in pyridine contributes one 2p electron to the π system. A lone pair of electrons on the sp^2-hybridized nitrogen atom of pyridine lies in the plane of the ring and does not contribute to the π system. In pyrrole, however, and in furan, and thiophene, the heteroatom contributes two electrons to the π system.

12.7 Polycyclic Aromatic Compounds

Polycyclic aromatic compounds that contain fused benzene rings contain $4n+2$ π electrons. Since they obey the Hückel rule, they are aromatic. For example, naphthalene contains 10 π electrons ($n = 2$) and anthracene has 14 π electrons ($n = 3$).

ANSWERS TO END-OF-CHAPTER EXERCISES
Criteria for Aromaticity

12.1   Determine whether each of the following is an aromatic compound.

Answers:

(a) Although there are six π electrons in the triene, it is not aromatic because the π bonds are not in a ring.

(b) There are six π electrons in the cyclic triene. However, there is an sp³-hybridized carbon atom interrupting the conjugation of the π bonds, so the compound is not aromatic.

(c) There are six p electrons in the bicyclic triene. However, only two of the π bonds are conjugated and each end of that system is separated from the third π bond by an sp³-hybridized carbon atom. The compound is not aromatic.

(d) There are eight π electrons, so the compound does not have 4n+2 π electrons. Two other facts tell us that it is not aromatic: (1) the π bonds are not in a continuous cyclic arrangement, and (2) there are intervening sp³-hybridized carbon atoms.

12.2   Determine whether each of the following is an aromatic compound.

Answers:

(a) There are six π electrons in the bicyclic compound. However, there is an sp³-hybridized carbon in the seven-membered ring. The compound is *not* aromatic.

(b) The π bonds are in a cyclic arrangement without any intervening sp³-hybridized carbon atoms. However, there are 8 π electrons, a number that is not consistent with the Hückel rule. The compound is not aromatic.

(c) The π bonds are in a cyclic arrangement without any intervening sp³-hybridized carbon atoms. However, there are 12 π electrons, a number that is not consistent with the Hückel rule. The compound is not aromatic.

(d) There are 14 π electrons, a number that is consistent with the Hückel rule for *n* = 3. However, there are two intervening sp³-hybridized carbon atoms. Therefore, compound is not aromatic.

12.3   Borazole is an aromatic compound. Explain why.

borazole

Answer:
Boron is a member of group III and has only six bonding electrons in its trivalent compounds. Thus, it is sp² hybridized and has a vacant 2p orbital. Nitrogen has an unshared pair of electrons in its trivalent compounds. The six electrons of nitrogen can be delocalized over the six atom ring using the 2p orbitals of both nitrogen and boron. Borazole is isoelectronic with benzene and is aromatic.

12.4   Is the following compound aromatic? Explain your answer.

Answer:
Boron is sp² hybridized and has a vacant 2p orbital. There is a network of 2p orbitals of carbon and boron that can form a delocalized π system. However, there are only four π electrons, indicated by the two carbon–carbon π bonds. This number does not fit the Hückel rule, and the compound is not aromatic.

12.5   Are the following compounds aromatic according to the Hückel rule?

Answers:
(a) No, because 8 π electrons does not fit the Hückel rule of 4n+2; second, there are four sp³-hybridzied carbon atoms.

(b) Yes, because 10 π electrons fits the Hückel rule.

(c) Yes, because oxygen contributes one of its two unshared electron pairs to the π system. The four electrons of the carbon–carbon π bond and the carbon–nitrogen π bond, and one unshared electron pair of electrons from oxygen give a total of 6 π electrons, which fits the Huckel rule. Note that the other electron pair of oxygen and that of nitrogen are in sp² hybrid orbitals that are perpendicular to the π system.

(d) No, because oxygen would contribute one of its two unshared electron pairs to the π system. The six electrons of the three carbon–carbon π bonds and one unshared electron pair from oxygen give a total of 8 π electrons, which does not fit the Hückel rule. Note that the other electron pair of oxygen cannot be included to give a total of 10 electrons because it is in an sp²-hybrid orbital perpendicular to the π system.

12.6 Are the following compounds aromatic according to the Hückel rule?

(a)

(b)

Answers:
(a) Yes, because the nitrogen atom bonded to a hydrogen atom contributes one of its unshared electron pairs to the π system. The four electrons of the two carbon–carbon π bonds and this one unshared electron pair from nitrogen give a total of 6 π electrons, which does fit the Hückel rule. Note that each of the other two nitrogen atoms contribute one electron to the π system as part of their double bonds. The remaining unshared electron pair of each nitrogen atom is in an sp^2 orbital that is perpendicular to the π system.

(b) No, because each nitrogen atom would contribute its unshared electron pair to the π system. The four electrons of the two carbon–carbon π bonds and the two unshared electron pairs give a total of 8 π electrons, which does not fit the Hückel rule.

Aromatic Ions

12.7 1,2,3,3-Tetrachlorocyclopropene reacts with one mole of $SbCl_5$, to give the $C_3Cl_3^+$ ion. Draw the structure of the ion and explain why it forms.

Answer:
The carbocation that results from abstraction of a chloride ion by the Lewis acid $SbCl_5$ is a cyclopropenium ion. It has two π electrons and fits the Hückel rule (with $n = 0$). The electrons can be delocalized over the three carbon atoms of the ring.

12.8 The dipole moment of dipropylcyclopropenone is 5 D. This value is significantly higher than that of acetone, which is 3 D. Write a resonance form that accounts for the larger dipole moment of the cyclic ketone.

Answer:
The dipolar resonance form of the carbonyl group places a negative charge on oxygen and a positive charge on one of the carbon atoms of the cyclopropene ring. The resulting cyclopropenium ion has two π electrons and fits the Hückel rule. The positive charge can be delocalized over the three carbon atoms of the ring. The increased polarity of the carbonyl group as a result of resonance increases the dipole moment.

12.9 Cyclooctatetraene reacts with potassium to give a stable dianion. Explain why. Inscribe the dianion in a circle, with one vertex pointed down, and draw a molecular orbital energy diagram for it.

Answer:
The cycloöctatetraenyl dianion has 10 π electrons, which fits the Hückel rule, so the dianion is aromatic.

12.10 The following hydrocarbon reacts with two moles of butyllithium to form the stable ion $C_8H_6^{2-}$. Draw the structure of this ion. Explain why it is stable.

Answer:
Loss of one proton from each of the methylene groups gives a dianion that has a total of 10 π electrons, which fits the Hückel rule. Therefore, the dianion is aromatic.

10 π electrons, aromatic

12.11 Is the following hydride ion transfer reaction favorable in the direction written?

Answer:
The cyclopropenium cation product is aromatic. Thus, its resonance stabilization is larger than the resonance stabilization of the allyl cation. The reaction is favorable in the direction written.

12.12 Is the product of the following reaction aromatic?

Answer:
Loss of a proton from the methylene group gives a carbanion that has 10 π electrons, which fits the Hückel rule. The cyclononatetraenyl anion is aromatic.

Polycyclic Aromatic Compounds

12.13   Consider the following resonance contributor of chrysene. Draw the most stable resonance contributor.

Answer:

chrysene   chrysene-most stable resonance form

The most stable resonance form has four "benzene-type" rings. Note that the points of fusion all have double bonds that are shared by two rings.

12.14   Draw the most stable resonance contributor of naphthacene.

Answer:
The resonance form on the left has only one "benzene-type" rings. The resonance form on the right has two "benzene-type" rings and is more stable.

Heterocyclic Aromatic Compounds

12.15   How many electrons does each heteroatom contribute to the π system in each of the following compounds?

Answers:
(a) Nitrogen contributes one electron from its 2p orbital, and oxygen contributes a lone pair to the aromatic sextet.

(b) Each nitrogen contributes one electron from a 2p orbital to the aromatic sextet.

(c) Nitrogen contributes one electron from its 2p orbital, and sulfur contributes a lone pair to the aromatic sextet.

12.16   How many electrons does each heteroatom contribute to the π system in each of the following compounds?

Answers:
(a) The nitrogen atom bonded to hydrogen contributes two electrons; the other nitrogen atom contributes one electron from its 2p orbital.

(b) The nitrogen atom bonded to hydrogen in the five-membered ring contributes two electron from its lone pair; the other three nitrogen atoms each contribute one electron from a 2p orbital.

(c) Each nitrogen atom contributes one electron to the aromatic sextet.

12.17 Identify the heterocyclic ring structure in each of the following compounds.

(a)

O_2N

H O
N—C—NH$_2$
CH=N

(b)

O_2N

NO$_2$

N

CH$_2$C—N(C$_2$H$_5$)$_2$
O

(c)

Cl

S

CH$_3$
C
O

Answers:

(a) furan   (b) pyrrole   (c) thiophene

12.18 Identify the aromatic heterocyclic ring structure contained in each of the following compounds.
     (a) tolmetin, a drug used to lower blood sugar levels

Answers:
(a) pyrrole

(a) CH$_3$—

O
C

N
H
CH$_2$CO$_2$H

(b) cephalothin sodium, a broad-spectrum antibacterial

(b) thiophene

S

O
H
N

S

N
O

CH$_2$—O—C—CH$_3$
O

CO$_2^-$ Na$^+$

(c) dantrolene, a muscle relaxant

(c) furan

O_2N—

O

CH=NH—N

O

NH

O

(d) ethionamide, an antitubercular agent

S
C
NH$_2$

N
C$_2$H$_5$

13

ELECTROPHILIC AROMATIC SUBSTITUTION

KEYS TO THE CHAPTER

13.1 Names of Benzene Derivatives

The derivatives of benzene have both common names and IUPAC names. Some monosubstituted benzene compounds have common names such as toluene, phenol, and aniline. Disubstituted compounds may be named using a numbering system or the prefixes *o-*, *m-*, and *p-* for *ortho*, *meta*, and *para*, respectively. However, compounds with three or more substituents must use numbers to locate the substituents. Substituents are named in alphabetical order.

13.2 Electrophilic Substitution Reactions

Aromatic rings are attacked by electrophiles, E^+, to give substituted aromatic compounds represented by Ar—E. The common reactions of aromatic compounds are designated by the type of group substituted on the aromatic ring. Thus, the term halogenation means that a halogen has substituted for hydrogen on the aromatic ring to give a product represented by Ar—X. In the case of chlorination, the product is Ar—CI.

      The mechanism of aromatic substitution consists of two steps. The first is electrophilic attack to give a carbocation intermediate. The second step is loss of a proton to regenerate the aromatic system. We discussed five types of electrophilic substitution reactions. They are summarized in the following table:

Reaction Type	Reagents
1. Halogenation	$FeBr_3/Br_2$ or $FeCl_3/Cl_2$
2. Nitration	Nitric acid/sulfuric acid
3. Sulfonation	Fuming sulfuric acid
4. Friedel–Crafts alkylation	$AlCl_3/CH_3Cl$
5. Friedel–Crafts acylation	Acyl chloride/ $AlCl_3$

13.3 Limitations of Friedel–Crafts Reactions

In a Friedel–Crafts alkylation, the electrophile is a carbocation that tends to rearrange to the most stable carbocation by either a hydride or methide shift. An alternate approach to obtaining the desired alkyl group on the aromatic ring is to acylate first and then reduce the carbonyl group.

      A second limitation reflects the effects of substituents already on the aromatic ring. Deactivating, *meta*-directing substituents preclude Friedel–Crafts alkylation and acylation.

13.4 Effects of Substitutents on the Reactivity of Aromatic Compounds

Substituents on the aromatic ring affect the rate of electrophilic substitution. Substituents that increase reactivity are activating groups; substituents that decrease the reactivity are deactivating groups. The degree to which the groups affect the reactivity is qualified by the adjectives "strongly" and "weakly" (see Table 13.1). The substituents already on the aromatic ring determine where the electrophile attacks. Substituents are either *ortho*, *para*, or *meta* directors. All activating groups are *ortho, para* directors, except halogens, which are weakly deactivating, but *ortho, para* directing. Other deactivating groups are *meta* directors.

13.5 Interpreting Rate Effects

Any group increases the electron density of the aromatic ring makes it more inviting to an electrophile, so the rate of reaction is faster. Conversely, groups that reduce the electron density decrease the rate of reaction. Groups can affect the electron density by an inductive effect, a resonance effect, or a combination of both.

Except for alkyl groups, all common substituents withdraw electron density from the ring by an inductive effect. This effects is greatest for groups with a formal positive charge (such as the nitro group), followed by groups with a partial positive charge (such as the carbonyl group).

Organic Chemistry Study Guide: Key Concepts, Problems, and Solutions http://dx.doi.org/10.1016/B978-0-12-801889-7.00013-3

Substituents that either have lone pair electrons or have multiple bonds to an electronegative atom directly bonded to the ring have a resonance interaction with the ring. Groups with lone pair electrons (such as hydroxyl and amino) increase the electron density of the ring. Groups with multiple bonds to an electronegative atom (such as the carbonyl group) withdraw electron density from the ring.

Halogens deactivate the ring toward electrophilic substitution but are *ortho–para* directors. The halogens withdraw electron density by an inductively effect but can donate electrons to the ring by resonance. Substituents such as the hydroxyl and amino groups donate electrons to the ring by resonance and overcome inductive electron withdrawal by their electronegative atoms.

The nitro group and cyano group withdraw electrons by a combination of both an inductive effect and a resonance effect, so they are strongly deactivating.

13.6 Interpreting Directing Effects

The location of the positive charge in the carbocation intermediate in electrophilic substituent determines whether *ortho*, *para*, or *meta* substitution occurs. The charge is always *ortho* and *para* to the position at which the electrophile enters. If the substituent can stabilize a positive charge by resonance, then substitution at the *ortho*, para sites is favored. *Meta* directors destabilize positive charge when an electrophile attacks *ortho* and *para* to them. As a consequence, the electrophile enters the *meta* position by default.

The location of the positive charge in the carbocation intermediate in electrophilic substituent determines whether *ortho*, *para*, or *meta* substitution occurs. The charge is always *ortho* and *para* to the position at which the electrophile enters. If the substituent can stabilize a positive charge by resonance, then substitution at the *ortho*, *para* sites is favored. *Meta* directors destabilize positive charge when an electrophile attacks *ortho* and *para* to them. As a result, the electrophile enters the meta position by default.

13.7 Functional Group Modification

Some functional groups can be introduced on the benzene ring indirectly when a functional group already on the ring is transformed by a chemical reaction. The major examples are
1. Oxidizing an alkyl group to a carboxylic acid
2. Reducing an acyl group to an alkyl group
3. Reducing a nitro group to an amino group. The amino group can be converted to other functional groups by converting the amino group to a diazonium ion. The groups that can replace the diazonium ion are:
1. halogens, using copper(I) halide
2. nitrile, using copper(I) cyanide
3. hydroxyl, using aqueous acid
4. hydrogen, using hypophosphorous acid

13.8 Synthesis of Substituted Aromatic Compounds

The *ortho*, *para*, or *meta*-directing properties of the each group added to the ring, and the chemical modifications that can be performed on these groups after they have been added dictate the order in which groups are added to an aromatic ring.

If two groups can be introduced by direct reaction, and their location is consistent with the directing characteristic of one of them, then the synthesis is straightforward. If the effect of two groups are cooperative—both *ortho*, *para* directing or both *meta*-directing—they act in concert. If the effects of the two groups are opposed, then the one with the greater *ortho*, *para* directing activity determines the distribution of products.

Summary of Reactions

1. Halogenation

2. Nitration

3. Sulfonation

4. Friedel–Crafts Alkylation

5. Friedel–Crafts Acylation

6. Side Chain Oxidation

7. Acyl Side Chain Reduction

8. Nitro Group Side Chain Reduction

(a)

$$C_6H_5NO_2 \xrightarrow{\text{Sn / HCl}} C_6H_5NH_2$$

(b)

$$C_6H_5NO_2 \xrightarrow{\text{H}_2 \text{ / Pd}} C_6H_5NH_2$$

9. Amino Group Side Chain Reactions via Diazonium Ions

$$C_6H_5NH_2 \xrightarrow[\text{H}_2\text{SO}_4]{\text{HO-N=O}} C_6H_5N_2^+ \quad \text{Diazonium Ion}$$

(a)

$$C_6H_5N_2^+ \xrightarrow{\text{CuCl}} C_6H_5Cl$$

(b)

$$C_6H_5N_2^+ \xrightarrow{\text{CuBr}} C_6H_5Br$$

(c)

$$C_6H_5N_2^+ \xrightarrow{\text{CuCN}} C_6H_5CN$$

(d)

$$C_6H_5N_2^+ \xrightarrow{\text{H}_3\text{O}^+} C_6H_5OH$$

(e)

$$C_6H_5N_2^+ \xrightarrow{\text{H}_3\text{PO}_2} C_6H_6$$

10. Benzylic Bromination with *N*-Bromosuccinimide (NBS)

$$C_6H_6 \xrightarrow{\text{NBS}} C_6H_5CH_2Br$$

SOLUTIONS TO END-OF-CHAPTER EXERCISES

Nomenclature of Aromatic Compounds

13.1 Identify each of the following as an *ortho*-, *meta*-, or *para*-substituted compound.

Answers:

(a) *para*

(b) *ortho*

(c) *meta*

13.2 Identify each of the following as an *ortho*-, *meta*-, or *para*-substituted compound.

Answers:

(a) *meta*

(b) *para*

(c) *ortho*

13.3 Name each of the following compounds.

Answers:
(a) ethylbenzene
(b) isopropylbenzene
(c) 1,4-diethylbenzene
(d) 1,3,5-trimethylbenzene

(a) —CH₂CH₃

(b)

(c) CH₃CH₂— —CH₂CH₃

(d)

13.4 Name each of the following compounds.

Answers:
(a) 1,2,4-trichlorobenzene
(b) 2,4-dibromophenol
(c) 3-bromoaniline
(d) 3,4-dichlorotoluene

(a)

(b)

(c)

(d)

13.5 Name each of the following compounds.

Answers:
(a) 4-chloro-3,5-dimethylphenol
(b) 2,6-dimethylaniline
(c) 4-chloro-2-phenylbenzene

(a)

(b)

(c)

13.6   Draw the structure of each of the following compounds.
        (a) 5-isopropyl-2-methylphenol, found in oil of marjoram
        (b) 2-isopropyl-5-methylphenol, found in oil of thyme
        (c) 2-hydroxybenzyl alcohol, found in the bark of the willow tree

Answers:

13.7   Draw the structure of 3,4,6-trichloro-2-nitrophenol, a lampricide used to control sea lampreys in the Great Lakes.

Answer:

13.8   N,N-Dipropyl-2,6-dinitro-4-trinuoromethylaniline is the IUPAC name for Treflan, a herbicide. Draw its structure. (The prefix N signifies the location of a substituent replacing hydrogen on a nitrogen atom.)

Answer:

Electrophiles

13.9   Some activated rings may be hydroxylated by reacting hydrogen peroxide (H_2O_2) with acid. What is the formula of the electrophile? How does it form?

Answer: The oxygen–oxygen bond of the conjugate acid of hydrogen peroxide cleaves heterolytically to give water and a hydroxyl cation, which is the electrophile.

13.10   Reactive aromatic rings can be iodinated using iodine monochloride (ICl). What is the electrophile?

Answer: The electrophile is I^+, which is produced by the heterolytic cleavage of the I—Cl bond.

13.11   Benzene reacts with mercuric acetate to give phenylmercuric acetate using perchloric acid ($HClO_4$) as a catalyst. What is the electrophile? How does it form?

Answer: The electrophile is $AcOHg^+$, which is formed by protonation of one of the acetate groups of the covalent $Hg(OAc)_2$ followed by heterolytic cleavage of the Hg—O bond. Acetic acid is the other product.

13.12 Treating an aromatic rings with *tert*-butyl alcohol, $(CH_3)_3COH$, in acid solution places a tertiary butyl group on the ring. What is the formula of the electrophile? How does it form?

Answer: The electrophile is $AcOHg^+$, which is formed by protonation of one of the acetate groups of the covalent $Hg(OAc)_2$ followed by protonation of the oxygen atom followed by heterolytic cleavage of the C—O bond. Water is the leaving group.

Properties of Ring Substituents

13.13 Some activated rings may be hydroxylated by reacting hydrogen peroxide (H_2O_2) with acid. What is the formula of the electrophile? How does it form?

Answer: The oxygen–oxygen bond of the conjugate acid of hydrogen peroxide cleaves heterolytically to give water and a hydroxyl cation, which is the electrophile.

13.14 Is the thiomethyl group, —S—CH₃, an activating or deactivating group. Will it be *ortho*, *para* directing or *meta* directing?

Answer: Both chlorine and sulfur are third row elements. However, since sulfur is less electronegative than chlorine, the ring is less deactivated by inductive electron withdrawal. Although third row elements are not effective electron donors by resonance, sulfur is a better donor of electrons than a chloro group because it is less electronegative. Therefore, the —SCH₃ group is an *ortho*, *para*-directing group.

13.15 The sulfonamide group is found in sulfa drugs. Is it an activating or deactivating group. Will it be *ortho*, *para*-directing or *meta* directing?

Answer: The sulfur atom is bonded to three electronegative atoms. Therefore, the sulfonamide group will inductively withdraw electron density from the aromatic ring and deactivate it toward electrophilic substitution. There are no lone pair electrons on sulfur to be donated by resonance. Thus, the sulfonamide group is a *meta* director.

13.16 Nitration of *N,N*-dimethylaniline, $C_6H_5N(CH_3)_2$, in 85% sulfuric acid gives a *meta* nitro compound as the major product. What is the structure of the ring substituent responsible for the orientation of the nitro product?

Answer: An ammonium ion is formed in acidic solution because the nitrogen atom is protonated. The conjugate acid of *N,N*-dimethylaniline has a positive charge on the nitrogen atom. Therefore, the group withdraws electron density from the aromatic ring and deactivates benzene toward electrophilic substitution. Since there are no lone pair electrons on the protonated nitrogen atom to be donated by resonance, the group is a meta director.

13.17 The percentages of *meta* nitro product formed in the nitration of benzene compounds containing CH₃—, CH₂Cl—, CHCl₂—, and CCl₃— groups are 5%, 16%, 34%, and 64%, respectively. Explain this trend in the data.

Answer: As the number of chlorine atoms bonded to the carbon atom increases, the inductive withdrawal of electron density from that carbon atom and the adjacent benzene ring increases. Since there are no lone pair electrons on the carbon atom to be donated by resonance, the groups become increasingly better *meta* directors by increased electron withdrawal.

Reagents for Electrophilic Substitution

13.18 What reagent is required for each of the following reactions? Write the structure of the major product(s) expected from each reaction.
(a) bromination of anisole   (b) sulfonation of toluene
(c) nitration of benzoic acid   (d) acetylation of bromobenzene

Answers: The reagents are as follows. The structures of the products are shown below.
(a) bromine with iron(lll) bromide
(b) fuming sulfuric acid (sulfur trioxide and sulfuric acid)
(c) nitric acid with sulfuric acid
(d) acetyl chloride with aluminum trichloride

13.19 What reagent is required for each of the following reactions? Write the structure of the principal product(s) expected from each reaction.

(a) chlorination of bromobenzene
(b) Friedel–Crafts methylation of anisole
(c) Friedel–Crafts acetylation of toluene
(d) nitration of trifluoromethylbenzene

Answers: The reagents are as follows. The structures of the products are shown below.
(a) chlorine with iron(lll) chloride
(b) methyl chloride with aluminum trichloride
(c) acetyl chloride with aluminum trichloride
(d) nitric acid with sulfuric acid

Friedel–Crafts Alkylation and Acylation

13.20 Write the structure of the product resulting from the Friedel–Crafts alkylation of benzene using chlorocyclohexane and aluminum trichloride.

Answer: Reaction of chlorocyclohexane with aluminum trichloride gives the cyclohexyl carbocation, which reacts with benzene to give cyclohexylbenzene.

13.21 What product results from the Friedel–Crafts alkylation of benzene using 1-chloro-2-methylpropane and aluminum trichloride?

Answer: Reaction of 1-chloro-2-methylpropane with aluminum trichloride gives a primary isobutyl carbocation, which rearranges to a *tert*-butyl carbocation by a 1,2-hydride shift. Thus, the product of the reaction is *tert*-butylbenzene.

13.22 Alkylation of benzene can be accomplished using an alkene such as propene and an acid catalyst. Identify the electrophile and the product.

Answer: Protonating C-1 of propene gives a secondary, isopropyl carbocation, which is the electrophile. The product is isopropylbenzene.

13.23 Write the structure of the product formed by alkylation of *p*-methylanisole using 2-methyl-l-propene and sulfuric acid.

Answer: The electrophile is the *tert*-butyl carbocation, $(CH_3)_3C^+$, which is formed by protonating the double bond. The *tert*-butyl carbocation adds *ortho* or *para* to the methyl group. The major product is the *para* isomer because steric hindrance diminishes attack at the *ortho* position.

13.24 Reaction of toluene with isopropyl alcohol, $(CH_3)_2CHOH$, using sulfuric acid gives a mixture of two isomers with the molecular formula $C_{10}H_{13}$. Write the structures of these compounds. How does the electrophile form?

Answer: The electrophile is the isopropyl carbocation, $(CH_3)_2CH^+$, which is formed by protonation of the oxygen atom followed by heterolytic cleavage of the C—O bond. Water is the leaving group. The isopropyl carbocation alkylates *ortho* or *para* to the methyl group. The major product will be the *para* isomer due to steric hindrance in the attack at the *ortho* position.

13.25 The following compound reacts with sulfuric acid to give a tricyclic hydrocarbon with molecular formula $C_{17}H_{18}$. Write its structure.

Answer:

13.26 4-Phenylbutanoyl chloride reacts in carbon disulfide with aluminum trichloride to give a ketone with molecular formula $C_{10}H_{10}O$. Write the structure of the product.

Answer: An intramolecular Friedel–Crafts acylation reaction occurs *ortho* to the butanoyl chain to give a cyclic ketone.

13.27 The following compound undergoes an intramolecular Friedel–Crafts acylation to give a cyclic ketone. Write the structure of the product.

Answer: An intramolecular Friedel–Crafts acylation reaction occurs on the ring containing the methoxy group because it is the more activated aromatic ring.

Electrophilic Aromatic Substitution Reactions

13.28 Indicate on which ring and at what position bromination of each compound will occur.

Answer: (a) Reaction occurs in the ring bonded to the nitrogen atom because that ring is activated. The other ring is deactivated by the carbonyl group. Reaction will occur at the *ortho* or *para* position.

Answer: (b) Reaction occurs in the ring bonded to the methylene group because it is slightly activated. The other ring is deactivated by the carbonyl group. Reaction will occur at the *ortho* or *para* position.

13.29 Indicate on which ring and at what position nitration of each compound will occur.

221

Answer: (a) Both rings are activated by the oxygen atom. However, the ring on the right is deactivated by the nitro group. Nitration will occur in the other ring at the *ortho* or *para* position.

Answer: (b) Both rings are slightly activated by the methylene group. However, the ring on the right is also activated by the hydroxyl group. Nitration will tend to occur in that ring *ortho* to the activating group. A smaller amount of product with the nitro group *ortho* to the methylene group also forms.

13.30   Write the structure of the major product of each of the following reactions, assuming that only monosubstitution occurs.

Answers: (a) The alkyl group is slightly activating and the bromo group is slightly deactivating. Bromination occurs *ortho* to the alkyl group.
(b) The bromo groups are deactivating but are *ortho, para*-directing groups. Nitration at C-2 between the two bromine atoms is sterically hindered. The other *ortho* positions at C-4- and C-6 are structurally equivalent, and only one product results.
(c) The dimethylamino group is strongly activating and the isopropyl group is slightly activating. Acetylation occurs *ortho* to the dimethylamino group.

13.31   Write the structure of the major product of each of the following reactions, assuming that only monosubstitution occurs.

Answers: (a) The amino group is strongly activating and the nitro group is strongly deactivating. Bromination occurs *ortho* to the amino group.
(b) The two carboxylic acid groups are deactivating and are meta directing groups. Nitration occurs at the C-5 position, which is meta to both groups.
(c) The isopropyl group is slightly activating and the bromo group is slightly deactivating. Acetylation may occur *ortho* or para to the diisopropyl group. The *ortho* position between the isopropyl and bromo group is sterically hindered. The other *ortho* position is substantially hindered by the isopropyl group. The *para* position with respect to the isopropyl group is also *ortho* with respect to the bromine. This is the most likely site for attack.

(a)

(b)

(c)

13.32 Potassium permanganate oxidizes alkyl groups to —CO_2H groups.

(a)

(b)

(c)

13.33 Write the product of the reaction of each of the following compounds with zinc-mercury amalgam and HCl.

Answer: Zinc-mercury amalgam and HCl convert carbonyl groups to methylene groups.

(a)

(b)

(c)

13.34 Write the product of the reaction of each of the following compounds with tin and HCl.

Answer: Tin and HCl reduces nitro groups to produce amino groups.

(a)

$$\xrightarrow{\text{Sn / HCl}}$$

(b)

$$\xrightarrow{\text{Sn / HCl}}$$

(c)

$$\xrightarrow{\text{Sn / HCl}}$$

13.35 Write the final product of the sequence of reactions for each of the following compounds.

Answers:

(a)

$$\xrightarrow[\text{HBr}]{\text{NaNO}_2} \quad \xrightarrow{\text{Cu}_2\text{Br}_2}$$

(b)

$$\xrightarrow[\text{H}_2\text{SO}_4]{\text{NaNO}_2} \quad \xrightarrow{\text{H}_3\text{O}^+}$$

(c)

$$\xrightarrow[\text{HBr}]{\text{NaNO}_2} \quad \xrightarrow{\text{H}_3\text{PO}_2}$$

13.36   Write the product of the following reactions.

Answers:

(a)

(b)

(c)

Synthesis of Aromatic Compounds

13.37   What reagent is required for each of the following reactions? Will an ortho and para mixture of products or the meta isomer predominate?
(a) nitration of bromobenzene   (b) sulfonation of nitrobenzene
(c) bromination of ethylbenzene   (d) methylation of anisole

Answers: (a) Nitric acid with sulfuric acid is required. The products are a mixture of *ortho* and *para* isomers of bromonitrobenzene, because the bromo group is *ortho, para* directing.
(b) Fuming sulfuric acid is the reagent. *m*-Nitrobenzenesulfonic acid is the product because the nitro group is *meta* directing.
(c) Bromine and iron(lll) bromide are required. The products are a mixture of *ortho* and *para* isomers of bromoethylbenzene because the ethyl group is *ortho, para* directing.
(d) Chloromethane and aluminum trichloride are required. The products are a mixture of *ortho* and *para* isomers of methylanisole because the methoxy group is *ortho, para* directing.

13.38   What reagent is required for each of the following reactions? Will an *ortho* and *para* mixture of products or the *meta* isomer predominate?
(a) bromination of benzoic acid   (b) acetylation of isopropylbenzene
(c) nitration of acetophenone   (d) nitration of phenol

Answers: (a) Bromine and iron(lll) bromide are required. The product is m-bromobenzoic acid, because the carboxylic acid group is meta directing.
(b) Acetyl chloride and aluminum trichloride are the reagents. The products are a mixture of *ortho* and *para* isomers of isopropylacetophenone because the isopropyl group is *ortho, para* directing.
(c) Nitric acid with sulfuric acid is required. The product is *m*-nitroacetophenone because the acetyl group is *meta* directing.
(d) Nitric acid alone is sufficient because the hydroxy group is strongly activating. The products are a mixture of *ortho* and *para* isomers of nitrophenol because the hydroxy group is *ortho, para* directing.

13.39   Starting with benzene, describe the series of reagents and reactions required to produce each of the following compounds.
(a) *p*-bromonitrobenzene   (b) *m*-bromonitrobenzene
(c) *p*-bromoethylbenzene   (d) *m*-bromoethylbenzene

Answers: (a) The bromo group is an *ortho, para* director, whereas the nitro group is a *meta* director, so bromination must occur first. Brominate benzene using bromine and iron(III) bromide. Nitrate the bromobenzene using nitric acid with sulfuric acid.
(b) The nitro group is a *meta* director, whereas the bromo group is an *ortho, para* director, so nitration must occur first. Nitrate benzene using nitric acid with sulfuric acid. Brominate the nitrobenzene using bromine and iron(III) bromide.
(c) The ethyl group is weakly activating, whereas the bromo group is weakly deactivating, so alkylation is done first. Alkylate benzene using bromoethane and aluminum tribromide. Brominate ethylbenzene using bromine and iron(III) bromide.
(d) Both an ethyl and a bromo group are *ortho, para* directors, but a *meta* director is required. The acetyl group is a *meta* director, so acylation must occur first. Acetylate benzene using acetyl chloride and aluminum trichloride. Then brominates acetophenone using bromine and iron(III) bromide. Finally, reduce the *m*-bromoacetophenone using zinc/mercury amalgam and HCl.

13.40   Starting with benzene, describe the series of reagents and reactions required to produce each of the following compounds.
(a) *m*-bromobenzenesulfonic acid   (b) *p*-bromobenzenesulfonic acid
(c) *p*-nitrotoluene   (d) *p*-nitrobenzoic acid

Answers: (a) The bromo group is an *ortho, para* director. The sulfonic acid group is a *meta* director. Thus, sulfonation must occur first. Sulfonate benzene using fuming sulfuric acid. Then brominate the benzenesulfonic acid using bromine and iron(III) bromide.
(b) Bromine is an *ortho, para* director, whereas the sulfonic acid group is a *meta* director. Thus, bromination must occur first. Brominate benzene using bromine and iron(III) bromide. Sulfonate bromobenzene using fuming sulfuric acid to give a mixture of isomers, and separate them by a physical method.
(c) The methyl group is an *ortho, para* director, whereas the nitro group is a *meta* director. Therefore, alkylation must occur first. Alkylate benzene using bromomethane and aluminum tribromide. Nitration of toluene using nitric acid with sulfuric acid gives a mixture of isomers.
(d) Both the nitro group and the carbonyl group are deactivating groups and *meta* directors. A methyl group is an *ortho, para* director. Therefore, alkylate benzene with bromomethane and aluminum tribromide. Then, convert toluene to *p*-nitrotoluene using nitric acid with sulfuric acid. Finally, oxidize *p*-nitrotoluene with potassium permanganate to convert the methyl group to —CO_2H.

13.41   Starting with either benzene or toluene, describe the series of reagents and reactions required to produce each of the following compounds.
(a) 3,5-dinitrochlorobenzene   (b) 2,4,6-trinitrotoluene   (c) 2,6-dibromo-4-nitrotoluene

Answers: (a) The nitro group is a *meta* director, but the chloro group is an *ortho, para* director. Therefore, nitrate benzene twice using nitric acid and sulfuric acid to give *m*-dinitrobenzene. Then chlorinate the *m*-dinitrobenzene using chlorine and iron(III) chloride. The chlorine is introduced *meta* to each of the nitro groups.
(b) The methyl group of toluene is an *ortho, para* director, but the nitro group is a *meta* director. Nitrate toluene three times using nitric acid with sulfuric acid. The reaction will become progressively more difficult as each nitro group further deactivates the products. The nitro groups are all *meta* to each other and *ortho* and *para* to the methyl group.
(c) The methyl group of toluene is an *ortho, para* director. The bromo group is an *ortho, para* director also, but the nitro group is a *meta* director. Nitrate toluene using nitric acid with sulfuric acid to obtain a mixture of the *ortho* and *para* isomers of nitrotoluene. Brominate twice using bromine and iron(III) bromide. The bromine atoms are introduced at the two equivalent positions that are *ortho* to the methyl group and *meta* to the nitro group.

13.42   Starting with either benzene or toluene, describe the series of reagents and reactions required to produce each of the following compounds.
(a) 2,4,6-tribromobenzoic acid   (b) 2-bromo-4-nitrotoluene   (c) l-bromo-3,5-dinitrobenzene

Answers: (a) The bromo group is an *ortho, para* director and the nitro group is a *meta* director, and both are deactivating. The methyl group is an *ortho, para* director. Nitrate toluene three times using nitric acid with sulfuric acid to give 2,4,6-tribromotoluene. Oxidize this compound using potassium permanganate to convert the methyl group into a carboxylic acid group.
(b) The methyl group of toluene is an *ortho, para* director. Nitrate toluene using nitric acid with sulfuric acid. Brominate this compound using bromine and iron(lll) bromide. The bromine is *ortho* to the methyl group and *meta* to the nitro group.

(c) The bromo group is an *ortho, para* director and the nitro group is a *meta* director. Thus, nitration must occur first. Nitrate benzene twice using nitric acid with sulfuric acid to give *m*-dinitrobenzene. Then brominate the *m*-dinitrobenzene using bromine and iron(III) bromide. The bromine is introduced at a position that is *meta* to each of the nitro groups.

13.43   Starting with either benzene or toluene, describe the series of reagents and reactions required to produce each of the following compounds.
(a) *m*-bromophenol   (b) *m*-bromoaniline   (c) *p*-methylphenol

Answers: (a) Both the bromo and hydroxyl groups are *ortho, para* directors. The nitro group is a *meta* director that can be converted into another functional group. Nitrate benzene to give nitrobenzene. Then brominate nitrobenzene to give *m*-bromonitrobenzene. Reduce   the nitro group to an amino group using tin and HCl. Convert the amino group into a diazonium group using nitrous acid, and react with hot aqueous acid to give *m*-bromophenol.
(b) Both the bromo and amino groups are *ortho, para* directors. The nitro group is a *meta* director. Nitrate benzene to give nitrobenzene. Then brominate nitrobenzene to give *m*-bromonitrobenzene. Reduce the nitro group to an amino group using Sn and HCl.
(c) The methyl group of toluene is an *ortho, para* director. Nitrate toluene to give a mixture of the *ortho* and *para* isomers of nitrotoluene. Reduce the nitro group to an amino group using tin and HCl. Convert the amino group into a diazonium group using nitrous acid and react with hot aqueous acid to replace the diazonium group with a hydroxyl group.

13.44   Starting with either benzene or toluene, describe the series of reagents and reactions required to produce each of the following compounds.
(a) *m*-bromochlorobenzene   (b) *p*-methylbenzonitrile   (c) 3,5-dibromotoluene

Answers: (a) The bromo and chloro groups are *ortho, para* directors, but they are *meta* to each other in the product. The nitro group is a *meta* director. Nitrate benzene using nitric acid with sulfuric acid. Brominate the nitrobenzene using bromine and iron(III) bromide to give the *meta* isomer. Reduce the nitro group to an amino group using tin and Hcl. Convert the amino group into a diazonium group using nitrous acid and react with copper(l) chloride
(b) Nitrate toluene to give a mixture of the *ortho* and *para* isomers of nitrotoluene. Separate the *para* isomer by a physical method. Reduce the nitro group to an amino group using tin and HCl. Convert the amino group into a diazonium group using nitrous acid and react with copper(l) cyanide to give the nitrile.

(c) The bromo and methyl groups are *ortho, para* directors, but they are *meta* to each other in the product. The nitro group is a versatile group that can be converted into other groups. Nitrate toluene to give *p*-nitrotoluene. The nitro group is a *meta* director, but the bromo groups must be *ortho* to it and *meta* to the methyl group. Reduce the nitro group to an *ortho, para* directing amino group using tin and HCl. The amino group is strongly activating, whereas the methyl group is only weakly activating. Brominate this compound twice using bromine and iron(lll) bromide. The bromine atoms are placed in the equivalent positions *ortho* to the amino group, since the *para* position is already occupied by —CH$_3$. Convert the amino group into a diazonium group using nitrous acid and then react with hypophosphorous acid to replace the diazonium group by hydrogen.

14

METHODS FOR STRUCTURE DETERMINATION NUCLEAR MAGNETIC RESONANCE AND MASS SPECTROMETRY

KEYS TO THE CHAPTER

14.1 Structure Determination

Although the identity of a molecule and its structure can be determined indirectly based on its chemical reactions, such methods destroy some portion of the sample of the compound. Spectroscopic methods determine molecular structure by physical methods. Therefore, the sample is not destroyed.

Even a relatively simple molecule can exist in many isomeric forms, and spectroscopic methods rapidly narrow the range of possibilities, greatly shortening the time required to determine a molecular structure.

14.2 Nuclear Magnetic Resonance

Many nuclei have a nuclear spin. A spinning nucleus generates a magnetic field, whose energy depends on the direction of spin in the presence of an applied magnetic field. The NMR method depends on detecting the absorption of energy required to change the direction of the spin of a nucleus.

Two nuclei that are important in the determination of the structures of organic compounds are ^1H and ^{13}C. The magnetic field strength required to "flip" the spin of various hydrogen atoms (or carbon atoms) within a molecule differs. The local magnetic fields differ throughout a molecule because the bonding characteristics differ. Thus, each hydrogen (or carbon) nucleus is unique, and distinct resonances are obtained for each structurally nonequivalent atom in a molecule.

14.3 The Chemical Shift

A spinning nucleus induces a small local magnetic field that opposes the applied magnetic field. This local field shields the nucleus from the applied field. A relative scale called the delta scale, in which one delta unit (δ) is 1 ppm of the applied magnetic field, is used to measure the chemical shift of hydrogen atoms. The resonance for the hydrogen atoms of tetramethylsilane, $(CH_3)_4Si$, is defined as $0\ \delta$. The delta scale is independent of the applied magnetic field. Shielded nuclei are found at high field and have small δ values. The chemical shifts for the hydrogens in organic compounds ranges from 0 to 10 δ.

Organic Chemistry Study Guide: Key Concepts, Problems, and Solutions http://dx.doi.org/10.1016/B978-0-12-801889-7.00014-5

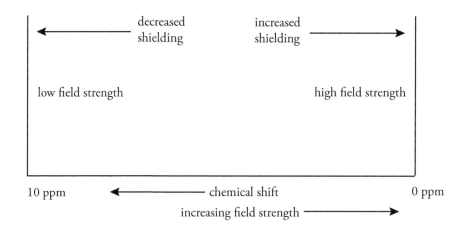

14.4 Detecting Sets of Nonequivalent Hydrogen Atoms

To understand the relationship between an NMR spectrum and the structure of a compound, we have to be able to recognize the equivalence of nuclei in the structure. The simplest examples, such as the six equivalent protons on the two methyl groups of 2-bromopropane, are straightforward. However, some nuclei that might look equivalent at first glance are actually nonequivalent. For example, both hydrogen atoms in 1-bromo-1-chloroethene are bonded to the same carbon atom, but they are not equivalent. One hydrogen atom is *cis* to the chlorine atom and the other is *cis* to the bromine atom. Replacing one hydrogen atom or the other by deuterium gives a set of diastereomers, and the hydrogen atoms are **diastereotopic**. Such hydrogen atoms usually have different chemical shifts.

Hydrogen atoms that are in mirror image environments are **enantiotopic** (Section 8.12). They have same chemical shifts. Replacing enantiotopic hydrogen atoms by deuterium atoms gives enantiomers, and their physical properties are identical, including their chemical shifts.

14.5 Structural Effects on Chemical Shifts

The chemical shifts of hydrogen atoms depend on the local electron density, which in turn affects the local magnetic field. Electronegative atoms deshield hydrogen atoms by an effect analogous to the inductive effect we have discussed many times. Deshielding results in a shift to lower field and larger δ values.

Electrons in π bonds are easily polarized, and they induce substantial local magnetic fields. These effects extend across three or four chemical bonds. Of particular importance is the deshielding of hydrogen atoms bonded to an aromatic ring. The chemical shifts of aromatic hydrogen atoms are large, meaning that they appear at low field with δ values of 7-8 ppm.

Table 14.5 gives the chemical shifts of hydrogen atoms in various structural environments.

14.6 Relative Peak Areas and Proton Counting

The NMR spectrum tells us how many sets of structurally nonequivalent hydrogen atoms are present in a molecule. Each set causes resonance absorptions in its own characteristic region. The area of each resonance peak is proportional to the relative number of hydrogen atoms of each kind. Therefore, NMR also tells us the ratios of the nonequivalent hydrogen atoms in a molecule. These ratios are proportional to the relative peaks heights of the resonances of the nonequivalent hydrogen atoms.

14.7 Spin-Spin Splitting

Multiple lines, called **multiplets**, are often observed for the absorptions of equivalent hydrogen atoms. This phenomenon is called **spin-spin splitting**. It results from the interaction of the nuclear spin of the hydrogen atoms on an adjacent carbon atoms. In general, sets of hydrogen atoms on nonequivalent neighboring carbon atoms couple with each other. If hydrogen atom A couples and causes splitting of the resonance for hydrogen atom B, then the resonance for hydrogen atom B is also split by hydrogen atom A. The separation of the components of the multiplet is the **coupling constant**, which is designated as *J*.

A set of one or more hydrogen atoms that has n equivalent neighboring hydrogen atoms has $n+1$ peaks in the NMR spectrum. Common multiplets include **doublets**, **triplets**, and **quartets**. The appearance of several sets of multiplets resulting from $n = 1$ to $n = 4$ is shown in Figure 14.15. The areas of the component peaks of a doublet are equal; the areas of the component peaks of other multiplets are not equal and are summarized in Table 14.16.

14.8 Structural Effects on Coupling Constants

The conformation of a molecule atom contributes to the magnitude of the coupling constant of vicinal hydrogen atoms. The coupling constant is largest for *anti* periplanar arrangements in saturated H—C—C—H compounds. The coupling constant for vinyl hydrogen atoms in an E arrangement is larger than for hydrogen atoms in an isomer with a Z configuration.

Coupling constants over more than three bonds are termed **long range**. These coupling constants can be used to determine the structures of isomeric aromatic compounds.

14.9 Ion Impact Mass Spectrometry

In **electron impact mass spectrometry**, the collision of a high energy electron with a sample molecule produces a **radical cation**, $M^{+\cdot}$. The first ion that forms in this process is the **parent ion**. Since the charge of the ion is +1, and since the mass of an electron is much smaller than the mass of a proton or neutron, the ratio of the mass of the parent ion, $M^{+\cdot}$, to its charge, m/z, equals the molecular mass of the compound.

The parent ion has very high energy, and it fragments in the instrument before it reaches the detector. The peak for the most abundant ion is assigned an arbitrary intensity of 100; this is the **base peak**. The parent ion, $M^{+\cdot}$, fragments two give two products: one is a cation or a radical cation, the other is neutral. We can identify the mass of the neutral particle that forms in a fragmentation reaction by subtracting the mass of the base peak from the mass of the parent ion.

One per cent of carbon atoms exist as the stable isotope ^{13}C. If the parent ion is sufficiently intense, then a **P+1 ion** will be present in the mass spectrum. The probability that a molecule will contain ^{13}C increases with the number of carbons in the compound.

Each type of functional group fragments with a characteristic pattern. The parent ion of a normal alkane fragments by forming a neutral methyl group, $CH_3\cdot$, and a primary radical cation. This species continues to fragment by losing successive $CH_2\cdot$ groups. The base peak for a normal alkane typically has a mass of $CH_3(CH_2)_n^{+\cdot}$. In contrast, the parent ion of a branched alkane usually does not have a parent ion. Instead, it fragments at a branch point to give a neutral fragment and a secondary or tertiary radical cation as the base peak.

Mass spectrometry allows us to chlorine and bromine by their isotopic abundances. Chlorine has two isotopes, ^{35}Cl and ^{37}Cl. The atomic mass of chlorine corresponds to an isotopic ratio $^{35}Cl/^{37}Cl$ of 3:1. The parent ion of a chloro compound has two peaks whose mass differ by two units having intensities are 3:1. It is also easy to identify bromine since the $^{79}Br/^{81}Br$ ratio is 1:1.

An alcohol ionizes in a mass spectrometer to produce a radical cation in which the oxygen atom bears the positive charge. The fragmentation of the base peak results from cleavage of the C—H bond rather than the C—O bond because the C—O bond is stronger.

The mass spectrum of a compound provides clues about the presence or absence of nitrogen. If a compound contains a single nitrogen atom, or any odd number of nitrogen atoms, its mass is an odd number. If the mass spectrum of a compound is an even number, then it has either an even number of nitrogen atoms, or none.

14.10 Effect of Dynamic Processes

Dynamic processes, such as rotation about a single bond, occur on such a rapid time scale that two seemingly nonequivalent hydrogen atoms have the same chemical shift. One such circumstance is found in cyclohexane because chair-chair interconversion exchanges the axial and equatorial hydrogen atoms so rapidly that they cannot be distinguished at room temperature. A second example is found in alcohols, in which the rapid exchange of protons between oxygen atoms of various alcohol molecules in a sample.

14.11 Carbon-13 NMR Spectroscopy

Carbon-13 (^{13}C) NMR spectroscopy permits the direct determination of the number of nonequivalent carbon atoms in a structure. If the carbon atoms are nonequivalent, each has a distinct resonance. More importantly, the count of the number of resonances gives the number of sets of equivalent carbon atoms that must be in the structure. The list of chemical shifts in Table 14.7 gives some idea about the identity of each carbon atom responsible for a resonance.

The intensity of the resonances of ^{13}C atoms is *not* proportional to the number of equivalent carbon atoms.

ANSWERS TO END-OF-CHAPTER EXERCISES

Infrared Spectroscopy

14.1   How can infrared spectroscopy be used to distinguish between propanone and 2-propen-l-ol?

Answer: The carbonyl group of propanone (acetone) has a strong absorption at 1749 cm^{-1}. 2-Propen-1-ol (allyl alcohol) has an absorption for the carbon-carbon double bond at 1645 cm^{-1} and an absorption for the oxygen-hydrogen bond at 3400 cm^{-1}.

14.2   How can infrared spectroscopy be used to distinguish between 1-pentyne and 2-pentyne?

Answer: 1-Pentyne is a terminal alkyne, so its sp-hybridized C—H bond has an absorption in the 3450 cm^{-1}, and another strong C≡C absorption at 2120 cm^{-1}. 2-Pentyne, which is an internal alkyne, does not have a C—H absorption at 3450 cm^{-1}. Also, the C≡C absorption is so weak that it is barely visible.

14.3   The carbonyl stretching vibration of ketones is at a longer wavelength than the carbonyl stretching vibration of aldehydes. Suggest a reason for this observation.

Answer: The longer wavelength absorption (smaller wavenumber) corresponds to a lower energy vibration. The dipolar resonance form of a ketone is more stable than that of an aldehyde because the extra alkyl group donates electron density. The increased contribution of the resonance form with a carbon-oxygen single bond means that the ketone carbonyl bond absorption requires less energy.

14.4   The carbonyl stretching vibrations of esters and amides occur at 1735 and 1670 cm^{-1}, respectively. Suggest a reason for this difference.

Answer: Both oxygen and nitrogen are inductively electron withdrawing, and they destabilize the dipolar resonance form of the carbonyl group. Since oxygen is more electronegative than nitrogen, this effect is larger for oxygen, so the dipolar resonance form of an ester is less stable that of an amide. The relative ability of the two atoms to donate electrons by resonance is also important. Because nitrogen donates electrons by resonance more effectively than oxygen, there is an increased contribution of a dipolar resonance form for the amide.

14.5   An infrared spectrum of a compound with molecular formula $C_4H_8O_2$ has an intense, broad band between 3500 and 3000 cm^{-1} and an intense peak at 1710 cm^{-1}. Which of the following compounds best fits this data?

   I: $CH_3CH_2CO_2CH_3$ II: $CH_3CO_2CH_2CH_3$ III: $CH_3CH_2CH_2CO_2H$

Answer: The absorptions correspond to an O—H and a carbonyl group, respectively. Only the carboxylic acid group of III has both structural features. The other two compounds are esters that would have an absorption corresponding to a carbonyl group but, because esters do not have an O—H group, would have no absorption in the 3500–3000 cm^{-1} region.

14.6   Explain why the carbonyl stretching vibrations of the following two esters differ.

$$CH_2=CH-CH_2-\overset{\overset{\displaystyle O}{\|}}{C}-O-CH_3 \qquad CH_3-CH=CH-\overset{\overset{\displaystyle O}{\|}}{C}-O-CH_3$$

$$1735 \text{ cm}^{-1} \qquad\qquad\qquad 1720 \text{ cm}^{-1}$$

Answer: The carbonyl group of the second compound is conjugated with a double bond. As a result, there is some contribution of a resonance form in which the carbon-oxygen bond has single bond character. The increased contribution of the resonance form with a carbon-oxygen single bond means that the carbonyl bond absorption requires less energy.

14.7   Explain how the two isomeric nitration products of isopropylbenzene can be distinguished using infrared spectroscopy.

Answer: The *ortho* nitro isomer has four adjacent C—H bonds, and the out-of plane bending of these bonds occurs at 748 cm^{-1}. The para nitro isomer has two sets of two adjacent C—H bonds, and the out-of plane bending occurs at 866 cm^{-1}.

14.8   Explain how the structures of the three isomeric trimethylbenzenes can be established using infrared spectroscopy.

Answer: The 1,2,3 isomer has three adjacent C—H bonds, and the out-of plane bending absorptions of these bonds occur in the 810–750 cm^{-1} region. There is another absorption in the 745–690 cm^{-1} region. The 1,2,4 isomer has two adjacent C—H bonds and one lone C—H bond. The absorptions for these bonds occur in the 860–800 and 900–860 cm^{-1} regions, respectively. The 1,3,5 isomer has three C—H bonds with no neighbors. The absorptions for these bonds occur in the 900–860 cm^{-1} region.

Calculation of Chemical Shift

14.9   The hydrogen NMR spectrum of CHCl$_3$, measured with a 360 MHz spectrometer, is a singlet that is 2622 Hz downfield from TMS. Calculate δ.

Answer: Divide 2622 Hz by (60×10^6) and multiply by 10^6 to obtain 7.28 δ.

14.10   The hydrogen NMR spectrum of CHI$_3$, measured with a 360 MHz spectrometer, is a singlet at 5.37 δ. Calculate the chemical shift in Hz relative to TMS.

Answer: Multiply 5.37 δ by 360 Hz because each ppm or δ unit equals 360 Hz. The chemical shift is 1933 Hz downfield from TMS.

Chemical Shifts and Structure

14.11   How many NMR signals should be observed for the hydrogen atoms in each of the following compounds?
(a) 2,2-dimethylpropane   (b) 2-methyl-l-propene   (c) 1,3,5-trimethylbenzene

Answers: (a) Only one, because all four methyl groups are equivalent. C-2 atom does not have a C—H.
(b) Two, because there are two equivalent methyl groups and two equivalent sp²-hybridized C—H bonds.
(c) Two, because there are three equivalent methyl groups and three equivalent sp²-hybridized C—H bonds on the benzene ring.
(d) There are four. The C-1 methyl group and the branching methyl group at C-2 are not equivalent! There are also resonances for sp²-hybridized C—H bond at C-3, and the C-4 methyl group.

14.12   How many NMR signals should be observed for the hydrogen atoms in each of the following compounds?
(a) 1,1-dichloroethene   (b) vinyl chloride (c) allyl bromide   (d) 1-bromo-l-chloroethene

(a) $Cl_2C{=}CH_2$ (b) $H(Cl)C{=}CH_2$ (c) $CH_2{=}CH{-}CH_2Br$ (d) $Cl(Br)C{=}CH_2$

Answers: (a) Only one, because there are two equivalent sp²-hybridized C—H bonds at C-2.
(b) The sp²-hybridized C—H bonds at C-2 are not equivalent. Thus, these two hydrogen atoms and the hydrogen at C-1 give three resonances.
(c) The sp²-hybridized C—H bonds at C-3 are not equivalent. Thus, these two hydrogen atoms, the hydrogen atom of the sp²-hybridized C—H bond at C-2, and the C-1 methylene hydrogen atoms give four resonances.
(d) The sp²-hybridized C—H bonds at C-2 are not equivalent, so each hydrogen atom has a different resonance.

14.13   How can the compounds of each pair be distinguished using hydrogen NMR spectroscopy?
(a) isopropyl ethyl ether and *tert*-butyl methyl ether (b) cyclohexane and *cis*-3-hexene
(c) 2,2-dimethyloxirane and *cis*-2,3-dimethyloxirane

Answers: (a) Each compound has resonances integrating as three hydrogen atoms in the 3.3-4.0 δ region associated with ethers. However, this signal for *tert*-butyl methyl ether is a singlet due to the three hydrogen atoms of the methyl group. Isopropyl ethyl ether has two resonances in this region. The resonance arising from the two hydrogen atoms of the methylene group is a quartet due to coupling of the methyl group. The other hydrogen resonance in this region is due to the methine hydrogen of the isopropyl group and is a septet because it is split by the two methyl groups.
(b) All 12 of the hydrogen atoms of cyclohexane are equivalent, and the resonance appears as a singlet at high field. The isomeric alkene has a multiplet in the 5.0-6.5 δ region due to the hydrogen atoms of the sp²-hybridized C—H bonds, as well as peaks due to the methylene and methyl hydrogen atoms.
(c) The resonance due to the hydrogen atoms of the equivalent methyl groups of the 2,2-dimethyloxirane is a singlet. The hydrogen atoms of the equivalent methyl groups of the isomeric compound are split by the hydrogen atom of the ring.

14.14   How can the compounds of each pair be distinguished using hydrogen NMR spectroscopy?
(a) 1,3-dibromopropane and 2,2-dibromopropane
(b) 1,1-dichlorobutane and 1,4-dichlorobutane
(c) *cis*-2-butene and 2-methyl-l-propene

Answers: (a) The equivalent C-1 and C-3 methylene groups of 1,3-dibromopropane give a triplet due to splitting by the hydrogen atoms at C-2. The methylene hydrogen atoms at C-2 give a quintet. The equivalent methyl groups of 2,2-dibromopropane give one peak, a singlet.
(b) The low-field portion of the spectrum of 1,1-dichlorobutane has an absorption due to the C-1 hydrogen atom. It is a triplet. The low-field portion of the spectrum of the isomeric 1,4-dichlorobutane has an absorption with an integrated intensity of four hydrogen atoms due to the C-1 and C-4 methylene groups. The signal is also a triplet.
(c) Both compounds have resonances in the 5.0-6.5 δ region due to the two hydrogen atoms of the sp²-hybridized C—H bonds. However, those of 2-methyl-1-propene have no nearest neighbor hydrogen atoms to split the signal. The signal for the isomeric *cis*-2-butene is split by the hydrogen atoms of the methyl groups.

14.15   Draw the structure of each of the following hydrocarbons whose hydrogen NMR spectrum consists of a singlet with the indicated chemical shift.
(a) C_5H_{10}; δ = 1.5   (b) C_8H_{18}; δ = 0.9   (c) $C_{12}H_{18}$; δ = 2.2   (d) C_8H_8; δ = 5.8

Answers: (a)   (b) $CH_3{-}C(CH_3)_2{-}C(CH_3)_2{-}CH_3$

(c) $CH_3{-}C(CH_3)_2{-}C{\equiv}C{-}C{\equiv}C{-}C(CH_3)_2{-}CH_3$ (d)

235

14.16   Draw the structure of each of the following halogen compounds whose hydrogen NMR spectrum consists of a singlet with the indicated chemical shift.

(a) $C_2H_3Cl_3$; $\delta = 2.7$ (b) $C_2H_4Cl_2$; $\delta = 3.7$ (c) C_4H_9Br; $\delta = 1.8$ (d) $C_3H_6Br_2$; $\delta = 2.6$

14.17   The hydrogen NMR spectrum of [18]annulene consists of signals at $\delta = 8.8$ ppm and $\delta = -1.9$ ppm. The negative value of δ corresponds to an "unusual" chemical shift that is upfield from TMS. The ratio of intensities of the 8.8 ppm to −1.9 ppm resonances is 2:1. Explain these data.

[18]annulene

Answer: The local magnetic field generated by the electrons of the ring is similar to that shown for benzene in Figure 14.12. The 12 hydrogen atoms extending away from the ring are deshielded, and their resonance occurs at 8.8 δ. There are also 6 hydrogen atoms inside the ring. The magnetic field experienced by these hydrogen atoms is reversed, and they are shielded. Thus, the resonance occur at high field. The negative value indicates that the resonance is at higher field than TMS.

14.18   The hydrogen NMR spectrum of [14]annulene consists of signals at $\delta = 7.8$ ppm and $\delta = -0.6$ ppm. Assign the resonances and predict the relative intensities of each.

[14]annulene

Answer: The 10 hydrogen atoms extending away from the ring are deshielded. Their resonance occurs at 7.8 δ. There are also 4 hydrogen atoms inside of the ring. The magnetic field experienced by these hydrogen atoms is reversed, and they are shielded. They are shielded, and their resonance occurs at −0.6 δ. The ratio of the low-field to high-field resonances is 5:2.

236

Multiplicity and Structure

14.19   Describe the multiplicity of each of the signals corresponding to a set of equivalent hydrogen atoms in each of the following ethers.
(a) $CH_3CH_2OCH_2CH_3$ (b) $CH_3OCH(CH_3)_2$ (c) $ClCH_2OCHClCH_3$ (d) $Cl_2CHOCHClCHCl_2$

Answers: (a) The resonance of the six hydrogen atoms of the two equivalent methyl groups is a triplet, and the resonance of the four hydrogen atoms of the two equivalent methylene groups is a quartet.
(b) The resonance of the hydrogen atoms of the methyl group bonded to the oxygen atom is a singlet. The resonance of the hydrogen atom of the other carbon atom bonded to the oxygen atom is a heptet. The resonance of the hydrogen atoms of the two equivalent methyl groups is a doublet.
(c) The resonance of the hydrogen atoms of the methylene group bonded to a chlorine atom is a singlet. The resonance of the hydrogen atom of the other carbon atom bearing a chlorine atom is a quartet. The resonance of the hydrogen atoms of the methyl group is a doublet.
(d) The resonance of the hydrogen atom at the carbon atom bearing two chlorine atoms and an oxygen atom is a singlet. The resonance of the hydrogen atom of the carbon atom bearing an oxygen atom and one chlorine atom is a doublet. The resonance of the third hydrogen atom is a doublet.

14.20   Describe the multiplicity of the lowest field resonance of each of the following alkyl halides.
(a) 1-chloropentane   (b) 1-chloro-2,2-dimethylpropane   (c) 3-chloropentane   (d) 1-chloro-2-methyl-2-butene

Answers: The lowest field resonance in each case is for hydrogen atom(s) at the carbon atom that is bonded to the chlorine atom.
(a) triplet, split by the two hydrogen atoms on C-2
(b) singlet, because there are no hydrogen atoms on C-2
(c) quintet, split by the four equivalent hydrogen atoms at C-2 and C-4
(d) singlet, because there are no hydrogen atoms on C-2

14.21   The chemical shifts of the C-1, C-2, and C-3 hydrogen atoms of 1,1,2-trichloropropane are 5.50, 4.22, and 1.20 ppm. The coupling constant of the C-2 and C-3 hydrogen atoms is 6.5 Hz and that of the C-2 and C-1 hydrogen atoms is 4.5 Hz. Draw the splitting diagram for the C-2 hydrogen atom.

Answer: The resonance for the hydrogen atom at C-2 is split into a quartet by the three hydrogen atoms at C-3 and is further split into doublets by the hydrogen atom at C-1.

237

14.22 Assume that the coupling constants for three nonequivalent hydrogen identified as H_a, H_b, and $H_{a,b}$ are $J_{a,b}$ = 6 Hz, $J_{a,c}$ =2 Hz, and $J_{b,c}$ = 6 Hz. Draw the splitting diagram for H_b. What is the appearance of this resonance?

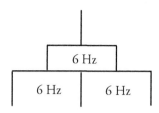

Answer: Because H_b is split by both H_a and H_c, its resonance should be a doublet of doublets. However, the coupling constants $J_{a,b}$ and $J_{b,c}$ are both 6 Hz. As a result, the inner lines of each doublet in the doublet of doublets overlap, so the resonance of H_b appears as a triplet.

14.23 Hydrogen bromide adds to 3-bromopropene under certain experimental conditions to give a compound whose NMR spectrum is a quintet at 2.10 δ and a triplet at 3.60 δ. The ratio of the total intensity of the quintet to that of the triplet is 1:2. What is the structure of the compound?

quintet

triplet ----- H   H   H

Br—C—C—C—Br

H   H   H

1,3-dibromopropane

Answer: The product is 1,3-dibromopropane. The resonance of the C-2 hydrogen atoms appears as a quintet. The resonance of the hydrogen atoms at C-1 and C-3 is a triplet.

14.24 The spectrum of a compound with molecular formula $C_3H_3Cl_5$ consists of a triplet at 4.5 δ and a doublet at 6.0 δ. The intensity ratio of the high-field to low-field signal is 1:2. What is the structure of the compound?

1H triplet

Cl   H   Cl

Cl—C—C—C—Cl

2H doublet ----- H   Cl   H

1,1,2,3,3-pentachloropropane

Answer: 1,1,2,3,3-Pentachloropropane has equivalent hydrogen atoms at C-1 and C-3 that give the low-field resonance which is split into a doublet by the hydrogen atom at C-2. The higher field resonance is due to the hydrogen atom at C-2, which is split into a triplet by the hydrogen atoms at C-1 and C-3. Note that 1,1,1,2,3-pentachloropropane also has two equivalent hydrogen atoms at C-3 another nonequivalent hydrogen atom at C-2. However, the low-field resonance would be for a single hydrogen atom at C-2, and it would be a triplet. The high-field resonance would be due to two hydrogen atoms at C-1, and it would be a doublet. These resonances are just the opposite of those in the spectrum described.

1H triplet

Cl   H   H

Cl—C—C—C—Cl

Cl   Cl   H ----- 2H doublet

1,1,1,2,3-pentachloropropane

Analysis of Spectra

14.25 Determine the structure of the compound corresponding to each of the following hydrogen NMR spectra.

(a) $C_3H_6Cl_2$

(b) C_4H_9Cl

The low-field resonance corresponds to the two hydrogen atoms at C-1. The nine line multiplet at 1.88 δ corresponds to the hydrogen atom at C-2, which is split by eight hydrogen atoms. However, the intensity of two of the "outer" lines is so small relative to the other seven lines that they are barely visible. The high-field doublet corresponds to the six hydrogen atoms in the two methyl groups.

14.26 Determine the structure of the compound corresponding to each of the following hydrogen NMR spectra.

(a) C_3H_8O

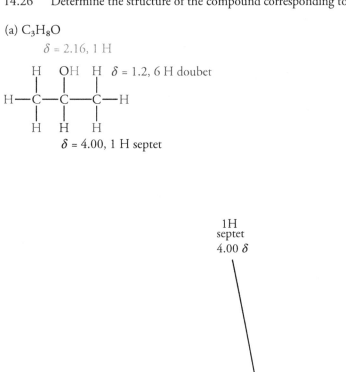

δ = 2.16, 1 H

H   OH   H   δ = 1.2, 6 H doubet

H—C—C—C—H

H   H   H

δ = 4.00, 1 H septet

1H
septet
4.00 δ

1H
2.16 δ

6H
1.2 δ

TMS

4   3   2   1   0

Chemical shift, δ (ppm)

(b) $C_3H_5Cl_3$

Cl   Cl   Cl   δ = 3.8, 4H doubet

H—C—C—C—H

H   H   H

δ = 4.4, 1 H quintet

J_{ab} = 5.278 Hz

4 H doubet ⟶

4.4 δ

1 H quintet

1.00

3.8 δ

4.00

TMS

5   4   3   2   1   0

Chemical shift, ppm (δ)

14.27   Determine the number of signals in the ^{13}C NMR spectrum of each of the following aromatic compounds.
(a) naphthalene   (b) 1,2,3-trimethylbenzene (c) 1,3,5-trimethylbenzene (d) 1,4-dimethylbenzene

Answers: (a) There are three sets of equivalent carbon atoms. One set is the C-1, C-4, C-5, and C-8 atoms. The second set is the C-2, C-3, C-6, and C-7 atoms. The third is the two carbon atoms at the points of fusion of the rings.
(b) There are six signals, because there are two sets of nonequivalent methyl groups and four sets of nonequivalent ring carbon atoms. The carbon atom of the methyl groups at C-1 and C-3 is equivalent and has a different resonance than the carbon atom of the C-2 methyl group. The C-1 and C-3 ring atoms are equivalent.
(c) The C-4 and C-5 ring atoms are equivalent. The C-2 and C-5 ring atoms each have their own resonance.
(d) There are three signals because there are a set of three equivalent methyl groups and two sets of nonequivalent ring carbon atoms. The C-1, C-3, and C-5 ring atoms are equivalent. The C-2, C-4, and C-6 ring atoms are equivalent.

14.28   Determine the number of signals in the ^{13}C NMR spectrum of each of the following ketones.

(a) CH_3—CH_2—$\overset{\overset{\displaystyle O}{\|}}{C}$—$CH_2$—$CH_3$ (b) CH_3—$\overset{\overset{\displaystyle O}{\|}}{C}$—$CH_2$—$CH_2$—$CH_3$

(c) CH_3—$\overset{\overset{\displaystyle O}{\|}}{C}$—$\underset{\underset{\displaystyle CH_3}{|}}{CH}$—$CH_3$

Answers: (a) There are three signals because there is one set of two equivalent methyl groups and one set of two equivalent methylene groups, as well as the carbonyl carbon atom.
(b) There are five signals because all carbon atoms are nonequivalent. The methyl carbon atoms are situated differently with respect to the carbonyl carbon atom, as are the two methylene carbon atoms.
(c) There are five signals. Only the two methyl carbon atoms bonded to the common carbon atom on the right of the structure are equivalent.

Dynamic Processes

14.29   Explain why the ^{13}C NMR spectrum of *trans*-1,4-dimethylcyclohexane contains only one methyl resonance.

Answer: Both methyl groups are equivalent, and they are in a diequatorial conformation in the most stable conformation of the *trans* isomer. A ring flip to give the diaxial conformation also has the methyl groups in equivalent environments.

14.30   The proton NMR of 2,2,3,3-tetrachlorobutane has a single resonance, a singlet, at 25 °C. Decreasing the temperature to −50 °C yields a spectrum with two singlets of unequal intensity. What structures account for the low-temperature spectrum?

Answer: At 25 °C, the rate of rotation about the carbon–carbon bond is rapid, and the NMR spectrum reflects the time average of two conformations. At the lower temperature, the spectrum is that of two conformations. Each conformation has one set of equivalent hydrogen atoms, and each of their resonances is a singlet.

14.31   The proton NMR of 1,1-dibromo-2,2-dichloroethane has a doublet of doublets with a coupling constant of 3.88 Hz. What does this coupling constant indicate about the predominant conformation?

Answer: The conformation with the hydrogen atoms *anti* to one another will have a large coupling constant, and the one with gauche hydrogen atoms will have a small coupling constant. Thus, the time average coupling constant is closer to the value for the gauche conformation because it is present in the larger amount.

15

ALCOHOLS: REACTIONS AND SYNTHESIS

KEYS TO THE CHAPTER

15.1 Overview Alcohol Reactions

Alcohols can react in several ways that differ in the number and type of bonds cleaved. These are:

1. Cleavage of the oxygen–hydrogen bond
2. Cleavage of the carbon–oxygen bond
3. Cleavage of the carbon–oxygen bond as well as the carbon–hydrogen bond at the carbon atom adjacent to the carbon atom bearing the hydroxyl group.
4. Cleavage of the oxygen–hydrogen bond as well as the carbon–hydrogen bond at the carbon atom bearing the hydroxyl group.

15.2 Converting Alcohols to Esters

Esters formed by the reaction of an alcohol with either an inorganic acid or a carboxylic acid have one feature in common. In both reactions, the oxygen atom bridging the alcohol and acid fragments is derived from the alcohol. The esterification reaction occurs by a substitution reaction mechanism in which a nucleophile attacks the carbonyl carbon atom to give a tetrahedral intermediate that subsequently ejects a leaving group. This process, known as nucleophilic acyl substitution, is another of the limited number of important mechanisms that dominate organic chemistry. We will meet this mechanism again in later chapters when we discuss the chemistry of acids and acid derivatives.

Hydroxide ion is a poor leaving group, and ester synthesis proceeds much more easily with a better leaving group such as the chloride ion. Thus, the formation of esters by reaction of alcohols with acid chlorides is a more favorable process than reaction of alcohols with the acids themselves.

15.3 Converting Alcohols to Haloalkanes

The conversion of an alcohol to a haloalkane can be done with thionyl chloride or phosphorus tribromide. In both cases, an intermediate is formed that converts the hydroxyl group into a better leaving group.

The reaction mechanism of the reaction of an alcohol with thionyl chloride occurs through a chlorosulfite intermediate. Subsequent displacement by chloride ion can occur with retention or inversion of configuration depending on the solvent. In pyridine as solvent, inversion occurs. In dioxane, retention occurs via a solvated ion pair, in which the chloride attacks by an internal return mechanism.

The reaction mechanism of the reaction of an alcohol with phosphorus tribromide occurs through a phosphite ester. The oxygen atom is now bound to phosphorus and as such is a better leaving group than the hydroxide ion. Nucleophilic substitution by bromide gives an alkyl bromide.

15.4 Oxidation of Alcohols

Alcohols are oxidized to carbonyl compounds by chromium(VI) compounds. The products depend on the structure of the substrate and the specific chromium reagent.

Alcohols are oxidized by the Jones reagent (CrO_3 in H_2SO_4). Primary alcohols are oxidized to aldehydes, which are further oxidized to carboxylic acids under the reaction conditions. The aldehyde usually cannot be isolated. Secondary alcohols are oxidized to ketones. Tertiary alcohols are not oxidized. Pyridinium chlorochromate (PCC), a reagent generated from CrO_3 and pyridine in CH_2Cl_2 as solvent, oxidizes alcohols only to carbonyl compounds. Secondary alcohols are oxidized to ketones, but primary alcohols are converted into aldehydes without being further oxidized to carboxylic acids.

Organic Chemistry Study Guide: Key Concepts, Problems, and Solutions http://dx.doi.org/10.1016/B978-0-12-801889-7.00015-7

15.5 Reactions of Vicinal Diols

Vicinal diols are oxidized by periodic acid, HIO_4, to give aldehydes if the two hydroxyl groups are secondary and ketones if they are tertiary. If one of the -OH groups is primary, the products are formic acid and an aldehyde.

15.6 Synthesis of Alcohols

Alcohols can be prepared by reduction of carbonyl compounds. Aldehydes yield primary alcohols; ketones yield secondary alcohols. The reduction of alcohols by hydrogen gas with a transition metal catalyst occurs more slowly than the reduction of alkenes, so both carbonyl groups and double bonds are reduced in compounds that contain both functional groups. Lithium aluminum hydride and sodium borohydride both reduce carbonyl compounds to alcohols without affecting carbon–carbon double bonds.

Alcohols with more complex hydrocarbon structures can be made by alkylation methods using a carbonyl compound and a Grignard reagent, which is carbanion-like reagent. This material is discussed in Section 15.10 and we will consider it again in Chapter 17.

15.7 Synthesis of Alcohols From Haloalkanes

The substitution of a halide ion by a hydroxide ion occurs in competition with an elimination reaction. We can circumvent this problem by using an oxygen-containing derivative that is a weak base as the nucleophile. The acetate ion is such a species. The product of the reaction of a haloalkane with acetate is an ester. Subsequent hydrolysis of the ester gives the alcohol and acetate ion. The hydrolysis occurs by nucleophilic acyl substitution.

15.8 Indirect Hydration Methods: Oxymercuration–Demercuration and Hydroboration-Oxidation

Direct hydration of an alkene to give an alcohol is limited by both rearrangement and the reversibility of the reaction. Two indirect methods for hydration are given in this section, each with different regioselectivity.

The first step in the conversion of an alkene to an alcohol by oxymercuration–demercuration resembles the addition of bromine in a mechanistic sense. Oxymercuration proceeds through a cyclic mercurinium ion. It does not rearrange because much of the positive charge is on the mercury atom. Subsequent attack by a nucleophile, water in this case, gives an oxymercury addition product. Treating the oxymercury adduct with sodium borohydride leads to the alcohol. The entire sequence, known as oxymercuration–demercuration, results in addition of water with the same regiospecificity as the direct hydration reaction. Thus, the process is a Markovnikov addition reaction.

Hydroboration–oxidation of an alkene gives an alcohol in a two-step sequence that is equivalent to an anti-Markovnikov addition of water. Hydroboration occurs by the addition of an H—B bond across the double bond by a cyclic four-center mechanism. The hydrogen atom that adds is not a proton but has the character of a hydride anion, and the boron is the "positive" part of the reagent. Thus, in a sense, the boron adds to the same carbon atom that adds a proton in the addition of reagents such as HBr.

The stereochemistry of the addition of an H—B bond across the double bond is *syn*. The subsequent replacement of the boron atom by oxygen using basic hydrogen peroxide occurs with retention of configuration.

15.9 Reduction of Carbonyl Compounds with Metal Hydrides

Both lithium aluminum hydride and sodium borohydride can reduce carbonyl compounds to alcohols without affecting carbon–carbon double bonds.

The mechanism of reduction of carbonyl compounds with hydride reagents occurs by attack of a nucleophilic hydride anion whose source is either borohydride or aluminum hydride on the carbonyl carbon. Lithium aluminum hydride must be used in an aprotic solvent such as ether to avoid its reactions with acidic hydrogens. Thus, this reagent shouldn't be used for reduction of carbonyl compounds that have hydroxyl groups or other acidic protons. Sodium borohydride can be used in protic solvents such as alcohols, and in fact, it is necessary to do so. The proton of the alcohol solvent is transferred to the oxygen atom, giving the alcohol product.

Lithium aluminum hydride and sodium borohydride have different reactivities: Lithium aluminum hydride reduces esters to alcohols; sodium borohydride does not.

15.10 Grignard Reagent

The Grignard reagent is a highly reactive organomagnesium compound formed by reacting a haloalkane with magnesium in an ether solvent. The carbon atom of a Grignard reagent has a partial negative charge. The Grignard reagent is a versatile material that can be used to form new carbon–carbon bonds. It acts as a nucleophile and attacks the carbonyl carbon atom to give an alkoxide which forms a salt with $(MgBr)^+$. Hydrolysis of the magnesium bromide salt gives the alcohol.

To determine how to combine two molecules to give an alcohol in a Grignard synthesis, examine the substituents bonded to the carbon bearing the hydroxyl group. One component of the new compound can come from the Grignard reagent. The other component must have been present in a carbonyl compound.

1. Primary alcohols are made from a Grignard reagent and formaldehyde.
2. Secondary alcohols can be made by reacting a Grignard reagent with an aldehyde.
3. Tertiary alcohols can be made by reacting a Grignard reagent with a ketone.

The use of the Grignard reagent is precluded if there is an acidic hydrogen in the substrate selected to react with the Grignard reagent. This acidic hydrogen destroys the Grignard reagent before it adds to the carbonyl group. If the substrate has a hydroxyl group, it can be protected by forming a trimethylsilyl ether.

15.11 Thiols

Thiols contain a sulfhydryl group (—SH). Thiols, also called mercaptans, have significantly different physical and chemical properties than alcohols. Thiols are lower boiling compounds because the —SH group does not form intramolecular hydrogen bonds. Thiols are stronger acids than alcohols. Like alcohols, thiols can be synthesized by displacement of a halide ion from haloalkanes. The SH⁻ ion is an excellent nucleophile. Oxidation of thiols produces disulfide bonds rather than analogs of aldehydes and ketones.

Summary of Reactions

1. Formation of Esters

2. Synthesis of Alkyl Halides

$$CH_3(CH_2)_4CH_2OH + SOCl_2 \longrightarrow CH_3(CH_2)_4CH_2Cl + HCl(g) + SO_2(g)$$

$$3\,R\!-\!OH + PBr_3 \longrightarrow R\!-\!Br + H_3PO_3$$

3. Oxidation of Alcohols

$$CH_3(CH_2)_4CH_2OH \xrightarrow{\text{Jones reagent}} CH_3(CH_2)_4CH_2CO_2H$$

4. Pinacol Rearrangement

2,3-dimethyl-2,3-butanediol
(pinnacol)

5. Oxidative Cleavage of Vicinal Diols

2-methyl-1-phenyl-1,2-butanediol → benzaldehyde + 2-butanone

cis-1,2-cyclohexanediol → 6-oxohexanal

6. Synthesis of Alcohols from Haloalkanes

7. Synthesis of Alcohols From Alkenes

A. Oxymercuration–Demercuration

B. Hydroboration–Oxidation

$$\xrightarrow[\text{2. } H_2O_2 \text{ / } OH^-]{\text{1. } BH_3}$$

anti-Markovnikov product

$$\xrightarrow[\text{2. } H_2O_2/HO^-]{\text{1. } BH_3/THF}$$

8. Reduction of Carbonyl Compounds

A. Catalytic Hydrogenation

$$(CH_3)_3\text{—}CH_2\text{—}\overset{\overset{\displaystyle O}{\|}}{C}\text{—H} \xrightarrow{\text{Ni / } H_2} (CH_3)_3\text{—}CH_2\text{—}\overset{\overset{\displaystyle OH}{|}}{\underset{\underset{\displaystyle H}{|}}{C}}\text{—H}$$

3,3-dimethylbutanal   3,3-dimethylbutanol

$$\xrightarrow[\text{100 atm}]{\text{Ni / } H_2}$$

At high pressure, both the vinyl and the carbonyl groups are reduced.

B. Sodium Borohydride Reduction

$$\xrightarrow[\text{2. } H_3O^+]{\text{1. } NaBH_4 \text{ /ethanol}}$$

Only the ketone is reduced.

C. Lithium Aluminum Hydride Reduction

$$\xrightarrow[\text{2. } H_3O^+]{\text{1. } LiAlH_4 \text{ /ether}}$$

methylbutanoate   1-butanol

9. Synthesis of Alcohols Using Grignard Reagents

1. Mg(s) / ether
2. HCHO
3. H_3O^+

1. Mg(s) / ether
2. C_6H_5CHO
3. H_3O^+

1. Mg(s) / ether
2. CH_3COCH_3
3. H_3O^+

Solutions to End-of-Chapter Exercises

Formation of Esters

15.1 Write the structural formula of each of the following esters.
(a) ethyl sulfate (b) dimethyl phosphate (c) propyl nitrate (d) 2-propylmethanesulfonate

Answers:

(a) CH_3CH_2—O—S—O—H (b) CH_3—O—P—O—CH_3

ethyl sulfate

dimethyl phosphate

(c) $CH_3CH_2CH_2$—O—N—O: (d) CH_3—S—O—$CH(CH_3)_2$

propyl nitrate

2-propylmethanesulfonate

15.2 Write the structural formula of each of the following esters.
(a) trimethyl phosphate (b) dipropyl sulfate (c) 2-propyl nitrate (d) l-butyl-*p*-toluenesulfonate

Answers:

(a) CH_3—O—P—O—CH_3 (b) $CH_3CH_2CH_2$—O—S—O—$CH_2CH_2CH_3$

trimethyl phosphate

dipropyl sulfate

(c) $(CH_3)_2CH$—O—N—O: (d) CH_3——S—O—$(CH_2)_3CH_3$

2-propyl nitrate

1-butyl-*p*-toluenesulfonate

15.3 Oxalyl chloride is a diacid chloride having the following structure. Draw the structure of the related diacid. Draw the structure of the product from reaction of one equivalent of benzyl alcohol with oxalyl chloride. Draw the structure of the product from the reaction of two equivalents of methyl alcohol with oxalyl chloride.

Answers:

Cl—C—C—Cl (a) HO—C—C—OH (b) ——CH_2—C—C—Cl

oxalyl chloride   oxalic acid

(c) CH_3—O—C—C—O—CH_3

15.4 What acid is contained in the following diester? Write the structure of an acid chloride that could be used to synthesize the diester.

Answer: The compound is a dimethyl ester of carbonic acid, which is an unstable add. The acid chloride is phosgene.

CH_3—O—C—O—CH_3

C
Cl   Cl
phosgene

$\xrightarrow{CH_3OH}$

CH_3—O   O—CH_3

15.5 The following diol reacts with one equivalent of tosyl chloride to give a single ester in good yield. Write the structure of the ester. Explain why the reaction is regioselective.

Answer: The ester is formed with the secondary alcohol, because its oxygen atom is a better nucleophile than the oxygen atom of the tertiary alcohol, which is more sterically hindered. The tertiary hydroxyl group does not react because only one equivalent of tosyl chloride is used.

15.6 The following diol reacts with one equivalent of tosyl chloride to give a single ester in good yield. Write the structure of the ester. Explain why the reaction is regioselective.

Answer: The equatorial hydroxyl group in the A ring of the steroid is less sterically hindered than the axial hydroxyl group of the C ring. Esterification occurs at the oxygen of the equatorial hydroxyl group because it is a more effective nucleophile.

Reactivity of Esters

15.7 Are alkyl esters of trifluoromethanesulfonic acid expected to be more or less reactive in S_N1 reactions than alkyl esters of methanesulfonic acid?

methanesulfonic acid   trifluoromethanesulfonic acid

Answer: The leaving group of the trifluoromethanesulfonic acid is a weak conjugate base of the more acidic trifluoromethanesulfonic acid. Weak bases are better leaving groups. Thus, alkyl esters of trifluoromethanesulfonic acid are more reactive than those of methanesulfonic acid because trifluoromethanesulfonic acid is the stronger acid and its conjugate base is the weaker base.

15.8 Describe the expected reactivity of the following compounds in S_N2 reactions compared to methanesulfonate esters.

(a) $R-\ddot{O}-\underset{\underset{:\ddot{O}:}{|}}{\overset{\overset{:\ddot{O}:}{|}}{Cl}}-\ddot{O}:$ (b) $R-\ddot{O}-N=\ddot{O}:$

Answer: The leaving groups are the perchlorate and nitrite ions, respectively. Perchloric acid is a strong acid and nitrous acid is a weak acid. Methanesulfonic acid is a strong acid. Thus, the perchlorate ion is comparable to the methanesulfonate ion as a leaving group. The nitrite ion is a much poorer leaving group.

15.9 The relative reactivities of alkyl p-nitrobenzoates and alkyl p-nitrobenzenesulfonates in S_N2 reactions relative to the reactivity of alkyl chlorides are 10^{-5} and 10^5, respectively. Explain the difference in the relative rates of reaction of the two esters.

an alkyl p-nitrobenzoate   an alkyl p-nitrobenzenesulfonate

Answer: Benzenesulfonic acids are strong acids, and benzoic acids are weak acids. Thus, esters of benzenesulfonic acids are more reactive because the benzenesulfonate ion is a weaker base than the benzoate ion.

15.10 Predict whether p-bromobenzenesulfonate is a better or worse leaving group than p-toluenesulfonate.

p-toluenesulfonate p-bromobenzenesulfonate

Answer: Bromine is an electron withdrawing group and makes the p-bromobenzenesulfonate ion a weaker base than p-toluenesulfonate. Thus, the bromine-substituted ion is a better leaving group than the tosylate ion.

Reactions of Alcohols with Acid

15.11 Explain why (R)-2-butanol in aqueous acid gradually loses its optical activity.

(R)-2-butanol   (S)-2-butanol

Answer: Protonation of the oxygen atom gives an oxonium ion, and water of the solvent acts as a nucleophile in an S_N2 reaction. The result is net inversion of configuration. Eventually, total racemization will occur.

15.12   Explain why l-phenyl-2-propen-1-ol rearranges to an isomer in the presence of a catalytic amount of H_2SO_4.

Answer: Protonation of the oxygen atom gives an oxonium ion which loses water from C-1 to give a resonance-stabilized allyl carbocation. Reaction of water at the original C-3 atom gives an alcohol in which the double bond is conjugated with the aromatic ring.

15.13   *cis*-2-Buten-1-ol isomerizes to form a mixture containing an isomeric alcohol when treated with dilute sulfuric acid. Write the structure of this alcohol.

Answer: Protonation of the oxygen atom gives an oxonium ion, and water can leave at C-1 giving a resonance-stabilized allyl carbocation. Reaction of water at the original C-3 atom gives 1-buten-3-ol.

15.14   When (*S*)-4-methyl-1,4-hexanediol is heated with acid, optically inactive 2-ethyl-2-methyltetrahydrofuran results. Write a mechanism for the reaction that accounts for the formation of the product and its lack of optical activity.

2-ethyl-2-methyltetrahydrofuran

Answer: Protonation of the tertiary alcohol at the C-4 atom and loss of water gives a tertiary carbocation. This carbocation is achiral, and subsequent attack of the oxygen atom of the hydroxyl group at C-1 can occur at either side of the plane to give the observed product, which is racemic.

Formation of Alkyl Halides

15.15 Draw the structure of the product of reaction for each of the following compounds with PBr_3.

Answers:

(a)

(b) $HOCH_2CH_2CHCH_2CH_2OH$ $\xrightarrow{PBr_3}$ $BrCH_2CH_2CHCH_2CH_2Br$ (with CH_3 substituents)

(c)

(d)   cyclopentane—CH_2CH_2OH $\xrightarrow{PBr_3}$ cyclopentane—CH_2CH_2Br

15.16 Draw the structure of the product of the reaction for each of the following compounds with $SOCl_2$ and pyridine.

Answers:

(a)

(b) $(CH_3)_2CH$—⟨benzene⟩—CH_2OH $\xrightarrow[\text{pyridine}]{SOCl_2}$ $(CH_3)_2CH$—⟨benzene⟩—CH_2Cl

(c)

(d)

15.17 Both 2-methyl-2-buten-1-ol and 3-methyl-2-buten-1-ol are converted to chlorides using concentrated Hcl. Which compound reacts at the faster rate?

2-methyl-2-buten-1-ol

3-methyl-2-buten-1-ol

Answer: Both compounds give resonance-stabilized allyl carbocations. However, the carbocation derived from 3-methyl-2-buten-1-ol has its positive charge distributed between a primary and a tertiary center. This carbocation is more stable than the carbocation derived from 2-methyl-2-buten-1-ol, which has its positive charge distributed between a primary and a secondary center. Thus, 3-methyl-2-buten-1-ol reacts at a faster rate because the reaction has a lower-energy transition state that resembles the carbocation is formed.

15.18 3-Methyl-3-cyclopentenol reacts with aqueous HBr to yield a mixture of two isomeric bromo compounds. Draw the structures of the two products. Predict the major isomer, assuming the reaction is not reversible. How might the data be different if the products can equilibrate?

major   minor

Answer: The allylic carbocation has its positive charge distributed between a secondary center at the original C-1 atom and a tertiary center at C-3. Under conditions of kinetic control, the bromine atom attacks the more positive center and gives 3-bromo-3-methyl-cyclopentene. If the products can equilibrate, the compound with the more substituted double bond would be favored. That isomer is 1-methyl-3-bromo-cyclopentene.

15.19 The yields of alkyl bromides obtained by reaction of an alcohol with PBr₃ are reduced if some of the HBr formed escapes from the reaction. In what alternate product would the alkyl groups be found under these conditions?

Answer: If the HBr escapes, then the nucleophilic bromide ion required to displace the substituted phosphite ion as the leaving group is lost. Thus, the phosphite ester remains and will regenerate the original alcohol when the reaction mixture is treated with water.

15.20 The yields of alkyl bromides in the reaction of alcohols with PBr₃ are increased if HBr is bubbled into the reaction vessel after the PBr₃ and alcohol are mixed. Explain why.

Answer: The extra HBr provides an added source of nucleophilic bromide ion to displace the substituted phosphite ion as the leaving group.

15.21 How could *trans*-4-*tert*-butylcyclohexanol be converted into *trans*-4-chloro-1-*tert*-butylcyclohexane? How could *trans*-4-*tert*-butylcyclohexanol be converted into *cis*-4-chloro-1-*tert*-butylcyclohexane?

Answer: Use thionyl chloride in dioxane, which gives a substitution product with retention of configuration by an internal return mechanism, to prepare the *trans* isomer. Use thionyl chloride in pyridine to achieve inversion of configuration to prepare the *cis* isomer.

15.22 What is the product of the reaction of (*R*)-2-octanol with thionyl chloride in pyridine? What is the product in diethyl ether as solvent?

Answer: Thionyl chloride in pyridine gives inversion and yields the (*S*) isomer. In diethyl ether, as in dioxane, the (*R*) isomer results by retention of configuration via an internal return mechanism.

15.23   Draw the structure of the product of the following series of reactions. What product would result if the alcohol reacted with HCl?

15.24   Draw the structure of the product of the following series of reactions. What product would result if the alcohol reacted with HBr?

Answer: Formation of a methanesulfonate followed by reaction with the nucleophilic chloride ion in an aprotic solvent tends to give substitution at the original C-1 atom, as shown above. In HCl, a resonance-stabilized carbocation forms which can react with chloride ion at C-1 to give the product shown above, or a rearrangement can occur to give 1-bromobicyclo[3.1.1]heptane.

15.25   Sterically hindered alcohols react with phosphorus tribromide but tend to give large quantities of rearranged product. The product mixture obtained from 2,2-dimethyl-l-propanol (neopentyl alcohol) contains 63% 1-bromo-2,2-dimethylpropane, 26% 2-bromo-2-methylbutane, and 11% 2-bromo-3-methylbutane. Explain the origin of the products. Why are sterically hindered alcohols more prone to give rearranged products?

Answer: Sterically hindered alcohols are less likely to undergo S_N2 reactions. The competing S_N1 reaction gives a primary carbocation that can undergo a methide shift, which gives the 2-bromo-2-methylbutane product. Although the resulting carbocation is less stable, a hydride shift of the carbocation resulting from the methide shift gives a secondary carbocation that leads to the 2-bromo-3-methylbutane product.

15.26   Both 2-chloropentane and 3-chloropentane are converted to a mixture of the two compounds in a concentrated HCl solution containing zinc chloride. The ratio of the 2-chloro to the 3-chloro compound is 2:1. Write a mechanism that explains how the zinc chloride accounts for the equilibration. Why does the observed ratio occur?

Answer: The zinc chloride acts as a Lewis acid that abstract a chloride ion to form a carbocation.

$$CH_3-\overset{\overset{\displaystyle Cl}{|}}{\underset{\underset{\displaystyle H}{|}}{C}}-\overset{\overset{\displaystyle H}{|}}{\underset{\underset{\displaystyle H}{|}}{C}}-\overset{\overset{\displaystyle H}{|}}{\underset{\underset{\displaystyle H}{|}}{C}}-CH_3 \xrightleftharpoons{ZnCl_2} CH_3-\overset{+}{C}-\overset{\overset{\displaystyle H}{|}}{\underset{\underset{\displaystyle H}{|}}{C}}-\overset{\overset{\displaystyle H}{|}}{\underset{\underset{\displaystyle H}{|}}{C}}-CH_3 + ZnCl_3^-$$

Answer: Sterically hindered alcohols are less likely to undergo S_N2 reactions. The competing S_N1 reaction gives a primary carbocation that can undergo a methide shift, which gives the 2-bromo-2-methylbutane product. Although the resulting carbocation is less stable, a hydride shift of the carbocation resulting from the methide shift gives a secondary carbocation that leads to the 2-bromo-3-methylbutane product.

$$CH_3-\overset{+}{C}-\overset{H}{\underset{H}{C}}-\overset{H}{\underset{H}{C}}-CH_3 \rightleftharpoons CH_3-\overset{H}{\underset{H}{C}}-\overset{+}{\underset{H}{C}}-\overset{H}{\underset{H}{C}}-CH_3$$

I   II

$$CH_3-\overset{H}{\underset{H}{C}}-\overset{H}{\underset{H}{C}}-\overset{+}{\underset{H}{C}}-CH_3$$

III

Answer: A hydride shift interconverts secondary carbocations with positive charge at C-2 and C-3. Since there are two equivalent carbocations (I and III) at C-2, and only one at C-3 (II), the ratio of 2-bromobutane to 3-bromobutane is 2:1.

Oxidation of Alcohols

15.27 Both 1-octanol and 2-octanol react with aqueous basic potassium permanganate. The product of the reaction of 2-octanol is not soluble in aqueous base, but the product of reaction of 1-octanol is soluble. What are the products? Explain the difference in solubility.

Answer: The product from 2-octanol is a ketone, which is insoluble in water because it is a nonpolar compound (even though there is a carbonyl group). The product from 1-octanol is a carboxylic acid, which reacts with base to form a carboxylate ion which is soluble in water.

$$CH_3(CH_2)_5-\overset{\overset{\displaystyle OH}{|}}{\underset{\underset{\displaystyle H}{|}}{C}}-CH_3 \xrightarrow{KMnO_4} CH_3(CH_2)_5-\overset{\overset{\displaystyle O}{\|}}{C}-CH_3$$

$$CH_3(CH_2)_6-\overset{\overset{\displaystyle OH}{|}}{\underset{\underset{\displaystyle H}{|}}{C}}-H \xrightarrow{KMnO_4} CH_3(CH_2)_6-\overset{\overset{\displaystyle O}{\|}}{C}-OH$$

15.28 Draw the structure of the product of each of the following reactions.

(a) $CH_3CH_2CH_2\overset{}{-}C\equiv C\overset{}{-}CH_2\overset{OH}{\underset{|}{CH}}CH_3$ $\xrightarrow[\text{reagent}]{\text{Jones}}$ $CH_3CH_2CH_2\overset{}{-}C\equiv C\overset{}{-}CH_2\overset{O}{\overset{||}{C}}CH_3$

(b) $CH_2\!\!=\!\!CH\!-\!CH\!\!=\!\!CH\!-\!CH_2CH_2OH$

$\downarrow$ PCC

$CH_2\!\!=\!\!CH\!-\!CH\!\!=\!\!CH\!-\!CH_2\overset{O}{\overset{||}{C}}\!-\!H$

(c) $H_3CO\!-\!\!\!\bigcirc\!\!\!-\!CH_2OH$ $\xrightarrow{\text{PCC}}$ $H_3CO\!-\!\!\!\bigcirc\!\!\!-\!\overset{O}{\overset{||}{C}}\!-\!H$

(d) (structure) $-CH_2OH$ $\xrightarrow{\text{Jones reagent}}$ (structure) $\overset{O}{\overset{||}{C}}\!-\!H$

15.29 Write the product formed from the oxidation of each of the compounds in Exercise 15.15 using PCC.

(a) (o-methylbenzene ring)$-CH_2CH_2OH$ $\xrightarrow{\text{PCC}}$ (o-methylbenzene ring)$-CH_2\overset{O}{\overset{||}{C}}\!-\!H$

(b) $HOCH_2CH_2\overset{CH_3}{\underset{|}{CH}}CH_2CH_2OH$ $\xrightarrow{\text{PCC}}$ $H\!-\!\overset{O}{\overset{||}{C}}CH_2\overset{CH_3}{\underset{|}{CH}}CH_2\overset{O}{\overset{||}{C}}\!-\!H$

(c) (cyclohexane)$\overset{OH}{\underset{|}{CH}}CH_3$ $\xrightarrow{\text{PCC}}$ (cyclohexane)$\overset{O}{\overset{||}{C}}CH_3$

(d) (cyclopentane)$-CH_2CH_2OH$ $\xrightarrow{\text{PCC}}$ (cyclopentane)$-CH_2\overset{O}{\overset{||}{C}}\!-\!H$

258

15.30 Write the product formed from the oxidation of each of the compounds in Exercise 15.16 using the Jones reagent.

(a)

(b) $(CH_3)_2CH$—⬡—CH_2OH $\xrightarrow[\text{reagent}]{\text{Jones}}$ $(CH_3)_2CH$—⬡—$\overset{\displaystyle O}{\underset{}{\overset{\|}{C}}}$—$OH$

(c) $\xrightarrow[\text{reagent}]{\text{Jones}}$ tertiary alcohol, no reaction

(d)

15.31 Write the product formed from the oxidation of the sex attractant of the Mediterranean fruit fly by PCC.

16.32 Write the product formed from the mosquito repellent by PCC.

Answer: The slower rate for the compound with deuterium on C-2 indicates that the C—D bond is cleaved in the rate-determining step.

15.33   Consider the relative rates of oxidation of the following three compounds by chromium(VI). What do these data reveal about the rate-determining step of the reaction?

Compound	$CH_3CH(OH)CH_3$	$CH_3CD(OH)CH_3$	$CD_3CH(OH)CH_3$
Relative rate	1.0	0.16	1.0

15.34 The rate of oxidation of *endo*-bicyclo[2.2.1]heptan-2-ol is faster than the rate of oxidation of the *exo* isomer. What does this fact indicate about the rate-determining step for the reaction?

endo-bicyclo[2.2.1]hepan-2-ol

exo-bicyclo[2.2.1]hepan-2-ol

Answer: The rate-determining step is not the formation of the chromium ester because the more hindered *endo* alcohol reacts at the slower rate. The rate-determining step is cleavage of the C—H bond, which is *exo* in the *endo* alcohol, and is sterically more accessible than the endo C—H bond in the *exo* alcohol.

15.35 Write a mechanism of the oxidation of an alcohol by chromium(VI) that uses only an intramolecular process for the abstraction of the a hydrogen atom. Considering the size of the ring in the cyclic process, how likely is it that this process will occur?

Answer: The reaction occurs via a five-atom transition state in an intramolecular process, which is both strain free and highly probable.

15.36 Which of the two sites within the following structure will be oxidized at the faster rate when only one equivalent of a chromium(VI) oxidizing agent is available?

Answer: The rate-determining step is cleavage of the C—H bond, as seen in Exercise 15.33, above. That bond is equatorial for the alcohol of the C ring of the steroid. It is oxidized faster than the equatorial alcohol of the A ring which has an axial C—H bond and is much more sterically hindered.

Reactions of Vicinal Diols

15.37 Draw two possible structures of products formed by treating the following vicinal diol with sulfuric acid.

Answer: A pinacol-type rearrangement occurs. Protonation occurs at either of the two equivalent oxygen atoms and water readily leaves to give a tertiary carbocation. Migration of either a methyl or an ethyl group from the adjacent carbon atom gives a protonated carbonyl compound that then loses a proton to give a mixture of two products. The mechanism for formation of one of them, by a methide shift, is shown below.

Answer: A pinacol-type rearrangement occurs. Protonation occurs at either of the two equivalent oxygen atoms and water readily leaves to give a tertiary carbocation. Migration of an ethyl group by a 1,2-ethide shift gives another carbocation that then loses a proton to give the product shown below.

15.38 Only one product is formed by treating the following vicinal diol with sulfuric acid. Draw its structure. Why is it formed rather an isomeric product?

Answer: A pinacol-type rearrangement occurs. Protonation occurs at the hydroxyl group of the benzyl carbon atom bonded to the two phenyl groups, and water readily leaves to give a tertiary carbocation that is stabilized by the phenyl groups. Migration of a methylene group of the cyclopentane ring gives a hydroxy carbocation that then loses a proton.

15.39 Draw the structure of the product(s) of the reaction of each of the following compounds with periodic acid.

15.40 Which compound in the following pairs reacts fastest with periodic acid?

(a)

or

(b)

or

(c)

or

Answer: Compounds with two hydroxyl groups in a *syn* periplanar arrangement or a *cis* arrangement can form the cyclic iodate ester more easily and then react to cleave the carbon–carbon bond. In (a), (b), and (c), the first compound (on the left) reacts at the faster rate.

15.41 The reaction of oleic acid ($C_{18}H_{34}O_2$) with osmium tetraoxide followed by reaction with periodate yields the following two compounds. Draw the structure of oleic acid.

Answer: Osmium tetroxide reacts with a double bond to give a vicinal diol. Subsequent reaction with periodate cleaves the carbon–carbon bond that was originally a double bond. The compound has a double bond between C-9 and C-10 of an unsaturated carboxylic acid and could be either *cis* or *trans*. The natural product is *cis*.

15.42 A hydrocarbon of molecular formula C_9H_{14} is found in sandalwood oil. Reaction of the compound with osmium tetraoxide followed by reaction with period ate yields the following compound. Draw the structure of the hydrocarbon.

Answer: Osmium tetroxide reacts with a double bond to give a vicinal diol. Subsequent reaction with periodate cleaves the carbon–carbon bond that was originally a double bond. The product is a dicarbonyl compound, which indicates that the original double bond was contained in a ring that has been cleaved.

Synthesis of Alcohols from Alkyl Halides

15.43   Which compound of each of the following pairs will react with ethanoate ion at the faster rate?
   (a) 1-iodohexane or l-bromohexane
   (b) l-bromo-l-phenylethane or l-bromo-2-phenylethane
   (c) l-bromo-2,2-dimethylpropane or 1-bromopentane

Answer: (a) Iodide ion is a better leaving group than bromide ion. Thus, 1-iodohexane reacts faster than 1-bromohexane with the same nucleophile.
   (b) A secondary benzylic carbocation results from 1-bromo-l-phenylethane which reacts faster than l-bromo-2-phenylethane, which is a primary halogen compound.
   (c) Both compounds are primary but 1-bromo-2,2-dimethylpropane is sterically hindered and reacts via an S_N2 mechanism at a much lower rate.

15.44   Would DMF or ethanol be the better solvent for the displacement of halide ions from alkyl halides by ethanoate ion?

Answer: DMF would be a better solvent because it is aprotic and does not decrease the nucleophilicity of the ethanoate ion as does the ethanol, which forms hydrogen bonds to the nucleophile.

15.45   Attempted synthesis of bicyclo[2.2.2]octan-l-ol by reaction of ethanoate with l-bromobicyclo[2.2.2]octane fails. Why?

1-bromobicyclo[2.2.2]octane

Answer: Reaction by an S_N2 mechanism is impossible because the nucleophilic ethanoate ion cannot approach from the back side of the C—Br bond. Although the compound is a tertiary halide, the carbocation required for the S_N1 mechanism cannot form because it cannot be planar due to restriction of the bicyclic ring. As a result, the substitution reaction cannot occur.

15.46   Predict the stereochemistry of the alcohols obtained by the reaction of *cis*-1-bromo-4-*tert*-butylcyclohexane with ethanoate followed by hydrolysis under basic conditions.

Answer: Displacement of the axial bromo group by attack of the ethanoate occurs with inversion of configuration to give the *trans* ester which has an equatorial C—O bond. Hydrolysis of the ester occurs with nucleophilic attack at the carbonyl carbon atom and releases the alkoxy portion of the ester. The product is the *trans*-4-*tert*-butylcyclohexanol.

(CH₃)₃C — [structure with Br, H, H, H]	1. CH₃CH₂O⁻ / ethanol → 1. H₂O /OH⁻
cis-1-bromo-4-*tert*-butylcyclohexane	(CH₃)₃C — [structure with H, OH, H, H] *trans*-4-*tert*-butylcyclohexanol

Hydration of Alkenes

15.47 Which of the isomeric $C_4H_{10}O$ alcohols can be produced by an acid-catalyzed hydration reaction of an alkene?

Answer: 2-Butanol can be prepared by acid-catalyzed hydration of 1-butene or either of the isomeric 2-butenes. 2-Methyl-2-propanol can be prepared from 2-methyl-1-propene. 1-Butanol cannot be prepared by acid-catalyzed hydration of an alkene.

(a) $CH_2{=}CH{-}CH_2{-}CH_3 \xrightarrow[H_2O]{H^+}$

$$CH_3{-}\overset{\overset{\displaystyle OH}{|}}{CH}{-}CH_2{-}CH_3$$

2-butanol

(b) trans-2-butene or cis-2-butene $\xrightarrow[H_2O]{H^+}$ 2-butanol

(c) 2-methyl-1-propene $\xrightarrow[H_2O]{H^+}$

$$CH_3{-}\underset{\underset{\displaystyle CH_3}{|}}{\overset{\overset{\displaystyle CH_3}{|}}{C}}{-}OH$$

2-methyl-2-propanol

15.48 The acid-catalyzed hydration of 3,3-dimethyl-1-pentene gives a mixture of two tertiary alcohols. Draw the structures and write a mechanism for their formation.

Answer: Addition of a proton to C-1 gives a secondary carbocation with an adjacent quaternary center. A shift of either a methyl or ethyl group can occur to give a tertiary carbocation. The products are 2,3-dimethyl-3-pentanol and 2,3-dimethyl-2-pentanol.

15.49 The acid-catalyzed hydration of 4-*tert*-butylcyclohexene gives a mixture of four secondary alcohols. Draw the structures and write a mechanism for their formation.

Answer: Addition of a proton can occur at either C-1 or C-2 to give secondary carbocations.

4-*tert*-butylcyclohexene

Addition of water then occurs at either the C-2 or C-1 center of the original compound. Attack from either side of each carbocation generates a mixture of *cis* and *trans* isomers. The products are *cis*- and *tert*-4-*tert*-butylcyclohexanol, as well as *cis*- and *trans*-3-*tert*-butylcyclohexanol.

trans-3-*tert*-butylcyclohexanol

trans-4-*tert*-butylcyclohexanol

Reduction of Carbonyl Compounds with Metal Hydrides

15.50 What is the product when each of the following reacts with lithium aluminum hydride?

(a)
1. LiAlH$_4$ / ether
2. H$_3$O$^+$

(b)
1. LiAlH$_4$ / ether
2. H$_3$O$^+$

(c)
1. LiAlH$_4$ / ether
2. H$_3$O$^+$

15.51 What is the product when each of the following reacts with sodium borohydride?

(a) CH_3CHCHO (CH_3) $\xrightarrow{NaBH_4 \text{ / ethanol}}$ CH_3CHCH_2OH (CH_3)

(b) CH_3CCH_2CHO (with CH_3 and CH_3 on central C) $\xrightarrow{NaBH_4 \text{ / ethanol}}$ $CH_3CCH_2CH_2OH$ (with CH_3 and CH_3)

(c) $CH_3C-C-CH_3$ (with O and CH_3, CH_3) $\xrightarrow{NaBH_4 \text{ / ethanol}}$ $CH_3C-C-CH_3$ (with OH, CH_3, H, CH_3)

15.52 The reduction of each of the following compounds by lithium aluminum hydride yields two products. Explain why.

(a)

(b)

Answer: Attack of hydride ion can occur from either side of the plane of the carbonyl group although not with equal probability. A mixture of geometric isomers occurs in both (a) and (b).

15.53 The reduction of each of the following compounds by sodium borohydride yields only one product. Explain why.

(a)

(b)

Answer: Each compound is symmetrical with respect to the plane of the carbonyl group. Only one alcohol can result from reduction, which occurs by attack of a hydride ion on the carbonyl carbon atom.

15.54 Assuming that steric factors control the reduction by sodium borohydride, what stereoisomer should predominate for the reduction of each of the following compounds?

(a) $\xrightarrow{\substack{NaBH_4 \\ \text{ethanol}}}$

(b) $\xrightarrow{\substack{NaBH_4 \\ \text{ethanol}}}$

Answer: Attack of the hydride ion should occur at the more sterically accessible position from "below" the average planes of the rings, which avoids the axial methyl groups. In the product from (a), the hydroxyl group is equatorial, and in the product from (b), it is axial.

266

15.55 What is the product of the reaction of each of the following compounds with sodium borohydride and also with lithium aluminum hydride?

(a)

1. LiAlH$_4$ / ether
2. H$_3$O$^+$

The ketone, the aldehyde, and the ester are reduced.

NaBH$_4$ / ethanol

Only the ketone and the aldehyde are reduced.

(b)

1. LiAlH$_4$ / ether
2. H$_3$O$^+$

The ketone and the ester are reduced.

NaBH$_4$ / ethanol

Only the ketone is reduced.

Oxymercuration–Demercuration

15.56 Name the final product of oxymercuration–demercuration of each of the following compounds.

(a)
$$\underset{\substack{H}}{\overset{\substack{H}}{C}} = \underset{\substack{CH_3}}{\overset{\substack{CH_3}}{C}} \quad \xrightarrow[\text{2. NaBH}_4]{\text{1. HgOAc}_2 / \text{H}_2\text{O}} \quad \underset{\substack{H}}{\overset{\substack{H}}{H-C}} - \underset{\substack{CH_3}}{\overset{\substack{OH}}{C}} - CH_3$$

2-methyl-2-propanol

(b)
$$\underset{\substack{H_3C}}{\overset{\substack{H_3C}}{C}} = \underset{\substack{CH_3}}{\overset{\substack{CH_3}}{C}} \quad \xrightarrow[\text{2. NaBH}_4]{\text{1. HgOAc}_2 / \text{H}_2\text{O}} \quad \underset{\substack{CH_3}}{\overset{\substack{CH_3}}{H-C}} - \underset{\substack{CH_3}}{\overset{\substack{OH}}{C}} - CH_3$$

2,3-dimethyl-2-butanol

(c)
$$\underset{\substack{H}}{\overset{\substack{H}}{C}} = \underset{\substack{H}}{\overset{\substack{CH_2CH_3}}{C}} \quad \xrightarrow[\text{2. NaBH}_4]{\text{1. HgOAc}_2 / \text{H}_2\text{O}} \quad \underset{\substack{H}}{\overset{\substack{H}}{H-C}} - \underset{\substack{H}}{\overset{\substack{OH}}{C}} - CH_2CH_3$$

2-butanol

(d)
$$\underset{\substack{H}}{\overset{\substack{H}}{C}} = \underset{\substack{H}}{\overset{\substack{CH_3}}{C}} \quad \xrightarrow[\text{2. NaBH}_4]{\text{1. HgOAc}_2 / \text{H}_2\text{O}} \quad \underset{\substack{H}}{\overset{\substack{H}}{H-C}} - \underset{\substack{H}}{\overset{\substack{OH}}{C}} - CH_3$$

2-propanol

15.57 Draw the structure of the oxymercuration–demercuration product of each of the following compounds.

(a)
$$\xrightarrow[\text{2. NaBH}_4]{\text{1. HgOAc}_2 / \text{H}_2\text{O}}$$

(b)
$$\xrightarrow[\text{2. NaBH}_4]{\text{1. HgOAc}_2 / \text{H}_2\text{O}}$$

(c)
$$\xrightarrow[\text{2. NaBH}_4]{\text{1. HgOAc}_2 / \text{H}_2\text{O}}$$

268

15.58 Draw the structure of the oxymercuration–demercuration product of each of the following compounds.

(a)

$C_6H_5CH_2CH=CH_2$

1. HgOAc$_2$ / H$_2$O
2. NaBH$_4$

$C_6H_5CH_2CH-CH_3$ with OH

(b)

1. HgOAc$_2$ / H$_2$O
2. NaBH$_4$

(c)

1. HgOAc$_2$ / H$_2$O
2. NaBH$_4$

15.59 How many products should be formed in the oxymercuration–demercuration of 4-*tert*-butylcyclohexene?

+ *cis* isomer

trans-3-*tert*-butylcyclohexanol

+ *cis* isomer

trans-4-*tert*-butylcyclohexanol

Hydroboration–Oxidation

15.60 Draw the final product of hydroboration–oxidation of each of the compounds in Exercise 15.56.

(a)

1. B$_2$H$_6$
2. H$_2$O$_2$ / HO$^-$

2-methyl-1-propanol

(b)

1. B$_2$H$_6$
2. H$_2$O$_2$ / HO$^-$

2,3-dimethyl-2-butanol

(c)

1. B$_2$H$_6$
2. H$_2$O$_2$ / HO$^-$

1-butanol

(d)

1. B$_2$H$_6$
2. H$_2$O$_2$ / HO$^-$

1-propanol

15.61 Draw the final product of hydroboration–oxidation of each of the compounds in Exercise 15.57.

(a)

$$\text{1. B}_2\text{H}_6 \quad \text{2. H}_2\text{O}_2 / \text{HO}^-$$

(b)

$$\text{1. B}_2\text{H}_6 \quad \text{2. H}_2\text{O}_2 / \text{HO}^-$$

(c)

$$\text{1. B}_2\text{H}_6 \quad \text{2. H}_2\text{O}_2 / \text{HO}^-$$

15.62 Draw the final product of hydroboration–oxidation of each of the compounds in Exercise 15.58.

(a)

$$\text{1. B}_2\text{H}_6 \quad \text{2. H}_2\text{O}_2 / \text{HO}^-$$

(b)

$$\text{1. B}_2\text{H}_6 \quad \text{2. H}_2\text{O}_2 / \text{HO}^-$$

(c)

$$\text{1. B}_2\text{H}_6 \quad \text{2. H}_2\text{O}_2 / \text{HO}^-$$

15.63 Draw the structure of the hyroboration–oxidation product of the following bicyclic hydrocarbon.

$$\text{1. B}_2\text{H}_6 \quad \text{2. H}_2\text{O}_2 / \text{HO}^-$$

15.64 Draw the structure of the hyroboration–oxidation product of the following bicyclic hydrocarbon.

$$\text{1. B}_2\text{H}_6 \quad \text{2. H}_2\text{O}_2 / \text{OH}^-$$

15.65   Can the following compound be synthesized by hydroboration–oxidation from 1-propylenecyclopentene?

Answer: No, because the indicated, tertiary alcohol requires Markovnikov addition of water, which is opposite that obtained from the hydroboration–oxidation method.

Grignard Reactions

15.66 What carbonyl compound and Grignard reagent are required to produce each of the following compounds?

(a) $CH_3CH_2\overset{OH}{\underset{|}{C}}HCH_2CH_3$ (b) $CH_3\overset{OH}{\underset{|}{C}}H\overset{}{\underset{|}{C}}HCH_3$ (c) $CH_3\overset{CH_3}{\underset{|}{C}}H\overset{}{\underset{|}{C}}HCH_2CH_3$ (d) $CH_3\overset{CH_3}{\underset{|}{C}}HCH_2\overset{OH}{\underset{|}{C}}HCH_2CH_3$

Answers: (a) Use the Grignard reagent of bromoethane and propanal.
   (b) Use the Grignard reagent of iodomethane and 2-methylpropanal or the Grignard reagent of 2-bromopropane and ethanal.
   (c) Use the Grignard reagent of bromoethane and 2-methylpropanal or the Grignard reagent of 2-bromopropane and propanal.
   (d) Use the Grignard reagent of bromoethane and 3-methylbutanal or the Grignard reagent of l-bromo-2-methylpropane and propanal.

15.67 What carbonyl compound and Grignard reagent are required to produce each of the following compounds?

Answers: (a) Use the Grignard reagent of bromocyclopentane and formaldehyde.
   (b) Use the Grignard reagent of iodomethane and 4,4-dimethylcyclohexanone.
   (c) Use the Grignard reagent of bromoethane and benzaldehyde or the Grignard reagent of bromobenzene and propanal.
   (d) Use the Grignard reagent of bromocyclohexane and 2-butanone or the Grignard reagent of iodomethane and 1-cyclohexyl-l-propanone or the Grignard reagent of iodoethane and 1-cyclohexyl-1-ethanone.

15.68   What carbonyl compound and Grignard reagent are required to produce the following compound?

Answers:

15.69   Outline all possible reaction sequences required to prepare Mestranol, a mild sleep-inducing agent, using the Grignard synthesis.

Answer: The ethynyl Grignard reagent adds to the carbonyl carbon atom from the least hindered side of the plane of the carbonyl group. Attack from the "top" of the ring is more hindered due to the axial methyl group at the ring juncture.

15.70 What carbonyl compound and Grignard reagent are required to produce each of the following compounds?

Answers: (a) Use the Grignard reagent of bromocyclopentane and formaldehyde.
           (b) Use the Grignard reagent of iodomethane and 4,4-dimethylcyclohexanone.
           (c) Use the Grignard reagent of bromoethane and benzaldehyde or the Grignard reagent of bromobenzene and propanal.
           (d) Use the Grignard reagent of bromocyclohexane and 2-butanone or the Grignard reagent of iodomethane and 1-cyclohexyl-l-propanone or the Grignard reagent of iodoethane and 1-cyclohexyl-1-ethanone.

15.71   Outline how each of the following alcohols could be made from the indicated starting material and other necessary compounds using the Grignard synthesis.
         (a) 2-cyclopentyl-2-propanol starting from bromocyclopentane
         (b) 1-cyclopentyl-1-ethanol starting from ethanal (CH$_3$CHO)
         (c) 1-nonanol starting from 1-bromooctane.
         (d) 3-heptanol starting from pentanal CH$_3$(CH$_2$)$_3$CHO

Answers: (a) Prepare the Grignard reagent of bromocyclopentane and add it to propanone, followed by hydrolysis.
           (b) Prepare the Grignard reagent of bromocyclopentane and add it to ethanal, followed by hydrolysis.
           (c) Prepare the Grignard reagent of 1-bromooctane and add it to formaldehyde, followed by hydrolysis.
           (d) Prepare the Grignard reagent of bromoethane and add it to pentanal, followed by hydrolysis.

15.72   Using bromobenzene, outline how each of the following alcohols could be made using the Grignard synthesis.
         (a) 1-phenylcyclopentanol
         (b) 3-phenyl-3-hexanol
         (c) 1-phenyl-l-octanol
         (d) benzyl alcohol

Answers: (a) Prepare the Grignard reagent of bromobenzene and add it to cyclopentanone, followed by hydrolysis.
           (b) Prepare the Grignard reagent of bromobenzene and add it to 3-hexanone, followed by hydrolysis.
           (c) Prepare the Grignard reagent of bromobenzene and add it to octanol, followed by hydrolysis.
           (d) Prepare the Grignard reagent of bromobenzene and add it to formaldehyde, followed by hydrolysis.

15.73 What carbonyl compound and Grignard reagent are required to produce each of the following compound?

$$CH_3CH_2-\overset{\overset{\displaystyle OH}{|}}{\underset{\underset{\displaystyle CH_3}{|}}{C}}-C\equiv CH$$

Answers: There are three possible combinations of Grignard reagent and a carbonyl compound.

(1) Prepare the Grignard reagent of an ethynyl group by reacting acetylene with a Grignard reagent such as methyl magnesium bromide. Then react it with 2-butanone, followed by hydrolysis.

(2) Prepare the Grignard reagent of iodoethane and react with it with S-butyn-2-one, followed by hydrolysis.

(3) Prepare the Grignard reagent of iodomethane and react with it with l-pentyn-3-one, followed by hydrolysis.

Synthetic Sequences

15.74 Write the structure of the final product of each of the following sequences of reactions.

(a) [cyclopentane]—$CH_2CH=CH_2$ → 1.$Hg(OAc)_2$ / H_2O / THF; 2. $NaBH_4$ → CrO_3 / H_2SO_4

(b) [benzene ring]—$CH=CH_2$ → 1. B_2H_6; 2. H_2O / OH^- → PCC

(c) [cyclobutane]—$CH_2CH=CH_2$ → 1. B_2H_6; 2. H_2O / OH^- → CrO_3 / H_2SO_4

Answers:

(a) [cyclopentane]—$CH_2\overset{\overset{\displaystyle O}{\|}}{C}-CH_3$

(b) [benzene ring]—$CH_2-\overset{\overset{\displaystyle O}{\|}}{C}-H$

(c) [cyclobutane]—$CH_2CH_2-\overset{\overset{\displaystyle O}{\|}}{C}-OH$

15.75 Write the structure of the final product of each of the following sequences of reactions.

(a)

(b)

(c)

15.76 Outline the steps required to convert testosterone into the indicated steroid structure.

Answer: Reduce the carbon–carbon double bond using hydrogen and a platinum catalyst. Then oxidize the secondary alcohol using PCC or the Jones reagent.

15.77 Outline the steps required to convert testosterone into the indicated steroid structure.

Answer: Reduce the carbon–carbon double bond using hydrogen and a platinum catalyst. Then reduce the ketone with sodium borohydride.

15.78 Propose a sequence of reactions that could be used to convert cyclohexanone into l-methylcyclohexene.

Answer: React the ketone with the methyl Grignard reagent to obtain 1-methylcyclohexanol. Dehydration of the alcohol using an acid catalyst yields 1-methylcyclohexene as the major product. Methylenecyclohexane is the minor product.

15.79 Propose a sequence of reactions that could be used to accomplish the following conversion.

274

Answer: React the ketone with the ethyl Grignard reagent to obtain cyclohexanol with an ethyl group bonded to the carbon bearing the hydroxyl group. Dehydration of the alcohol using catalyst yields an alkene which, when hydrogenated using hydrogen and a platinum catalyst, gives the desired product.

Sulfur Compounds

15.80 There are four isomeric compounds $C_4H_{10}S$ with an —SH group. Draw the structures of the compounds.

$$CH_3—CH_2—CH_2—CH_2—SH$$

1-butanethiol

$$CH_3—\overset{\overset{\displaystyle SH}{|}}{CH}—CH_2—CH_3$$

$$CH_3—\overset{\overset{\displaystyle}{|}}{\underset{\underset{\displaystyle CH_3}{|}}{CH}}—CH_2—SH$$

2-methyl-1-propanethiol

$$CH_3—\overset{\overset{\displaystyle SH}{|}}{\underset{\underset{\displaystyle CH_3}{|}}{C}}—CH_3$$

2-methyl-2-propanethiol

15.81 There are three isomeric compounds C_3H_8S. Draw their structures.

$$CH_3—CH_2—CH_2—SH$$
1-propanethiol

$$CH_3—\overset{\overset{\displaystyle SH}{|}}{CH}—CH_3$$
2-propanethiol

$$CH_3—CH_2—S—CH_3$$
ethylmethylsulfide

275

15.82   Draw the structure of each of the following compounds.
(a) 1-propanethiol
(b) 2-methyl-3-pentanethiol
(c) cyclopentanethiol

Answers: (a) CH_3—CH_2—CH_2—SH
1-propanethiol

(b) CH_3—$\overset{\overset{\displaystyle SH}{|}}{CH}$—$\overset{\overset{\displaystyle CH_3}{|}}{CH}$—$CH_2$—$CH_3$
2-methyl-3-pentanethiol

(c) —SH

cyclopentanethiol

15.83 Draw the structure of each of the following compounds.
(a) 2-propanethiol
(b) 2-methyl-1-propanethiol
(c) cyclobutanethiol

Answers:

(a) CH_3—$\overset{\overset{\displaystyle SH}{|}}{CH}$—$CH_3$ (b) CH_3—$\overset{\overset{\displaystyle CH_3}{|}}{CH}$—$CH_2$—SH   (c)

2-propanethiol   2-methyl-1-propanethiol   cyclobutanethiol

15.84 Adding sodium hydroxide to an aqueous solution of $CH_3CH_2CH_2SH$ eliminates the odor. Explain why.

Answer: The strong base reacts with the thiol to give a thiolate ($CH_3CH_2CH_2S^-$) that is nonvolatile.

15.85 The boiling points of ethanethiol and dimethyl sulfide are 35 and 37 °C, respectively. Why are the boiling points similar? What types of intermolecular forces are responsible for this similarity?

CH_3CH_2SH CH_3—S—CH_3
ethanethiol   dimethyl sulfide

Answer: The S—H group does not form hydrogen bonds. As a result, the intermolecular attractive forces of both the thioether and the thiol are similar. There are similar van der Waals forces in each of the compounds because they are isomers.

15.86 Indicate two methods to produce the scent marker of the red fox using a thiol as one of the reactants.

Answer: Use either of the following combinations of a thiol and a haloalkane in the presence of base to generate the thiolate anion.

16

ETHERS AND EPOXIDES

KEYS TO THE CHAPTER

The chemistry of ethers has substantially less variety than the chemistry of alcohols because several of the reactions of alcohols involve the O—H bond, namely dehydration and oxidation reactions are not possible for ethers. However, if comparable reactions are considered, such as substitution, then alcohols and ethers have similar reactivities. The reactions of epoxides are the result of ring strain, which leads to the formation of ring-opened products.

16.1 Structure of Ethers

Ethers contain an oxygen atom bonded to two alkyl groups, two aryl groups, or one of each. The geometry of ethers resembles that of alkanes, with the substitution of a methylene carbon atom by an sp²-hybridized oxygen atom. Conformations of ethers resemble those of alkanes. The two nonbonding electron pairs of the ether oxygen are directed to the corners of a tetrahedron.

16.2 Nomenclature of Ethers

The common names of simple ethers are based on the names of the alkyl or aryl groups bonded to the oxygen atom. The name results from listing the alkyl (or aryl) groups in alphabetical order and appending the name ether.

The IUPAC name is based on the longest carbon chain bonded to the oxygen atom. The smaller group bonded to the oxygen atom is named as an alkoxy group and is regarded as a substituent on the longer chain.

The three-, five-, and six-membered cyclic ethers have common names. Three-membered ring compounds are called epoxides of the corresponding alkene from which they may be synthesized. The common names of five- and six-membered ring compounds are called tetrahydrofurans and tetrahydropyrans, respectively. In the IUPAC system, each ring size has a specific name. The names for cyclic ethers having three-, four-, five-, and six-membered rings are oxirane, oxetane, oxolane, and oxane, respectively. The oxygen atom in each ring is assigned the number 1, and the ring is numbered in the direction that gives the lowest numbers to substituents.

16.3 Physical Properties of Ethers

Ethers have two polar C—O bonds and have substantial dipole moments. They are more polar than alkanes, but less polar than alcohols. Ethers do not have an O—H bond and cannot act as hydrogen bond donors. They can, however, serve as hydrogen bond acceptors, which makes the low-molecular-weight ethers soluble in water. Because they cannot form intermolecular hydrogen bonds, ethers have boiling points that are substantially lower than those of alcohols of comparable molecular weight. The boiling points of ethers are very close to the boiling points of alkanes of similar molecular weight. Ethers are excellent solvents for both nonpolar and nonpolar solutes. They are aprotic, so they are used as solvents for reagents that react with acidic protons, as is the case for the Grignard reagent. Polyethers readily dissolve polar compounds and hydrogen bond donors.

16.4 Polyether Antibiotics

Polyether antibiotics act by selectively bind to cations, typically sodium or potassium. They disrupt the ion balance in bacterial cells, and therefore kill bacteria.

16.5 Ether synthesis by Alkoxymercuration–Demercuration of Alkenes

Alkoxymercuration occurs by the same mechanism as oxymercuration. The only difference is the alkyl group bonded to the oxygen atom of an alcohol compared to the hydrogen atom bonded to the oxygen atom of water. The regiospecificity of the reaction corresponds to net Markovnikov addition.

Organic Chemistry Study Guide: Key Concepts, Problems, and Solutions http://dx.doi.org/10.1016/B978-0-12-801889-7.00016-9

16.6 The Williamson Ether Synthesis

The Williamson ether synthesis is an S_N2 reaction in which an alkoxide ion is a nucleophile that displaces a halide ion from an alkyl halide to give an ether. The reaction occurs with inversion of configuration at chiral centers and can be limited by possible competing elimination reactions.

Intramolecular Williamson ether synthesis occurs at rates that depend on the number of atoms in the transition state. The rates are affected by the probability of the alkoxide approaching the carbon atom bearing the halide ion, as well as the strain of the resulting ring compound. The observed order of reactivity in terms of the ring size is 3 > 5 > 6 > 4.

16.7 Reactions of Ethers

Both ethers and alcohols can act as bases because they have two lone pairs of electrons on the oxygen atom. They are both very weak bases and can only be protonated to form the conjugate acid, an oxonium ion, by a strong acid. The formation of an oxonium ion is analogous to the reaction of water with a strong acid to give the hydronium ion. Oxonium ions are intermediates in many reactions catalyzed by strong acids in the reactions of both ethers and alcohols.

Ethers react with strong acids such as HBr and HI to give cleavage products. The reaction proceeds by a two-step process in which first the oxygen atom is protonated and then the halide ion attacks one of the alkyl groups to displace an alcohol by an S_N2 process. The alkyl group attacked by the halide ion is controlled by the reactivity order 1° > 2° > 3°. The product alcohol can react with additional HX to give a second mole of a haloalkane.

16.8 Ethers as Protecting Groups

Protecting groups are provided by reagents that easily form derivatives of the functional group to be "protected," but can also be easily removed when required. A protecting group is used to render a functional group unreactive toward specific reagents that are required to transform a second functional group in the molecule.

Alcohols are protected by preparing silyl ethers of the general formula $R'—O—SiR_3$. The silyl ether is obtained by reacting an alcohol $R'—OH$ with a chlorosilane $Cl—SiR_3$. The silyl ether is cleaved by fluoride ion to liberate the alcohol after other transformation are completed.

16.9 Synthesis of Epoxides

Epoxidation is the reaction of an alkene with certain peroxyacids such as MCPBA or MMPP. The stereochemistry of the groups of the alkene is retained in the epoxide. Halohydrins undergo an intramolecular Williamson ether synthesis. The reaction occurs inversion of configuration at the center where the halide ion is displaced. Halohydrins are formed by addition of halogen to a double bond in aqueous solution.

16.10 Reactions of Epoxides

The ring strain of the three-membered epoxide ring results in ring-opening reactions in which one of the C—O bonds breaks. The regiochemistry of the ring opening and the stereochemistry of the product depend on whether the reaction occurs under basic or acidic conditions.

Nucleophiles displace an alkoxide ion, a reaction not observed in acyclic compounds. As in the case of other S_N2 reactions, the order of reactivity for a substrate with a nucleophile is primary > secondary > tertiary. The resulting product has the nucleophile and the hydroxyl group on adjacent carbon atoms. The nucleophile is on the less substituted carbon atom. Grignard reagents react with epoxides to give an alcohol containing two additional carbon atoms between the alkyl group and the carbon atom bearing the hydroxyl group in the product. Thus, for the reaction of a Grignard reagent, $RMgBr$, with ethylene oxide, the product is RCH_2CH_2OH.

Under acidic conditions, the oxygen atom is protonated to give a cyclic intermediate that resembles the bromonium and mercurinium ions we have encountered in other mechanisms. Subsequently, the nucleophilic reagent attacks the more substituted carbon atom because it has the greater partial positive charge. Thus, the nucleophile is bonded to the more substituted carbon atom in the product.

The stereochemistry of the ring opening of an epoxide is easily predicted. Nucleophilic attack of a reagent such as methoxide ion occurs with inversion of configuration at the least substituted carbon atom. The stereochemistry of the other carbon atom of the original epoxide is unchanged because no bond to that center is broken in the reaction. In a substituted epoxide, the ring opening reaction under acidic conditions occurs by inversion of configuration at the more substituted center where the nucleophile attacks. The configuration at the least substituted center is unchanged.

16.11 Reactions of Epoxides

The sulfur analogs of ethers are sulfides. They are named using alkylthio groups as substituents to the parent chain. Cyclic sulfides have common names.

Sulfides are prepared by reaction of a thiolate, the conjugate base of a thiol, with an alkyl halide. Because the thiolate ion is less basic and is a better nucleophile than an alkoxide. The sulfur analog of a Williamson synthesis has fewer complications due to elimination reactions.
Sulfide are oxidized to sulfoxides and then to sulfones.

16.12 Spectroscopy of Compounds with C—O and C—S Bonds

Infrared spectroscopy is usually not used to confirm the presence of either C—O or C—S bonds because the stretching vibrations occur in a region complicated by other absorptions. The O—H stretching vibration of alcohols is easily seen as a strong broad absorption in the spectrum in the 3400–3600 cm^{-1} region.

The chemical shift of hydrogen atoms bonded to the carbon atom bearing the oxygen atom of either alcohols or ethers occurs in the 3–4 δ region. The chemical shift of hydrogen atoms bonded to the carbon atom bearing the sulfur atom of either thiols or sulfides is less deshielded and occurs in the 2.5 δ region. Both O—H and S—H hydrogen atoms have variable chemical shifts depending on concentration.

The a carbon atom of an alcohol or an ether has a chemical shift that reflects the deshielding of the electronegative oxygen atom. The deshielding due to a sulfur atom is smaller.

Summary of Reactions

1. Synthesis of Ethers by Addition of an Alcohol to an Alkene

$$\text{cyclopentane}=CH_2 + CH_2OH \xrightarrow{H^+} \text{product with } OCH_3, CH_3$$

2. Synthesis of Ethers by Alkoxymercuration–Demercuration of Alkenes

$$\text{cyclopentene}-CH_3 + Hg(OAc)_2 \xrightarrow[\text{2. } NaBH_4 / OH^-]{\text{1. } CH_3CH_2OH} \text{product with } OCH_2CH_3, CH_3$$

3. Williamson Ether Synthesis

$$CH_3CH_2CH_2\overset{\overset{\displaystyle CH_3}{|}}{C}HCH_2OH \xrightarrow[\text{2. } CH_3CH_2I]{\text{1. } NaH} CH_3CH_2CH_2\overset{\overset{\displaystyle CH_3}{|}}{C}HCH_2O-CH_2CH_3$$

4. Cleavage of Ethers

5. Silyl Ethers as Protecting Groups

Add protecting group

$$CH_3CH_2\overset{\overset{\displaystyle CH_3}{|}}{C}H\overset{\underset{\displaystyle Br}{|}}{C}HCH_2OH \xrightarrow[\text{pyridine}]{(CH_3)_3SiCl} CH_3CH_2\overset{\overset{\displaystyle CH_3}{|}}{C}H\overset{\underset{\displaystyle Br}{|}}{C}HCH_2O-Si(CH_3)_3$$

Remove protecting group

$$CH_3CH_2\overset{\overset{\displaystyle CH_3}{|}}{C}H\overset{\underset{\displaystyle Br}{|}}{C}HCH_2O-Si(CH_3)_3 \xrightarrow[\text{THF}]{(C_4H_9)_4N^+F^-} CH_3CH_2\overset{\overset{\displaystyle CH_3}{|}}{C}H\overset{\underset{\displaystyle Br}{|}}{C}HCH_2OH + (CH_3)_3SiF$$

6. Synthesis of Epoxides

$$\xrightarrow[CH_2Cl_2]{MCPBA}$$

$$\xrightarrow[\text{ethanol}]{MMPP}$$

7. Ring Cleavage of Epoxides

Base catalyzed

(S)-2-methyloxirane → 1.CH₃O⁻ / CH₃OH, 2. H₃O⁺ → (S)-1-methoxy-2-propanol

$$\text{(S)-2-methyloxirane} \xrightarrow[\text{2. H}_3\text{O}^+]{\text{1. CH}_3\text{O}^- / \text{CH}_3\text{OH}} \text{(S)-1-methoxy-2-propanol}$$

Acid catalyzed

$$\text{(S)-2-methyloxirane} \xrightarrow[\text{H}^+]{\text{CH}_3\text{OH}} \text{(R)-2-methoxy-1-propanol (major product)} + \text{(S)-1-methoxy-2-propanol (minor product)}$$

8. Synthesis of Sulfides

$$\text{CH}_3\text{CH}_2\text{CH}_2\text{CHCH}_2\text{SH} \xrightarrow[\text{2. CH}_3\text{CH}_2\text{I}]{\text{1. NaOH}} \text{CH}_3\text{CH}_2\text{CH}_2\text{CHCH}_2\text{S—CH}_2\text{CH}_3$$

(with CH₃ substituent on the CH)

9. Oxidation of Sulfides

Sulfide to Sulfoxide

$$\text{CH}_3\text{CH}_2\text{CH}_2\text{CHCH}_2\text{S—CH}_3 \xrightarrow{\text{H}_2\text{O}_2} \text{CH}_3\text{CH}_2\text{CH}_2\text{CHCH}_2\overset{\text{O}}{\text{S}}\text{—CH}_3$$

Sulfoxide to Sulfone

$$\text{CH}_3\text{CH}_2\text{CH}_2\text{CHCH}_2\overset{\text{O}}{\text{S}}\text{—CH}_3 \xrightarrow{\text{CH}_3\text{CO}_3\text{H}} \text{CH}_3\text{CH}_2\text{CH}_2\text{CHCH}_2\overset{\text{O}}{\underset{\text{O}}{\text{S}}}\text{—CH}_3$$

Exercises

Ether Isomers

16.1   Draw the structures of the isomeric ethers with the following characteristics.
      (a)   molecular formula $C_4H_{10}O$
      (b)   a methyl ether with molecular formula $C_5H_{12}O$
      (c)   a saturated ether with the molecular formula C_3H_6O

Answers:

    (a) $CH_3CH_2CH_2\!\!-\!\!O\!\!-\!\!CH_3$ (b) $CH_3\!\!-\!\!\underset{\underset{\displaystyle CH_3}{|}}{\overset{\overset{\displaystyle CH_3}{|}}{C}}\!\!-\!\!O\!\!-\!\!CH_3$ (c)

16.2   Draw the structures of the following ethers.
      (a)   oxane and two methyl oxolane
      (b)   2-methoxypropene
      (c)   *cis*-2,3-dimethyloxetane

Answers:

   (a)

Ether Nomenclature

16.3   Give the common name of each of the following compounds.

Answers:
(a) dicyclopentyl ether
(b) phenyl propyl ether
(c) cyclopentyl propyl ether

16.4   Give the common name of each of the following compounds.

Answers:
(a) *tert*-butyl phenyl ether
(b) benzyl *tert*-butyl ether
(c) cyclohexyl propargyl ether

16.5   Assign the IUPAC name of each of the following compounds.

Answers:
(a) 2-methoxypentane
(b) 2-methoxy-4-methylpentane
(c) 2,4-diethoxyheptane

282

16.6   Assign the IUPAC name of each of the following compounds.

Answers:
(a) 1,1-dimethoxypentane
(b) 2,4-dimethoxypentane
(c) 3-ethoxyhexane

(a) $CH_3CH_2CH_2\overset{\overset{\displaystyle OCH_3}{|}}{C}H OCH_3$ (b) $CH_3\overset{\overset{\displaystyle OCH_3}{|}}{C}HCH_2\overset{\overset{\displaystyle OCH_3}{|}}{C}HCH_3$ (c) $CH_3CH_2CH_2\overset{\overset{\displaystyle OCH_2CH_3}{|}}{C}HCH_2CH_3$

16.7   Draw the structure of each of the following general anesthetics.
        (a)   1,1,1,3,3,3-hexafluoroisopropyl methyl ether (isoindoklon)
        (b)   2-chloro-1,1,2-trifluoro-1-(difluoromethoxy)ethane (enflurane)

Answers:

16.8   What is the common name of each of the following anesthetics?
        (a) $CH_2=CH-O-CH=CH_2$
        (b) $CF_3CHCl-O-CHF_2$

Answers:
(a) divinyl ether   (b) difluoromethyl 1-chloro-2,2,2-trifluoroethyl ether

16.9   Draw the structure of each of the following compounds.
        (a)   trans-4-methoxycyclohexanol
        (b)   3-ethoxy-1,1-dimethylcyclohexane
        (c)   12-crown-4

Answers:

16.10   Draw the structure of each of the following compounds.
        (a)   3-methoxyoxolane
        (b)   trans-2-chloro-l-methoxycyclobutane
        (c)   cis-2-ethoxy-3-methyloxirane
        (d)   15-crown-5

Answers:

Properties of Ethers

16.11 Explain why 1,4-dioxane is more soluble in water (it is miscible) than diethyl ether.

Answer: Diethyl ether is somewhat soluble in ether. The added oxygen atom of 1,4-dioxane increases the extent of hydrogen bonding with water, leading to increased solubility. The ratio of carbon to oxygen atoms decreases from 4:1 in diethyl ether to 2:1 in 1,4-dioxane.

16.12 Explain why *p*-ethylphenol is more soluble in water than ethoxybenzene.

Answer: Ethoxybenzene is only a hydrogen bond acceptor, but *p*-ethylphenol is both a hydrogen bond acceptor and a hydrogen bond donor.

16.13 The boiling points of dipropyl ether and diisopropyl ether are 91 and 68 °C, respectively. Explain why the boiling points of these isomeric ethers differ.

Answer: Diisopropyl ether has a more compact conformation and has smaller London forces than the cylindrical conformation of dipropyl ether.

16.14 The boiling points of 1-ethoxypropane and 1,2-dimethoxyethane are 64 and 83 °C, respectively. Explain why.

Answer: Although they have similar molecular weights, 1,2-dimethoxyethane has an additional oxygen atom and is more polar.

16.15 Explain why dipropyl ether is soluble in concentrated sulfuric acid, whereas heptane is insoluble.

Answer: The oxygen atom of dipropyl ether can accept a proton from sulfuric acid. The conjugate base is an alkoxoxonium ion and is soluble in the polar medium. Heptane cannot be protonated.

16.16 Explain why aluminum trichloride dissolves in tetrahydropyran, releasing heat.

Answer: Aluminum trichloride is a Lewis base. It forms a coordinate covalent bond with lone pair electrons on the oxygen atom of tetrahydrofuran. The process is exothermic.

16.17 Explain why some potassium compounds dissolve in 18-crown-6, but the related rubidium compounds do not.

Answer: The atomic radius of rubidium is larger than the atomic radius of potassium. 18-crown-6 has a cavity the right size to solvate potassium, which means that it would be too small for the rubidium cation.

16.18 Some sodium compounds dissolve in 15-crown-5. Would the related lithium compounds be more or less likely to be soluble in 18-crown-6 or 12-crown-4?

Answer: The atomic radius of lithium is smaller than the atomic radius of sodium. 15-Crown-5 has a cavity the right size to solvate the sodium cation, which means that a smaller ring such as 12-crown-4 would be suitable for the lithium cation.

16.19 Draw the stable conformation of 1,4-dioxane.

16.20 Explain why the most stable chair conformation of 1,3-dioxan-5-ol has an axial hydroxyl group, but the most stable chair conformation of 5-methoxy-1,3-dioxane has an equatorial methoxy group.

Answer: The methoxy group should occupy the equatorial position of 1,3-dioxane much like it does in cyclohexane. The hydroxyl group occupies the axial position of 1,3-dioxane because it forms hydrogen bonds with the ring oxygen atoms.

Synthesis of Ethers

16.21 Write a mechanism for the formation of 1,4-dioxane from ethylene glycol catalyzed by sulfuric acid.

16.22 Write a mechanism for the formation of 2,2-dimethyloxolane from 4-methyl-1,4-pentanediol and sulfuric acid. Which of the two oxygen atoms remains in the ether?

Answer: The oxygen atom bonded to C-1 remains in the ether.

16.23 Reaction of 1-hexene with mercuric acetate in methanol as solvent followed by reduction of the intermediate product with sodium borohydride yields 2-methoxyhexane. What is the structure of the intermediate product? How is it formed?

Answer: Electrophilic attack of $Hg(OAc)_2$ gives a mercurinium ion that then reacts with the nucleophilic methanol at the carbon atom with the larger positive charge, which is the more substituted carbon atom.

16.24 Reaction of 4-penten-l-ol with mercuric acetate followed by reduction with sodium borohydride yields 2-methyl-oxolane. Write the structure of the intermediate and explain why this ether forms.

Answer: Electrophilic attack of $Hg(OAc)_2$ gives a mercurinium ion that reacts with the nucleophilic oxygen atom of the alcohol at C-1. Attack occurs at the carbon atom with the larger positive charge, which is the more substituted carbon atom, C-4.

16.25 Reaction of 5-chloro-2-pentanol with sodium hydride yields 2-methyloxolane. Write the structure of the intermediate and explain why this ether forms.

Answer: An alkoxide ion, formed by reaction of the alcohol with the hydride ion, displaces chloride in an intramolecular S_N2 reaction.

16.26 Treatment of 3,4-dibromo-l-butanol with sodium hydroxide yields a cyclic ether. What is the structure of the ether? What alternate ether could form, and why is it not produced?

Answer: The alkoxide ion displaces a bromide ion from C-4, giving an oxolane. Displacement of a bromide ion from C-3 would give a more highly strained oxetane.

16.27 Which of the following compounds can be synthesized in good yield using the Williamson method? Explain why the method would fail for the remaining compounds.
(a) ethyl cyclopentyl ether   (b) 1-methyl-l-methoxycyclohexane   (c) tert-butyl cyclohexyl ether
(d) di-sec-butyl ether   (e) 2-methyl-3-phenoxyhexane

Answers:
(a) Reaction of the alkoxide of cyclopentanol with iodoethane gives a good yield because the S_N2 reaction occurs at a primary center.
(b) Although the alkoxide of 1-methylcyclohexanol is sterically hindered, the reaction of this nucleophile with iodomethane gives a good yield because the alkyl halide is unhindered, and it cannot undergo an elimination reaction.
(c) The reaction of the alkoxide of cyclohexanol with 2-bromo-2-methylpropane will give only elimination product. The reaction of tert-butoxide with bromocyclohexane may give some ether, but the major product will be cyclohexene, an elimination product.
(d) The reaction of the alkoxide of 2-butanol with 2-bromobutane gives largely elimination products.
(e) The reaction of the phenoxide with 3-bromo-2-methylhexane gives largely elimination products.

16.28 In each case, select a primary alkyl halide in order to avoid a competing elimination product. The combination of alkyl halide and the alcohol required to form the alkoxide ion is shown below.

(a) [structure] —CH$_2$CH$_2$Br + $^-$O— [structure] ⟶ [structure] —CH$_2$CH$_2$—O— [structure]

(b) $^-$O■ [structure] ■O$^-$ + 2 CH$_3$I ⟶ CH$_3$O■ [structure] ■OCH$_3$

(c) [structure] + BrCH$_2$CH$_2$CH$_2$CH$_3$ ⟶ [structure]

16.29 (R)-2-Octanol reacts with sodium hydride followed by treatment with iodoethane to give 2-ethoxyoctane. Based on the mechanism of this reaction, what is the configuration of the product?

Answer: The hydroxyl group at C-2 is converted to an alkoxide by NaH. The C—C bond of the alkoxide of 2-octanol is unaffected, so the configuration at the C-2 chiral center is unchanged in the subsequent S$_N$2 reaction with iodoethane.

16.30 (R)-2-Octanol with p-toluenesulfonyl chloride to give a tosylate. Subsequent reaction of the tosylate with ethanol gives 2-ethoxyoctane. Based on the mechanism of these reactions, what is the configuration of the tosylate.

Answer: The tosylate forms by reaction of the oxygen atom of the alcohol with the sulfur atom of p-toluenesulfonyl chloride. Hence, the tosylate has the same configuration as the original alcohol. The subsequent displacement reaction occurs at the carbon atom bearing the tosylate group. The C—C bond is broken in an S$_N$2 reaction when ethoxide ion displaces the tosyl group. Thus, the ether is formed with inversion of configuration.

16.31 The following compound reacts with sodium hydroxide to produce 1,4-dioxane. Write the sequence of reactions leading to this product.

$$Cl—CH_2CH_2—O—CH_2CH_2—Cl$$

[reaction scheme] ⟶ OH$^-$ ⟶ [structure] ⟶ OH$^-$ ⟶ [structure] ⟶ —Cl$^-$ ⟶ [structure]

Answer: Displacement of one halide ion by hydroxide ion gives an alcohol that can exchange a proton with hydroxide ion to give an alkoxide ion. This ion displaces chloride to give dioxane.

16.32   Determine the best choice of reactants to synthesize each of the following ethers using the Williamson method.

(a)

(b)

(c)

Answer: Each compound can be made by an intramolecular Williamson synthesis using a halogen substituted alcohol and a strong base such as sodium hydride. In both (a) and (b), two reactants can be considered. However, the isomer that has a primary halide is the best choice to preclude a competing elimination reaction.

16.33   Determine the best choice of reactants to synthesize each of the following ethers using the Williamson method.

dibucaine

Answer: React the conjugate base of the phenol with 1-bromobutane.

16.34　What reactants might be used to produce the antihistamine diphenylpyraline using the Williamson synthesis? What difficulties might be encountered?

diphenylpyraline

elimination

Answer: (a) The reagents will give the desired product. (b) These reagents give only an elimination reaction.

Synthetic Sequences

16.35　Draw the structure of the final product of each of the following reaction sequences.

(a)

$CH_2CH=CH_2$

1. Hg(OAc)$_2$　3. NaH
2. NaBH$_4$　4. CH$_3$I

OCH$_3$
CH$_2$CHCH$_3$

(b)

—CH=CH$_2$

1. B$_2$H$_6$　3. NaH
2. H$_2$O$_2$ / OH$^-$　4. CH$_3$CH$_2$Br

—CH$_2$CH$_2$OCH$_2$CH$_3$

(c)

—CH=CH$_2$

1. B$_2$H$_6$
2. H$_2$O$_2$ / OH$^-$
3. NaH
4. ethylene oxide

CH$_2$CH$_2$CH$_2$OCH$_2$CH$_2$OH

Answers: (a) Addition of mercuric acetate followed by reduction with sodium borohydride gives the product of indirect addition of water in a Markovnikov sense. Formation of the alkoxide using sodium hydride followed by displacement of iodide from methyl iodide is a Williamson synthesis of an ether.
(b) Addition of borane followed by oxidation with hydrogen peroxide gives the product of "indirect" addition of water in an anti-Markovnikov sense. Formation of the alkoxide using sodium hydride followed by displacement of bromide from ethyl bromide is a Williamson synthesis of an ether.
(c) Addition of borane followed by oxidation with hydrogen peroxide gives the product of "indirect" addition of water in an anti-Markovnikov sense. Formation of the alkoxide using sodium hydride followed by nucleophilic attack on ethylene oxide opens the ring to give an ether product that is also an alcohol.

16.36 Draw the structure of the final product of each of the following reaction sequences.

Answers: (a) Reduction with sodium borohydride gives a primary alcohol. Reaction of the alcohol with the active metal potassium gives an alkoxide that displaces iodide ion from methyl iodide in a Williamson ether synthesis.

(b) Reduction with lithium aluminum hydride gives a secondary alcohol. Reaction of the alcohol with the strong base sodium hydride gives an alkoxide that displaces iodide ion from ethyl iodide in a Williamson ether synthesis.

(c) Reduction with sodium borohydride gives a primary alcohol. Reaction of the alcohol with the active metal sodium gives an alkoxide that attacks the primary carbon atom of the oxirane to give an ether that is also an alcohol.

Reactions of Ethers

16.37 Write the structure of a compound with the given molecular formula that reacts with HI to yield the following iodo compound(s).

(a) $C_5H_{12}O_2$ yields a mixture of iodomethane, iodoethane, and 1,2-diiodoethane
(b) $C_5H_{12}O_2$ yields a mixture of iodomethane and 1,3-diiodopropane
(c) $C_5H_{10}O$ yields only 1,5-diiodopentane
(d) $C_4H_8O_2$ yields only 1,2-diiodoethane

Answers:

(a) CH_3—O—CH_2CH_2—O—CH_2CH_3 (b) CH_3O—$CH_2CH_2CH_2$—O—CH_3

16.38 Anisole can be cleaved by LiI in dimethylformamide to give iodomethane and phenoxide ion. What is the mechanism of the reaction? Explain why the reaction can occur without a source of acid.

Answer: Iodide ion in the polar, aprotic solvent dimethylformamide is an excellent nucleophile. This is an S_N2 reaction in which iodide attacks at the methyl group. Phenoxide ion, which is resonance stabilized, is a good leaving group. Phenol is a stronger acid than an alcohol, and therefore, phenoxide ion is a weaker base than an alkoxide ion. We recall weak bases are better leaving groups than strong bases.

16.39 Write the product of the reaction of the following bicyclic ether with HBr.

Answer: Protonation of the ether followed by an S_N2 displacement by bromide ion gives a bromo alcohol. The reaction occurs with inversion of configuration at the center attacked by the bromide ion.

16.40 A compound with the molecular formula $C_{10}H_{18}O$ reacts with HCl. What is the structure of the compound?

Answer: The reactant is a bicyclic ether. Protonation of the ether, followed by cleavage of either of the two C—O bond, occurs by an S_N1 mechanism because both centers are tertiary. Capture of the carbocation by chloride gives a chloro alcohol with the OH group on a tertiary carbon atom. Reaction of that alcohol with HCl gives the dichloro product.

Synthesis of Epoxides

16.41 Draw the structure of the product of the following compound with sodium hydroxide.

Answer: The alkoxide ion derived from the alcohol is *trans* to the C—Br bond, and a Williamson synthesis reaction occurs to give an epoxide.

16.42 The reaction of *trans*-2-chlorocyclohexanol with sodium hydroxide yields an epoxide, but the *cis* isomer does not. Explain this difference.

Answer: The alkoxide ion derived from the alcohol is *trans* to the C—Cl bond in a diaxial conformation, which is formed by a ring flip of the more stable diequatorial conformation. A Williamson synthesis reaction readily occurs to give an epoxide. In the *cis* isomer, a *trans* arrangement of the C—Cl and C—O bonds is not possible.

16.43 Write the structure of the bromohydrin formed by adding aqueous bromine to (*E*)-2-butene. What is the stereochemistry of the epoxide formed from this bromohydrin?

Answer: The *anti* addition reaction of bromine and water followed by loss of a proton gives a bromohydrin. Intramolecular displacement of bromide by an alkoxide in a base-catalyzed reaction gives an epoxide. The *trans* groups in the (*E*)-2-butene are also *trans* in the epoxide.

16.44 Write the structure of (2*S*,3*S*)-3-bromo-2-butanol. What is the stereochemistry of the epoxide formed from this bromohydrin?

Answer: The displacement of bromide by an alkoxide derived from the alcohol inverts the configuration at the C-3 center, so the product is (2*S*,3*R*), which is *meso*.

16.45 Two products can result from the epoxidation of the following bicycloalkene with MCPBA. Draw their structures.

Answer: The epoxide ring could form on the same side of the decalin ring as the methyl group or on the opposite side.

16.46 What alkene is required to synthesize disparlure, the sex attractant of the gypsy moth using MCPBA?

Answer: The alkyl groups of the epoxide are *cis*, so the alkyl groups in the alkene must also be *cis*.

16.47 Draw the structure of a bromo alcohol that could yield the following epoxide. Assign the configuration of each stereogenic center of this bromo alcohol.

Answer: The required compound is (2R,3S)-3-bromo-2-butanol.

16.48 Draw the structure of two possible bromo alcohols that could yield the following epoxide. Which one should give the higher yield? Assign the configuration of each stereogenic center of this bromo alcohol.

Answer: Compound I gives the higher yield because it reacts fastest in a displacement of a halide ion from a primary center.

Reactions of Epoxides

16.49 Each of the following compounds is a commercial product that is made from ethylene oxide. Propose a synthesis of each compound.
   - (a) ethyl cellosolve, CH_3CH_2—O—CH_2CH_2—OH
   - (b) ethyl carbitol, CH_3CH_2—O— CH_2CH_2—O—CH_2CH_2—OH
   - (c) diethanolamine, HO—CH_2CH_2—NH—CH_2CH_2—OH

Answers: (a) A ring opening reaction of ethylene oxide with ethanol under either acidic or basic conditions is required.
   - (b) React the product of (a) with a second mole of ethylene oxide in acid or base. The hydroxyl group of ethyl cellosolve is the nucleophile that opens the epoxide ring.
   - (c) First, a ring opening reaction of ethylene oxide with ammonia gives 2-aminoethanol. Reaction of this compound with ethylene oxide leads to ring opening as the result of nucleophilic attack of the nitrogen atom rather than the oxygen atom.

16.50   Divinyl ether is prepared by an elimination reaction of the following compound. Propose a synthesis of divinyl ether starting from ethylene oxide.

$$Cl—CH_2CH_2—O—CH_2CH_2—Cl$$

Answer: React ethylene glycol with hydroxide to obtain the conjugate base of ethylene glycol. React this conjugate base with ethylene glycol. Convert the diol to a dichloro compound by reacting it with thionyl chloride. Dehydrohalogenate the dichloro compound using sodium hydroxide.

16.51   Write the product of reaction of cyclopentene oxide with aqueous base.

16.52   The reaction of methyllithium with an epoxide followed by neutralization gives an alcohol. Write the product of the reaction of cyclohexene oxide with methyl lithium, showing the structure in its most stable conformation.

Answer: Reaction of the methyl carbanion inverts the configuration at the center where it attacks. Thus, the methyl group and the hydroxyl group are *trans* in the product. The most stable conformation is diequatorial.

16.53   Write the structure of the amino alcohol formed in the reaction of (2S,3R)-2,3-dimethyloxirane with aqueous ammonia. Explain why this compound forms in preference to a diol. Assign the stereochemistry of each chiral center in the product.

Answer: Ammonia is a better nucleophile than water. The reaction occurs with inversion of configuration at the center attacked by the nucleophilic ammonia. Attack at one carbon atom gives the (2R,3R) product, and attack at the other carbon atom gives the (2S,3S) product. Thus, a racemic mixture of enantiomers results.

16.54   Epoxide rings can be cleaved by phenoxides. Propose a synthesis of the muscle relaxant methocarbamol using this fact.

methocarbamol

Answer: A ring opening reaction of the following epoxide and the conjugate base of the phenol will occur by nucleophilic attack at the primary carbon atom of the epoxide.

16.55   A mixture of 2,2-dimethyloxirane and ethanethiol is treated with sodium hydroxide. Write the structure of the expected product.

Answer: The ring opening reaction of the epoxide by the nucleophilic thiolate ion occurs at the primary carbon atom of the epoxide.

16.56   Epoxide rings can he cleaved by metal hydrides, which serve as a source of hydride ion. Write the product of the reaction of l-methylcyclohexene oxide and $LiAID_4$.

Answer: The deuteride ion attacks at the less hindered secondary carbon atom rather than the tertiary carbon atom. Thus, the deuteride ion and the hydroxyl group formed will be *trans*. The most stable conformation of the product has the methyl group in the equatorial position because the conformational preference of a methyl group is larger than for a hydroxyl group.

16.57   The reaction of (2S,3R)-2,3-dimethyloxirane with aqueous acid gives a mixture of enantiomers. Explain why a mixture is obtained. What are the configurations of all stereogenic centers of both isomers?

Answer: The reaction occurs with inversion of configuration at the center of the conjugate acid of the epoxide attacked by the nucleophilic water. Attack at one of the secondary carbon atoms gives the (2R,3R) product, and attack at the other secondary carbon atom gives the (2S,3S) products. Thus, a racemic mixture of enantiomers results.

16.58   The reaction of (2R,3R)-2,3-dimethyloxirane with aqueous acid gives a single optically inactive product. Explain the origin of this product. What are the configurations of all stereogenic centers of the product?

Answer: The reaction occurs with inversion of configuration at the center of the conjugate acid of the epoxide attacked by the nucleophilic water. Attack at one carbon atom gives a product designated (2S,3R), and attack at the other carbon atom gives a product designated (2R,3S). However, the two centers in the diol product are equivalent, and thus only a single product, which is *meso*, obtained.

16.59 Sodium ethoxide in ethanol reacts with 1-(chloromethyl)oxirane containing a ^{14}C label at the position shown by the asterisk. Write a two-step mechanism that explains why the label is at the indicated position in the product.

Answer: Ring opening of the epoxide as the result of nucleophilic attack by ethoxide ion yields an alkoxide that displaces the primary chloride and forms an epoxide ring.

16.60 The following compound isomerizes in aqueous NaOH to give a tetrahydropyran. What is the structure of the product? Write a mechanism for its formation.

Answer: The alkoxide ion formed by proton exchange with base is a nucleophile. It attacks the primary carbon atom of the epoxide to open the three-membered ring. The result of the intramolecular substitution reaction is a six-membered ring.

16.61   Deduce the structure of a compound with the molecular formula $C_4H_6Cl_2O$ based on the following proton NMR spectrum.

6.00

5.1 5.0 4.9 4.8 4.7 4.6 4.5 4.4 4.3 4.2 4.1 4.0 3.9 3.8 3.7 3.6

Answer: The spectrum contains a low-field singlet at ~5 δ and two triplets centered at 3.85 δ. The two identical triplets correspond to two methylene groups. One of them is bonded to a choline atom, the other to an oxygen atom. The low-field singlet also contains two hydrogen atoms. The compound is $Cl-CH_2-O-CH_2CH_2-Cl$.

16.62   Deduce the structure of a compound with the molecular formula C_4H_8O based on the following proton NMR spectrum plus the information that the ^{13}C NMR spectrum has two peaks whose chemical shifts are 30 and 45 ppm.

Answer: The formula of the compound tells us that it is a ring. There are two kinds of sp³-hybridized carbon atoms based on the chemical shifts of the two peaks in the C-13 NMR spectrum. The proton NMR spectrum tells us that there are two kinds of protons. The integration of the peaks tells us that each of these have the same number of hydrogen atoms. Since there are four carbons, each contains two hydrogen atoms. The signal near 3.9 δ is due to a pair of methylene groups adjacent to oxygen. The peak at 1.9 δ is due to two hydrogen atoms bonded to the carbon atom farthest from the oxygen atom. The compound is tetrahydrofuran.

1.00

4   3   2   1

17

ORGANOMETALLIC CHEMISTRY OF TRANSITION METAL ELEMENTS AND INTRODUCTION TO RETROSYNTHESIS

KEYS TO THE CHAPTER

17.1 Transition Metal Complexes

Transition metal complexes consist of a transition metal bonded to molecules or ions called **ligands** in a **coordination complex**. In the formation of a coordination complex, the metal ion acts as a Lewis acid and the ligands act as Lewis bases. A coordination complex in solution is in equilibrium with its component metal atom (or ion) and its ligands.

The number of ligands that form coordinate covalent bonds in a transition metal complex is called the **coordination number** of the complex. The most common coordination numbers are 2, 4, and 6. The metal atom (or ion) and its bonded ligands constitute the **coordination sphere** of the complex.

A covalent bond in which one of the bonding groups contributes both bonding electrons to the metal is called a **coordinate covalent bond**.

Palladium metal, Pd(0), has a $[Kr]4d^{10}$ electron configuration. Coordination complexes in which Pd(0) has four ligands have tetrahedral geometry. Pd(0) is sp^3 hybridized in these complexes.

Pd(II) has a $[Kr]4d^8$ electron configuration. It has square planar geometry in which Pd(II) is dsp^2 hybridized. This geometry places the electron pairs in the coordinate covalent bonds, and therefore the ligands as far apart as possible.

Coordination complexes of palladium and other metals having six ligands usually have octahedral geometry. This geometry minimizes steric repulsion. The bonding in an octahedral complex is d^2sp^3.

17.2 Gilman Reagents

A **Gilman reagent** is an organometallic compound that has the general formula R_2CuLi, where the R group is derived from an alkyl halide. Gilman reagents undergo a wide variety of substitution reactions in which the R group of the reagent acts as a nucleophile that replaces a halogen in an alkyl, alkenyl, or aryl halide (Figure 17.2).

A Gilman reagent is prepared from an alkyl halide in a two-step process. First, lithium reacts with haloalkanes to give organolithium reagents. The organolithium compound then reacts with Cu(I) iodide to give a lithium dialkylcuprate.

Gilman reagents are nucleophiles that react by either an S_N2 mechanism or a more complex pathway of oxidative addition followed by reductive elimination to give products in which the alkyl group of the Gilman reagent couples with a substrate molecule.

Organic Chemistry Study Guide: Key Concepts, Problems, and Solutions http://dx.doi.org/10.1016/B978-0-12-801889-7.00017-0

17.3 Organopalladium-Catalyzed Cross-Coupling Reactions

Organopalladium complexes catalyze cross-coupling reactions between two aryl groups—Suzuki coupling; between an aryl group and an alkenyl group—the Heck reaction; and between an aryl group and an alkyne— the Sonogashira reaction.

17.4 The Suzuki Coupling Reaction

The Suzuki coupling reaction couples an aryl or vinyl halide to an aryl or vinyl boronic acid. The Suzki catalyst is $Pd(PPh_3)_4$.

Alkenyl and aryl boronic acids can be prepared by the reaction of a Grignard reagent with trimethoxyborane, $B(OCH_3)_3$ to give a dimethoxyaryl borane, $ArB(OCH_3)_2$. Hydrolysis of the borane gives the aryl boronic acid, $ArB(OH)_2$.

The vinyl halide adds to the Suzuki catalyst in an oxidative addition reaction in the first step of the catalytic cycle. Then, hydroxide displaces iodide. The base that is present in the reaction medium activates the boronic acid. The aryl group of the borate anion then adds to the catalyst. Cross-coupling of the aryl and vinyl groups occurs in a reductive elimination step. The palladium atom of the catalyst returns to its original oxidation state, Pd(0), and the cycle continues.

17.5 The Heck Reaction

The Heck reaction couples aryl halides and alkenes. The product of the reaction is a conjugated aryl alkene.

17.6 The Sonogashira Reaction

The Sonogashira reaction couples the sp-hybridized carbon of a terminal an alkyne to a wide range of other substrates, including aryl and alkenyl halides. The reaction requires a palladium catalyst and a catalytic amount of CuI. It is carried out in an amine solvent.

17.7 The Wilkinson Catalyst: Homogeneous Catalytic Hydrogenation

Wilkinson's catalyst, $Ru[(PPh_3)_3Cl$, catalyzes homogeneous hydrogenation. The reaction is regiospecific. Alkenes are reduced in the following order.

Relative rates of hydrogenation by Wilkinson's catalyst

17.8 Noyori Asymmetric Reduction of Ketones

The Noyori asymmetric hydrogenation of ketones uses chiral ruthenium catalysts for the stereospecific hydrogenation of ketones. One of the most common ligands for the Noyori reaction is BINAP, a chiral catalyst. BINAP stereoisomers result from restricted rotation around single bonds. They are **atropisomers**. The chiral ruthenium catalysts carry out stereospecific **chiral synthesis** reactions.

17.9 The Grubbs Metathesis Reaction

The Grubbs reaction exchanges the groups attached to the double bond of alkenes. The two alkenes exchange partners to give two new products in which neither one is oxidized or reduced. This process is a metathesis reaction. Most of the time both reactants for the Grubbs are terminal alkenes. When two terminal alkenes react, they exchange methylene groups. Thus, one of the products is ethene, which escapes from the solution as a gas, converting a reversible process to one that is effectively irreversible.

Ethene escapes as a gas, pulling the reaction to completion.

17.10 Introduction to Retrosynthesis

The design of a series of steps leading from simple, readily available starting materials to a more complex product that contains new functional groups and carbon–carbon bonds requires careful thought. One way to imagine a complex synthetic procedure is to go backward from the desired end-product to the initial reactants. This mode of thinking backward is called **retrosynthesis**. Consider the following synthetic scheme, viewed from right to left. The molecules leading to the **target molecule, T**, are called **precursor molecules**. In this scheme, there are four precursor molecules. Each double arrow pointing to the left is a **reverse step**.

Retrosynthetic Scheme I

Each precursor molecule is also a target molecule, but one with a simpler structure than the one preceding it. In terms of structural complexity, then, A < 4 < 3 < 2 < 1 < T. In each reverse step, we imagine how a bond can break to give two fragments; this is called **disconnection**.

The fragments of a hypothetical bond dissociation are called **synthons**. To have synthetic value, a synthon must correspond to an actual molecule called a **synthetic equivalent** that can be used in a synthetic reaction.

SUMMARY OF REACTIONS

1. Preparation of Gilman Reagents

$$CH_3{-}Br \ + \ 2\,Li \ \longrightarrow \ CH_3{-}Li \ + \ LiBr$$

methyl lithium

$$2\,CH_3{-}Li \ + CuI \ \longrightarrow \ \left[CH_3{-}\bar{C}u{-}CH_3\right]Li^+ \ + \ LiI$$

alkyl lithium · · · · · · · · lithium dimethylcuprate
(a Gilman reagent)

2. Reactions of Gilman Reagents

3. Types of Cross-Coupling Reactions

4. Synthesis of Aryl Boronic Acids for the Suzuki Reaction

trimethoxyborane dimethoxyphenylborane

phenylboronic acid

5. Suzuki Coupling Reactions

Suzuki coupling reactions

6. The Heck Reaction

7. The Sonogashira Reaction

8. Wilkinson's Catalyst: Homogeneous Catalytic Hydrogenation

$$\xrightarrow[\text{H}_2]{\text{Rh(PPh}_3)_3\text{Cl}}$$

9. Noyori Asymmetric Reduction of Ketones

$$\xrightarrow[\text{H}_2]{(R)\text{-[BINAP]Ru(OAc)}_2}$$

R isomer

$$\xrightarrow[\text{H}_2]{(S)\text{-[BINAP]Ru(OAc)}_2}$$

S isomer

10. The Grubbs Metathesis Reaction

$$\xrightarrow{\text{Grubbs Catalyst}}$$

+ $CH_2\!\!=\!\!CH_2$

(*E*)-2-pentene

304

ANSWERS TO EXERCISES

Gilman Reagent

17.1 What is the product of the following reaction?

Answer:

17.2 What is the product of the following reaction?

Answer:

17.3 What is the product of the following reaction?

Answer:

17.4 What is the product of the following reaction?

Answer:

(S)-2-bromobutane   (R)-2-phenylbutane

17.5 What is the product of the following reaction?

Answer:

Suzuki Coupling

17.6 What is the product of the following reaction?

Answer:

17.7 What is the product of the following reaction?

Answer:

17.8 What is the product of the following reaction?

Answer:

Heck Reaction

17.9 What is the product of the following reaction?

Answer:

17.10 What is the product of the following reaction?

Answer:

17.11 What is the product of the following reaction?

Answer:

Sonogashira Coupling

17.12 What is the product of the following reaction?

Answer:

Wilkinson's Catalyst

17.13 What is the product of the following reaction?

Answer:

17.14 What is the product of the following reaction?

Answer:

Grubbs Alkene Metathesis

17.15 What is the product of the following reaction?

Grubbs catalyst → ?

Answer:

Grubbs catalyst →

17.16 What is the product of the following reaction?

Grubbs catalyst → ?

Answer:

Grubbs catalyst →

17.17 What is the product of the following reaction?

Grubbs catalyst
?

Answer:

Grubbs catalyst

+

17.18 What is the product of the following reaction?

Grubbs catalyst
?

Answer:

Grubbs catalyst

+

17.19 What is the product of the following reaction?

Grubbs catalyst
?

Answer:

Grubbs catalyst

+

17.20 What is the product of the following reaction?

Answer:

18

KEYS TO THE CHAPTER

ALDEHYDES AND KETONES

The chemistry of carbonyl compounds will occupy us for most of the rest of this text. The key points of this chapter revolve around the structure of the carbonyl group and the influence of the structure of the carbonyl bond on the physical and chemical properties of aldehydes and ketones.

18.1 The Carbonyl Group

The carbonyl group is C=O with the carbon atom bonded to two other atoms. Carbonyl compounds with only hydrogen, alkyl, or aryl groups bonded to the carbonyl carbon atom are aldehydes or ketones. Aldehydes have one hydrogen atom and one alkyl or aryl group bonded to the carbonyl carbon atom. Ketones have only alkyl or aryl groups bonded to the carbonyl carbon atom.

The carbonyl carbon and oxygen are sp²-hybridized. The reactivity of the carbonyl group is interpreted based on its π electrons and the two sets of nonbonded electrons. In addition, pay particular attention to the dipolar structure that is a contributing resonance structure for the carbonyl group. The structure of the carbonyl bond affects both the stability of carbonyl compounds and their reactivity.

18.2 Physical Properties of Aldehydes and Ketones

Carbonyl compounds are polar compounds, and as a result, they have higher boiling points than alkanes of similar molecular weight. The lower-molecular-weight carbonyl compounds are soluble in water because water can form hydrogen bonds to the carbonyl oxygen atom. Alkanes are insoluble in water.

Carbonyl compounds cannot form intermolecular hydrogen bonds like alcohols can, because they lack a hydrogen atom bonded to an electronegative atom such as oxygen. Thus, they have lower boiling points than alcohols of similar molecular weight.

18.3 Redox Reactions of Carbonyl Compounds

Aldehydes are easily oxidized by a variety of oxidizing agents, such as Tollens reagent, Benedict's solution, and Fehling's solution. A reaction is detected by the formation of metallic silver with Tollens reagent or a red precipitate of Cu_2O with Benedict's or Fehling's solution. Ketones are not oxidized by these reagents. The product of oxidation is a carboxylate ion which yields the carboxylic acid when neutralized. Recall that stronger oxidizing agents such as the Jones reagent (Chapter 15) oxidize primary alcohols to carboxylic acids. Thus, aldehydes are also oxidized by this reagent.

Aldehydes and ketones are reduced to primary and secondary alcohols, respectively. Hydrogen gas with a platinum catalyst may be used as a reducing agent, but high pressures are required, and any carbon–carbon double bond would be reduced first. Lithium aluminum hydride and sodium borohydride reduce carbonyl groups without affecting carbon–carbon double bonds. Aldehydes yield primary alcohols; ketones yield secondary alcohols.

The carbonyl group in both aldehydes and ketones can be reduced to a methylene group in a Clemmensen reduction with Zn/Hg and HCl, or with NH_2NH_2 and KOH in a Wolff–Kishner reduction.

18.4 Synthesis of Carbonyl Compounds

An effective synthesis of *any* type of functional group depends on the type of substrates available, and on the specificity of several possible reagents that could be used. We have already described four general methods to prepare carbonyl compounds that have already been presented.

1. Oxidation of alcohols
2. Friedel–Crafts acylation of aromatic compounds
3. Ozonolysis of alkenes
4. Oxidative cleavage of diols
5. Hydroboration–oxidation of alkynes

Primary alcohols are oxidized by PCC to give aldehydes. Secondary alcohols are oxidized by the same reagent to give ketones. The Jones reagent (CrO_3 in acetone/H_2SO_4) further oxidizes primary

Organic Chemistry Study Guide: Key Concepts, Problems, and Solutions http://dx.doi.org/10.1016/B978-0-12-801889-7.00018-2

alcohols beyond the aldehyde state to give carboxylic acids. The Jones reagent reacts with secondary alcohols to give ketones, which are not further oxidized.

Friedel–Crafts acylation gives a carbonyl compound with the carbonyl group directly attached to an aromatic ring. The reaction is limited to aromatic compounds lacking strongly deactivating groups. Two variations are an intramolecular cyclization reaction of carboxylic acids using HF, and the Gatterman–Koch reaction using CO and HCl, which behave together like formyl chloride to give an aldehyde.

Ozonolysis of an alkene gives two carbonyl fragments. If the fragments are identical or if one of the two fragments can be isolated, then the method may be useful. For example, the ozonolysis of a methylenecycloalkane give formaldehyde and a cycloalkanone. This method depends on the availability of the alkene, which in turn must usually be prepared by an elimination reaction of another substrate.

The oxidative cleavage of diols also gives two carbonyl fragments. As a synthesis of carbonyl compounds, this reaction is limited because usually a synthetic method focuses on formation of a single product. Furthermore, the diol must be prepared by oxidation of an alkene.

Indirect hydration of alkynes can be controlled to add one mole of a borane. The resulting enol has a hydroxyl group bonded to an unsaturated carbon atom. The enol rapidly rearranges to the isomeric ketone. Because an alkyne can react twice with any reagent, the addition reaction of a borane is controlled by using disiamylborane, which has two sterically hindering 1,2-dimethylpropyl groups. Only one mole of the borane compound adds to the double bond at the least hindered position.

18.5 A Preview of Carbonyl Synthesis

Each of the synthetic methods given in this section depends on the chemistry of functional groups we will encounter in future chapters.

Acid chlorides can be prepared by the reaction of carboxylic acids with thionyl chloride. They are reduced to aldehydes by either of two methods. The Rosenmund reduction uses hydrogen gas and a modified palladium catalyst. Lithium tri(*tert*-butoxy)aluminum hydride also converts acid chlorides to aldehydes. The aldehyde is not further reduced by this hydride reagent as it would be by LiAlH$_4$. Esters are reduced to aldehydes using diisobutylaluminum hydride (DIBAL). The aldehyde itself is not formed in the reaction but is formed in the aqueous workup reaction to give a hemiacetal which decomposes to the aldehyde. We will discuss these latter reactions in Chapter 18.

Carboxylic acids and acid chlorides react with certain organometallic compounds to give ketones. Addition of an organolithium compound to a carboxylic acid gives a salt of a geminal diol that gives a diol upon hydrolysis. This rapidly eliminates water to give a ketone. Reaction of an acid chloride with a Gilman reagent gives a ketone directly. The product does not react with the Gilman reagent.

Nitriles are reduced by modified hydride reagents to give imine anion intermediates. Upon aqueous workup, the imine hydrolyzes to give the more stable carbonyl compound, which is an aldehyde. Grignard reagents react with nitriles to give a salt of an imine. It is converted to an imine and then to a ketone in the aqueous workup.

18.6 Spectroscopy of Aldehydes and Ketones

The infrared spectra of aldehydes and ketones have strong carbonyl group absorptions in the neighborhood of 1700 cm^{-1}. Aldehydes absorb at slightly higher wavenumber than ketones. The aldehyde C—H bond also has a characteristic absorption at 2710 cm^{-1}. The position of the carbonyl absorption occurs at lower wavenumber with conjugation due to a greater contribution of the dipolar resonance form because it decreases the double bond character of the carbonyl group. The position of the carbonyl absorption of cycloalkanones depends on ring size.

The proton NMR of aldehydes has a characteristic absorption near 10 δ due to the aldehyde C—H bond. The a hydrogen atoms of both aldehydes and ketones have NMR absorptions in the 2.0–2.5 δ region

The α carbon atoms of both aldehydes and ketones have C-13 NMR absorptions in the 30–50 δ region. The carbonyl carbon atom is easily identified by its absorption in the 190–220 δ region.

Summary of Reactions

1. Oxidation of Aldehydes

Naphthalene-CHO $\xrightarrow{\text{Ag(NH}_3)_2{}^+ \text{ (aq)}}$ Naphthalene-COOH + Ag(s)

cyclohexyl-CHO $\xrightarrow{\text{Cu}_2{}^+ / \text{HO}^-}$ cyclohexyl-COOH + Cu$_2$O

2. Reduction of Aldehydes and Ketones to Alcohols

cyclopentyl-C(CH$_3$)(H)-CHO $\xrightarrow[\text{ethanol}]{\text{NaBH}_4}$ cyclopentyl-C(CH$_3$)(H)-CH$_2$OH

cyclopentyl-CO-CH$_2$CH$_3$ $\xrightarrow[\text{2. H}_3\text{O+}]{\text{1. LiAlH}_4}$ cyclopentyl-C(OH)(H)-CH$_2$CH$_3$

3. Reduction of Aldehydes and Ketones to Methylene Groups

cyclohexanone $\xrightarrow{\text{Zn(Hg) / HCl}}$ cyclohexane

C$_6$H$_5$-CO-CH$_2$CH$_3$ $\xrightarrow{\text{H}_2\text{NNH}_2/ \text{KOH}}$ C$_6$H$_5$-CH$_2$-CH$_2$CH$_3$

4. Synthesis of Aldehydes and Ketones by Oxidation of Alcohols

(a) CH$_3$O—C$_6$H$_4$—CH$_2$OH $\xrightarrow{\text{PCC}}$ CH$_3$O—C$_6$H$_4$—CHO

(b) CH$_2$—CH$_2$—C(OH)(H)—CH$_3$ $\xrightarrow{\text{Jones reagent}}$ CH$_2$—CH$_2$—CO—CH$_3$

315

5. Synthesis of Aryl Ketones by Friedel–Crafts Acylation

6. Synthesis of Carbonyl Compounds by Ozonolysis of Alkenes

7. Synthesis of Carbonyl Compounds by Oxidative Cleavage of Vicinal Diols

2-methyl-1-phenyl-1,2-butanediol   benzaldehyde   2-butanone

cis-1,2-cyclohexanediol   6-oxohexanal

8. Synthesis of Carbonyl Compounds by Hydration of Alkynes

$$\text{C}_6\text{H}_5-\text{C}\equiv\text{C}-\text{H} + \text{H}-\text{B}\begin{pmatrix}\text{CH(CH}_3)_2\\ \text{CH(CH}_3)_2\end{pmatrix} \longrightarrow$$

disiamylborane

$$\xrightarrow[\text{NaOH}]{\text{H}_2\text{O}_2}$$

9. Synthesis of Carbonyl Compounds by Reduction of Acid Derivatives

$$\xrightarrow{\text{H}_2 \,/\, \text{Pd-C}}$$

$$\text{CH}_3(\text{CH}_2)_5\text{CH}_2-\overset{\overset{\text{O}}{\|}}{\text{C}}-\text{Cl} \xrightarrow[\text{2. H}_3\text{O}^+]{\text{1. Li}^+[\text{AlH(O(CH}_3)_3]^-} \text{CH}_3(\text{CH}_2)_5\text{CH}_2-\overset{\overset{\text{O}}{\|}}{\text{C}}-\text{H}$$

octanoyl chloride   octanal

$$\text{CH}_3(\text{CH}_2)_5\text{CH}_2-\overset{\overset{\text{O}}{\|}}{\text{C}}-\text{Cl} \xrightarrow[\text{2. H}_3\text{O}^+]{\text{1. LiAlH}_4} \text{CH}_3(\text{CH}_2)_5\text{CH}_2-\overset{\overset{\text{OH}}{|}}{\underset{\underset{\text{H}}{|}}{\text{C}}}-\text{H}$$

octanoyl chloride   1-octanol

$$\text{CH}_3(\text{CH}_2)_{10}\text{CH}_2-\overset{\overset{\text{O}}{\|}}{\text{C}}-\text{OCH}_3 \xrightarrow[\text{2. H}_3\text{O}^+]{\text{1. DIBAL}} \text{CH}_3(\text{CH}_2)_{10}\text{CH}_2-\overset{\overset{\text{O}}{\|}}{\text{C}}-\text{H}$$

methyl hexadecanoate   hexadecanal

10. Reactions of Organometallic Reagents with Acid Derivatives

$$\xrightarrow{\text{LiOH}} \qquad \xrightarrow{\text{CH}_3\text{Li}}$$

$$+\; [(\text{CH}_3)_2\text{CH}]_2\text{CuLi} \longrightarrow$$

11. Formation of Carbonyl Compounds from Nitriles

$$\text{C}_6\text{H}_5\text{—CH}_2\text{—C}{\equiv}\text{N} \xrightarrow[\text{2. H}_3\text{O}^+]{\text{1. LiAlH(OCH}_2\text{CH}_3)_3} \text{C}_6\text{H}_5\text{—CH}_2\text{—C}\overset{\text{O}}{\underset{\text{H}}{\|}}$$

$$\text{C}_6\text{H}_5\text{—CH}_2\text{—C}{\equiv}\text{N} \xrightarrow[\text{2. H}_3\text{O}^+]{\text{1. CH}_3\text{MgBr}} \text{C}_6\text{H}_5\text{—CH}_2\text{—C}\overset{\text{O}}{\underset{\text{CH}_3}{\|}}$$

SOLUTIONS TO EXERCISES

Nomenclature of Aldehydes and Ketones

18.1 Write the structure for each of the following compounds.

(a) 2-methylbutanal   (b) 3-ethylpentanal   (c) 2-bromopentanal

(d) 3,4-dimethyloctanal   (e) 1-bromocyclobutanecarbaldehyde

Answers:

$$\underset{\overset{|}{CH_3}}{(a)\ CH_3CH_2CHCHO} \qquad \underset{\overset{|}{CH_2CH_3}}{(b)\ CH_3CH_2CHCH_2CHO} \qquad \underset{\overset{|}{Br}}{(c)\ CH_3CH_2CH_2CHCHO}$$

$$\underset{\underset{CH_3}{|}}{\overset{\overset{CH_3}{|}}{(d)\ CH_3CH_2CH_2CH_2CHCHCH_2CHOCHO}}$$

(e) [cyclobutane ring with Br and CHO substituents]

18.2 Write the structure of each of the following compounds.

(a) 3-bromo-2-pentanone   (b) 2,4-dimethyl-3-pentanone   (c) 4-methyl-2-pentanone

(d) 3,4-dimethyl-2-pentanone   (e) 2-methyl-1,3-cyclohexanedione

Answers:

$$\underset{\overset{|}{Br}\ \ \overset{||}{O}}{(a)\ CH_3CH_2CHCCH_3} \qquad \underset{\overset{CH_3}{|}\ \ \underset{||}{\overset{CH_3}{|}}\atop O}{(b)\ CH_3CHCCHCH_3} \qquad \underset{\overset{CH_3}{|}\ \ \overset{O}{||}}{(c)\ CH_3CHCH_2CCH_3}$$

$$\underset{\underset{CH_3}{|}}{\overset{\overset{CH_3}{|}\ \ \overset{O}{||}}{(d)\ CH_3CHCHCCH_3}} \qquad (e)\ [\text{cyclohexanedione ring with CH}_3]$$

18.3 Give the IUPAC name of each of the following compounds.

$$(a)\ CH_3CH_2CH_2CHO \quad \underset{\underset{CH_3}{|}}{(b)\ CH_3\overset{\overset{CH_3}{|}}{C}-CH_2CHO} \quad \underset{\overset{CH_3}{|}}{(c)\ CH_3CHCHO} \quad \underset{\underset{CH_2H_3}{|}}{(d)\ CH_3CH_2\overset{\overset{CH_3}{|}}{CH}CHCHO}$$

Answers:

(a) butanal   (b) 3,3-dimethylbutanal   (c) 2-methylpropanal   (d) 2-ethyl-3-methylpentanal

18.4 Give the IUPAC name of each of the following compounds.

$$(a)\ CH_3CH_2\overset{\overset{O}{||}}{C}CH_2CH_3 \quad (b)\ CH_3\overset{\overset{O}{||}}{C}-\underset{\underset{CH_3}{|}}{\overset{\overset{CH_3}{|}}{C}}-CH_3 \quad (c)\ CH_3\overset{\overset{CH_3}{|}}{CH}\underset{\underset{O}{||}}{C}CH_2CH_3 \quad (d)\ CH_3\overset{\overset{CH_3}{|}}{CH}CH_2\overset{\overset{O}{||}}{C}CH_3$$

Answers:

(a) 3-pentanone   (b) 3,3-dimethyl-2-butanone   (c) 2-methyl-3-pentanone   (d) 5-methyl-3-hexanone

18.5   Give the IUPAC name of each of the following compounds.

(a)

(b) CH_3

(c)

(d)

Answers:

(a) 4-chloro-2,3-dimethylheptanal   (b) 6-ethyl-3-methyl-2-nonanone   (c) 4-ethyl-2,3,5-trimethylheptanal   (d) 8-methyl-4-nonanone

18.6   Give the IUPAC name of each of the following compounds.

(a)

(b)

(c)

(d)

Answers:

(a) 2-methylcyclobutanone (b) 5-chlorocyclodecanone   (c) 2-ethylcyclohexanone   (d) 1-cyclohexyl-1-pentanone

18.7   Many aldehydes and ketones are better known by their common names. Draw the structural formula of each of the following carbonyl compounds. Their common names are given within parentheses.
(a)   2,2-dimethylpropanal (pivaldehyde)
(b)   2-hydroxy-1,2-diphenyl-1-ethanone (benzoin)
(c)   2-propenal (acrolein)
(d)   4-methyl-3-penten-2-one (mesityl oxide)
(e)   5,5-dimethyl-1,3-cyclohexanedione (dimedone)

(a) $CH_3\overset{\overset{\displaystyle CH_3}{|}}{\underset{\underset{\displaystyle CH_3}{|}}{C}}CHO$ (b)

(c) $CH_2{=}CHCHO$

(d) $CH_3\overset{\overset{\displaystyle CH_3}{|}}{C}{=}CH\overset{\overset{\displaystyle O}{||}}{C}CH_3$ (e)

320

18.8 Draw the structural formula of each of the following carbonyl compounds. The common name of each compound is given within parentheses.

(a) 3,3-dimethyl-2-butanone (pinacolone) (b) 4-hydroxy-4-methyl-2-pentanone (diacetone alcohol)
(c) (E)-2-butenal (crotonaldehyde) (d) 1,3-diphenyl-2-buten-l-one (dypnone)
(e) 2,3-butanedione (biacetyl)

Answers:

Properties of Aldehydes and Ketones

18.9 The H—C—H bond angle of formaldehyde is 116.5°. The H—C—C bond angle of acetaldehyde is 118.2°. Explain this difference.

Answer: Although sterically small, the methyl and hydrogen atoms bonded to the carbonyl carbon atom occupy more space than two hydrogen atoms of formaldehyde and the bond angle widens to accommodate the methyl group.

18.10 The C=C bond length in alkenes and the C=O bond length in aldehydes are 134 and 123 pm, respectively. Explain this difference.

Answer: The atomic radius of oxygen is smaller than the atomic radius of carbon, and the bond length between two atoms reflects their respective atomic radii.

18.11 The dipole moments of acetone and isopropyl alcohol are 2.7 and 1.7 D, respectively. Explain this difference.

Answer: The oxygen atom of the "two" carbon–oxygen bonds of the carbonyl group pulls electron density away from the carbonyl carbon atom more than the single oxygen atom of the carbon–oxygen bond of the alcohol.

18.12 The dipole moments of propanal and propenal are 2.52 and 3.12 D, respectively. Consider the resonance forms of these compounds and explain the difference in their dipole moments.

Answer: An additional dipolar resonance form of propenal gives added stabilization of the positive charge and increases the polarity of the molecule.

18.13 The boiling points of butanal and 2-methylpropanal are 75 and 61 °C, respectively. Explain this difference.

Answer: 2-Methylpropanal is a more compact molecule that has a more spherical shape than the cylindrically shaped butanal, so its London forces are smaller.

18.14 The boiling points of 2-heptanone, 3-heptanone, and 4-heptanone are 151, 147, and 144 °C, respectively. What is responsible for this trend?

Answer: As the polar carbonyl group moves to the interior of the molecule, the compound more closely resembles an alkane structure and the intermolecular attractive forces are smaller.

18.15 The boiling points of 2-hydroxy- and 3-hydroxybenzaldehydes are 197 and 240 °C, respectively. Suggest a reason for this difference.

Answer: As the polar carbonyl group moves to the interior of the molecule, the compound more closely resembles an alkane structure and the intermolecular attractive forces are smaller.

18.16   The boiling points of 2-hydroxy- and 3-hydroxyacetophenones are 218 and 296 °C, respectively. Suggest a reason for this difference.

Answer: 2-Hydroxyacetophenone can form intramolecular hydrogen bonds, which decrease the number of intermolecular hydrogen bonds that would cause a higher boiling point like that of 3-hydroxyacetophenone.

18.17   The solubilities of butanal and 1-butanol in water are 7 and 9 g/100 mL, respectively. Explain this difference.

Answer: Butanal is a hydrogen bond acceptor of water molecules but 1-butanol is both a hydrogen bond acceptor and a hydrogen bond donor, so it is somewhat more soluble.

18.18   The solubilities of butanal and 2-methylpropanal in water are 7 and 11 g/100 mL, respectively. Explain this difference.

Answer: 2-Methylpropanal is a more compact molecule that has a more spherical shape than the cylindrically shaped butanal, so it interferes less with the hydrogen bonding arrays of water.

Oxidation and Reduction of Carbonyl Compounds

18.19   What is observed when an aldehyde reacts with Benedict's solution? What is observed when an aldehyde reacts with Tollens reagent?

Answer: A red precipitate of Cu_2O forms when an aldehyde reacts with Benedict's solution. A precipitate of silver usually seen as a silver mirror forms when an aldehyde reacts with Tollens reagent.

18.20   What class of compounds results from the reduction of ketones with sodium borohydride? What class of compounds results from the reduction of aldehydes by lithium aluminum hydride?

Answer: Ketones yield secondary alcohols when reduced using sodium borohydride. Aldehydes yield primary alcohols when reduced using lithium aluminum hydride.

18.21   Draw the structure of the product of each of the following reactions.

Answers:
(a) Oxidation of an aldehyde by Tollenss reagent yields a carboxylate ion in solution. Upon neutralization, the isolated product is the carboxylic acid.
(b) Oxidation of an aldehyde by Benedict's solution or Fehling's solution yields a carboxylate ion in solution. Upon neutralization, the isolated product is the carboxylic acid.
(c) Oxidation of a primary alcohol by CrO_3 in sulfuric acid yields a carboxylic acid.

18.22   Determine the best choice of reactants to synthesize each of the following ethers using the Williamson method.

(a)

(b)

(c)

Answers:
(a) Oxidation of an aldehyde by Benedict's solution or Fehling's solution yields a carboxylate ion in solution. Upon neutralization, the isolated product is the carboxylic acid.
(b) Oxidation of an aldehyde by Tollenss reagent yields a carboxylate ion in solution. Upon neutralization, the isolated product is the carboxylic acid.
(c) Oxidation of a primary alcohol by PCC acid yields an aldehyde

18.23   What is the product when each of the following reacts with lithium aluminum hydride?

(a)

(b)

(c)

Answer: Lithium aluminum hydride reduces aldehydes to primary alcohols and ketones to secondary alcohols. Carbon–carbon double bonds and aromatic rings are unaffected.

18.24   What is the product when each of the following reacts with sodium borohydride?

(a) HOCH₂CH₂— [cyclohexane ring] —C(=O)—CH₃ → HOCH₂CH₂— [cyclohexane ring] —C(OH)(H)—CH₃

(b)

(c)

Answer: Sodium borohydride reduces aldehydes to primary alcohols and ketones to secondary alcohols. Carbon–carbon double bonds and aromatic rings are unaffected.

18.25   Explain why the reduction of carvone by lithium aluminum hydride yields two products.

carvone

Answer: The hydride reagent can attack from either face of the carbonyl group, so both *cis* and *trans* isomers may form.

18.26   Explain why the reduction of the following compound by sodium borohydride yields two products.

Answer: The hydride reagent can attack from either face of the carbonyl group, so both *cis* and *trans* isomers may form.

324

18.27   What is the product of each of the following reactions?

(a) H_2NNH_2 /KOH

(b) $CH_3-\overset{O}{\overset{||}{C}}$ ⬡ $\overset{O}{\overset{||}{C}}-H$ Zn(Hg) / HCl CH_3CH_2-⬡$-CH_3$

(c) H_2 (1 atm) / Pd

Answers:
(a)   The Wolff–Kishner reduction converts the ketone into a methylene group.
(b)   The Clemmensen reduction converts the ketone into a methylene group and the aldehyde group into a methyl group.
(c)   At atmospheric pressure the carbon–carbon double bond is reduced.

18.28   What is the product of each of the following reactions?

(a) H_2NNH_2 / KOH

(b) Zn(Hg) / HCl

(c) H_2 (1 atm) / Pd

Answers:
(a)   The Wolff–Kishner reduction converts the ketone into a methylene group.
(b)   The Clemmensen reduction converts the ketone into a methylene group.
(c)   At atmospheric pressure, the carbon–carbon double bond is reduced and the carbonyl group is unaffected.

Synthesis of Carbonyl Compounds

18.29 What is the product of each of the following reactions?

(a)

$\xrightarrow{\text{PCC}}$

(b)

$\xrightarrow{\text{HF}}$

(c)

$\xrightarrow[\text{2. Zn/ } H_3O^+]{\text{1. } O_3}$

(d)

$\xrightarrow[\text{2. } HIO_4]{\text{1. } OsO_4}$

(e) $CH_3CH_2\overset{\overset{\displaystyle CH_3}{|}}{C}HCH_2C\equiv CH$ $\xrightarrow[\text{2. } H_2O_2 \text{ / } HO^-]{\text{1. disiamylborane}}$ $CH_3CH_2\overset{\overset{\displaystyle CH_3}{|}}{C}HCH_2CH_2CHO$

Answers:
(a)   PCC oxidizes primary alcohols to aldehydes.
(b)   Carboxylic acids are acylating agents of aromatic rings in the presence of HF. The cyclization occurs at C-1 benzylic position, which is more reactive than C-3.
(c)   Ozonolysis of the alkene yields formaldehyde and a ketone.
(d)   Reaction with osmium tetroxide gives a diol that is subsequently cleaved by periodic acid.
(e)   Reaction of a terminal alkyne with disiamylborane followed by hydrogen peroxide yields an aldehyde.

18.30　What is the product of each of the following reactions?

(a)

(b)

(c)

(d)

(e)

Answers:
(a)　PCC oxidizes primary alcohols to aldehydes.
(b)　Carboxylic acids are acylating agents of aromatic rings in the presence of HF. The cyclization occurs at either of two *ortho* positions with respect to the side chain containing the carboxyl group.
(c)　Ozonolysis of the alkene contained in a ring gives a dicarbonyl group which in this case is a dialdehyde.
(d)　Reaction with osmium tetroxide gives a diol that is subsequently cleaved by periodic acid.
(e)　Reaction of an internal alkyne with disiamylborane followed by hydrogen peroxide yields a ketone with the carbonyl group at the least hindered carbon atom of the original alkyne.

18.31   What is the product of each of the following reactions?

(a)

H_2 / Pd

quinoline

(b)

1. DIBAL

2. H_3O^+

(c)

1. CH_3Li

2. H_3O^+

(d)

1. $LiAlH(OCH_2CH_3)_3$

2. H_3O^+

Answers:
(a)   The Rosenmund reaction reduces acid chlorides to form aldehydes.
(b)   DIBAL reduces esters to form aldehydes.
(c)   Methyl lithium reacts with carboxylic acids to give ketones by adding a methyl group to the carbonyl carbon atom of the carboxylic acid.
(d)   Reduction of nitriles by complex hydride reagents containing alkoxy groups gives aldehydes.

18.32 What is the product of each of the following reactions?

(a) $\xrightarrow[\text{quinoline}]{\text{H}_2\text{ / Pd}}$

(b) $\xrightarrow[\text{2. H}_3\text{O}^+]{\text{1. DIBAL}}$

(c) $-\text{CH}_2\text{CO}_2\text{H}$ $\xrightarrow[\text{2. H}_3\text{O}^+]{\text{1. C}_6\text{H}_5\text{Li}}$ $-\text{CH}_2\overset{\text{O}}{\overset{\|}{\text{C}}}-$

(d) $\xrightarrow[\text{2. H}_3\text{O}^+]{\text{1. CH}_3\text{CH}_2\text{MgBr}}$

Answers:
(a) The Rosenmund reaction reduces acid chlorides to form aldehydes.
(b) DIBAL reduces esters to form aldehydes.
(c) Phenyl lithium reacts with carboxylic acids to give ketones by adding a phenyl group to the carbonyl carbon atom of the carboxylic acid.
(d) Reaction of nitriles Grignard reagents gives ketones whose carbonyl group is the carbon of the original nitrile.

Biochemical Reductions

18.33 Reduction of 5-chloro-2-pentanone by NADPH in an enzyme-catalyzed reaction yields (S)-5-chloro-2-pentanol. From which face does the hydrogenation occur?

$\xrightarrow[\text{dehydrogenase}]{\text{NAPH}}$

(S)-5-chloro-2-pentanol

Answer: The S-enantiomer can only be produced if the NADPH adds hydride to the *si* face of the carbonyl group.

18.34 Reduction of the ketone group of the following keto ester occurs stereospecifically at the *re* face using NADH in an enzyme-catalyzed reaction. Draw the structure of the product and assign its configuration.

$\xrightarrow[\text{dehydrogenase}]{\text{NADH}}$

Answer: The R-enantiomer can only be produced if the NADH adds hydride to the *re* face of the carbonyl group.

Spectroscopy of Aldehydes and Ketones

18.35 Explain why the IR absorption of the C=C bond of 1-butene (1642 cm^{-1}) is at higher wavenumber than the C=C bond absorption of 3-buten-2-one (1613 cm^{-1}).

Answer: The double bond of 3-buten-2-one is conjugated to the carbonyl group. One of the resonance structures has a double bond between C-3 and C-4. Therefore, the C-3—C-4 bond has some double bond character, which decreases the energy required for bond-stretching.

18.36 Explain why the carbonyl stretching absorption of cyclohexanones are shifted approximately 20 cm^{-1} to higher wavenumber when a bromine atom is substituted in the equatorial position at the α carbon atom.

Answer: The bromine atom withdraws electrons from the carbonyl carbon by an inductive effect. This increases the polarity of the carbonyl bond and, therefore, increases the energy required to stretch the carbonyl bond.

18.37 Based on the following C-13 NMR data, deduce the structures of isomeric compounds having the molecular formula C_4H_8O.
  (a)   25.7 ppm, 68.0 ppm
  (b)   15.5 ppm, 41.0 ppm, 204.9 ppm
  (c)   7.9 ppm, 29.4 ppm, 36.9 ppm, 209.2 ppm
  (d)   13.7 ppm, 15.7 ppm, 45.8 ppm, 207.6 ppm

Answer: Compounds with the formula C_4H_8O must contain a ring or a double bond. Possibilities include cyclic ethers, cyclic alcohols, unsaturated alcohols, unsaturated ethers, and ketones.

  (a) A compound with only two C-13 resonance must has two sets of nonequivalent carbon atoms. The signals correspond to an internal methylene group, δ = 25.7 ppm, and a methylene group bonded to oxygen, δ = 68.0 ppm. The compound is tetrahydrofuran.

  (b) Since there are three resonances and four carbon atoms, two of the carbons must be equivalent. The carbon with a chemical shift of 204.9 ppm is an aldehyde carbon. The resonance at 41.0 ppm corresponds to a methylene group bonded to the carbonyl carbon; the carbon with a chemical shift of 15.5 ppm corresponds to two equivalent methyl group. The compound is 2-methylpropanal.

  (c) Since there are four resonances and four carbon atoms, there must be four nonequivalent carbon atoms. The carbons with chemical shifts of 29.4 and 36.9 ppm are alpha to a carbonyl carbon with a chemical shift of 209.2 ppm. The resonance at 7.9 ppm corresponds to a methyl group bonded to an alpha carbon. The compound is 2-butanone.

  (d) Since there are four resonances and four carbon atoms, there must be four nonequivalent carbon atoms. The carbons with chemical shifts of 45.8, 15.7, and 13.7 are C-21, C-3, and C-4; C-1 is an aldehyde carbon. The compound is butanal.

18.38 Based on the following carbon NMR data, deduce the structures of isomeric ketones having the molecular formula $C_5H_{10}O$.
(a)   18.1 ppm, 27.3 ppm, 41.5 ppm, 211.7 ppm
(b)   13.5 ppm, 18.5 ppm, 29.3 ppm, 45.2 ppm, 206.6 ppm
(c)   7.3 ppm, 35.3 ppm, 209.3 ppm

Answer: Compounds with the formula $C_5H_{10}O$ must contain a ring or a double bond. Possibilities include cyclic ethers, cyclic alcohols, unsaturated alcohols, unsaturated ethers, and ketones.

  (a) The compound has four nonequivalent carbon atoms. The carbons with chemical shifts of 27.3 and 41.5 ppm are alpha to a carbonyl carbon with a chemical shift of 211.7 ppm. The signal with a chemical shift of 18.1 ppm corresponds to two equivalent methyl groups. The compound is 3-methyl-2-butanone.

  (b) Since there are five resonances and five carbon atoms, there must be five nonequivalent carbon atoms. The carbons with chemical shifts of 29.3 and 45.2 ppm are alpha to a carbonyl carbon with a chemical shift of 209.2 ppm. The resonances at 18.5 and 13.5 ppm corresponds to a methylene group and a methyl group. The compound is 2-pentanone.

  (c) Since there are three resonances and five carbon atoms, there must be three nonequivalent carbon atoms. A carbonyl carbon with a chemical shift of 209.3 ppm is bonded to two equivalent ethyl groups. The compound is 3-pentanone.

18.39   Deduce the structure of isomeric ketones having the molecular formula $C_{10}H_{12}O$ based on the following proton NMR spectra.

(a)

Answer: The 5H multiplet corresponds to the monosubstituted benzene ring; the 2H singlet corresponds to a methylene group with no hydrogens on neighboring carbon atomss; the 2H quartet and 3H triplet are the classical signature of an ethyl group.

(b)

Answer: The doublet of doublets corresponds to the *para*-substituted benzene ring; the 3H singlet corresponds to a methylene group with no hydrogens on bonded carbons; the 2H quartet and 3H triplet are the classical signature of an ethyl group.

4-phenyl-2-butanone

3H singlelet

2H triplets

TMS

5H multiplet

Chemical shift, ppm (δ)

Answer: The 5H multiplet corresponds to the mono-substituted benzene ring; the 3H singlet corresponds to a methyl group with no hydrogens on bonded carbons; the 2H triplets correspond to the two methylene groups. The methylene group at 2.8 δ corresponds to the methylene group bonded to the carbonyl carbon.

18.40   Deduce the structure of isomeric ketones having the molecular formula $C_9H_{10}O_2$ based on the following proton NMR spectra.

p-methoxyacetophenone

3H singlet

3H singlet

TMS

2H   2H

Chemical shift, ppm (δ)

Answer: The doublet of 2H doublets between 7 and 8 δ corresponds to the *para*-substituted benzene ring; the 3H singlet corresponds at 2.2 δ corresponds to a methyl group bonded to the carbonyl carbon; the 3H singlet at 3.8 δ corresponds to the methyl group bonded to an oxygen atom.

phenoxyacetone

2H singlet →

← 3H singlet

5H

TMS

Chemical shift, ppm (δ)

Answer: The 5H multiplet between corresponds to the mono-substituted benzene ring; the 3H singlet corresponds at 2.0 δ corresponds to a methyl group bonded to the carbonyl carbon; the 2H singlet at 4.2 δ corresponds to the methylene group.

p-ethoxybenzaldehyde

1H singlet

3H triplet

2H   2H

2H quartet

TMS

Chemical shift, ppm (δ)

Answer: The 1H singlet at 9.9 ppm corresponds to the aldehyde hydrogen; the doublet of doublets (2H each) corresponds to the two sets of nonequivalent protons on the *para*-substituted benzene ring; the 2H quartet corresponds at 4.1 δ corresponds to a methylene group bonded to oxygen, and the 3H triplet corresponds to the methyl group.

19

ALDEHYDES AND KETONES: NUCLEOPHILIC ADDITION REACTIONS

KEYS TO THE CHAPTER

The chemistry of carbonyl compounds depends on the bonding characteristics of the carbonyl group. A dipolar resonance form contributes to the carbonyl group. Much of the chemistry of carbonyl compounds involves addition of a nucleophile to the partially positive carbon atom of the carbonyl group. Aldehydes and ketones undergo addition reactions with unsymmetrical reagents such as H—Nu. The nucleophilic part of the reagent adds to the carbonyl carbon, and the electrophilic part adds to the oxygen atom. These products are obtained in both acid- and base-catalyzed reactions.

19.1 Synthesis of Cyanohydrins

Hydrogen cyanide adds to an aldehyde or ketone to give a compound called a **cyanohydrin**. The equilibrium constant for the addition of HCN is much more favorable for aldehydes than for ketones. Thus, as we noted above, because ketones are more stable than aldehydes, the addition reactions of ketones are less favorable (have smaller equilibrium constants) than addition reactions of aldehydes.

19.2 Hydration of Carbonyl Compounds

The equilibrium constant for the addition of water to a carbonyl compound is greater than one for only a few compounds such as formaldehyde. Although hydration is not a useful synthetic reaction, the equilibrium constant for hydration provides a basis for evaluating the effect of structure on carbonyl addition reactions in general.

The equilibrium constants for hydration formation is greater for aldehydes than for ketones because the carbonyl carbon of an aldehyde has a greater partial positive charge than the carbonyl carbon of a ketone. Nucleophiles are attracted to a carbonyl carbon atom because of its partial positive charge. The partial positive charge is larger in an aldehyde, which has only one alkyl group bonded to the carbonyl carbon atom, than in a ketone, which has two alkyl groups. There is also a steric effect. The carbonyl carbon atom has a larger bond angle than the tetrahedral product. As a result of increased steric hindrance, the addition product is less stable for ketones than for aldehydes.

Groups bonded to the α carbon atom can affect the electron density of the carbonyl carbon atom by either resonance or inductive effects. Resonance stabilization of the carbonyl group decreases the equilibrium constant for addition reactions. Inductive electron withdrawal by electronegative groups such as halogen atoms destabilizes the carbonyl group and increases the equilibrium constant for addition reactions.

19.3 Mechanisms of Carbonyl Compounds Addition Reactions

Addition reactions to carbonyl groups can occur by either of two mechanisms depending on whether the reaction conditions are acidic or basic. In both cases, a nucleophile adds to the carbonyl carbon atom, and an electrophile adds to the oxygen carbon. The difference between the mechanisms for addition under acidic or basic conditions is in the sequence of two possible reactions. Under acidic conditions, a proton adds to the carbonyl oxygen atom to give a conjugate acid that has a greater positive charge on the carbonyl carbon atom, which then reacts with a nucleophile. Under basic conditions, the nucleophile attacks the carbonyl carbon atom to give an alkoxide ion, which subsequently reacts with the electrophile—usually a proton.

Organic Chemistry Study Guide: Key Concepts, Problems, and Solutions http://dx.doi.org/10.1016/B978-0-12-801889-7.00019-4

19.4 Kinetic Effects in Addition Reactions

The same structural features that control the equilibrium constant for the addition reaction of carbonyl compounds also affect the rate of the reaction. In the case of the rate of a reaction, the stabilization of the partial positive charge on the carbonyl carbon diminishes its attraction to a nucleophile. Thus, aldehydes react faster than ketones. The steric features of the alkyl groups in aldehydes, and to a greater degree in ketones, affect the reaction by decreasing the rate of reaction with increasing size of the groups. The groups bonded to the carbonyl carbon sterically hinder the approach of the nucleophile in the same way observed in S_N2 reactions.

19.5 Addition of Alcohols to Carbonyl Compounds

Alcohols add to aldehydes to give acetals and to ketones to give ketals. The equilibrium constant is less than one. However, if an intramolecular acetal or ketal can form, then the equilibrium constant is greater than one. There is very little difference in the bond energies of the reactants and products in either case. However, there is an entropy change for the intermolecular reaction since two moles of reactants give one mole of products. For the intramolecular reaction, the entropy change is nearly zero because there is no difference in the number of moles of reactant and product.

Hemiacetals result from the reaction of an aldehyde, RCHO, with an alcohol, R'OH. In a hemiacetal a hydroxyl group, an —OR' group from the alcohol, an —R group from the aldehyde, and a hydrogen atom all bonded to the same carbon atom. Hemiketals result from the reaction of a ketone, R_2CO, with an alcohol, R'OH. In a hemiketal, the original carbonyl carbon is bonded to an —OH group an —OR' group from the alcohol, and two —R groups from the ketone. The —R groups may be alkyl or aryl groups in either hemiacetals or hemiketals.

19.6 Formation of Acetals and Ketals

Hemiacetals can be converted into acetals and hemiketals to ketals. An acetal results from the reaction of a hemiacetal with an alcohol in acidic solution. An acetal has two —OR' groups from the alcohol, an —R group from the original aldehyde, and a hydrogen atom bonded to the original carbonyl carbon. A ketal results from the reaction of a hemiketal with an alcohol in acidic solution. A ketal has two —OR' groups from the alcohol and two —R groups from the ketone bonded to the original carbonyl carbon. The —R groups may be alkyl or aryl groups

The formation of acetals from aldehydes and alcohols is driven to completion by removing the water that forms during the reaction.

The mechanism of conversion of a hemiacetal into an acetal (or a hemiketal into a ketal) proceeds by protonation of the alcohol to convert it into a better leaving group. The resulting oxocarbocation is resonance stabilized. The oxocarbocation subsequently reacts with the nucleophilic oxygen atom of the alcohol, and the last step is proton transfer from the oxonium ion. The reaction requires acidic conditions. As a result, acetals (or ketals) can regenerate the carbonyl compound under acidic conditions in the presence of water. However, both acetals and ketals are stable in basic solutions.

19.7 Acetals and Ketals as Protecting Groups

A protecting group must easily form a derivative of a functional group and easily removed to regenerative it when required. A protecting group makes a functional group unreactive toward specific reagents that are required to transform a second functional group in the molecule. Acetals and ketals are ideal protecting groups because they are easily formed in acidic solution and easily removed when the compound is again exposed to acid.

Cyclic acetals and ketals derived from ethylene glycol are used as protecting groups for carbonyl compounds. In dicarbonyl compounds, it is possible to selectively form a derivative of one carbonyl group if the structural features between the two are sufficiently different.

Alcohols can also be protected by their incorporation into an acetal or ketal. Both 1,2-diols and 1,3-diols react with acetone to give a ketal. Alcohols react under acid-catalyzed conditions with dihydropyran to give a THP derivative which is easily hydrolyzed.

19.8 Thioacetals and Thioketals

The sulfur analogs of acetals and ketals are synthesized using the dithiols 1,2-ethanedithiol or 1,3-ethanedithiol. Thioacetals and thioketals are stable under acidic and basic conditions. Unprotecting the carbonyl group requires mercury(II) chloride. Thioacetals and thioketals can be reduced to methylene compounds using Raney nickel. Thus, the sequence of thioacetal formation followed by reduction is another method to reduce carbonyl compounds to hydrocarbons.

19.9 Addition of Nitrogen Compounds

Amines and other nitrogen derivatives that are sufficiently nucleophilic attack the carbonyl carbon atom to give a tetrahedral addition product, called a hemiaminal, but it is unstable with respect to dehydration. As a result, an imine forms. The overall reaction is an addition–elimination reaction In every case, the net result of adding a nitrogen compound, R—NH$_2$, to an to an aldehyde or ketone gives a product in which the original carbonyl carbon is bonded to the nitrogen with a double bond.

19.10 The Wittig Reaction

Alkenes can be synthesized from carbonyl compounds by reaction with a phosphorus ylide in the Wittig reaction. It is produced by the reaction of a primary or secondary alkyl halide with triphenylphosphine to obtain a phosphonium ion. This compound is then treated with a strong base such as butyl lithium, which removes a proton from the alkyl group. The negatively charged carbon atom of the ylide reacts as a nucleophile, and attacks the carbonyl carbon atom to give an intermediate that triphenylphosphine oxide, leaving the desired alkene.

Summary of Reactions

1. Formation of Cyanohydrins

2. Formation of Acetals and Ketals

$$CH_3-CH_2-CH_2-\overset{\overset{\displaystyle \ddot{O}:}{\|}}{C}-H + 2\ CH_3OH \xrightleftharpoons{H^+} CH_3-CH_2-CH_2-\overset{\displaystyle OCH_3}{\underset{\displaystyle OCH_3}{C}}-H$$

butanal

methyl acetal of butanal

$$CH_3-\overset{\overset{\displaystyle \ddot{O}:}{\|}}{C}-CH_2CH_3 + 2\ CH_3CH_2OH \xrightleftharpoons{H^+} CH_3-\overset{\displaystyle OCH_2CH_3}{\underset{\displaystyle OCH_2CH_3}{C}}-CH_2CH_3$$

2-butanone

ethyl acetal of 2-butanone

$$\xrightarrow[\text{TsOH / benzene}]{HOCH_2CH_2OH}$$

$+\ H_2O$

$$\xrightarrow[\text{TsOH / benzene}]{HOCH_2CH_2OH}$$

$+\ H_2O$

3. Protection of Alcohols by Acetal Formation

$$\xrightarrow{H^+}$$

$+\ H_2O$

$$CH_3-C\equiv C-\overset{\displaystyle CH_2OH}{\underset{\displaystyle H}{C}}-CH_2OH + O{=}C\overset{\displaystyle CH_3}{\underset{\displaystyle CH_3}{}} \xrightarrow{H^+} CH_3-C\equiv C-$$

$+\ H_2O$

$$BrCH_2CH_2CH_2-\ddot{\underset{..}{O}}H + \xrightarrow{H^+} BrCH_2CH_2CH_2-\ddot{\underset{..}{O}}$$

4. Thioacetals and Thioketals as Protecting Groups

5. Reduction of Thioacetals and Thioketals to Methylene Groups

339

6. Addition of Nitrogen Compounds to Carbonyl Compounds

$\overset{\cdot\cdot}{N}H_2$—OH
hydroxyl amine

cyclohexanone oxime

+ H_2O

$\overset{\cdot\cdot}{N}H_2$—NH—$\overset{O}{\overset{\|}{C}}$—NH$_2$
semicarbazide

cyclopentanone semicarbazone

+ H_2O

benzaldehyde

+

$H_2\overset{\cdot\cdot}{N}$—NH

O_2N

NO_2

NO_2

NO_2

+ H_2O

benzaldehyde 2,4-DNP

7. Synthesis of Alkenes With the Wittig Reaction

$(C_6H_5)_3\overset{+}{P}$—CH$_3$ I$^-$

butyllithium

C=P(C$_6$H$_5$)$_3$ ⟷ ^-C—$\overset{+}{P}$(C$_6$H$_5$)$_3$

+ C=P(C$_6$H$_5$)$_3$ ⟶ C + O=P(C$_6$H$_5$)$_3$

EXERCISES

Reactivity of Carbonyl Compounds

19.1   Which member of each of the following pairs of compounds reacts faster with sodium borohydride?
(a)   cyclopropanone or cyclopentanone
(b)   acetophenone or benzaldehyde
(c)   acetone or 3,3-dimethyl-2-butanone

Answers:
(a) Cyclopropanone is the more reactive because of its greater ring strain. Thus, cyclopropanone reacts faster with sodium borohydride.
(b) Both compounds have resonance-stabilized carbonyl groups. However, acetophenone also has a methyl group bonded to the carbonyl carbon atom and is a ketone. Thus, acetophenone has a more stabilized carbonyl group and is less reactive. The carbonyl group of acetophenone is also more sterically hindered than the carbonyl group of benzaldehyde. Thus, benzaldehyde reacts faster with sodium borohydride.
(c) Both compounds are ketones, and the inductive effects of the alkyl groups are similar in both. However, one of the alkyl groups of 3,3-dimethyl-2-butanone is a *tert*-butyl group. Therefore, the carbonyl group of this compound is more sterically hindered than the carbonyl group of acetone. As a result, acetone reacts faster with sodium borohydride.

19.2   Which member of each of the following pairs of compounds reacts faster with sodium borohydride?
(a)   benzaldehyde or acetaldehyde
(b)   cyclopentanone or cyclohexanone
(c)   *p*-trifluoromethylbenzaldehyde or benzaldehyde

Answers:
(a) Benzaldehyde has a resonance-stabilized carbonyl group and is less reactive than an aldehyde such as acetaldehyde that does not have such stabilization. The carbonyl group of benzaldehyde is also more sterically hindered than the carbonyl group of acetaldehyde. As a result, acetaldehyde reacts faster with sodium borohydride.
(b) Reduction of cyclopentanone gives an alcohol with additional steric hindrance. In the case of cyclohexanone, the product does not have additional steric hindrance because the C—H and C—OH bonds are each staggered with respect to bonds on adjacent carbon atoms. Thus, cyclohexanone reacts faster with sodium borohydride.
(c) The trifluoromethyl group is inductively electron withdrawing, and it destabilizes the dipolar resonance form of the carbonyl group. As a result, *p*-trifluoromethylbenzaldehyde is more reactive than benzaldehyde.

Equilibrium Constants of Hydration Reactions

19.3   The equilibrium constants for the formation of hydrates of acetaldehyde and chloroacetaldehyde are 1 and 37, respectively. Explain whether you expect the equilibrium constant for formation of the hydrate of trichloroacetaldehyde to be greater or less than 37.

Answer:
Each chlorine atom withdraws electron density from the carbonyl carbon by an inductive effect, and therefore increases the equilibrium constant for the reaction. The cumulative effect is huge: if each additional chlorine atom increase the equilibrium constant by a factor of 37, the estimated equilibrium constant is $(37)^3$, which is approximately 5×10^4.

19.4   The equilibrium constant for formation of a hydrate of acetone is 1.4×10^{-3}. Explain whether you expect the equilibrium constant for formation of the hydrate of 1,3-dichloroacetone to be greater or less than 1.4×10^{-3}.

Answer:
As explained in exercise 19.3, each additional chlorine atom increases the equilibrium constant for the reaction. Therefore, the equilibrium constant for formation of the hydrate of 1,3-dichloroacetone is greater than 1.4×10^{-3}.

19.5   Explain why the methoxy group of *p*-methoxybenzaldehyde decreases the equilibrium constant for hydration relative to benzaldehyde, whereas the methoxy group of *m*-methoxybenzaldehyde increases the equilibrium constant.

Answer:
The *p*-methoxy group stabilizes the carbonyl group by a resonance interaction in which the methoxy group releases electrons through the benzene ring to the carbonyl group. The *m*-methoxy group destabilizes the carbonyl group somewhat by an inductive effect in which the methoxy group withdraws electron density. No donation of electrons by resonance is possible for two groups located in *meta* positions.

19.6   The equilibrium constants for the formation of hydrates of acetone, acetophenone, and benzophenone are 1.4×10^{-3}, 6.6×10^{-6}, and 1.7×10^{-7}, respectively. Explain why the second phenyl group of benzophenone has a much smaller effect on the equilibrium constant than the phenyl group of acetophenone compared to acetone.

Answer:
   One phenyl group can stabilize the carbonyl group because it can exist in a coplanar conformation with the carbonyl group, allowing the π electrons to overlap. Two phenyl groups cannot both be coplanar with the carbonyl group because there is a steric repulsion between the *ortho* hydrogen atoms of the two rings. The *ortho* hydrogen atoms of the two rings would have to occupy the same space if rings were coplanar. Rotation of the ring about the bond to the carbonyl carbon atom relieves this repulsion but eliminates the resonance interaction with the carbonyl group. The resonance interaction between the carbonyl group and the first phenyl group stabilizes acetophenone compared to acetaldehyde, but the second phenyl group has only a small effect.

One phenyl group rotates out of the plane, so the
π electrons in the two benzene rings do not overlap.

19.7   Explain why the equilibrium constant for hydration of cyclopropanone is significantly larger than for hydration of cyclopentanone.

Answer:
   Cyclopropanone has a larger equilibrium constant because it has more ring strain cyclopentanone. The hydrate of cyclopropanone is less strained, so the reaction is favored.

19.8   Considering the role of torsional strain in determining cycloalkane stability, predict the order of the equilibrium constants for hydration of cyclopentanone and cyclohexanone.

Answer:
   Hydration of cyclopentanone gives a hydrate with additional torsional strain between the C—OH bonds and the adjacent C—H bonds. The hydrate of cyclohexanone does not have added torsional strain because the C—OH bonds are staggered with respect to adjacent C—H bonds. Thus, the equilibrium constant for formation of the hydrate of cyclohexanone is larger.

Formation of Cyanohydrins

19.9   Explain why hydrogen cyanide reacts with 2-propanone to give a good yield of an addition product, but 2,2,4,4-tetramethyl-3-pentanone gives a poor yield in the same reaction.

Answer:
   The carbonyl carbon atom of 2,2,4,4-tetramethyl-3-pentanone has two *tert*-butyl groups bonded to it at 120° angles. Formation of the cyanohydrin decreases the bond angle between the two *tert*-butyl groups to 109°, resulting in increased steric crowding. Thus, the reaction is less favored than the reaction with 2-propanone.

19.10   Explain why the equilibrium constant for formation of a cyanohydrin of cyclohexanone is much larger than the equilibrium constant for formation of a cyanohydrin of cyclopentanone.

Answer:
   Adding HCN to cyclopentanone gives a cyanohydrin with additional torsional strain between the C—OH and C—CN bonds and the adjacent C—H bonds. The cyanohydrin of cyclohexanone does not have added torsional strain because the C—OH and C—CN bonds are each staggered with respect to adjacent C—H bonds. Thus, cyclohexanone has the larger equilibrium constant.

19.11   Explain why the equilibrium constant for formation of a cyanohydrin of butanone is much larger than the equilibrium constant for the formation of a cyanohydrin of 3,3-dimethylbutanone.

Answer:
   The carbonyl carbon atom of 3,3-dimethyl-2-butanone has a *tert*-butyl and a methyl group bonded to it at 120° angles. Formation of the cyanohydrin decreases the bond angle between the two groups to 109°, resulting in increased steric strain. The alkyl groups in 2-butanone are methyl and ethyl, so the steric repulsion is far less. Thus, the reaction with 3,3-dimethylbutanone is less favored than the reaction with butanone.

19.12 Is the equilibrium constant for formation of a cyanohydrin of *p*-methoxylbenzaldehyde larger or smaller than the equilibrium constant for benzaldehyde?

Answer:
The *p*-methoxy group stabilizes the carbonyl group by a resonance interaction in which the methoxy group releases electrons through the benzene ring to the carbonyl group. Thus, its conversion to a cyanohydrin, in which this resonance stabilization is lost, is unfavorable. Therefore, the equilibrium constant for cyanohydrin formation is smaller for *p*-methoxylbenzaldehyde than for benzaldehyde.

19.13 Is the equilibrium constant for formation of a cyanohydrin of *p*-methylbenzaldehyde larger or smaller than the equilibrium constant for benzaldehyde?

Answer:
The *p*-methyl group donates electron density through the aromatic ring to the carbonyl group; hence, it stabilizes the carbonyl group. Thus, the conversion of *p*-methylbenzaldehyde to a cyanohydrin, in which this stabilization is lost, is less favorable than for benzaldehyde. The equilibrium constant for formation of the cyanohydrin of *p*-methylbenzaldehyde is smaller.

19.14 Is the equilibrium constant for formation of a cyanohydrin of *p*-ethoxybenzaldehyde larger or smaller than the equilibrium constant for *p*-dimethylaminobenzaldehyde?

Answer:
Nitrogen is a better donor of electrons by resonance than oxygen. Therefore, the dimethylamino group stabilizes the carbonyl group by a resonance interaction more effectively than the *p*-ethoxy group. Thus, the conversion of *p*-dimethylaminobenzaldehyde to a cyanohydrin, in which this stabilization is lost, is less favorable than for *p*-ethoxybenzaldehyde.

19.15 Explain why the equilibrium constant for formation of a cyanohydrin of 3,3,5-trimethylcyclohexanone is smaller than the equilibrium constant for cyclohexanone.

Answer:
Conversion into a cyanohydrin must place either a hydroxyl or a cyano group in an axial position, so there will be a 1,3-diaxial interaction with the axial methyl group at C-3. This steric repulsion between groups decreases the stability of the product and therefore decreases the equilibrium constant compared to that of cyclohexanone.

19.16 Two cyanohydrins of 4-*tert*-butyl-butylcyclohexanone exist. Which is the kinetic product? Which is the thermodynamic product?

Answer:
Attack of the cyanide occurs fastest in the equatorial direction to form the cyanohydrin with the hydroxyl group in the axial position. This is the kinetic product. However, the hydroxyl group has a larger conformational preference for the equatorial position than the cyano group. As a consequence, the reversible formation of the cyanohydrin under equilibrium conditions favors formation of the compound with the cyano group axial and the hydroxyl group equatorial.

19.17 When benzaldehyde is heated with acetone cyanohydrin and a catalytic amount of base, the following reaction occurs. Explain whether you expect the equilibrium constant to be greater or less than 1.0. Write a mechanism for the reaction.

Answer:
There is no direct reaction between the two reactants. Each carbonyl compound can exist in equilibrium with its cyanohydrin and HCN. Thus, the HCN is simply distributed between the two carbonyl compounds. The mechanism is the standard one for addition of HCN to a carbonyl compound. The equilibrium constants for formation of the cyanohydrin for benzaldehyde and propanone are 210 and 30, respectively. Thus, the reaction as written has $K = 7$.

19.18   Write the structure of the product for the reaction of one equivalent of HCN with each of the following compounds.

Answers:
(a) Both carbonyl groups are equivalent, and only one reacts with one equivalent of HCN.

(b) The equilibrium constant for the saturated aldehyde side chain is larger than for benzaldehyde, so reaction occurs at that site.

(c) The ketone group that is para to the methoxy group is resonance stabilized, so the equilibrium constant for its reaction with HCN is smaller than for the keto group that is *meta* to the methoxy group. The methoxy group is inductively electron withdrawing, and it destabilizes the dipolar resonance form of the carbonyl group *meta* to it, so it is more reactive.

(d) The cyclic ketone is sterically hindered by the two methyl groups at the adjacent carbon atom. Thus, the cyanohydrin formed occurs at the methyl ketone of the chain bonded to the ring.

Addition of Alcohols to Carbonyl Compounds

19.19   Identify each of the following as a hemiacetal, hemiketal, acetal, or ketal.

Answers:
(a) acetal
(b) ketal
(c) acetal
(d) hemiacetal

(a) $CH_3CH_2CH(OCH_3)_2$ (b) $CH_3CH_2C(OCH_3)_2CH_3$

(c) $CH_3\overset{\overset{\displaystyle OCH_2CH_3}{|}}{C}HOCH_2CH_3$ (d) $CH_3CH_2\overset{\overset{\displaystyle OCH_3}{|}}{C}HOH$

19.20   Identify each of the following as a hemiacetal, hemiketal, acetal, or ketal.

Answers:
(a) ketal
(b) diacetal   (a) CH₃CH₂C(OCH₃)₂CH₃
(c) hemiketal
(d) ketal

(a) $CH_3CH_2C(OCH_3)_2CH_3$ with OCH₃ groups above and below

(b) $(CH_3O)_2CHCH_2CH(OCH_3)_2$

(c)

(d)

19.21   Identify each of the following as a hemiacetal, hemiketal, acetal, or ketal.

Answers:
(a) acetal
(b) hemiacetal (a)
(c) hemiketal
(d) ketal

(b)

(c)

(d)

19.22   Identify each of the following as a hemiacetal, hemiketal, acetal, or ketal.

Answers:
(a) hemiacetal
(b) ketal   (a)
(c) acetal
(d) acetal

(b)

(c)

(d)

19.23   Identify the functional groups in talaromycin A, a substance found in the fungus that grows in poultry litter.

Answers:
(a) primary alcohol
(b) secondary alcohol
(c) ketal

talaromycin A

19.24   Identify the functional groups in daunosamine, a component of Adriamycin, used in cancer chemotherapy.

Answers:
(a) secondary alcohol
(b) amino group
(c) hemiacetal

daunosamine

19.25   Is the equilibrium constant for the following reaction greater than or less than 1.0?

$$CF_3CHO + CH_3CH(OCH_3)_2 \rightleftharpoons CH_3CHO + CF_3CH(OCH_3)_2$$

Answer: The equilibrium constants for the formation constants for acetals parallel those for the formation of hydrates. The fluoro groups favors formation of the acetal of trifluoroethanal compared to ethanal. Thus, the equilibrium constant is greater than one.

19.26   Which compound should exist to the larger extent as a hemiacetal, 4-hydroxybutanal or 5-hydroxypentanal?
Answer:
Formation of a five-membered hemiacetal ring of 4-hydroxybutanal is less favored than formation of a six-membered hemiacetal ring of 5-hydroxypentanal because there is torsional strain in the five-membered ring. The six-membered ring has no torsional strain and is favored.

19.27   Benzaldehyde reacts with 1,2-propanediol to give two isomeric acetals. Draw their structures.
Answer:
The benzene ring and the methyl group can be *cis* or *trans* in the cyclic acetal.

19.28   Acetone reacts with 1,2,3-propanetriol to give two isomeric ketals. Draw their structures.
Answer:
Acetone can form a ketal with a six-membered ring by reacting with the C-1 and C-3 hydroxyl groups, or a ketal with a five-membered ring by reacting with the C-1 and C-2 hydroxyl groups.

19.29   2-Oxopropanal reacts with excess methanol in an acid-catalyzed reaction to give a compound with molecular formula $C_5H_{10}O_3$. Draw its structure.
Answer:
An acetal forms. The ketone is less reactive, so ketal formation is not favored.

19.30   2-Oxopropanal reacts with excess ethylene glycol in an acid-catalyzed reaction to give a compound with molecular formula $C_7H_{12}O_4$. Draw its structure. What is the difference between the product of this reaction and the product in Exercise 19.29?
Answer:
Both an acetal forms at C-1 and a ketal forms at C-2 because the equilibrium constants for formation of cyclic derivatives are more favorable than those for reaction with two moles of an alcohol to form acyclic derivatives.

19.31 What carbonyl compound and alcohol are required to form each of the following compounds?

(a)

$HOCH_2CH_2OH$

(b)

$HOCH_2CH_2OH$

(c)

CH_3OH

(d)

CH_3CH_2OH

19.32 What carbonyl compound and alcohol are required to form each of the following compounds?

(a)

$HOCH_2CH_2OH$

(b)

$HOCH_2CH_2OH$

(c)

$HOCH_2CH_2OH$

Using Protecting Groups in Synthesis

19.33 Outline a series of steps to carry out the following syntheses.

(a)

Answers:
(a) Step 1. Prepare a cyclic ketal of the ketone using ethylene glycol. Step 2. Reduce the ester with lithium aluminum hydride to give the primary alcohol. Step 3. Hydrolyze the ketal with dilute acid.

(b) Step 1. Prepare a cyclic acetal of the aldehyde using ethylene glycol. Step 2. Oxidize the secondary alcohol to a ketone using PCC. Step 3. Add the Grignard reagent prepared from iodomethane. Step 4. Hydrolyze the acetal with dilute acid, which will also convert the magnesium alkoxide to a tertiary alcohol.

CHO

(b)

H   OH

1. HOCH$_2$CH$_2$OH
2. PCC

3. CH$_3$MgI
4. H$_3$O$^+$

CHO

CH$_3$ OH

(c) Prepare a cyclic ketal of the ketone using ethylene glycol. Reduce the ester with lithium aluminum hydride to give the primary alcohol. Hydrolyze the ketal with dilute acid.

(c)

O

O

O   OCH$_3$
H

1. HOCH$_2$CH$_2$OH
2. LiAlH$_4$ / ether

3. H$_3$O$^+$

O

CH$_2$OH

H

19.34   Outline a series of steps to carry out the following syntheses.

Answers: (a) Step 1. Prepare either an acetal of the aldehyde with methanol or a cyclic acetal using one mole of ethylene glycol. The ketone is less reactive, and it remains unprotected. Step 2. Reduce the ketone with sodium borohydride to give a secondary alcohol. Step 3. Hydrolyze the acetal with dilute acid.

(a)

O

O

H

1. HOCH$_2$CH$_2$OH
(1 equivalent)

2. NaBH$_4$
3. H$_3$O$^+$

OH

O

H

(b) Step 1. Prepare a THP derivative of the alcohol using dihydropyran. Step 2. Add the Grignard reagent prepared from iodomethane. Step 3. Hydrolyze the THP derivative with dilute acid, which will also convert the magnesium alkoxide to a secondary alcohol.

CHO

(b)

H   OH

1. dihydropyran
(DHP)

2. CH$_3$MgI / ether
3. H$_3$O$^+$

HO   CH$_3$

H   OH

(c) Step 1. Prepare a THP derivative of the alcohol using dihydropyran. Step 2. Add sodium amide to make the acetylide ion of the acetylene. Step 3. Add bromoethane to alkylate the acetylide ion. Step 4. Hydrolyze the THP derivative with dilute acid. Step 5. Oxidize the secondary alcohol to a ketone using PCC.

(c)

1. dihydropyran
(DHP, gives THP derivative)
2. NaH (gives acetylide)
3. CH_3CH_2Br
4. H_3O^+
5. PCC

19.35   Outline a series of steps to carry out the following syntheses.

Answers:(a) Step 1. Prepare a cyclic ketal using ethylene glycol. Step 2. Obtain the secondary alcohol by hydroboration–oxidation of the carbon–carbon double bond. Step 3. Hydrolyze the ketal with dilute acid.

(a)

1. $HOCH_2CH_2OH$
2a. BH_3
2b. H_2O_2 / OH^-
3. H_3O^+

(b) Step 1. Prepare a THP derivative of the alcohol using dihydropyran. Step 2. Form the Grignard reagent of this protected bromo compound. Step 3. React the Grignard reagent with ethanal. Step 4. Hydrolyze the THP derivative with dilute acid.

(b) $BrCH_2CH_2CH_2OH$

1. dihydropyran
2. Mg / ether
3. CH_3CHO
4. H_3O^+

$HO-\overset{\overset{\displaystyle CH_3}{|}}{C}HCH_2CH_2OH$

(c) Step 1. Oxidize the secondary alcohol using PCC. Step 2. Prepare a ketal using ethylene glycol. Step 3. Add sodium amide to make the acetylide ion of the acetylene. Step 4. React the acetylide ion with ethanal to give the acetylenic alcohol. Step 5. Hydrolyze the ketal derivative with dilute acid.

(c) Step 1. Oxidize the secondary alcohol using PCC. Step 2. Prepare a ketal using ethylene glycol. Step 3. Add sodium amide to make the acetylide ion of the acetylene. Step 4. React the acetylide ion with ethanal to give the acetylenic alcohol. Step 5. Hydrolyze the ketal derivative with dilute acid.

(c)

1. PCC
2. HOCH$_2$CH$_2$OH
3. NaH
4. CH$_3$CHO
5. H$_3$O$^+$

19.36 Outline a series of steps to carry out the following syntheses.

Answers: (a) Step 1. Prepare either an acetal of the aldehyde with methanol or a cyclic acetal using one mole of ethylene glycol. The ketone is less reactive, and it remains unprotected. Step 2. Reduce the ketone with sodium borohydride to give a secondary alcohol. Step 3. Hydrolyze the acetal with dilute acid.

(a)

1. HOCH$_2$CH$_2$OH (1 equiv.)
2. NaBH$_4$
3. H$_3$O$^+$

(b) Step 1. Prepare a cyclic ketal of the diol using acetone. Step 2. Form the Grignard reagent of this protected bromo compound. Step 3. React the Grignard reagent with ethanal. Step 4. Hydrolyze the cyclic ketal with dilute acid.

(b)   HOCH$_2$CHCH$_2$CH$_2$CH$_2$Br

1. CH$_3$COCH$_3$
2. Mg / ether
3. CH$_3$CHO
4. H$_3$O$^+$

HOCH$_2$CHCH$_2$CH$_2$CHCH$_3$

(c) Step 1. Prepare a THP derivative of the alcohol using dihydropyran. Step 2. Add sodium amide to make the acetylide ion of the acetylene. Step 3. React the acetylide ion with ethanal to give the acetylenic alcohol. Step 4. Hydrolyze the THP derivative with dilute acid.

(c)

1. dihydropyran
2. NaH
3. CH_3CHO
4. H_3O^+

19.37   Reduction of the following compound by the Wolff–Kishner method gives $C_{10}H_{20}O_2$, but Clemmensen reduction gives C_8H_{16}. Why do the products differ?

(a)

H_2NNH_2 / KOH

Wolff–Kishner

(b)

Zn / HCl

Clemmensen

Answer: (a) Under the basic conditions of the Wolff–Kishner reduction, the ketone is reduced to a methylene group and the cyclic ketal is not affected. (b) Under the acidic conditions of the Clemmensen reduction, the ketal hydrolyzes and both ketones of the product are reduced.

19.38 (a) Draw the structure of the product obtained from the following reactant using Wolff–Kishner conditions. (b) What product would be observed for a Clemmensen reduction?

Answers: (a) Under the basic conditions of the Wolff–Kishner reduction, the ketone is reduced to a methylene group and the cyclic ketal is not affected. (b) Under the acidic conditions of the Clemmensen reduction, the ketal hydrolyzes and both ketones of the product are reduced.

(a)

(b)

Thioacetals and Thioketals

19.39 Draw the structure of the product of each of the following reactions.

Answers:
(a) The aldehyde reacts with two moles of the thiol to form a thioacetal.
(b) The ketone reacts with one mole of the 1,2-dithiol to form a cyclic thioketal.
(c) The ketone reacts with one mole of the 1,3-dithiol to form a cyclic thioketal.

19.40 Thiols react with dihydropyran in an acid-catalyzed reaction to form an addition product. Write the mechanism for the reaction and draw the structure of the addition product.

Answers: (a) Dihydropyran is protonated. (b) The nucleophilic sulfur of the thiol reacts with the carbocation to give the addition product.

19.41 Aldehydes and ketones react with 2-thioethanol to give cyclic derivatives. Draw the structures of two possible products from the reaction of 4-*tert*-butylcyclohexanone with this reagent.

Answer: The five-membered cyclic product can have either an equatorial or an axial oxygen atom.

353

19.42 Predict the product of the reaction of one equivalent of 1,2-ethanedithiol with the following steroid.

Answer: The cyclohexanone ring reacts because it produces a derivative with no torsional strain. The cyclic thioketal of the cyclopentanone would have torsional interactions between the C—S and C—H bonds on adjacent carbon atoms.

Addition of Nitrogen Compounds

19.43 Write the structure of the product for each of the following combinations of reactants.
   (a)   ethanal and methylamine
   (b)   acetone and ethylamine
   (c)   benzaldehyde and hydrazine
   (d)   3-pentanone and hydroxylamine
   (e)   1-phenyl-2-propanone and semicarbazide

Answers:

19.44   What reactants are required to form each of the following structures?

(a) (b) (c)

Answers:

(a) NH_2OH

(b) $C_6H_5NHNH_2$

(c) $NH_2NHCONH_2$

19.45   What reactants are required to form each of the following imines?

(a)   (b)

Answers:

(a)   + NH_2 →

(b)   +   →

19.46 What reactants are required to form each of the following imines?

(a) (b)

Answers:

(a)

(b)

19.47 Explain why reaction of cyclohexanone with hydroxylamine yields a single product. However, cyclopentanecarbaldehyde yields two isomeric oximes.

Answer: Geometric isomers are possible about the carbon–nitrogen double bond if the two groups bonded to the sp²-hybridized carbon atom are nonequivalent. (a) The two methylene groups of the ring in cyclohexanone oxime are equivalent. (b) In the oxime of cyclopentanecarbaldehyde, there is a hydrogen atom and a cyclopentane ring, so *cis* and *trans* isomers can form.

(a)

(b)

19.48 Draw the structure of the product of reaction of hydrazine with two molar equivalents of benzaldehyde.

Answer:

356

19.49 2,4-Pentanedione reacts with hydroxylamine to yield 3,5-dimethylisoxazole. Write a mechanism for the reaction.

3,5-dimethylisoxazole

Answer: An oxime forms with one of the two equivalent carbonyl groups. The hydroxyl group of the oxime adds to the other carbonyl carbon atom, giving a hemiacetal that subsequently dehydrates to give a heterocyclic aromatic compound.

19.50 2,4-Pentanedione reacts with hydrazine to yield 3,5-dimethylpyrazole, not an isomeric diimine. Explain why the pyrazole forms.

Answer: A hydrazone forms with one of the two equivalent carbonyl groups. The amino group of the hydrazone adds to the other carbonyl carbon atom, giving a hemiaminal. Hemiaminals normally dehydrate. However, in this case, dehydration involving the carbon atom adjacent to the carbon atom bearing the hydroxyl group gives a more stable aromatic compound.

3,5-dimethylpyrazole

not formed

3,5-dimethylpyrazole

Ylide Chemistry and the Wittig Reaction

19.51   Draw the structure of the product of each of the following reactions.

(a)

$(C_6H_5)_3P=CH_2$ →

(b)

$(C_6H_5)_3P=CHC_6H_5$ →

(c)   CHO $=P(C_6H_5)_3$ →

19.52   Outline a synthesis of each of the following compounds using a Wittig reaction.
(a) ethylidenecyclopentane (b) 2-ethyl-1-pentene   (c) 4-propyl-3-heptene

Answers:
(a) Step 1. Prepare the ethyltriphenylphosphonium salt. Step 2. Add butyllithium. Step 3. React the ylide with cyclopentanone.

(b) Step 1. Prepare the methyltriphenylphosphonium salt. Step 2. Add butyllithium. Step 3. React the ylide with 3-hexanone.

(c) Step 1. Prepare the propyltriphenylphosphonium salt. Step 2. Add butyllithium. Step 3. React the ylide with 4-heptanone.

19.53   Draw the structures of the two products formed in the following reaction.

Answer:

CH_3CH_2—$\overset{O}{\overset{\|}{C}}$—H + $CH_3CH_2\overset{CH_3}{\overset{|}{C}}$=$P(C_6H_5)_3$ →

$\underset{CH_3CH_2}{\overset{H}{}}$C=C$\underset{CH_2CH_3}{\overset{CH_3}{}}$

+

$\underset{H}{\overset{CH_3CH_2}{}}$C=C$\underset{CH_2CH_3}{\overset{CH_3}{}}$

19.54 Draw the structure of the ylide formed by reacting 1-bromo-2-butyne with triphenylphosphine, followed by reaction with sodium ethoxide. Explain why such a relatively weak base is sufficient to generate the ylide.

Answer: The negative charge on C-1 of the original halogen compound is delocalized by resonance with the triple bond. Thus, a weaker base than butyllithium is sufficient to abstract the proton.

$$CH_3-C\equiv C-CH_2Br \ + \ (C_6H_5)_3P: \longrightarrow \ CH_3-C\equiv C-CH_2-\overset{+}{P}(C_6H_5)_3$$

$$CH_3-C\equiv C-CH_2-\overset{+}{P}(C_6H_5)_3 \xrightarrow[\text{ethanol}]{CH_3CH_2O^-} \ CH_3-C\equiv C-\overset{-}{C}H-\overset{+}{P}(C_6H_5)_3$$

19.55 Suggest a mechanism for the following reaction of a sulfur ylide with a ketone.

+ $(CH_3)_2S\!=\!\!=\!CH_2$
dimethylsulfonium
methylide

Answer: The dipolar resonance form of dimethylsulfonium methylide has a negative charge on the methylene carbon atom, which is a nucleophilic site. It attacks the carbonyl carbon atom giving an alkoxide, which then forms the epoxide via an intramolecular Williamson synthesis. The leaving group is dimethyl sulfide.

19.56 Acetone reacts with diazomethane to yield 2,2-dimethyloxirane Write a mechanism for this reaction.

$$:N\equiv\overset{+}{N}-CH_2^-$$
diazomethane

Answer: The dipolar resonance form of diazomethane has a negative charge on the methylene carbon atom, which is a nucleophilic site. It attacks the carbonyl carbon atom, giving an alkoxide which then forms the epoxide via an intramolecular Williamson synthesis. The leaving group is nitrogen.

19.57 The reaction of dimethylsulfonium methylide, $(CH_3)_2S=CH_2$, with 4-*tert*-butylcyclohexanone can potentially give two isomeric epoxide products. Draw their structures. What factors may control the relative amounts of the two compounds formed?

Answer: The initial attack of the negatively charged methylene group is affected by the steric environment of the carbonyl carbon atom. Attack along an equatorial pathway is less sterically hindered that attack from an axial direction. Thus, the methylene group of the more favored epoxide product is in an equatorial-like position.

4-*tert*-butylcyclohexanone

19.58 Suggest a mechanism for the following reaction of cyclohexanone with triethylphosphonoacetate. Draw the structure of the by-product of the reaction.

Answer: Sodium hydride abstracts a proton from the central methylene group. The resulting negative charge is resonance stabilized by conjugation with the carbonyl group. The dipolar resonance form of the phosphorus grouping has a positive charge on the phosphorus atom. Thus, the conjugate base of the compound resembles the ylide used in the Wittig reaction.

CARBOXYLIC ACIDS

20.1 The Carboxyl and Acyl Groups

The reactivity of the carboxyl group depends on the interplay of its π electrons, its nonbonded electron pairs, and the hydroxyl group. The carboxyl group is resonance stabilized, with two contributing dipolar resonance structures. The lone-pair electrons of the hydroxyl oxygen atom are donated to the electron-deficient carbon atom in one dipolar resonance structure. As a result, the electron density of the carbonyl carbon atom in a carboxylic acid is larger than for aldehydes and ketones.

Carboxylic acids are the "parents" of acid derivatives. The acyl group, RCO, is bonded to an electronegative atom in carboxylic acids and their derivatives. The acyl group of an ester is bonded to the oxygen atom of an alkoxy group. The acyl group of an amide is bonded to a nitrogen atom. Cyclic esters and amides are lactones and lactams, respectively. The acyl group is bonded to a chlorine atom, a carboxyl group, and a thiolate group in acid chlorides, acid anhydrides, and thioesters, respectively.

20.2 Nomenclature of Carboxylic Acids

The common names of the unbranched carboxylic acids do not provide information about the number of carbon atoms contained in the chain, so the common names simply have to be learned. Carbon atoms in the parent chain are named with Greek lower case letters (α, β, γ, etc.) starting with the carbon atom directly bonded to the carboxyl carbon atom. Branches on the chain are indicated using these Greek letters to identify the location of substituents or alkyl groups.

IUPAC names of carboxylic acids are based on the alkane names, for which the suffice -oic acid replaces the final -e in the name of the alkane. The carbon chain is numbered starting with the carboxyl carbon atom, but the number 1 is not included in the name, since C-1 is automatically reserved for the carboxyl group. The carboxyl group takes priority over halogen atoms, alkyl groups, double or triple bonds, and other functional groups containing a carbonyl group. Other wise, the names of carboxylic acids follow the rules for the other functional groups we have considered.

The names of salts of carboxylic acids begin with the name of the metal ion followed by the name of the acid, modified by changing the suffice -oic acid to -oate. For example, the sodium salt of propanoic acid is sodium propanoate.

20.3 Physical Properties of Carboxylic Acids

Carboxylic acid molecules form intermolecular hydrogen bonds. Carboxylic acids exist as dimers, and consequently, they have higher boiling points than other hydrocarbons of similar molecular weight. The carboxyl group forms hydrogen bonds with water molecules, and low-molecular-weight carboxylic acids are soluble in water.

20.4 Acidity of Carboxylic Acids

Carboxylic acids are more acidic than alcohols. Loss of a proton from the carboxyl group leaves the carboxylate anion, which is stabilized by delocalization of two electron pairs over the carbon atom and two oxygen atoms. The formation of stable conjugate bases enhances the acidity of carboxylic acids.

Electronegative groups attached to the α-carbon atom pull electron density away from the carboxyl group. This inductive effect increases the acidity of carboxylic acids. Increased acidity increases the acid ionization constant, K_a, and decreases the pK_a.

Both the methoxy group and the nitro group of substituted ethanoic acids increase their acidity by an inductive effect. The oxygen atom of a methoxy group is more electronegative than carbon, hence it withdraws electron density carboxyl group. The nitrogen atom of the nitro group, which has a formal positive charge, is also electron withdrawing.

The acidity of a substituted benzoic acid is affected by the change in the electron density of the aromatic ring. The substituents do not interact by resonance with the carboxylate group, nor does the carboxylate group interact by resonance with the aromatic ring.

Organic Chemistry Study Guide: Key Concepts, Problems, and Solutions http://dx.doi.org/10.1016/B978-0-12-801889-7.00020-0

20.5 Carboxylate Ions

In water, unsubstituted carboxylic acids are present largely as the nonionized form. However, if the pH of a solution is greater than the pK_a of the acid, then the carboxylic acid exists as its conjugate base. At pH 7, physiological pH, carboxylic acids exist predominantly as carboxylate ions. The solubility of carboxylate anions in water provides an easy way to separate them form other, water-insoluble organic compounds.

20.6 Synthesis of Carboxylic Acids

Carboxylic acids are prepared from substrates with the proper hydrocarbon skeleton by oxidation of either an alcohol or an aldehyde. Because special methods are required to prepare aldehydes, the more common substrate for preparation of a carboxylic acid is the structurally related alcohol. The Jones reagent is used as the oxidizing agent

Oxidation of an alkylbenzene by $KMnO_4$ produces a benzoic acid. The haloform reaction oxidizes a methyl ketone by removing the methyl group. The reagent is a halogen in basic solution. Preparation of carboxylic acids with one more carbon atom than the substrate is accomplished by two alternate procedures, both of which commence with an alkyl halide. Conversion of an alkyl halide into a Grignard reagent followed by reaction with carbon dioxide gives a carboxylic acid. A haloalkane, or an aryl halide, can be used. The second method is based on nucleophilic substitution of a halide ion by a cyanide ion. The resulting nitrile is hydrolyzed to form a carboxylic acid. The limitation on the reaction is the first step, which is effective for primary haloalkanes and to a limited extent for secondary haloalkanes. The competing reaction is dehydrohalogenation of the haloalkane.

20.7 Reduction of Carboxylic Acids

It is more difficult to reduce carboxylic acids than to reduce aldehydes or ketones. The acidic carboxyl proton reacts with the metal hydride reagents commonly used for reduction of carbonyl groups. Only lithium aluminum hydride is sufficiently reactive to reduce a carboxylic acid to a primary alcohol. Sodium borohydride does not reduce carboxylic acids.

Diborane reduces carboxylic acids to give primary alcohols. Diborane does not readily reduce other unsaturated functional groups such as nitrile or nitro groups. Diborane cannot be used if the substrate contains a double bond because hydroboration would occur.

20.8 Decarboxylation Reactions

Loss of a carboxyl group and replacement by an atom such as hydrogen is decarboxylation. Carboxylic acids containing a β keto group undergo decarboxylation in a cyclic mechanism. Both β keto acids and malonic acids undergo this reaction.

The Hunsdiecker reaction replaces a carboxyl group by a halogen. The reagent is bromine, which reacts with the silver salt of the carboxylic acid. Mercury(II) oxide and bromine also can be used with the carboxylic acid. An unsaturated carboxylic acid cannot be used because bromine adds to double bonds.

20.9 Reactions of Carboxylic Acids and Derivatives—A Preview

The salts of carboxylic acids are weak nucleophiles but can be used to displace halide ion from a haloalkane to give esters. The most common reaction of carboxylic acids and their derivatives is attack of nucleophiles at the carbonyl carbon atom. However, in contrast to aldehydes and ketones, the tetrahedral intermediate is unstable, and it ejects a leaving group. This overall process is nucleophilic acyl substitution.

Although nucleophilic acyl substitution has the same stoichiometry as nucleophilic substitution at a saturated carbon atom, the mechanisms are very different. Nucleophilic acyl substitution occurs in two steps via a tetrahedral intermediate.

20.10 Conversion of Carboxylic Acids Into Acyl Halides

Carboxylic acids react with thionyl chloride to give acyl chlorides. The mechanism resembles the reaction of alcohols with thionyl chloride. The hydroxyl group acts as a nucleophile to displace a chloride ion from thionyl chloride. However, the subsequent displacement of the sulfur-containing moiety occurs by a nucleophilic substitution in the case of alcohols, and nucleophilic acyl substitution in the case of acids.

20.11 Ester Synthesis

The synthesis of esters can be carried out in several ways.
1. Alkylation of a carboxylate ion using a haloalkane, which resembles the Williamson ether synthesis. It occurs by a S_N2 mechanism and therefore is limited to primary haloalkanes.
2. Methylation of a carboxylic acid using diazomethane. The reaction is largely limited to this simplest diazo compound but will occur with other diazo compounds.
3. Reaction of acyl chlorides with alcohols, which occurs by nucleophilic acyl substitution. The reaction is general for all alcohols and any acyl chloride.
4. The Fischer esterification, which uses a carboxylic acid and an alcohol, is an acid-catalyzed equilibrium reaction that is driven to product by removal of the by-product, water.

20.12 Mechanism of Esterification

The Fischer esterification reaction occurs by nucleophilic attack of the alcohol on the carbonyl carbon atom of the carboxylic acid to give a tetrahedral intermediate. The acid catalyst protonates the carbonyl oxygen atom to increase the partial positive charge of the carbonyl carbon atom. Isotope studies show that the oxygen atom of the alcohol is the bridging atom of the ether. The configuration of a chiral alcohol is retained in this esterification reaction because the bond to the oxygen atom of the alcohol is not cleaved.

20.13 Brief Synthetic Review

The reactions we discussed in this chapter and in Chapter 19 can be combined in many ways in organic synthesis. Our brief review reminds us that there is almost always more than one way to convert one functional group into another. Sometimes a single regioselective reaction will act on one functional group and leave the other one unaltered. Other times, potentially reactive groups have to be protected.

20.14 Spectroscopy of Carboxylic Acids

Carboxylic acids are characterized by the strong absorption due to the carbonyl group in the infrared spectra of these compounds. The absorption occurs in the same region as the carbonyl groups of aldehydes and ketones, but the absorption for carboxylic acids occurs at slightly higher wavenumber, and tends to be somewhat broadened. The O—H bond of carboxylic acids absorbs in the same region as that for alcohols. However, the absorption is very much broader for carboxylic acids, and it overlaps the C—H absorptions.

The proton NMR spectra of carboxylic acids show a characteristic absorption in the 9–12 δ region for the strongly deshielded carboxyl proton. The α hydrogen atoms of carboxylic acids occur in the 2.0–2.5 δ region, which is the same region for the α hydrogen atoms of aldehydes and ketones.

The α carbon atoms of carboxylic acids have C-13 NMR absorptions in the 20 d region, which is at slightly higher field than for aldehydes and ketones. The carbonyl carbon atom is easily identified by its absorption in the 200 δ region.

Summary of Reactions

1. Synthesis of Carboxylic Acids by Oxidative Methods

$$\text{(cyclopentyl)}-CH_2OH \xrightarrow[\text{Jones Reagent}]{CrO_3 / H_2SO_4} \text{(cyclopentyl)}-CO_2H$$

$$-CHO \xrightarrow{Ag(NH_3)_2^+} -CO_2H$$

$$\xrightarrow{KMnO_4}$$

CH₃ ... OCH₂CH₃ → CO₂H ... OCH₂CH₃

$$\xrightarrow{Br_2 / OH^-} \quad + \ CHBr_3$$

2. Synthesis of Carboxylic Acids from Haloalkanes

$$\text{(phenyl)}-Br \xrightarrow[\text{2. } CO_2 \\ \text{3. } H_3O^+]{\text{1. Mg / ether}} \text{(phenyl)}-CO_2H$$

$$CH_3CH_2\overset{\overset{\displaystyle CH_3}{|}}{C}HCH_2-Br \xrightarrow[\text{2. } H_3O^+]{\text{1. NaCN / DMF}} CH_3CH_2\overset{\overset{\displaystyle CH_3}{|}}{C}HCH_2-\overset{\overset{\displaystyle O}{||}}{C}-OH$$

3. Reduction of Carboxylic Acids

$$-CO_2H \xrightarrow[\text{2. } H_3O^+]{\text{1. LiAlH}_4 \text{/ether}} -CH_2OH$$

$$\xrightarrow[\text{diglyme}]{B_2H_6}$$

CH₃ ... CO₂H → CH₃ ... CH₂OH

364

4. Decarboxylation of Carboxylic Acids

$$\text{heat}$$

$$+ \; CO_2$$

$$\frac{\text{1. } Ag_2O}{\text{1. } Br_2}$$

$$+ \; CO_2 + AgBr$$

5. Synthesis of Acyl Halides

$$SOCl_2$$

6. Synthesis of Esters

$$CH_3CH_2\!-\!Cl \; + \; {}^-O\!-\!\overset{\displaystyle O}{\overset{\|}{C}}\!-\!CH_2\overset{\displaystyle CH_3}{\overset{|}{C}}HCH_3 \longrightarrow CH_3CH_2\!-\!O\!-\!\overset{\displaystyle O}{\overset{\|}{C}}\!-\!CH_2\overset{\displaystyle CH_3}{\overset{|}{C}}HCH_3$$

$$+ \; CH_2N_2 \longrightarrow$$

$$+ \; N_2$$

$$\overset{\displaystyle CH_3}{\overset{|}{C}}H_3CHCH_2\!-\!OH \; + Cl\!-\!\overset{\displaystyle O}{\overset{\|}{C}}\!-\!CH_2CH_3 \longrightarrow \overset{\displaystyle CH_3}{\overset{|}{C}}H_3CHCH_2\!-\!O\!-\!\overset{\displaystyle O}{\overset{\|}{C}}\!-\!CH_2CH_3$$

$$\overset{\displaystyle CH_3}{\overset{|}{C}}H_3CHCH_2\!-\!OH \; + HO\!-\!\overset{\displaystyle O}{\overset{\|}{C}}\!-\!CH_2CH_3 \xrightarrow{H^+} \overset{\displaystyle CH_3}{\overset{|}{C}}H_3CHCH_2\!-\!O\!-\!\overset{\displaystyle O}{\overset{\|}{C}}\!-\!CH_2CH_3$$

Solutions to Exercises

Nomenclature

20.1 Give the common name for each of the following acids.
(a) $CH_3CH_2CO_2H$ (b) HCO_2H (c) $CH_3(CH_2)_4CO_2H$
(d) $CH_3(CH_2)_{10}CO_2H$ (e) $CH_3(CH_2)_{16}CO_2H$ (f) $CH_3(CH_2)_{14}CO_2H$

Answers:
(a) propionic acid (b) formic acid   (c) caproic acid   (d) lauric acid   (e) stearic acid   (f) palmitic acid

20.2 Give the common name for each of the following acids.

Answers:
(a) β-chlorobutyric acid
(b) α-bromopropionic acid
(c) β-bromo-γ-methylvaleric acid

(d) α-chloro-γ,γ-dimethylvaleric acid

(a) $\overset{\overset{\displaystyle Cl}{|}}{CH_3CHCH_2CH_2CO_2H}$

(b) $\overset{\overset{\displaystyle CH_3}{|}}{BrCHCO_2H}$

(c) $CH_3\overset{\overset{\displaystyle CH_3}{|}}{CH}\overset{\underset{\displaystyle Br}{|}}{CH}CH_2CO_2H$

(d) $CH_3\overset{\overset{\displaystyle CH_3}{|}}{\underset{\underset{\displaystyle CH_3}{|}}{C}}CH_2\overset{\overset{\displaystyle Cl}{|}}{CH}CO_2H$

20.3 Give the IUPAC name for each of the following acids.

Answers:
(a)   *trans*-2-hydroxycyclohexanecarboxylic acid
(b)   (*E*)-3-chloro-3-phenyl-2-propenoic acid
(c)   5-hexynoic acid
(d)   cyclodecanecarboxylic acid

(a)

(b)

(c) $HC\equiv C-CH_2CH_2CH_2CO_2H$ (d)

20.4 Give the IUPAC name for each of the following acids.

Answers:
(a)   3-methoxybenzoic acid
(b)   2-(3-methoxyphenyl)propanoic acid
(c)   3-ethyl-2-pentenoic acid
(d)   cyclopentylethanoic acid

(a)

(b)

(c)

(d)

20.5 The IUPAC name of ibuprofen, the analgesic in Motrin, Advil, and Nuprin, is 2-(4-isobutylphenyl)propanoic acid. Draw its structure.

Molecular Formulas

20.6 What is the general molecular formula for each of the following classes of compounds?
  (a) saturated acyclic carboxylic acid
  (b) saturated acyclic dicarboxylic acid
  (c) saturated monocyclic carboxylic acid
  (d) monounsaturated acyclic carboxylic acid

Answers:
(a) $C_nH_{2n}O_2$
(b) $C_nH_{2n-2}O_4$
(c) $C_nH_{2n-2}O_2$
(d) $C_nH_{2n-2}O_2$

20.7 Draw the structure of two isomers having the following characteristics
  (a) dicarboxylic acids with molecular formula $C_4H_4O_4$
  (b) carboxylic acids with molecular formula $C_4H_8O_2$
  (c) saturated carboxylic acids with molecular formula $C_5H_8O_2$

Answers:

(a) $HO_2CCH_2CH_2CO_2H$ HO_2CCHCO_2H with CH_3 substituent

(b) $CH_3CH_2CH_2CO_2H$ CH_3CHCO_2H with CH_3 substituent

(c) cyclopropane with CH_2CO_2H ; cyclobutane with CO_2H

20.8 10-Undecenoic acid is the anti fungal agent contained in Desenex and Cruex. Write the structure.

Answer: $CH_2{=}CH(CH_2)_8CO_2H$

Properties of Acids

20.9 Explain why 1-butanol is less soluble in water than butanoic acid.
Answer:
  Since butanoic acid has one more oxygen atom than butanol, it has more unshared pairs of electrons that act as hydrogen bond acceptors with water molecules. Therefore, it is more soluble than 1-butanol in water.

20.10 Explain why adipic acid is much more soluble in water than hexanoic acid.

Answer: Adipic acid has two carboxylic acid groups that act as hydrogen bond acceptors with water molecules, as well as another hydrogen atom bonded to oxygen that is a hydrogen bond donor to a water molecule. Therefore, it is much more soluble in water than hexanoic acid.

20.11 Explain why the boiling point of decanoic acid is higher than that of nonanoic acid.

Answer:Decanoic acid has one more methylene group than nonanoic acid, so it has stronger van der Waals attractive forces than nonanoic acid and a higher boiling point.

20.12 Explain why the boiling point of 2,2-dimethylpropanoic acid (164 °C) is lower than that of pentanoic acid (186 °C).

Answer:The shape of 2,2-dimethylpropanoic acid is approximately spherical, so it has a smaller surface area than the cylindrically shaped pentanoic acid. Thus, the van der Waals forces are smaller for 2,2-dimethylpropanoic acid, and its boiling point is lower.

20.13 Explain why the boiling point of 4-methoxybenzoic acid (278 °C) is higher than that of 2-methoxybenzoic acid (200 °C).

Answer: 4-Methoxybenzoic acid has a high boiling point because it can form intermolecular hydrogen bonds between carboxylic acid groups, forming a dimer.

20.14 Explain why the boiling point of *trans*-2-butenoic acid (185 °C) is higher than that of *cis*-2-butenoic acid (169 °C).

The *trans* isomer has a larger dipole moment than the *cis* isomer due to the contribution of both the methyl and carboxyl groups.

Acidity of Carboxylic Acids

20.15 The K_a of methoxyacetic acid is 2.7×10^{-4}. Explain why this value differs from the K_a of acetic acid (1.8×10^{-5}).

Answer: The methoxy group of methoxyacetic acid inductively attracts electron density, which stabilizes the conjugate base. Therefore, it is more acidic (has a larger K_a) than acetic acid.

20.16 The K_a values of benzoic acid and *p*-nitrobenzoic acid are $6\ 3 \times 10^{-5}$ and $3\ 8 \times 10^{-4}$, respectively. Explain why these values differ.

Answer: The higher K_a for *p*-nitrobenzoic acid means than the nitro group stabilizes its carboxylate ion compared to benzoate ion. The nitro group inductively attracts electron density, but there is a long distance between it and the carboxylate ion. The stabilization results from electron withdrawal from the aromatic ring by resonance. As a consequence of lower electron density at C-1 in the resonance-stabilized structure, the electrons of the carboxylate ion are inductively pulled toward the aromatic ring.

20.17 Estimate the pK_a values of the two carboxyl groups in 3-chlorohexanedioic acid.

Answer: The location of the chlorine atom with respect to the C-1 carboxyl corresponds to that in 3-chlorobutanoic acid, so its pK_a must be close to 4.06. The location of the chlorine atom with respect to the carboxylic acid of the C-6 carboxyl group corresponds to that in 4-chlorobutanoic acid, so its pK_a must be close to 4.52.

20.18 The pK_a of 3-cyanobutanoic acid is 4.44. Using the pK_a values of chlorine-substituted butanoic acids as a guide, estimate the pK_a of 2-cyanobutanoic acid.

Answer: The pK_a increases from 4.06 to 2.84 for 3-chlorobutanoic acid and 2-chlorobutanoic acid, respectively. A similar increase would be expected for the cyano compounds. The estimated pK_a is 3.22.

20.19 The pK_a for the first dissociation of dicarboxylic acids levels off at approximately 4.85. The pK_a of long-chain carboxylic acids levels off at approximately 4.55. What relationship exists between these two numbers? What structural features are responsible for this difference?

Answer: They differ by 0.3 units on the log scale, which corresponds to a factor of 2 in K_a. The long chain dicarboxylic acids are twice as acidic as the long chain carboxylic acids because there are two hydrogen atoms per molecule in the dicarboxylic acids. The difference reflects a statistical factor, not an influence of structure.

20.20 The difference between the pK_a values for dissociation of the first and second protons of the long-chain dicarboxylic acids is about 1 unit. The difference between the pK_a values for both oxalic and malonic acids is about 3 units. Explain these data, focusing on the pK_a for the second ionization step.

Answer:

The ionization constant for the transfer of a second proton from an acid to water is expected to be smaller than for the first proton because it is more difficult to remove a proton from a negatively charged ion than from a neutral molecule. In the case of the oxalate ion and the malonate ion, the carboxylate group and carboxylic acid group are close enough to form a hydrogen bond. As a consequence, more energy is required to remove the proton, and the acid is weaker.

20.21 The methoxy group is an effective donor of electrons and as a consequence is an activating group in electrophilic aromatic substitution. Explain why the pK_a of methoxyacetic acid (3.5) is less than that of acetic acid (4.7).

Answer:

The methoxy group cannot donate electrons by resonance in methoxyacetic acid as it does in an aromatic ring because there is an intervening methylene group in the acetic acid. The smaller pK_a of methoxyacetic acid indicates that the methoxy group stabilizes the carboxylate ion, but this stabilization is the result of inductive electron withdrawal by the oxygen atom.

20.22 The pK_a values of cyanoacetic acid and nitroacetic acid are 2.45 and 1.65, respectively. What do these data indicate about the substituent properties of —CN and —NO_2?

Answer:

The groups cannot affect the acidity of the acetic acids by resonance because there is an intervening methylene group. Therefore, the difference indicates that the nitro group is inductively more electron withdrawing than the cyano group. The nitro group has a formal positive charge on nitrogen. The carbon atom of the cyano group is partially positive as a result of the electronegativity difference between carbon and nitrogen atom. However, it does not have a formal charge.

20.23 The substituent effects of the hydroxyl and methoxy groups are quite similar, as evidenced by the pK_a values of *p*-hydroxy- and *p*-methoxybenzoic acids, which are 4.48 and 4.47, respectively. However, the pK_a values of *o*-hydroxy- and *o*-methoxybenzoic acids are 2.97 and 4.09, respectively. Explain why the values for the *ortho* isomers are so different.

Answer:

The enhanced acidity of the *o*-hydroxy compound compared to the *p*-methoxy compound indicates that its carboxylate group is stabilized. The hydrogen atom of the hydroxyl group can form a hydrogen bond with the oxygen atom of the carboxylate group. This type of interaction does not exist in the *p*-methoxy compound.

20.24 The pK_a values of *para*-substituted benzoic acids for the —PCl_2 and —$Si(CH_3)_3$ groups are 3.6 and 4.3, respectively. Based on these data, determine whether these groups are activating or deactivating in electrophilic aromatic substitution.

Answer:

The pK_a value of the phosphorus compound is smaller than that of benzoic acid. Thus, the phosphorus group withdraws electron density from the aromatic ring and stabilizes the carboxylate ion. The phosphorus group should be a deactivator in aromatic substitution reactions. The pK_a value of the silicon compound is slightly larger than that of benzoic acid. Thus, it donates electron density to the aromatic ring and destabilizes the carboxylate ion. The silicon group is a weak activator in aromatic substitution reactions.

20.25 *p*-Methoxybenzoic acid is a weaker acid than benzoic acid, but *p*-(methoxymethyl)benzoic acid is a stronger acid than *p*-methylbenzoic acid. Why does the methoxy group have opposite effects in these two cases?

Answer:

The smaller K_a for *p*-methoxybenzoic acid means that the methoxy group destabilizes its carboxylate ion. The methoxy group inductively attracts electron density and could stabilize the ion, but there is a long distance between it and the carboxylate group. The destabilization results from electron donation to the aromatic ring by resonance. As a consequence of higher electron density at C-1 in the resonance-stabilized structure, the carboxylate ion is destabilized. In the case of the methoxymethyl group, there is no resonance interaction between the oxygen atom and the aromatic ring. A methyl group is inductively electron donating, but the added methoxy group makes the methoxymethyl group somewhat inductively electron withdrawing. The net effect is a decrease in electron density at C-1, thus stabilizing the carboxylate ion.

20.26 The van del Waals radii of fluorine and hydrogen atoms arc similar. The pK_a values of o-, m-, and p-fluoro benzoic acids are 4.1, 3.9, and 3.3, respectively. The pK_a value of benzoic acid is 4.2. Explain the order of the pK_a values of the fluorobenzoic acids. Estimate the contribution of fluorine as an electron donor in terms of resonance.

Answer:

The pK_a values of o-, m-, and p-fluorobenzoic acid are 3.3, 3.9, and 4.1, respectively. All of the fluorine compounds are stronger acids than benzoic acid, suggesting that in each case the fluorine atom withdraws electron density from the ring and stabilizes the carboxylate ion. The effect is largest for the *ortho* compound and decreases with distance between the fluorine atom and the carboxylate ion, which is typical of an inductive effect. If the fluorine atom effectively donated electrons by resonance, the acidity of the *para* substituted compound would be greater than that of benzoic acid.

20.27 Compare the pK_a values of biphenyl-3-carboxylic acid (4.14) and biphenyl-4-carboxylic acid (4.21) to that of benzoic acid (4.20) and explain the different effect of the phenyl group on the pK_a values.

Answer:

The two phenyl rings do not interact by resonance because the *ortho* hydrogen atoms do not allow the two rings to be coplanar. As a result, the observed effect is that of an sp^2-hybridized carbon atom of the phenyl group, which is inductively electron withdrawing. That effect stabilizes the carboxylate ion. The effect, although small, is greater in the case of the 3-carboxylic acid because the phenyl ring is closer to the carboxyl group.

20.28 Which is the stronger acid in each of the following pairs of aromatic carboxylic acids? What accounts for the difference in acid strength?

Answers:

(a) The fluoro compound is more acidic because the fluorine atoms inductively withdraw electron density from the aromatic ring and stabilize the carboxylate ion.

(b) Oxygen is more electronegative than nitrogen. As a result of inductive electron withdrawal, the carboxylate ion of the oxygen compound is more stable than the carboxylate ion of the nitrogen compound. The furan derivative is more acidic.

(c) The inductive electron withdrawal by oxygen decreases with distance from the carboxylate ion. Thus, the first compound is more acidic.

20.29 The pK_a of benzoic acid is 4.2. The pK_a of probenecid is 3.4. Explain why probenecid is the stronger acid. (Probenecid is a drug that is used to treating gout and hyperuricemia.)

probenecid

Answer:

The nitrogen atom and the two oxygen atoms bonded to the sulfur atom make this group strongly electron withdrawing. As a result, the carboxylate ion is stabilized by a decrease in electron density at C-1.

20.30   Predict the pK_a of indomethacin, an anti-inflammatory agent.
Answer:

There are no strongly electron withdrawing groups near the carboxyl group. However, there is an sp²-hybridized carbon atom bonded to the spa methylene group, and it can withdraw electron density from that group, which is bonded to the carboxyl group. The effect on the acidity should be small, but the compound should be a stronger acid than acetic acid. Its pK_a is less than 4.72.

indomethacin

Carboxylate Anions

20.31   The pK_a of penicillin G is 2.8. Is it more soluble in stomach acid (pH ~2) or in blood (pH =7.4)?

Answer:
Since the pH of the stomach is less than the pK_a of the acid, it will be predominantly protonated at pH 2. However, in blood it will be predominantly a carboxylate ion. So penicillin G is more soluble in blood than in the stomach.

penicllin G

20.32   Sodium benzoate is used as a preservative in foods, but only if the pH is greater than 5. In what form is the compound present at pH 7?
Answer:
Carboxylic acids exist to a larger degree as carboxylate salts in a more basic medium and hence are more soluble. Thus, sodium benzoate is more soluble in foods whose pH is greater than 5.

20.33   Explain why benzoic acid with an ¹⁸O isotopic label in the hydroxyl oxygen atom can be prepared, but that it cannot be used in mechanistic studies in aqueous solutions.
Answer:
As the carboxylic acid, the two oxygen atoms are nonequivalent. However, in aqueous solution the carboxylate ion forms, and the two oxygen atoms become equivalent.

Synthesis of Carboxylic Acids
20.34   Outline the steps required to prepare cyclohexanecarboxylic acid from each of the following reactants.
(a) bromocyclohexane   (b) cyclohexanol   (c) cyclohexene   (d) vinylcyclohexane   (e) cyclohexylmethanol
Answers:
(a)   Prepare the Grignard reagent using magnesium and ether and then add it to carbon dioxide, followed acidification in workup.
(b)   Convert the alcohol into bromocyclohexane using PBr₃ and then proceed as in part (a).
(c)   Add HBr to the double bond to form bromocyclohexane and then proceed as in part (a).
(d)   Use ozone under oxidative workup conditions.
(e)   Oxidize the alcohol to the carboxylic acid using the Jones reagent

20.35   Outline the steps required to prepare hexanoic acid from each of the following reactants.
(a) 1-chloropentane   (b) 1-hexanol   (c) hexanal   (d) 1-hexene   (e) 1-heptene

Answers:
(a)   Prepare the Grignard reagent using magnesium and ether and then add it to carbon dioxide, followed by acidification in workup.
(b)   Oxidize the alcohol to the carboxylic acid using the Jones reagent.
(c)   Oxidize the aldehyde to the carboxylic acid using the Jones reagent.
(d)   Prepare 1-hexanol using B₂H₆ followed by treatment with basic hydrogen peroxide (the hydroboration–oxidation procedure). Then oxidize the alcohol using the Jones reagent.
(e)   Use ozone under oxidative workup conditions.

20.36 Outline the steps required to convert methylenecyclohexane to each of the following compounds.
(a) cyclohexanecarboxylic acid   (b) cyclohexylacetic acid   (c) 1-methylcyclohexanecarboxylic acid
Answers:
(a) Prepare cyclohexylmethanol using B_2H_6 followed by treatment with basic hydrogen peroxide (the hydroboration–oxidation procedure). Then oxidize the alcohol to the carboxylic acid using the Jones reagent.
(b) Add HBr in the presence of peroxide to give a primary halogen compound (anti-Markovnikov addition). Prepare the Grignard reagent using magnesium and ether and then add it to carbon dioxide, followed by acidification in workup. Alternatively, displace bromide ion by cyanide ion, followed by hydrolysis of the nitrile.
(c) Add HBr to give a tertiary halogen compound. Prepare the Grignard reagent using magnesium and ether and then add it to carbon dioxide, followed by acidification in workup.

20.37 Outline the steps required to convert p-ethylanisole into each of the following compounds.
(a) p-methoxybenzoic acid (b) 2-(p-methoxyphenyl)propanoic acid   (c) 3-(p-methoxyphenyl)butanoic acid

Answers:
(a) Oxidize the ethyl group to a carboxylic acid using a strong oxidizing agent such as potassium permanganate.
(b) Use NBS to prepare the secondary bromo compound of the ethyl side chain. Then, prepare the Grignard reagent using magnesium and ether. Third, add it to carbon dioxide, followed by acidification in workup.
(c) Use NBS to prepare the secondary bromo compound of the ethyl side chain. Second, prepare the Grignard reagent using magnesium and ether; third, add it to ethylene oxide to give 3-(p-methoxyphenyl)butanol. Oxidize the alcohol using the Jones reagent.

20.38 Fatty acids from natural sources are long-chain unbranched carboxylic acids that contain an even number of carbon atoms. Outline steps to convert the readily available dodecanoic acid (lauric acid) into the rare tridecanoic acid.
Answer:
First, reduce the carboxylic acid to dodecanol. Second, use PBr_3 to give 1-bromododecane. Third, prepare the Grignard reagent using magnesium and ether. Fourth, add the Grignard reagent to carbon dioxide, followed by acidification in workup.

20.39 Pivalic acid, $(CH_3)_3CCO_2H$, can be prepared from tert-butyl chloride. What method should be used?
Answer:
First, prepare the Grignard reagent using magnesium and ether. Second, add the Grignard reagent to carbon dioxide, followed by acidification in workup. The alternate method of displacing bromide ion by cyanide ion to give a nitrile cannot be accomplished with a tertiary halide.

20.40   Draw the structure of the product of each of the following reactions.

Answers:

(a)
$Ag(NH_3)^+$

(b)
$KMnO_4$

(c)
CH_3 $I_2/$ OH^-

(d)
Br
1. Mg (ether)
2. CO_2
3. H_3O^+
$-CO_2H$

(e)
CH_2Br
1. CN^-
2. H_3O^+
CH_2CO_2H

373

20.41 Draw the structure of the product of each of the following reactions.

Answers:

(a)

(b)

(c)

(d)

(e)

Reduction of Carboxylic Acids

20.42 Metal hydride reductions occur by nucleophilic attack at the carbonyl carbon atom of acyl derivatives. Reduction of carboxylic acids with hydride reagents occurs slowly, but reduction by diborane occurs rapidly. Based on the structure of BH_3, the active reagent in diborane reductions suggests the structure of the first intermediate formed in the reaction.

Answer:

Boron is electron deficient in BH_3 and can form a coordinate covalent bond to the oxygen atom of the carbonyl group. As a result, the hydrogen atom can be transferred to the carbonyl carbon atom. The cyclic intermediate would resemble that of the reaction of BH_3 with an alkene in the hydroboration reaction.

20.43 Diborane slowly reduces nitriles to amines but rapidly reduces aldehydes and ketones. Using the structure of BH_3, and the mechanism your wrote in Exercise 20.42, explain why nitriles react more slowly than aldehydes and ketones.

Answer:
The unshared pair of electrons of the nitrile can coordinate to the boron atom, but the linear geometry of the nitrile group does not allow close approach of the hydrogen atom to the carbon atom.

20.44 Lithium borohydride is a more active reducing agent than sodium borohydride, but less active than lithium aluminum hydride. Lithium borohydride reduces the ester group of the following compound selectively. Explain this selectivity.

Answer: Aluminum is more electropositive than boron. As a result, the electron pair in an Al—H bond is more available to the hydrogen atom, which can depart as a hydride ion in a reaction. Lithium borohydride is more active than sodium borohydride because the coordination of the smaller lithium ion is stronger than that for the larger sodium ion. Coordination of the metal ion polarizes the carbonyl bond and increases the partial positive charge on the carbonyl carbon atom. As a result, the center is more reactive toward hydride ion derived from the borohydride ion.

20.45 Draw the structure of the product of the following reaction. What relationship exists between this compound and the product of the reaction of Exercise 20.44?

Answer: BH_3 reduces the carboxylic acid group, but the ester group is unchanged. This compound is the methyl ester of the enantiomer of the product in Exercise 20.44.

Decarboxylation

20.46 Could the Hunsdiecker reaction be used to decarboxylate an unsaturated carboxylic acid?

Answer: No, because the bromine would react with the unsaturated carbon–carbon π bond and give an addition product.

20.47 Which carboxylic acid should decarboxylate the more easily in a Hunsdiecker reaction, benzoic acid or cyclohexanecarboxylic acid?

Answer: Cyclohexanecarboxylic acid decarboxylates more readily because the carboxyl carbon atom is bonded to an sp^3-hybridized carbon atom, which is a weaker bond than the bond to the sp^2-hybridized carbon atom of benzoic acid.

20.48 The following β-keto acid does not decarboxylate on heating. Based on the mechanism for the reaction, explain this observation.

Answer: The cyclic mechanism for decarboxylation of β-ketoacids forms an enol that isomerizes to a carbonyl compound. The double bond at the bridgehead atom for the enol in the bicyclic compound would be highly strained.

20.49   Saturated carboxylic acids do not decarboxylate, but β,γ-unsaturated carboxylic acids do. Explain why, using a mechanism to show the decarboxylation of 3-butenoic acid. Use your mechanism to predict the product of decarboxylation of (E)-4-methyl-3-pentenoic acid.

Answer: A cyclic concerted mechanism gives an alkene with a double bond between the original α and β carbon atoms of 3-butenoic acid. The product of the decarboxylation of (E)-4-methyl-3-pentenoic acid is 3-methyl-l-butene.

Acyl Halides

20.50   Acyl halides can be prepared by reaction of a carboxylic acid with one equivalent of oxalyl chloride. The by-products of the reaction are HCl, CO_2, and CO. Write a mechanism for this reaction.

Answer: The carboxylic acid first reacts with oxalyl chloride to form HCl and an anhydride, which then decomposes by a cyclic mechanism in which the chlorine atom of the intermediate attacks the carbonyl carbon atom of the original carboxylic acid. A concerted process in the third step releases the acyl chloride, CO_2, and CO.

20.51   Explain why acyl halides of hydroxy acids cannot be prepared using thionyl chloride.

Answer: Thionyl chloride will also react with the hydroxyl group.

Synthesis of Esters

20.52   Draw the structure of the product of each of the following reactions.

Answers:

(a) The nucleophilic carboxylate group displaces a halide ion to form an ester. The iodide ion is a better leaving group than the chloride ion. The resulting ester product retains the chlorine atom.

(b) Diazomethane reacts with the acid to give a carboxylate ion and the methyldiazonium ion. Subsequent reaction of these two ions gives a methyl ester.

(c) The reaction of an acid chloride with an alcohol occurs by attack of the nucleophilic oxygen atom of the alcohol on the carbonyl group. The diol has a tertiary and a primary alcohol. The primary alcohol is less sterically hindered and reacts at a faster rate to produce an ester.

(d) A Fischer esterification reaction requires that the equilibrium be shifted toward the ester product. The resonance stabilization of the phenol favors the reactant side of the reaction, so esterification of phenols is not favorable. Esterification of the benzyl alcohol occurs.

(a) [structure: cyclobutane-CO_2^-] + $ClCH_2CH_2I$ (one equivalent) $\longrightarrow$ [ester product with Cl]

(b) $HO-\overset{O}{\underset{\|}{C}}-$[benzene ring]$-CH_2OH$ $\xrightarrow{CH_2N_2}$ $CH_3O-\overset{O}{\underset{\|}{C}}-$[benzene ring]$-CH_2OH$

(c) $CH_3-\overset{O}{\underset{\|}{C}}-Cl$ + $CH_3-\overset{OH}{\underset{CH_3}{\overset{|}{\underset{|}{C}}}}-CH_2CH_2CH_2OH$

$\downarrow$ pyridine

$CH_3-\overset{OH}{\underset{CH_3}{\overset{|}{\underset{|}{C}}}}-CH_2CH_2CH_2O-\overset{O}{\underset{\|}{C}}-CH_3$

(d) [naphthalene with CH_2OH and OH] $\xrightarrow[CH_3CO_2H]{H_3O^+}$ [naphthalene with $CH_2O-\overset{O}{\underset{\|}{C}}-CH_3$ and OH]

20.53   Draw the structure of the product of each of the following reactions.

Answers:
(a) The reagent resembles diazomethane and has phenyl groups in place of hydrogen atoms. It reacts to give a carboxylate ion and a diphenylmethyldiazonium ion. Subsequent reaction of these two ions gives a diphenylmethyl ester.
(b) The nucleophilic carboxylate group displaces a halide ion to form an ester. Although the bromide ion is a better leaving group than the chloride ion, the reaction does not occur at that site because the carbon atom is sp^2 hybridized. The ester product results from nucleophilic attack at the spa carbon atom to give a product that retains the bromine atom.
(c) A Fischer esterification reaction occurs via an equilibrium process that is easily pushed toward the ester product by adding excess methanol.
(d) The reaction of an acid bromide with an alcohol occurs via nucleophilic attack of the nucleophilic oxygen atom of the alcohol on the carbonyl group. The diol has a secondary and a primary alcohol. The primary alcohol is less sterically hindered and reacts at a faster rate to produce an ester.

(a) $HO-\overset{O}{\overset{\|}{C}}-\langle C_6H_4 \rangle-\overset{O}{\overset{\|}{C}}-CH_3$ +

  $\downarrow (C_6H_5)_2CN_2$

  $(C_6H_5)_2CHO-\overset{O}{\overset{\|}{C}}-\langle C_6H_4 \rangle-\overset{O}{\overset{\|}{C}}-CH_3$

(b) $CH_3CH_2-\overset{O}{\overset{\|}{C}}-O^-$ + $\underset{H}{\overset{ClCH_2CH_2}{}}C=C\underset{Br}{\overset{CH_3}{}}$

  $\downarrow$

  $CH_3CH_2-\overset{O}{\overset{\|}{C}}-O-CH_2CH_2\ \underset{H}{}C=C\underset{Br}{\overset{CH_3}{}}$

(c) $HC\equiv C-CH_2CO_2H \xrightarrow[H_3O^+]{CH_3OH} HC\equiv C-CH_2-O-\overset{O}{\overset{\|}{C}}-OCH_3$

(d) $CH_3-\underset{H}{\overset{OH}{C}}-CH_2CH_2CH_2OH$ + $\langle C_6H_5 \rangle-\overset{O}{\overset{\|}{C}}-Br$

  $\downarrow$

  $CH_3-\underset{H}{\overset{OH}{C}}-CH_2CH_2CH_2O-\overset{O}{\overset{\|}{C}}-\langle C_6H_5 \rangle$

20.54 Outline the steps necessary to prepare the following compound from cyclohexanecarboxylic acid.

Answers:
(a) Divide the reactant into two portions.
(b) Reduce one portion of the carboxylic acid using lithium aluminum hydride to obtain cyclohexylmethanol.
(c) Convert the second portion of the carboxylic acid to an acid chloride using thionyl chloride.
(d) React the acid chloride with the alcohol in the presence of pyridine to neutralize the Hcl formed.

20.55 Outline the steps necessary to prepare the following compound from benzoic acid.

Answers:
(a) Divide the reactant into two portions.
(b) Prepare 2-phenylethanol from one portion of benzoic acid by the Hunsdiecker reaction to give bromobenzene.
(c) Make the Grignard reagent of bromobenzene.
(d) Add ethylene oxide to the Grignard reagent to make the alcohol.
(e) React the alcohol with another portion of benzoic acid in a Fischer esterification reaction.

20.56 What alcohol and acid are required to form each of the following esters by Fischer esterification?

Answers:
(a) butanoic acid and 1-butanol
(b) butanoic acid and methanol
(c) butanoic acid and ethanol

(a) $CH_3CH_2CH_2CH_2$—$\overset{\displaystyle O}{\overset{\|}{C}}$—O—$CH_2CH_2CH_3$ (b) CH_3—$\overset{\displaystyle O}{\overset{\|}{C}}$—O—$CH_2CH_2CH_3$

(c) CH_3CH_2—$\overset{\displaystyle O}{\overset{\|}{C}}$—O—$CH_2CH_2CH_3$

20.57 What alcohol and acid or acyl derivative are required to form each of the following esters?

(a) $CH_2CH_2CO_2CH(CH_3)_2$

(b) $CO_2CH(CH_3)_2$

(c)

Answers:
(a) 3-phenylpropanoic acid and 2-propanol
(b) cycloheptanecarboxylic acid and 2-propanol
(c) m-bromobenzoic acid and cyclobutanol
(d) p-methoxybenzoic acid and cyclopentanol

Multistep Synthesis

20.58 Write the structure of the final product of each of the following sequences of reactions.

Answers:
(a) Oxidation gives a carboxylic acid, which is converted into an acid chloride by $SOCl_2$. Subsequent reaction with methanol gives a methyl ester.
(b) Reduction by lithium aluminum hydride gives a primary alcohol, which is then converted to a chloro compound by $SOCl_2$. Formation of a Grignard reagent followed by addition to carbon dioxide gives a carboxylic acid with one more carbon atom than the starting material.
(c) Oxidation by PCC gives an aldehyde. Subsequent reaction with methanol gives an acetal.

20.59 Write the structure of the final product of each of the following sequences of reactions.

Answers:
(a) Oxidation gives a carboxylic acid. Subsequent reaction with methanol in a Fischer esterification reaction gives a methyl ester.
(b) Reduction by lithium aluminum hydride give a secondary alcohol, which is then converted to an acetate ester by Fischer esterification.
(b) Reduction by the hydride reagent gives an aldehyde. Subsequent reaction with ethylene glycol gives a cyclic acetal.

20.60   Each of the following compounds has a resonance due to a single hydrogen atom at a lower field position than 10 δ. Based on the molecular formula, and the indicated remaining resonances, propose a structure for each compound. The number of hydrogen atoms and multiplicity are given in parentheses.

(a) $C_5H_{10}O_2$, 1.25 ppm (9H singlet)

(b) $C_3H_5ClO_2$, 1.75 ppm (3H doublet), 4.45 ppm (1H quartet)

(c) $C_8H_8O_2$, 1.4 ppm (3H singlet), 7.25 ppm (2H singlet), 8.0 ppm (2H doublet)

(d) $C_3H_6O_3$, 3.4 ppm (2H singlet), 4.0 ppm (3H singlet)

(e) $C_9H_{10}O_3$, 2.7 ppm (2H triplet), 4.2 ppm (2H triplet), 7.4 ppm (5H complex multiplet)

Answers:

Each molecular formula given has at least two oxygen atoms, so the resonance lower than 10 δ reveals the presence of a carboxyl group proton in each compound.

(a) The singlet of intensity 9H at 1.25 ppm is due to the nine equivalent protons of a tertiary butyl group. The compound is 2,2-dimethylpropanoic acid.

(b) The doublet of intensity 3H and the quartet of intensity 1H indicate three protons and one proton that split each other on neighboring carbon atoms. The resonance at 1.75 ppm is due to a methyl group, and the resonance at 4.45 ppm is due to a proton bonded to a carbon atom which also has a chlorine atom bonded to it. The compound is 2-chloropropanoic acid.

(c) The molecular formula and the resonances at 7.25 and 8.0 indicate that a benzene ring is present. The singlet of intensity 3H at 1.4 ppm indicates a methyl group with no neighboring protons. The doublets of intensity 2 at 7.25 and 8.0 ppm, respectively, indicate two sets of equivalent protons on a benzene ring. The compound is p-methylbenzoic acid.

(d) The "extra" oxygen atom in the molecular formula could indicate an alcohol or an ether group, but there is no resonance of intensity 1H, so it cannot be an alcohol. The singlet of intensity 2H at 3.4 ppm and the singlet of intensity 3 at 4.0 ppm are due to a methylene and a methyl group, respectively, bonded to an oxygen atom. The compound is methoxyethanoic acid.

(e) The molecular formula and the multiplet at 7.4 ppm indicate that a benzene ring is present. The "extra" oxygen atom in the molecular formula could indicate an alcohol or an ether group, but there is no resonance of intensity 1, so it cannot be an alcohol. The triplets of intensity 2 at 2.7 and 4.2 ppm are due to neighboring methylene groups that are also bonded to a carbonyl carbon atom and an oxygen atom, respectively. The complex multiplet of intensity 5 at 7.4 ppm is due to a phenyl group. The compound is 3-phenoxypropanoic acid.

$$\text{C}_6\text{H}_5\text{—O—CH}_2\text{CH}_2\text{CH}_2\text{CO}_2\text{H}$$

20.61   Each of the following compounds has a resonance due to a single hydrogen atom at a lower field position than 10 δ. Based on the molecular formula, and the indicated remaining resonances, propose a structure for each compound. The number of hydrogen atoms and multiplicity are given in parentheses.
(a) $C_6H_{12}O_2$, 1 07 ppm (9H singlet), 2.21 ppm (2H singlet)
(b) $C_3H_5ClO_2$, 2.85 ppm (2H triplet), 3.80 ppm (2H triplet)
(c) $C_8H_8O_2$, 3.6 ppm (2H singlet), 7.25 ppm (5H singlet)
(d) $C_4H_8O_3$ 1 27 ppm (3, triplet), 3.55 ppm (2H quartet), 4.13 ppm (2H singlet)
(e) $C_9H_{10}O_3$, 1.72 ppm (3H doublet), 4.95 ppm (2H quartet), 7.4 ppm (5H complex multiplet)

Answers:
Each molecular formula given has at least two oxygen atoms, so the resonance lower than 10 δ reveals the presence of a carboxyl group proton in each compound.
(a) The singlet of intensity 9H is due to the nine equivalent protons of a *tert*-butyl group. The singlet of intensity 2H at 2.21 ppm is due to a methylene group that is bonded to a carbonyl carbon atom and has no neighboring protons. The compound is 3,3-dimethylbutanoic acid.

$$\text{CH}_3\text{—}\overset{\displaystyle \text{CH}_3}{\underset{\displaystyle \text{CH}_3}{\overset{|}{\underset{|}{\text{C}}}}}\text{—CH}_2\text{CO}_2\text{H}$$

(b) The triplets of intensity 2H at 2.85 and 3.80 ppm are due to neighboring methylene groups that are also bonded to a carbonyl carbon atom and a chlorine atom, respectively. The compound is 3-chloropropanoic acid.

$$\text{ClCH}_2\text{—}\overset{\displaystyle \text{H}}{\underset{\displaystyle \text{H}}{\overset{|}{\underset{|}{\text{C}}}}}\text{—CO}_2\text{H}$$

(c) The molecular formula and the 5H resonances at 7.25 ppm that a benzene ring is present and that it has one substituent. The singlet of intensity 2 at 3.6 ppm indicates a methylene group with no neighboring protons. The compound is phenylethanoic acid.

$$\text{C}_6\text{H}_5\text{—CH}_2\text{CO}_2\text{H}$$

(d) The extra oxygen atom in the molecular formula could indicate an alcohol or an ether group, but there is no resonance of intensity 1H, so it cannot be an alcohol. The triplet of intensity 3H at 1.27 ppm and the quartet of intensity 2H at 3.55 ppm indicate three protons and two protons, respectively, splitting each other on neighboring carbon atoms, with the methylene group also bonded to an oxygen atom. The singlet of intensity 2H at 4.13 ppm is due to a methylene group that is bonded to a carbonyl carbon atom and an oxygen atom. The compound is ethoxyethanoic acid.

$$\text{CH}_3\text{CH}_2\text{O—}\overset{\displaystyle \text{H}}{\underset{\displaystyle \text{H}}{\overset{|}{\underset{|}{\text{C}}}}}\text{—CO}_2\text{H}$$

(e) The molecular formula and complex multiplet of intensity 5H at 7.4 ppm indicate that a phenyl group is present. The "extra" oxygen atom in the molecular formula could indicate an alcohol or an ether group, but there is no singlet of intensity 1H, so it cannot be an alcohol. The doublet of intensity 3 at 1.72 ppm and the quartet of intensity 1 at 4.95 ppm indicate three protons and one proton, respectively, splitting each other on neighboring carbon atoms. The resonance of intensity 1 at 4.95 ppm also indicates that the proton is bonded to a carbon atom that is bonded to an oxygen atom and one other atom. The compound is 2-phenoxypropanoic acid.

20.62 Deduce the structure of each of the following compounds based on the molecular formula and the carbon-13 NMR data.
(a) $C_6H_{12}O_2$, 9.3 ppm, 24.6 ppm, 33.5 ppm, 42.7 ppm, 185.5 ppm
(b) $C_6H_6O_2$, 128.7 ppm, 129.6 ppm, 131.2 ppm, 133.0 ppm, 167.7 ppm
(c) $C_4H_8O_2$, 13.4 ppm, 18.5 ppm, 36.3 ppm, 179.6 ppm

Answers:
(a) The compound has six carbon atoms, but five signals, so two of the carbon atoms are equivalent. The quartets at 9.3 and 24.6 ppm are due to methyl groups isolated from and close to the carbonyl carbon atom. The triplet at 33.5 ppm is due to a methylene group. The singlet at 42.7 ppm must be due to a quaternary carbon atom because it is not split by any bonded hydrogen atoms. The singlet at 185.5 ppm must be due to a carboxyl carbon atom. The compound is 2,2-dimethylbutanoic acid.

(b) The molecular formula, $C_7H_6O_2$, indicates that a benzene ring is possible. The compound has seven carbon atoms, but five signals, so there must be two sets of two equivalent carbon atoms or one set of three equivalent carbon atoms. The doublets at 128.7, 129.6, and 131.2 ppm are due to carbon atoms in a benzene ring. The singlet at 133.0 ppm is due to a carbon atom with a double bond to another carbon atom, but no hydrogen atom, so it indicates a connecting carbon atom in a benzene ring. The singlet at 167.7 ppm must be due to a carboxyl carbon atom. The compound is benzoic acid. Note that the compound contains two sets of two equivalent carbon atoms in the benzene ring, C-2 and C-6 in one set, and C-3 and C-5 in the other.

(c) The compound has four nonequivalent carbon atoms with signals corresponding to a methyl carbon at 13.4 ppm, methylene carbons at 18.5 and 36.3 ppm, and carbonyl carbon at 179.6 ppm. The compound is butanoic acid.

$CH_3CH_2CH_2CO_2H$

20.63 Deduce the structure of each of the following compounds based on the molecular formula and the carbon-13 NMR data.
(a) $C_5H_{10}O_2$, 13.5 ppm, 22.0 ppm, 27.0 ppm, 34.1 ppm, 179.7 ppm
(b) $C_7H_6O_3$, 115.8 ppm, 121.9 ppm, 132.7 ppm, 162.5 ppm, 169.0 ppm

Answers:
(a) The compound has five nonequivalent carbon atoms with signals corresponding to methyl (quartet at 13.5 ppm), methylene carbons at 22.0, 27.0, and 34.1 ppm, and a carbonyl carbon 179.7 ppm). The compound is pentanoic acid.

$CH_3CH_2CH_2CH_2CO_2H$

(b) The molecular formula, $C_7H_6O_3$, indicates that a benzene ring is possible. The "extra" oxygen atom in the molecular formula could indicate an alcohol or an ether group. The compound has seven carbon atoms, but five signals, so there must be two sets of two equivalent carbon atoms, or one set of three equivalent carbon atoms. The signals at 115.8 and 121.9 ppm indicate carbon atoms in a benzene ring. The signals at 132.7 and 162.5 also indicate connecting carbon atoms in a benzene ring, one of which is shifted further down field because it is deshielded by an oxygen atom. The signal at 169.0 ppm must be due to a carboxyl carbon atom. The compound is *p*-hydroxybenzoic acid. *p*-Hydroxybenzoic acid contains two sets of two equivalent carbon atoms in the benzene ring, C-2 and C-6 in one set, and C-3 and C-5 in the other. The hydroxyl group would easily be detected in an proton NMR spectrum.

21 CARBOXYLIC ACID DERIVATIVES

KEYS TO THE CHAPTER

21.1 Nomenclature of Carboxylic Acid Derivatives

The names of the acid derivatives with the exception of nitriles resemble those of the structurally related carboxylic acid. The acid halides are named using *-oyl halide* in place of *-oic acid* of carboxylic acids. Compounds with an acid halide bonded to a ring carbon atom are named by appending *carbonyl halide* to the name of the alkane. The acid anhydrides are named using *-oic anhydride* in place of *-oic acid* of carboxylic acids.

Esters are named by the name of the alkyl group bonded to the bridging oxygen atom followed by the name of the acid, with *-oate* replacing *-oic acid*. Compounds with the acid portion bonded to a ring carbon atom are named by appending *carboxylate* to the name of the alkane. Lactones are named by adding *lactone* as a separate word to the name of the hydroxy acid.

Amides are named using *amide* in place of *-oic acid* of carboxylic acids. Compounds with an amide bonded to a ring carbon atom are named by appending *carboxamide* to the name of the alkane. The prefix *N-* is used to identify alkyl or aryl groups bonded to the nitrogen atom of the amide. Lactams are named by adding *lactam* as a separate word to the name of the amino acid. Nitrites are named by adding *nitrile* to the name of the related alkane. The name includes the carbon of the cyano group.

21.2 Physical Properties of Acyl Derivatives

Acid halides and acid anhydrides are polar compounds and have physical properties that resemble those of structurally similar carbonyl derivatives of similar molecular weight.

Esters are polar molecules, but they cannot form intermolecular hydrogen bonds like carboxylic acids do. Hence, esters have lower solubility in water and have lower boiling points than carboxylic acids.

The lower-molecular-weight amides are soluble in water as a result of hydrogen bonding to water molecules. The melting points of primary and secondary amides are higher than those for alkanes of similar molecular weight because of intermolecular hydrogen bonding. Tertiary amides have lower melting points than isomeric primary and secondary amides because they do not have an N—H bond to form intermolecular hydrogen bonds.

Nitriles are very polar acid derivatives due to the large bond moment of the triple bond between carbon and nitrogen. Lower-molecular-weight nitriles are soluble in water.

21.3 Basicity of Acyl Derivatives

Protonation of the carbonyl oxygen atom is a reaction common to all acid-catalyzed reactions of acid derivatives. The lone-pair electrons of oxygen are in sp^2-hybridized orbitals, and the oxygen atom is less basic than the oxygen atom of alcohols or ethers, whose lone-pair electrons are in sp^3-hybridized orbitals.

The basicity of acid derivatives is related to the ability of the electronegative atom bonded to the carbonyl carbon atom to stabilize the conjugate acid by resonance donation of electron density from that atom. Amides are the most basic because nitrogen is an effective donor of electrons. Esters are less basic than amides because oxygen is more electronegative and is a less effective electron donor. Acid chlorides are very much less basic because chlorine is not effective at stabilizing a positive charge.

Nitriles are very weak bases because the lone-pair electrons of the nitrogen atom are in an sp-hybridized orbital. Also, there is no alternative stabilized resonance form for the conjugate acid.

21.4 Nucleophilic Acyl Substitution

Nucleophilic acyl substitution (acyl transfer reaction) occurs by a two-step mechanism. First, attack of the carbonyl carbon atom of an acyl derivative by a nucleophile yields a tetrahedral intermediate. The tetrahedral intermediate can then eject a leaving group. The net result is a substitution reaction. The first step is rate determining. Thus, the rate of the reaction depends on the effect of the hydrocarbon group and the attached electronegative atom on the stability of the acid derivative and the transition state, not on the leaving group characteristics of the group displaced by the nucleophile.

Organic Chemistry Study Guide: Key Concepts, Problems, and Solutions http://dx.doi.org/10.1016/B978-0-12-801889-7.00021-2

The acid-catalyzed reaction of acid derivatives occurs by protonation of the carbonyl oxygen atom, which increases the electrophilicity of the carbonyl carbon atom. Attack of a nucleophile gives a tetrahedral intermediate that subsequent ejects the leaving group. Loss of a proton from the carbonyl oxygen atom completes the reaction.

The base-catalyzed reaction of acid derivatives occurs by abstraction of a proton from the nucleophile H—Nu, which prepares the nucleophile for attack of the carbonyl oxygen atom. The attack of the nucleophile gives a tetrahedral intermediate that has a negative charge on the oxygen atom. Return of an electron pair from that oxygen atom to reform the carbon–oxygen double bond results in ejection of the leaving group.

The different effects of the electronegative atoms on the reactivity of acid derivatives are explained using resonance and inductive effects on the stability of the reactant. The transition state resembles the tetrahedral intermediate, which cannot be stabilized by resonance. The stability of the transition states of all acyl derivatives for a reaction with the same nucleophile is approximately the same. Thus, the stability of the reactant controls the rate of its reaction. The most stable acyl derivatives are the least reactive. Amides are the most stable, and the least reactive, because nitrogen is an effective donor of electrons to the carbonyl group. Anhydrides and esters are somewhat less stable, because oxygen is more electronegative than nitrogen and is a less effective donor of electrons. Anhydrides are less stable because the donation of electrons to one carbonyl group is in competition with the donation of electrons to the second carbonyl group. Thus, in comparison to esters, where the oxygen atom need only stabilize one carbonyl group, anhydrides are more reactive than esters. Acid chlorides are very much less stable because chlorine is not effective at stabilizing positive charge by donation of electron density by resonance.

21.5 Hydrolysis of Acyl Derivatives

Acyl halides and acid anhydrides react readily with water to give carboxylic acids. Esters react with water in an equilibrium reaction to give an alcohol and a carboxylic acid. Amides are stable to water under neutral conditions.

The degree of completion of the reaction of an ester with water is increased by use of an equivalent amount of hydroxide ion. The saponification reaction is spontaneous because the product is a carboxylate ion that is a weaker base than hydroxide ion.

Amides are difficult to hydrolyze, and an equivalent amount of acid or base must be used. In the case of acid, the products are the carboxylic acid and the conjugate acid of the amine. In the case of base, the products are the amine and the conjugate base of the carboxylic acid.

Nitriles hydrolyze with difficulty using either concentrated acid or base. In the case of acid, the product is the ammonium ion and the carboxylic acid. In the case of base, the product is the conjugate base of the carboxylic acid and ammonia.

21.6 Reaction of Acyl Derivatives with Alcohols

The reactivity of acyl derivatives with alcohols parallels their reactivity with water. The reaction of either acid chlorides or acid anhydrides with alcohols is an excellent way to prepare esters.

Esters react with alcohols to interchange alkoxy groups in a *transesterification* reaction. The reaction is one to be avoided. Reactions of esters in an alcohol solvent are selected so that the alcohol and the ester contain the same alkoxy group.

21.7 Reaction of Acyl Derivatives with Amines

Amines (or ammonia) are better nucleophiles than alcohols (or water), so the reactions of amines with acyl derivatives are faster than the corresponding reaction with alcohols. Acid chlorides react with ammonia, primary amines, and secondary amines to produce primary, secondary, and tertiary amides. A second mole of the nitrogen compound is required to neutralize the HCl formed. Usually, pyridine is added to conserve the amine and allow the use of only one mole of the amine.

Acid anhydrides react with ammonia, primary amines, and secondary amines to produce primary, secondary, and tertiary amides. The by-product of the reaction is one mole of the carboxylic acid related to the acid anhydride.

Esters react similarly with ammonia, primary amines, and secondary amines to give an alcohol as the by-product.

21.8 Reduction of Acyl Derivatives

The salts of carboxylic acids are weak nucleophiles but can be used to displace halide ion from a haloalkane to give esters. The most diverse reaction of carboxylic acids and their derivatives is attack of nucleophiles at the carbonyl carbon atom. However, in contrast to aldehydes and ketones, the tetrahedral intermediate is unstable, and it ejects a leaving group, This overall process is nucleophilic acyl substitution, or an acyl transfer reaction.

Esters are reduced only by lithium aluminum hydride to give a primary alcohol related to the acid portion of the ester, with the alcohol of the original ester as a by-product. The process occurs by nucleophilic attack of hydride at the carbonyl carbon atom to give a tetrahedral intermediate that subsequently ejects an alkoxide ion. An aldehyde results, which is then rapidly reduced to the primary alcohol. Diisobutylaluminum hydride (DIBAL) reacts with esters to give the aldehyde.

Amides are reduced by lithium aluminum hydride to give amines containing the groups originally bonded to the nitrogen atom and an alkyl group derived from the acid minus its oxygen atom.

Nitriles are reduced by a variety of reagents. Raney nickel and hydrogen at high pressures give primary amines. Lithium aluminum hydride also reduces nitriles to amines.

21.9 Reaction of Acyl Derivatives with Organometallic Reagents

The carbanion derived from an organometallic compound is a nucleophile that can attack the carbonyl carbon atom of acyl derivatives. Ejection of a leaving group from the tetrahedral intermediate gives the final product, which may or may not react a second time with the organometallic compound.

Acid chlorides react with Grignard reagents in a manner similar to that for esters to give an intermediate aldehyde that subsequently reacts further to give an alcohol. However, this reaction is not used because esters are better starting materials. Reaction of acid chlorides with the Gilman reagent gives ketones that do not react further.

Esters react with Grignard reagents to add an alkyl (or aryl) group to the carbonyl carbon atom and eject an alkoxide ion. The resulting aldehyde reacts further with the reagent to give a tertiary alcohol containing two equivalents of the alkyl (or aryl) group of the Grignard reagent.

21.10 Spectroscopy of Acid Derivatives

The infrared spectra of nitriles contain an absorption for the carbon–nitrogen triple bond in the 2200–2250 cm^{-1} region. The absorption is more intense than that for carbon–carbon triple bond in the 2100–2200 cm^{-1} region.

Acyl derivatives are characterized by a strong absorption due to the carbonyl group in the infrared spectra of these compounds. The absorption occurs in the same region as for the carbonyl groups of aldehydes and ketones. However, the position of the absorption is strongly affected by the electronegative atom and its contribution to the stability of the dipolar resonance form by a combination of inductive and resonance affects. Esters have absorptions at 1735 cm^{-1}, but that position is affected by conjugation of double bonds with the acyl group. The absorptions of lactones show the same changes with ring size as for cycloalkanones.

Acid chlorides absorb at 1800 cm^{-1} because the chlorine atom cannot stabilize the dipolar resonance form by donation of electrons. In fact, the inductive electron withdrawal of electrons destabilizes that resonance form and increases the double bond character of the carbonyl group.

Amides very effectively donate electrons to the carbonyl carbon atom by resonance. Hence, the double bond character of the carbonyl group is reduced, and the resulting absorption of the carbonyl group occurs in the 1650–1655 cm^{-1} region.

The proton NMR of the a hydrogen atoms of acyl derivatives occurs in the 2 δ region, which is the same region for the α hydrogen atoms of aldehydes and ketones. The chemical shift of the hydrogen atoms on the carbon atom bonded to the oxygen atom of the alkoxy part of esters are at somewhat lower field (4 δ) than alcohols. The chemical shift of the hydrogen atoms bonded to the alkyl carbon atom bearing the nitrogen atom in amides occur in the 2.6–3.0 δ region.

The absorption of a carbon atoms of acid derivatives have ^{13}C NMR absorptions in the 20 δ region. The carbonyl carbon atom is easily identified by its absorption in the 165–180 δ region. The absorptions of the carbon atom bonded to the oxygen atom of the alkoxy part of an ester are in the 60 δ region.

Summary of Reactions

1. Hydrolysis of Acid Derivatives

2. Reactions of Acid Derivatives with Alcohols and Phenols

3. Reactions of Acid Derivatives with Amines

$$CH_3CH_2\overset{\overset{\displaystyle CH_3}{|}}{C}HCH_2-\overset{\overset{\displaystyle O}{\|}}{C}-Cl \xrightarrow[\text{pyridine}]{CH_3CH_2NH_2} CH_3CH_2\overset{\overset{\displaystyle CH_3}{|}}{C}HCH_2-\overset{\overset{\displaystyle O}{\|}}{C}-NHCH_2CH_3$$

o-methoxyaniline $+$ $CH_3-\overset{\overset{\displaystyle O}{\|}}{C}-O-\overset{\overset{\displaystyle O}{\|}}{C}-CH_3 \longrightarrow$ o-methoxy-acetanilide ($NH-\overset{\overset{\displaystyle O}{\|}}{C}-CH_3$)

$$CH_3CH_2\overset{\overset{\displaystyle CH_3}{|}}{C}HCH_2-\overset{\overset{\displaystyle O}{\|}}{C}-OH + CH_3NH_2 \longrightarrow CH_3CH_2\overset{\overset{\displaystyle CH_3}{|}}{C}HCH_2-\overset{\overset{\displaystyle O}{\|}}{C}-NHCH_3$$

4. Reduction of Acid Chlorides

$$CH_3CH_2\overset{\overset{\displaystyle CH_3}{|}}{C}HCH_2-\overset{\overset{\displaystyle O}{\|}}{C}-Cl \xrightarrow[\text{2. H}_3\text{O}^+]{\text{1. LiAlH}_4} CH_3CH_2\overset{\overset{\displaystyle CH_3}{|}}{C}HCH_2-\overset{\overset{\displaystyle H}{\underset{\displaystyle H}{|}}}{C}-OH$$

$$CH_3CH_2\overset{\overset{\displaystyle CH_3}{|}}{C}HCH_2-\overset{\overset{\displaystyle O}{\|}}{C}-Cl \xrightarrow[\text{2. H}_3\text{O}^+]{\text{1. LiH(OC(CH}_3))_3} CH_3CH_2\overset{\overset{\displaystyle CH_3}{|}}{C}HCH_2-\overset{\overset{\displaystyle O}{\|}}{C}-H$$

5. Reduction of Esters

benzoic acid methyl ester (OCH_3) $\xrightarrow[\text{2. H}_3\text{O}^+]{\text{1. LiAlH}_4}$ benzyl alcohol (CH_2OH) $+$ CH_3OH

benzoic acid methyl ester (OCH_3) $\xrightarrow[\text{2. H}_3\text{O}^+]{\text{1. DIBAL}}$ benzaldehyde (H) $+$ CH_3OH

6. Reduction of Amides

naphthalene amide ($NHCH_3$) $\xrightarrow[\text{2. H}_3\text{O}^+]{\text{1. LiAlH}_4}$ naphthalene-CH_2-$NHCH_3$

7. Reduction of Nitriles

8. Reaction of Acid Derivatives with Organometallic Reagents

SOLUTIONS EXERCISES

Nomenclature

21.1 Give the IUPAC name for each of the following carboxylic acid derivatives.

Answers:
(a) ethyl phenylethanoate
(b) 3-cyclohexylbutanenitrile
(c) *N,N*-diethylcyclobutanecarboxamide
(d) 2-bromoethyl 3-bromobenzoate
(e) 3,4-dimethoxybenzoyl chloride

21.2 Give the IUPAC name for each of the following carboxylic acid derivatives.

Answers:
(a) cyclohexyl benzoate
(b) 4-methylbenzonitrile
(c) *N*-cyclohexyl-2-fluoroethanamide
(d) 3,4-dimethylbenzoyl chloride
(e) 2-methylbutyl 5-cyclopentylpentanoate

21.3 Write the structure of each of the following compounds.
(a) phenyl octanoate   (b) butanoic anhydride   (c) *N*-ethyl-4,4-dimethylcyclohexanecarboxamide
(d) 2-bromo-3-methylbutanoyl chloride   (e) *trans*-4-methylcyclohexanecarbonitrile

Answers:

21.4 Write the structure of each of the following compounds.
   (a) 2-chloropropyl 3-bromobutanoate   (b) 4-methoxyphthalic anhydride (c) N,N-dimethyl-3-cyclopropylpentanamide
   (d) cyclobutanecarbonyl bromide   (e) (R)-2-methylbutanenitrile

Answers:

(a) structures

(b) structure

(c) structure (d) structure (e) structure

21.5 The common name of the vasodilator cyclandelate is 3,5,5-trimethylcyclohexyl mandelate. Give the structure and name of the acid contained in the ester.

mandelic acid

21.6 Hydrolysis in the body is required for diloxanide furanoate to be effective against intestinal amebiasis. What is the acid component of the drug? Considering the name of the drug, name the acid.

diloxanide furanoate

furanoic acid

Answer: The carboxylic acid is derived from the part of the compound that contains the carboxylate group. The name of the acid is furanoic acid, which is obtained by replacing *-oate* with *-ic acid*.

Cyclic Acyl Derivatives

21.7 (a) Identify the oxygen-containing functional group in the following structure, a pheromone of the female Japanese beetle. (b) What is the configuration around the alkene moiety?

Answers: (a) The functional group is an ester contained within a ring and is therefore a lactone. (b) The configuration about the double bond is (Z).

21.8 Identify the nitrogen-containing functional group within the four-membered ring of cephalosporin C, antibiotic.

cephalosporin C

Answer: The functional group within the four-membered ring is an amide and is therefore a lactam.

21.9 Name each of the following lactones.

Answers:
(a) 5-hydroxypentanoic acid lactone
(b) 9-hydroxynonanoic acid lactone
(c) 4-hydroxypentanoic acid lactone

(a)   (b)   (c)

21.10 Draw the structure of each of the following lactams.
(a) 3-aminopropanoic acid lactam   (b) 4-aminopentanoic acid lactam   (c) 5-aminopentanoic acid lactam

Answers:
(a)   (b)   (c)

21.11 Which of the following compounds are lactones?

(a)   (b)   (c)

Answer: Compound (a) has two carbonyl groups joined by a common oxygen atom and is a cyclic anhydride. Only (b) and (c) are lactones.

21.12 Which of the following compounds are lactams?

(a) (b) (c)

Answer: In compound (b), a methylene group separates the nitrogen atom and the carbonyl group. It is not a lactam. Only (a) and (c) are lactams.

Properties of Acid Derivatives

21.13 The boiling points of methyl pentanoate and butyl ethanoate are 126 and 125 °C, respectively. Explain the similarity of these boiling points.

Answer: The compounds are isomers and have similar polarities. They also have similar molecular shapes. Thus, the dipole–dipole as well as London forces are similar, and as a result, they have similar boiling points.

21.14 The boiling points of methyl pentanoate and methyl 2,2-dimethylpropanoate are 126 and 102 °C, respectively. Explain why these values differ.

Answer: The compounds are isomeric methyl esters and thus have similar polarities. However, they have different molecular shapes. The acid portion of methyl 2,2-dimethyipropanoate has a nearly spherical shape, but the acid moiety of methyl pentanoate has a cylindrical shape. Thus, the van der Waals forces of methyl 2,2-dimethylpropanoate are smaller, and the boiling point is lower.

21.15 The boiling points of acetonitrile and 1-propyne are 81.5 and −23 °C, respectively. Account for this difference in boiling point between two compounds with similar molecular weights.

Answer: The molecular weights differ only by 1 amu, and both compounds have similar cylindrical shapes. However, the nitrile has a polar carbon–nitrogen triple bond and as a result has larger dipole–dipole forces than propyne, which has a much less polar carbon–carbon triple bond.

21.16 The boiling points of acetamide and acetic acid are 221 and 118 °C, respectively. Account for this difference in boiling point between two compounds with similar molecular weights.

Answer: Both compounds can form intermolecular hydrogen bonds between a hydrogen atom bonded to an electronegative atom on one molecule and the carbonyl group on a neighboring molecule. However, the N—H group is a better hydrogen bond donor, and it forms stronger hydrogen bonds, so the amide has a higher boiling point.

21.17 Explain why protonation of *N,N*-dimethylformamide occurs at the oxygen atom rather than the nitrogen atom.

Answer: The amide functional group is resonance stabilized. As a result, the oxygen atom has increased electron density and the nitrogen atom has decreased electron density.

Protonation at the electron pair of nitrogen would give a conjugate acid in which there is no resonance stabilization of the carbonyl group. Resonance stabilization of the carbonyl group is still possible in the conjugate acid obtained by protonating oxygen.

no resonance stabilization resonance stabilization

21.18   The rotational barrier around the nitrogen-carbonyl carbon bond of *N,N*-dimethylformamide is approximately 87 kJ mole^{-1}. Why is this energy barrier substantially higher than values for other single bonds?

Answer: The carbon–nitrogen bond has some double bond character as a result of resonance stabilization of the carbonyl group by donation of electrons from the nitrogen atom, as shown in the answer to Exercise 21.17.

Cyclic Acyl Derivatives

21.19   Indicate whether each of the following reactions will occur.

(a) $CH_3-\overset{\overset{\displaystyle O}{\|}}{C}-Cl$ + CH_3OH $\longrightarrow$ $CH_3-\overset{\overset{\displaystyle O}{\|}}{C}-O-CH_3$ + HCl

(b) $CH_3-\overset{\overset{\displaystyle O}{\|}}{C}-NH_2$ + CH_3OH $\longrightarrow$ $CH_3-\overset{\overset{\displaystyle O}{\|}}{C}-O-CH_3$ + NH_3

(c) $CH_3-\overset{\overset{\displaystyle O}{\|}}{C}-O-CH_3$ + CH_3NH_2 $\longrightarrow$ $CH_3-\overset{\overset{\displaystyle O}{\|}}{C}-\overset{\overset{\displaystyle H}{}}{N}-CH_3$ + CH_3OH

Answers:
(a) Esters are more stable than acid chlorides because the carbonyl group is more stabilized by donation of electrons by resonance from oxygen than from chlorine. The reaction occurs to give the more stable ester.
(b) Amides are more stable than esters because the carbonyl group is more stabilized by donation of electrons by resonance from nitrogen than from oxygen. The reaction will not occur.
(c) Amides are more stable than esters because the carbonyl group is more stabilized by donation of electrons by resonance from nitrogen than from oxygen. The reaction will occur.

21.20   Indicate whether each of the following reactions will occur.

(a) $CH_3-\overset{\overset{\displaystyle O}{\|}}{C}-O-\overset{\overset{\displaystyle O}{\|}}{C}-CH_3$ + NH_3 $\longrightarrow$ $CH_3-\overset{\overset{\displaystyle O}{\|}}{C}-NH_2$ + CH_3CO_2H

(b) $CH_3-\overset{\overset{\displaystyle O}{\|}}{C}-O-\overset{\overset{\displaystyle O}{\|}}{C}-CH_3$ + HCl $\longrightarrow$ $CH_3-\overset{\overset{\displaystyle O}{\|}}{C}-Cl$ + $CH_3-\overset{\overset{\displaystyle O}{\|}}{C}-OH$

(c) $CH_3-\overset{\overset{\displaystyle O}{\|}}{C}-O-\overset{\overset{\displaystyle O}{\|}}{C}-CH_3$ + CH_3OH $\longrightarrow$ $CH_3-\overset{\overset{\displaystyle O}{\|}}{C}-O-CH_3$ + CH_3CO_2H

Answers:
(a) Amides are more stable than acid anhydrides because the carbonyl group is more stabilized by donation of electrons by resonance from nitrogen than from oxygen. The reaction will occur.
(b) Acid anhydrides are more stable than acid chlorides because the carbonyl group is more stabilized by donation of electrons by resonance from oxygen than from chlorine. The reaction will not occur.
(c) Esters are more stable than acid anhydrides because the carbonyl group is more stabilized by donation of electrons by resonance from oxygen bonded to an alkyl group than from oxygen bonded to another carbonyl group. The reaction will occur.

21.21   Considering the stability of the reactant, explain why thioesters react more readily than esters in acyl substitution reactions.

Answer: Esters are more stable than thioesters because the carbonyl group is more stabilized by donation of electrons by resonance from oxygen than from sulfur. The electrons of sulfur are in the third energy level, and overlap of 3p and 2p orbitals in thioesters is less effective than overlap of 2p and 2p orbitals in esters.

21.22 Considering the stability of the reactant, explain why thioesters are less reactive than acid chlorides in acyl substitution reactions.

Answer: Both sulfur and chlorine are third row elements and neither contributes electrons very effectively by resonance. However, to the extent that they do, sulfur is a better donor of electrons by resonance because it is less electronegative than chlorine. Thus, the resonance effect makes the thioester more stable than the acid chloride. There is also a stronger inductive electron withdrawal of electrons by the more electronegative chlorine, which destabilizes the carbonyl group leading to increased reactivity of the acid chloride.

21.23 Explain the order of reactivity in a saponification reactions of each of the following pairs of compounds.

(a)

(b)

(c)

(d)

Answers:
(a) The *tert*-butyl group of the alcohol portion of the ester on the right sterically hinders approach of a nucleophile to the carbonyl group. Thus, the compound on the left is the more reactive.
(b) The resonance interaction of the nonbonded electrons of the oxygen atom with the aromatic ring of the phenolic portion of the ester decreases the electron density at oxygen. Thus, there is a decreased availability of electrons to the carbonyl group, the carbonyl group is not as stable for the compound on the left and it is more reactive
(c) The axial carbomethoxy group of the compound on the right is sterically hindered and approach of a nucleophile is also sterically hindered. Thus, the compound on the left is the more reactive.
(d) The carbonyl group of the six-membered lactone can be approached in an equatorial direction. The five-membered lactone has its carbonyl group in the plane of the ring, so approach of a nucleophile must occur perpendicular to that plane and is sterically hindered.

21.24 Explain the order of reactivity in saponification reactions of each of the following pairs of compounds.

(a)

(b)

(c)

(d)

396

Answers Problem 21.24:

(a) The *tert*-butyl group bonded to the carbonyl carbon atom of the compound on the right sterically hinders approach of a nucleophile. Thus, the compound on the left is the more reactive

(b) The resonance interaction of the nonbonded electrons of the oxygen atom is decreased by the resonance electron withdrawal of those electrons by the *p*-nitro group. As a result, the carbonyl group is not as stable for the compound on the left and it is more reactive

(c) The trifluoromethyl group bonded to the carbonyl carbon atom inductively withdraws electron density. As a result, the carbonyl group is not as stable in the compound on the left, and it is more reactive toward nucleophiles.

(d) Ring strain of the compound is increased by the incorporation of an sp^2-hybridized carbon atom. In the transition state, that carbon atom becomes sp^3-hybridized, and the ring strain is decreased. Thus, the activation energy is smaller for the compound on the left.

21.25   Explain the position of the following equilibrium.

Answer:
Esters are more stable than thioesters because the carbonyl group is more stabilized by donation of electrons by resonance from oxygen than from sulfur. The electrons of sulfur are in the third energy level, and overlap of 3p and 2p orbitals in thioesters is less effective than overlap of 2p and 2p orbitals in esters. Therefore, the equilibrium favors the more resonance stabilized ester.

21.26   Which equilibrium constant for the following reactions is larger, K_I or K_{II}?

Answer:
In both reactions, a more stable amide is formed from an ester. The resonance interaction of the nonbonded electrons of the oxygen atom with the aromatic ring of the phenolic moiety of the aromatic ester decreases the electron density at oxygen. Thus, there is a decreased availability of electrons to the carbonyl group. Therefore, the carbonyl group is not as stable for the phenyl ester. Thus, the equilibrium constant is larger for the reaction of the phenyl ester than for the cyclohexyl ester.

21.27   Explain why the tautomeric equilibrium between an imidic acid and an amide lies on the side of the amide.

imidic acid   amide

Answer:
The carbon–oxygen double bond is more polar and more stable than the carbon–nitrogen double bond. Also, the nitrogen atom of the amide stabilizes the carbonyl group by resonance donation of electrons. Because the oxygen atom of the imidic acid is more electronegative than a nitrogen atom, it is less effective in donation of electrons by resonance. Therefore, the amide is more stable than the imidic acid.

21.28 Explain why esters react with hydroxylamine (NH$_2$OH) to give hydroxamic acids rather than O-acyl hydroxyl amines.

(an hydroxamic acid)   (an O-acyl hydroxyl amine)

Answer:
The carbonyl group of the hydroxamic acid is stabilized by donation of electrons of nitrogen by resonance. Because the oxygen atom of the O-acyl hydroxylamine is more electronegative than a nitrogen atom, it is less effective in donation of electrons by resonance.

21.29 One equivalent of methylamine reacts with S-ethyl-O-methylthiocarbonate as shown by the following equation. What other product is possible? Explain the observed selectivity of the reaction.

Answer:
The other possible product results from displacement of the ethoxy group as ethanol. The activation energy of the second step of the two step nucleophilic acyl substitution reaction is affected by the identity of the leaving group. Thiols are stronger acids than alcohols. Hence, the thiolate ion is a weaker base than the alkoxide ion. Weaker bases are generally better leaving groups, so thiolate is a better leaving group than methoxide ion.

21.30 Methanol reacts with glutaric anhydride to give a good yield of a monomethyl ester. Explain why the diester does not form.

Answer:
The other possible product results from displacement of the ethoxy group as ethanol. The activation energy of the second step of the two step nucleophilic acyl substitution reaction is affected by the identity of the leaving group. Thiols are stronger acids than alcohols. Hence, the thiolate ion is a weaker base than the alkoxide ion. Weaker bases are generally better leaving groups, so thiolate is a better leaving group than methoxide ion.

21.31 Explain why alcohols react with the following mixed anhydride to give good yield of acetate esters.

Answer:
The trifluoroacetate moiety of the compound withdraws electrons from the carbonyl carbon atom on the right to a greater extent than the acetate moiety of the compound withdraws electrons from the carbonyl carbon atom on the left. Thus, the nucleophile attacks the carbonyl group on the left to release the trifluoroacetate group.

21.32 Ethanol reacts with the following mixed anhydride to give two esters in a 36:64 ratio. Which of the two possible esters forms in the larger amount?

Answer:

A methyl group is sterically smaller than an ethyl group. Therefore, attack of a nucleophile is more favorable at the carbonyl group on the left to release a propanoate ion. The major ester is ethyl ethanoate; the minor ester is ethyl propanoate.

21.33 *p*-Hydroxyaniline reacts with acetic anhydride to give *N*-(4-hydroxyphenyl)acetamide. Explain why the reaction is selective, and acetylation does not occur at oxygen.

Answer:

The nitrogen atom of an amino group is more basic than the oxygen atom of a hydroxyl group. For atoms in the same period, the more basic atom is the better nucleophile.

21.34 Explain why the following bicyclic lactam hydrolyzes at a significantly faster rate than 5-aminopentanoic acid lactam.

5-aminopentanoic acid lactam

Answer:

The amide cannot be resonance stabilized by donation of electrons from nitrogen to the carbonyl group. Such a resonance contributor would have a carbon–nitrogen double bond at the bridgehead of the bicyclic structure, which would be highly strained. Resonance stabilization is possible for 5-aminopentanoic acid lactam, so it is more stable, and it hydrolyzes more slowly. The product of the hydrolysis reaction of 5-aminopentanoic acid lactam is shown below.

21.35 Mevalonic acid readily forms a lactone when heated. Draw two possible structures for the lactone. Which of the two is formed?

Answer:
The lactone is formed from the primary alcohol to give a six-membered ring. Reaction with the tertiary alcohol would give a more strained, four-membered lactone.

21.36 Explain why the rate of acid-catalyzed esterification of 2,2-dimethylcyclohexanecarboxylic acid is slower than that of cyclohexanecarboxylic acid.

2,3-dimethylcyclohexanecarboxylic acid

Answer:
The approach of the nucleophile to the carboxyl carbon atom is sterically hindered by the quaternary center at C-2.

Reactions of Acyl Derivatives

21.37 Draw the structures of the products of hydrolysis of each of the following esters.

Answers:

(a) $CH_3-\overset{O}{\overset{\|}{C}}-O-CH_2CH_2CH_3 \xrightarrow{H_2O} CH_3-\overset{O}{\overset{\|}{C}}-OH + HO-CH_2CH_2CH_3$

(b) $CH_3(CH_2)_2CH_2-\overset{O}{\overset{\|}{C}}-O-CH_2(CH_2)_2CH_3$
$\downarrow H_2O$
$CH_3(CH_2)_3-\overset{O}{\overset{\|}{C}}-OH + HO-CH_2(CH_2)_2CH_3$

(c) $CH_3CH_2-O-\overset{O}{\overset{\|}{C}}-CH_2CH_3 \xrightarrow{H_2O} CH_3CH_2-OH + HO-\overset{O}{\overset{\|}{C}}-CH_2CH_3$

(d) $CH_3(CH_2)_2CH_2-O-\overset{O}{\overset{\|}{C}}-CH_2CH_2CH_3$
$\downarrow H_2O$
$CH_3(CH_2)_2CH_2-OH + HO-\overset{O}{\overset{\|}{C}}-CH_2CH_2CH_3$

21.38   Draw the structures of the hydrolysis products of each of the following esters.

Answers:

(a)

$$\xrightarrow{\text{H}_2\text{O}}$$

+ $HOCH_2CH_3$

(b)

$$\xrightarrow{\text{H}_2\text{O}}$$

+ $HOCH(CH_3)_2$

(c)

$$\xrightarrow{\text{H}_2\text{O}}$$

+

(d)

$$\xrightarrow{\text{H}_2\text{O}}$$

+ 2 $HOCH_3$

(e)

$$\xrightarrow{\text{H}_2\text{O}}$$

+ $HOCH(CH_3)_2$

(f)

$$\xrightarrow{\text{H}_2\text{O}}$$

CO_2H + $HOC(CH_3)$

21.39   Hydrolysis of ambrettolide, contained in hibiscus, yields (*E*)-16-hydroxy-7-hexadecenoic acid. Draw the structure of ambrettolide.

Answer:
Since there is only one product, ambrettolide must be a lactone.

ambrettolide

21.40 Hydrolysis of beeswax gives a mixture containing unbranched acids with 26 and 28 carbon atoms and unbranched alcohols with 30 and 32 carbons atoms. Draw the structures of all possible components of beeswax

Answers:

$$CH_3(CH_2)_{24}CH_2-\overset{\overset{\displaystyle O}{\|}}{C}-O-CH_2(CH_2)_{29}CH_3$$

$$CH_3(CH_2)_{26}CH_2-\overset{\overset{\displaystyle O}{\|}}{C}-O-CH_2(CH_2)_{29}CH_3$$

$$CH_3(CH_2)_{24}CH_2-\overset{\overset{\displaystyle O}{\|}}{C}-O-CH_2(CH_2)_{31}CH_3$$

$$CH_3(CH_2)_{26}CH_2-\overset{\overset{\displaystyle O}{\|}}{C}-O-CH_2(CH_2)_{31}CH_3$$

21.41 Draw the structures of the hydrolysis products of each of the following compounds.

Answers:

(a)

(b)

(c)

(d)

21.42   Draw the structures of the acid-catalyzed hydrolysis products of each of the following compounds.

Answers:

(a)   H₃O⁺

(b) —CN   H₃O⁺   —CO₂H

(c) —C(=O)—NHCH(CH₃)₂   H₃O⁺   —C(=O)—OH + H₂NCH(CH₃)₂

(d) CH₃——CO₂CH₃   H₃O⁺   CH₃——CO₂H + CH₃OH

21.43   Draw the structure of the product of each of the following reactions.

Answers:

(a)

pyridine

(b) + CH_3CH_2OH $\xrightarrow{H^+}$

(c)

CH_3CH_2OH | H^+

(d)

404

21.44 Draw the structure of the product of each of the following reactions.

Answers:

(a)

CH_3CH_2OH | pyridine

(b)

$+$ CH_3CH_2OH $\xrightarrow{H^+}$

(c)

CH_3CH_2OH | H^+

$+$ $HO-CH_2CH_2-$

(d)

$+ (CH_3CO_2)_2O \longrightarrow$

21.45   Draw the structure of the product of each of the following reactions.

Answers:

(a)

pyridine

(b)

+ CH$_3$CH$_2$NH$_2$ →   HO—⋯—NHCH$_2$CH$_3$

(c)

+

↓

(d)

+ H—N⟩   →

406

21.46   Draw the structure of the product of each of the following reactions.

Answers:

(a)

pyridine

(b)

(c)

(d)

Reduction of Acyl Derivatives

21.47 Draw the structures of the products of each of the following reactions.

Answers:

21.48 Draw the structure of the product of each of the following reactions.

Answers:

(a)

1. LiAlH₄
2. H₃O⁺

(b)

1. LiAlH₄
2. H₃O⁺

(c) CH₃—

1. LiAlH(t-OBu)₃
2. H₃O⁺

(d)

1. DIBAL
2. H₃O⁺

(e)

1. LiAlH₄
2. H₃O⁺

(f)

1. LiAlH₄
2. H₃O⁺

21.49 A compound obtained from the wax of the sperm whale has the molecular formula $C_{32}H_{64}O_2$. Reduction by LiAlH₄ gives 1-hexadecanol. Draw the structure of the compound.

Answer:
Since only one 16-carbon product is obtained, the compound must be an ester with 16 carbon atoms in the acyl group and in the alkyl group.

$$CH_3(CH_2)_{14}CH_2—\overset{\displaystyle O}{\overset{\|}{C}}—O—CH_2(CH_2)_{15}CH_3$$

21.50 A compound obtained from hibiscus has the molecular formula $C_{16}H_{28}O_2$. Reduction by $LiAlH_4$ gives (*E*)-7-hexadecen-1,16-diol. Draw the structures of two possible compounds that could yield this diol.

Reactions with Organometallic Compounds

21.51 What ester is required to produce alcohols of the general structure R_2CHOH using a Grignard reagent.

Answer:

The two *R* groups are derived from the Grignard reagent. The third group on the carbon atom bearing oxygen is hydrogen and was originally bonded to the carbonyl carbon atom of the ester. Thus, the ester must be a derivative of formic acid, such as ethyl formate. The alkyl group in the ester is not part of the alcohol product.

21.52 Dimethyl carbonate reacts with Grignard reagents to give tertiary alcohols with the general structure R_2CHOH. Write the structures of the intermediates formed after the addition of one and two moles of the Grignard reagent.

Answer:

The first mole of Grignard reagent reacts to give a tetrahedral intermediate, which loses methoxide ion and gives a methyl ester. After this point, the subsequent reactions are those of an ester with a Grignard reagent. The second intermediate is a ketone.

21.53 Butanoyl chloride reacts at −78 °C with one equivalent of the Grignard reagent derived from 1-iodopropane in THF to give 4-heptanone. Why doesn't a second equivalent of the Grignard reagent react?

Answer:

The acid chloride is much more reactive toward the Grignard reagent than the product ketone. At the low temperature of the reaction, the ketone does not react. At a higher temperature, the Grignard reagent would add to the carbonyl group.

21.54 Suggest two possible synthetic routes to prepare the following ketone using a lithium dialkylcuprate.

Answer:
React an acid chloride with a lithium dialkylcuprate. The acid chloride consists of the carbonyl carbon atom and one of the two alkyl groups bonded to it in the product. The lithium dialkylcuprate contains the other alkyl group. One possible combination is cyclopentylethanoyl chloride and lithium di(2-methyl propyl)cuprate. This combination is shown below. The second combination that gives the same compound is 3-methylbutanoyl chloride and lithium di(cyclopentylmethyl)cuprate.

Multistep Synthesis

21.55 Draw the structures of the products of each of the following reactions.

(a) CH_2CH_2OH 1. CrO_3
    2. PCl_3
    3. $(CH_3)_2CuLi$

(b)   1. $LiAlH_4$
    OCH_3 2. H_3O^+
    3. $SOCl_2$
    4. Mg / ether
    5. $CH_3CO_2CH_3$

(c)   1. $LiAlH(t\text{-}OBu)_3$
    2. H_3O^+
    3. CH_3OH / H^+

Answers:
(a) Oxidation gives a carboxylic acid, which is then converted into an acid chloride by PCl_3. Reaction with the lithium dialkylcuprate gives a ketone.
(b) Reduction by lithium aluminum hydride gives an alcohol. Reaction with $SOCl_2$ gives a chloro compound, which when converted to a Grignard reagent adds to the methyl acetate to give a tertiary alcohol.
(c) Reduction by the complex aluminum hydride gives an aldehyde. Reaction with methanol gives an acetal.

21.56 Draw the structure of the final product of each of the following reaction sequences.

(a)

CHO
1. CrO₃
2. SOCl₂
3. CH₃OH
(pyridine)

(b)

1. CH₃NH₂
2. LiAlH₄

(c)

1. LiAlH(t-OBu)₃
2. H₃O⁺
3. Zn(Hg) / HCl

Answers:
(a) Oxidation gives a carboxylic acid, which is then converted into an acid chloride. Reaction with methanol gives an ester.
(b) Reaction with the amine gives an amide. Reduction by lithium aluminum hydride gives an amine.
(c) Reduction by the lithium aluminum tri-*tert*-butoxide hydride gives an aldehyde. A Clemmensen reduction then gives an alkane.

Spectroscopy of Acid Derivatives

21.57 Would you expect the carbonyl stretching absorption of acyl bromides to occur at higher or lower wavenumber than the carbonyl stretching absorption of acyl chlorides?

Answer: The bromine atom cannot effectively contribute electrons by resonance to the carbonyl group. It is less electronegative than chlorine and does not destabilize the dipolar resonance form as much as chlorine does. Thus, the carbonyl group will have more single bond character in the acyl bromide, and the bond will require less energy to stretch. This effect results in an absorption at a lower wavenumber for the acyl bromide.

21.58 Explain why the carbonyl stretching absorption of thioesters occurs at 1690 cm⁻¹, whereas that of acyl chlorides occurs at 1800 cm⁻¹.

Answer: Neither the sulfur nor the chlorine atom can effectively contribute electrons by resonance to the carbonyl group. Sulfur is less electronegative than chlorine and does not destabilize the dipolar resonance form as much as chlorine does. Thus, the carbonyl group of the thioester has more single bond character than the acyl chloride and requires less energy to stretch. This effect results in an absorption at a lower wavenumber for the thioester.

21.59 A compound with molecular formula C_4H_5N has a strong absorption at 2250 cm⁻¹. Suggest two possible structures and explain how they could be distinguished by other infrared absorptions.

Answer:
The absorption is due to a nitrile group. There is an additional deficiency of two hydrogen atoms, which could be due to a double bond or a ring. The stretching frequency is at the upper value of the range for nitriles (2200–2250 cm⁻¹). Thus, the double bond cannot be in conjugation with the nitrile group because such a resonance stabilized functional group would absorb at lower wavenumber. Only one unsaturated nitrile is possible, and only one cycloalkylnitrile is possible.

$CH_2{=}CHCH_2{-}C{\equiv}N$

$-C{\equiv}N$

21.60 A compound with molecular formula C_4H_5N has a strong absorption at 2250 cm⁻¹. Suggest two possible structures and explain how they could be distinguished by other infrared absorptions.

I   II   III

Answer:

The absorption of compound II occurs at lower wavenumber than compound I because conjugation of the carbon–carbon double bond with the carbonyl group increases the single bond character of the carbonyl group. The absorption of compound III occurs at higher wavenumber than compound I because the electron withdrawal of the carbon–carbon double bond decreases the electron density of the oxygen atom, and hence the availability of electrons to stabilize the dipolar resonance form of the carbonyl group. Thus, the order for compounds I, II, and III is 1775, 1741, and 1806 cm⁻¹, respectively.

21.61 Deduce the structure of the following compound based on the molecular formula and the following proton NMR spectrum.

9H
1.42 δ

$C_6H_{11}ClO_2$

2H
4.14 δ

$ClCH_2 \!-\! \overset{\displaystyle O}{\overset{\|}{C}} \!-\! O \!-\! \overset{\displaystyle CH_3}{\underset{\displaystyle CH_3}{\overset{|}{\underset{|}{C}}}} \!-\! CH_3$

tert-butylchloroethanoate

4   3   2   1
Chemical shift, δ (ppm)

Answer:

The singlet of intensity 9H at 1.5 ppm is due to nine equivalent protons, a good indicator of a *tert*-butyl group. The singlet of intensity 2H at 3.9 ppm is due to two equivalent protons on a carbon atom bonded to an oxygen atom or a chlorine atom. The compound is *tert*-butyl chloroethanoate. Note that, although chloromethyl 2,2-dimethylpropanoate might seem to fit the spectrum also, the singlet due to the protons in the —CH_2Cl group would be shifted further down field because they would be deshielded by both the ester oxygen atom and the chlorine atom.

21.62   Deduce the structure of the following compound based on the molecular formula, $C_5H_{10}O_3$, and the following proton NMR spectrum.

$C_5H_{10}O_3$

2H
4.27 δ

3H
1.26 δ

Chemical shift, δ (ppm)

Answer:

The "extra" oxygen atom in the molecular formula could indicate an alcohol or an ether group in addition to an ester. The triplet of intensity 6H indicates six equivalent protons (two methyl groups) with a neighboring methylene group. The quartet of intensity 4H indicates four equivalent protons (two methylene groups) with a neighboring methyl group; the signal's position at 4.2 ppm indicates that the protons are bonded to a carbon atom that is also bonded to an oxygen atom. Thus there are two equivalent —CH_2CH_2 groups in the compound. The compound is diethylcarbonate.

$$CH_3CH_2-O-\overset{\overset{\textstyle O}{\|}}{C}-O-CH_2CH_3$$

diethylcarbonate

21.63   Deduce the structure of the following compound based on the molecular formula and the following proton NMR spectrum.

ethyl 2-phenylethanoate

$C_{10}H_{12}O_2$

Answer:

The singlet of intensity 5H at 7.1 ppm is due to a phenyl group. The triplet of intensity 3H at 1.2 ppm and the quartet of intensity 2 at 4.1 ppm indicate three protons and two protons, respectively, splitting each other on neighboring carbon atoms; the methylene group is bonded to an oxygen atom. The singlet of intensity 2H indicates a methylene group with no neighboring protons; its position at 3.6 ppm indicates that the protons are bonded to a carbon atom that is also bonded to an oxygen atom or a benzene ring. The compound is ethyl 2-phenylethanoate.

21.64   Deduce the structure of the following compound based on the molecular formula and the following proton NMR spectrum.

chloromethyl 2,2-dimethylpropanoate

Answer:
The formula is $C_7H_{11}ClO_2$. The singlet of intensity 9H at 1.3 ppm is due to nine equivalent protons, a good indicator of a *tert*-butyl group. The singlet of intensity 2H is due to two equivalent protons; its position at 5.7 ppm indicates that the protons are bonded to a carbon atom bonded to an oxygen atom and a chlorine atom. The compound is chloromethyl 2,2-dimethylpropanoate.

21.65   Deduce the structure of the following compound based on the molecular formula and the following proton NMR spectrum.

Answer:
The singlet of intensity 3H at 2.1 ppm is due to a methyl group bonded to the carbonyl carbon atom. The triplets of intensity 2H at 3.5 and 4.4 ppm are due to neighboring methylene groups that are also bonded to an oxygen atom and a bromine atom, respectively. The compound is bromoethyl ethanoate.

21.66 Deduce the structure of the following compound based on the molecular formula and the following proton NMR spectrum.

ethyl *p*-methoxybenzoate

$C_{10}H_{12}O_3$

Answer:
The "extra" oxygen atom in the molecular formula could indicate an alcohol or an ether group and an ester. The doublets of intensity 2H at 6.9 and 7.9 ppm are due to two sets of equivalent protons on a benzene ring. The triplet of intensity 3H at 1.15 ppm indicates a methyl group split by a neighboring methylene group. The triplet of intensity 2H at 4.25 ppm indicates a methylene group split by a neighboring methyl group and bonded to an oxygen atom. The singlet of intensity 3H at 3.84 ppm is due to a methyl group with no neighboring hydrogens. The compound is ethyl *p*-methoxyoxybenzoate.

22

CONDENSATION REACTIONS OF CARBONYL COMPOUNDS

Keys to the Chapter

22.1 The α Carbon Atom of Carbonyl Compounds

The carbon atom bonded to the carbonyl carbon atom is known as the α carbon atom. It is a reactive site because its hydrogen atom is acidic. Extraction of the α hydrogen atom results in formation of a carbanion that serves as a nucleophile. The pK_a of the α hydrogen atom is approximately 18, which means that the K_a is approximately 30 powers of 10 larger than the K_a for hydrocarbons. The increased acidity is the result of resonance stabilization of the enolate ion. One of the two resonance forms of the enolate ion has the negative charge on the α carbon atom; the other resonance form has the negative charge on the oxygen atom.

Enolates are formed by reaction of a carbonyl compound with a base. The concentration of enolate formed depends on the K_b of the base, which in turn is related to the K_a of the conjugate acid of the base. Sodium hydroxide is a weaker base than the enolate ion, and it is not sufficiently basic to give a high concentration of enolate. Sodium amide is a much stronger base, and it quantitatively converts carbonyl compounds to their enolates.

Enolates can react as nucleophiles that attack an electrophilic center at the oxygen atom or the carbon atom of a second enolate. Although the electron density of the enolate is highest at the oxygen atom, the most common reaction site of enolates is at the carbon atom. This selectivity is related to the bonds formed in the transition state. Reaction at the oxygen atom forms an enol product that contains a carbon–carbon double bond. Reaction at the carbon atom forms a keto product that contains a carbon–oxygen double bond. The greater stability of the carbonyl group favors formation of the keto product.

22.2 Basicity of Acyl Derivatives

Aldehydes and ketones both exist in two isomeric forms known as keto and enol tautomers. They differ in the location of a hydrogen atom and the type of double bond. The keto form is more stable than the enol form. Tautomerization is a net process by which protons are transferred from one site to another by a series of steps in which the solvent is an intermediary. In acidic solution, the steps are protonation of the carbonyl oxygen atom by an acid to give a conjugate acid, followed by deprotonation of the α carbon atom by the conjugate base of the acid. The acid and base are the hydronium ion and water. In basic solution, the steps are deprotonation of the α carbon atom to give an enolate ion, followed by protonation of the oxygen atom by the conjugate acid of the base. The base and acid are the hydroxide ion and water.

The stability of an enol is reflected in its concentration in equilibrium with the keto form. Ketones have a smaller concentration of enol than aldehydes. This fact reflects the greater stability of ketones compared to aldehydes as a result of electron donation of alkyl groups to the carbonyl carbon atom. The stability of isomeric enols from a ketone reflects the stability due to the degree of substitution of the double bond. Conjugation of the double bond of the enol increases its stability.

22.3 Consequences of Enolization

The hydrogen atom of the α carbon atom of a carbonyl compound is called an **enolizable hydrogen atom**. If the α carbon atom is chiral, the formation of the isomeric enol results in loss of optical activity because a racemic mixture is formed when the keto form is regenerated. If the keto–enol equilibrium occurs in a protic solvent containing deuterium in place of hydrogen, then the α hydrogen atoms are exchanged by deuterium.

Organic Chemistry Study Guide: Key Concepts, Problems, and Solutions http://dx.doi.org/10.1016/B978-0-12-801889-7.00022-4

22.4 α-Halogenation Reactions of Aldehydes and Ketones

The α hydrogen atoms of carbonyl compounds can be replaced by halogen atoms. The regioselectivity and the number of hydrogen atoms substituted depend on whether acidic or basic conditions are used. Under acidic conditions, one hydrogen atom is substituted without the complication of multiple substitution. Under basic conditions, multiple substitution occurs.

Acid-catalyzed halogenation occurs by halogenation of the enol, whose formation is the rate determining step. The double bond of the enol is attacked by the halogen in a reaction similar to the electrophilic attack of a simple alkene. However, the resulting intermediate is not the bromonium ion, but an oxocarbocation. Loss of a proton from the hydroxyl group gives the halogenated product. Multiple substitution is disfavored because the halogen atom makes the carbonyl oxygen atom less basic and decreases the rate of formation of the enol. The halogen also destabilizes the conjugate acid because it tends to withdraw electron density from the oxocarbocation. In ketones with two nonequivalent α carbon atoms, the more substituted carbon atom is halogenated. This regioselectivity reflects the greater stability of the enol with the more substituted carbon–carbon double bond.

Halogenation under basic conditions occurs by nucleophilic attack of the enolate at the halogen molecule. Because the halogen atom is inductively electron withdrawing, the α hydrogen atom of the halogenated ketone is more acidic. Therefore, not only does multiple substitution occur, but it continues at the α carbon atom originally substituted in preference to a second α carbon atom. Continued halogenation of a methyl ketone forms a trihalomethyl derivative that is cleaved into a carboxylate and a haloform as the result of nucleophilic attack of the carbonyl carbon atom.

22.5 Alkylation of Enolate Ions

Enolates, formed by the abstraction of the α hydrogen atom by a strong base, are nucleophiles. Lithium diisopropylamide (LDA) or sodium hydride are required as bases. The site of proton abstraction is related to the acidity of the two possible α hydrogen atoms, which is in the order primary > secondary > tertiary. Reaction of the enolate with an alkyl halide forms a alklylated ketones. Multiple alkylation can occur as the result of proton exchange between the original enolate and the alkylated ketone, followed by alkylation of that enolate ion.

22.6 The Aldol Condensation of Aldehydes

The aldol condensation is the reaction of two moles of an aldehyde to form a β-hydroxyaldehyde, or aldol, in the presence of a base. The product is formed by addition of the enolate, formed by abstraction of an a hydrogen atom of one aldehyde by hydroxide ion, to the carbonyl carbon atom of the second aldehyde. Protonation of the alkoxide by exchange of a proton from water gives the aldol. The first step is an addition reaction to form a tetrahedral product. Subsequent dehydration in the reaction mixture often occurs to give an α,β-unsaturated aldehyde. The combination of the two steps constitutes a condensation reaction.

Under the basic conditions of the aldol condensation reaction, a dehydrated product forms. As a result of this step, the formation of an α,β-unsaturated aldehyde is favorable.

22.7 Mixed Aldol Condensation

A mixed aldol condensation is the formation of an aldol incorporating two different aldehydes. The reaction gives a mixture of four possible products if both aldehydes have α hydrogen atoms. If only one has α hydrogen atoms, then two products can result. If the carbonyl group of the aldehyde without α hydrogen atoms is more reactive toward nucleophiles, then one product results.

22.8 Intramolecular Aldol Condensation

Intramolecular aldol condensations are more favorable than intermolecular aldol condensations. Cyclization occurs if the α carbon atom and the second carbonyl carbon atom can bond to form a five- or six-membered ring. If two or more reactions can yield these rings, it is necessary to consider which process is favored. The various possible enolates exist in low concentration under equilibrium conditions. Thus, the enolate that is the better nucleophile attacks the more reactive carbonyl carbon atom and dominates the product formed. In general, for example, intramolecular aldol condensations where the enolate attacks the carbonyl carbon atom of an aldehyde are favored over addition to the carbonyl carbon atom of a ketone.

22.9 Conjugation in α,β Unsaturated Aldehydes and Ketones

The carbon–carbon double bond in an α,β-unsaturated aldehyde or ketone affects the stability of the carbonyl group, and hence its reactivity. The positive charge on the carbonyl carbon atom in the dipolar resonance form can be stabilized by donation of π electrons from the carbon–carbon double bond. Thus, there is some partial positive charge at the β carbon atom. This contributing resonance structure decreases the reactivity of the carbonyl carbon atom toward nucleophiles and offers an alternate site for reactivity at the β carbon atom.

22.10 Conjugate Addition Reactions

Addition of compounds represented by H—Nu to the carbon–oxygen double bond of an O-unsaturated aldehyde or ketone gives a 1,2-addition product. Addition of the nucleophilic portion of the reagent at the β carbon atom and a hydrogen atom at the carbonyl oxygen atom is a 1,4-conjugate addition.

Strong nucleophiles such as the hydride ion of metal hydrides, and the carbanion of a Grignard reagent, react to give 1,2-addition products with α,β-unsaturated aldehydes and ketones. Weak electrophiles such as cyanide ion, amines, alcohols, and thiols give 1,4-addition products. The Gilman reagent also gives 1,4-addition products.

22.11 The Michael Reaction and Robinson Annulation

The reaction of an enolate with α,β-unsaturated aldehydes and ketones is the **Michael reaction**. The enolate is called a Michael donor, and the α,β-unsaturated carbonyl compound is called a Michael acceptor. The carbonyl group of the Michael donor remains in the condensation product and may undergo an intramolecular aldol condensation reaction with the α carbon atom of the original Michael acceptor. This process, which forms a ring, is termed the **Robinson annulation**.

22.12 α Hydrogen Atoms of Acid Derivatives

The acidity of α hydrogen atoms of acid derivatives is affected by the amount of positive charge on the carbonyl carbon atom, which in turn is affected by the stabilization of that charge by the electronegative atom bonded to the carbonyl carbon atom. The acidity of α hydrogen atoms of esters (pK_a = 25) is less than that of aldehydes and ketones (pK_a = 20) because the oxygen atom of the alkoxy group supplies electrons by resonance to the partially positive carbonyl carbon atom. The resulting delocalization places some positive charge on oxygen and decreases the pK_a. As a result, the acidity of α hydrogen atoms of esters is less than that of aldehydes and ketones.

Enolates of esters can be prepared in low concentration at equilibrium by using the alkoxide ion in an alcohol corresponding to the alkoxy group contained in the ester. The equilibrium constant for the reaction is approximately 10^{-9} because the pK_a values of the ester and the alcohol differ by 10^9. High concentrations of the ester enolate are prepared by using LDA, which is a poor nucleophile and a strong base. The pK_a of diisopropyl amine is 40. Thus, the equilibrium constant for the reaction is approximately 10^{19}.

Dicarbonyl compounds such as ethyl acetoacetate and dimethyl malonate are significantly stronger acids (the pK_a values are 11 and 13, respectively). The increased acidity results from delocalization of negative charge in the conjugate base by the additional carbonyl group.

22.13 Reactions at the α Carbon Atom of Acid Derivatives

A hydrogen atom bonded to the α carbon atom of an ester is an **enolizable hydrogen atom**. If the α carbon atom is chiral, the formation of the ester enolate results in loss of optical activity because a racemic mixture is formed when the ester is regenerated. If the equilibrium occurs in a protic solvent containing deuterium in place of hydrogen, then the α hydrogen atoms are exchanged by deuterium.

Ester enolates, formed by the abstraction of the α hydrogen atom by a strong base, are nucleophiles. Lithium diisopropylamide (LDA) is required as the base. Reaction of the ester enolate with an alkyl halide forms α alkylated esters. Only primary haloalkanes can be used.

The α hydrogen atoms of carboxylic acids can be replaced by a single halogen atom using bromine and a small amount of phosphorus in the **Hell–Volhard–Zelinsky reaction**. The reaction occurs by bromination of a small amount of the acyl bromide. Under the reaction conditions, the bromoacyl bromide is converted into the bromocarboxylic acid. If one equivalent of PBr_3 is used with the Br_2, an a bromoacyl bromide is formed. Reaction of this compound with an alcohol gives an α-bromo ester.

22.14 The Claisen Condensation

The reaction of two moles of an ester in the presence of the alkoxide base corresponding to the alkoxyl group of the ester produces a β-keto ester. The ester must have two α hydrogen atoms, and one equivalent of base is required. The Claisen reaction occurs in four steps.

1. Abstraction the α hydrogen atom by the alkoxide ion.
2. Attack of the carbonyl carbon atom of the ester by an ester enolate.
3. Ejection of an alkoxide ion from the conjugate base of a hemiketal.
4. Abstraction of an a hydrogen atom of the keto ester by the alkoxide ion.

The addition of dilute acid at the end of the reaction protonates the conjugate base of the keto ester. Proton exchange between the β-keto ester and the alkoxide provides the driving force for the reaction.

The **Dieckmann condensation** is an intramolecular variation of the Claisen condensation. The ester must have two α hydrogen atoms and one equivalent of base is required. The intramolecular reaction is favorable because one mole of the diester gives one mole of keto ester and one mole of alcohol.

A mixed Claisen condensation is the formation of a keto ester incorporating two different esters. The reaction gives a mixture of four possible products if both esters have α hydrogen atoms. If only one has α hydrogen atoms, then two products can result. If the carbonyl group of the ester without α hydrogen atoms is more reactive toward nucleophiles, then one product results.

Nonenolizable esters react with the enolate of ketones to give β-diketones. The enolate of the ketone attacks the carbonyl carbon atom of the ester in a reaction similar to that of the Claisen condensation.

22.15 Aldol-Type Condensations of Acid Derivatives

Aldol-type condensations occur between a carbonyl compound and the enolate of an ester. The product is either a β-hydroxy ester or its related α,β-unsaturated ester. The **Knoevenagel condensation** occurs between a malonate ester and an aldehyde or ketone. This aldol-type reaction gives an α,β-unsaturated ester. The **Reformatskii reaction** occurs between a zinc enolate of an ester and an aldehyde or ketone to produce a β-hydroxy ester.

22.16 β-Dicarbonyl Compounds In Synthesis

The alkylation of acetoacetate or malonate esters is a useful synthetic process that is synthetically equivalent to the direct alkylation of a ketone or an ester. The acidity of both compounds is higher than that of ketones and esters and allows the abstraction of the a proton by an alkoxide ion to quantitatively form the conjugate base. Alkylation of the conjugate base occurs by nucleophilic attack on a haloalkane. Hydrolysis of the alkylated acetoacetate leads to decarboxylation of the keto acid and formation of a ketone. Hydrolysis of the alkylated malonate ester leads to decarboxylation of the diacid, and formation of an acid.

22.17 Michael Condensation of Acid Derivatives

The reaction of an enolate with α,β-unsaturated aldehydes and ketones is termed the **Michael reaction**. The enolate is called a Michael donor, and the α,β-unsaturated carbonyl compound, such as 3-buten-2-one, is called a Michael acceptor. 1,3-Dicarbonyl derivatives, such as dimethyl malonate, easily form enolates that act as Michael donors. Subsequent hydrolysis of the addition product and decarboxylation yields 1,5-dicarbonyl compounds.

Summary of Reactions

1. Exchange of α Hydrogen Atoms

2. Isomerization of Carbonyl Compounds

3. α Halogenation of Carbonyl Compounds

4. Aldol Condensation

5. Conjugate Addition Reactions of Carbonyl Compounds

6. Michael Condensation of Carbonyl Compounds

7. Exchange of α Hydrogen Atoms of Esters

8. α Alkylation of Esters

9. α Halogenation of Carboxylic Acids (Hell–Volhard–Zelinsky Reaction)

10. Claisen Condensation

11. Aldol-Type Condensations of Acid Derivatives

12. Acetoacetate Ester Synthesis

13. Malonate Ester Synthesis

14. Michael Condensation of Acid Derivatives

Solutions to Exercises

Acidity of α Hydrogen Atoms

22.1 The pK_a of 2,4-pentanedione is 9. Calculate the equilibrium constant for the acid–base reaction of 2,4-pentanedione with sodium ethoxide. The pK_a of ethanol is 15.9.

Answer: The pK_a of 2,4-pentanedione is 1×10^{-9}, so it is a stronger acid than ethanol, whose K_a is approximately 1.3×10^{-16}. Thus, the reaction of 2,4-pentanedione with sodium ethoxide has $K_a = 7.7 \times 10^6$.

22.2 The pK_a of acetonitrile, CH_3CN, is 25. Calculate the equilibrium constant for the acid–base reaction of acetonitrile with LDA. The pK_a of isopropylamide is 40.

Answer: The K_a values of acetonitrile and diisopropylamine are 1×10^{-25} and 1×10^{-49}, respectively. Thus, acetonitrile is a stronger acid than diisopropylamine, and the reaction of acetonitrile with lithium diisopropylamide has $K_a = 1 \times 10^{15}$.

22.3 The pK_a of acetophenone is 16. Calculate the equilibrium constant for the acid–base reaction of acetophenone with LDA.

Answer: The K_a values of acetophenone and diisopropylamine are 1×10^{-16} and 1×10^{-49}, respectively. Thus, acetophenone is a stronger acid than diisopropylamine, and the reaction of acetophenone with lithium diisopropylamide has $K_a = 1 \times 10^{24}$.

22.4 The pK_a of nitromethane is 10.2. Calculate the equilibrium constant for the acid–base reaction of nitromethane with sodium ethoxide. The pK_a of ethanol is 15.9.

Answer: The K_a of nitromethane is 6.3×10^{-11}, so it is a stronger acid than ethanol, whose K_a is approximately 1.3×10^{-15}. Thus, the reaction of nitromethane with sodium ethoxide has $K_{eq} = 4.8 \times 10^5$.

22.5 The pK_a values of acetone and 3-pentanone, as measured in DMSO, are 26.5 and 27.1, respectively. Explain this order of values.

Answer: Acetone is the stronger acid, so its conjugate base is more stable. The conjugate base of acetone has a contributing resonance form with a negative charge on a primary carbon atom. For 3-pentanone, the charge of the carbanion is on a secondary carbon atom. Because primary carbanions are more stable than secondary carbanions, acetone gives a more stable conjugate base than 3-pentanone.

22.6 The pK_a values of acetone and 1-phenyl-2-propanone, as measured in DMSO, are 26.5 and 19.8, respectively. Explain this order of values.

Answer: 1-Phenyl-2-propanone is the stronger acid, so its conjugate base is more stable. The conjugate base of 1-phenyl-2-propanone has a contributing resonance form with a negative charge on a secondary carbon atom that is also benzylic. Thus, the charge is delocalized, and this conjugate base is resonance stabilized. For the conjugate base of acetone, the charge is localized on a primary carbon atom, so it is less stable.

Stability of Enols

22.7 Which ketone has the larger percent enol at equilibrium, cyclohexanone or cyclobutanone?

Answer: Both enols have the double bond within their respective rings in the enol form, and both have the same degree of substitution. However, the double bond of the enol of cyclobutanone increases the strain energy of the small ring, so this enol is less stable than the enol of cyclohexanone. Therefore, cyclohexanone has a larger percent of enol at equilibrium.

22.8 Which ketone has the larger percent enol at equilibrium, 1,3-cyclohexanedione or 1,4-cyclohexanedione?

Answer: Both enols have the double bond within a six-membered ring in the enol form. However, the double bond of the enol of 1,3-cyclohexanedione is conjugated with the second carbonyl group and is resonance stabilized. The enol of 1,4-cyclohexanedione is not resonance stabilized. Therefore, 1,3-cyclohexanedione has a larger percent of enol at equilibrium.

22.9   (a) Write the structures of the isomeric enols of 2,2-dimethyl-3-pentanone, and (b) rank them in order of their stability.

Answer: There is only one α-hydrogen atom, and it is located at C-4. However, two enols can form that are geometric isomers. The most stable enol has the large *tert*-butyl *trans* to the methyl group.

more stable, *trans* enol

22.10   (a) Write the structures of the isomeric enols of 2-methylcyclopentanone and (b) rank them in order of relative stability.

Answer: There are two α-carbon atom: C-2 and C-5. The double bond at C-2 is more substituted, and more stable.

more substituted,
more stable enol.

22.11   Which ketone has the larger percent enol at equilibrium, 1,2-diphenylethanone or 1,3-diphenyl-3-propanone?

Answer: The enol of 1,2-diphenylethanone has extended conjugation between the two phenyl rings through the carbon–carbon double bond. The enol of 1,3-diphenyl-3-propanone has only one ring only conjugated to the carbon–carbon double bond and is therefore less stable.

22.12   Write the structure for the enol tautomer of the following molecule. What structural features contribute to its stability?

Answer: The keto form has its carbonyl group conjugated with a benzene ring. However, the enol tautomer has a double bond that is part of the aromatic ring system of phenanthrene. The resonance stabilization of the fused ring system favors formation of the enol, which is a phenol.

Enolates

22.13 Write the resonance form with a negative charge on the oxygen atom for the enolates derived from each of the following compounds.
(a) 3,3-dimethyl-2-butanone   (b) acetophenone   (c) 2,2-dimethylcyclohexanone

Answer: In each case, there is only one α-carbon atom with enolizable hydrogen atoms, so only one enolate ion forms.

22.14 Write the resonance form with a negative charge on the oxygen atom for all possible enolates derived from each of the following compounds. Which enolate is the most stable in each case?
(a) 2-pentanone   (b) 1-phenyl-2-propanone   (c) 1,3-cyclohexanedione

Answer: (a) There are two α-carbon atoms with enolizable hydrogen atoms. C-1 gives a single enolate. C-3 gives a pair of enolates that are geometric isomers. The more substituted double bond of the enolates at C-3 is more stable than the enolate with a double bond at C-1. Of the two enolates involving at C-3, the isomer with the alkyl groups *trans* to one another is more stable.

enols of 2-pentanone in order of decreasing stability

Answer: (b) There are two α-carbon atoms with enolizable hydrogen atoms. C-3 gives a single enolate. C-1 gives a pair of enolates that are geometric isomers. The more substituted double bond of the enolates at C-1 is more stable than the enolate with a double bond at C-3 because the double bond is conjugated with the benzene ring. Of the two enolates involving the C-3 atom, the isomer with the alkyl and aryl groups *trans* to one another is more stable.

enols of acetophenone in order of decreasing stability

Answer: (c) There are two enolates. The enolate at C-6 has a localized double bond. The enolate at C-2 is conjugated with the second carbonyl group and is resonance stabilized.

enols of 1,3-cyclohexanedione in order of decreasing stability

22.15 Write the contributing resonance forms of the conjugate base of (a) acetonitrile (CH_3CN) and (b) nitromethane (CH_3NO_2).

Answers: (a) $:\bar{C}H_2—C\equiv N:$ ⟷ $CH_2=C=\ddot{N}:^-$

(b)

22.16 Write the contributing resonance forms for all possible conjugate bases of 3,6,6-trimethyl-2-cyclohexenone.

Answers:

22.17 The following ketone gives a mixture of two enolates in approximately equal amounts (53:47). (a) Write the structures of the enolates and (b) explain why they are of comparable stability.

Answer: (a)

Answer: (b) The double bond of each enolate has the same degree of substitution.

22.18 2-Methylcyclopentanone gives a mixture of two enolates in a 94:6 ratio. (a) Write their structures and (b) assign their relative stabilities.

Answer: (a)

Answer: (b) There are two α-carbon atoms with enolizable hydrogen atoms. They are located at C-2 and C-5. The double bond of the enolate to C-2 has a higher degree of substitution and is more stable. It constitutes 94% of the mixture of enolates.

22.19   3-Pentanone gives a mixture of two enolates in a 84:16 ratio. (a) Write their structures and (b) assign their relative stabilities.

Answer: (a)

enols of 3-pentanone in order of decreasing stability

Answer: (b) There are two α-carbon atoms with enolizable hydrogen atoms, but they are equivalent. The C-2 atom gives a pair of enolates that are geometric isomers. Of the two enolates, the isomer with *trans* alkyl groups is more stable. It constitutes 84% of the enolate mixture.

22.20   2,2-Dimethyl-3-pentanone gives a mixture of two enolates. Based on the data in Exercise 22.19, predict how the ratio of the amounts of the two enolates would differ from the ratio for 1-pentanone.

Answer: (a)

enols of 2,2-dimethyl-3-pentanone in order of decreasing stability

Answer: (b) There is only one α-carbon atom with enolizable hydrogen atoms. The C-3 atom gives a pair of enolates that are geometric isomers. Of the two enolates, the isomer with *trans* alkyl groups is more stable. However, in this compound the alkyl groups are a *tert*-butyl and a methyl group, so the steric hindrance is larger than for the ethyl and methyl groups of the enolate of 3-pentanone. Thus, the ratio of the two isomers is much larger, and the *trans* isomer constitutes more than 84% of the enolate mixture.

22.21   (a) Write the mechanism for the following isomerization reaction, which occurs using sodium ethoxide in ethanol. (b) Predict which isomer is more stable.

Answer: (a)

Answer: (b) Formation of an enolate occurs by abstraction of a proton from the bridgehead position by ethoxide ion. Protonation in a reverse step by ethanol can occur to give either the original ketone or its isomer. The isomer is a ketone derived from *trans*-decalin, which has its rings fused by diequatorial bonds, and is more stable than *cis*-decalin.

22.22 Write the mechanism for the following isomerization reaction, which occurs using sodium ethoxide in ethanol. Predict which isomer is more stable.

Answer: (a)

$CH_3CH_2O^-$ / CH_3CH_2OH

$CH_3CH_2O^-$ / CH_3CH_2OH

Answer: (b) The enolate is resonance stabilized, with charge distributed between a secondary and a tertiary carbon atom. Protonation can occur to give the original ketone or the isomer. This isomeric ketone is resonance stabilized by the extended conjugation of two double bonds with the carbonyl group, so it is more stable than the original ketone, whose carbonyl group is not conjugated with a double bond.

22.23 Write a mechanism for the base-catalyzed isomenzation of 3-cyclohexenone to 2-cyclohexenone.

Answer: (a)

$CH_3CH_2O^-$ / CH_3CH_2OH

$CH_3CH_2O^-$ / CH_3CH_2OH

Answer: (b) The enolate formed is resonance stabilized, with charge distributed between two secondary carbon atoms. Protonation can occur to give the original ketone or the isomer. This isomeric ketone is not resonance stabilized by conjugation of the ketone with the double bond, so it is less stable than the original conjugated ketone. Nevertheless, at equilibrium there is a low concentration of the less stable 2-methyl-3-cyclopentenone.

22.24 Write a mechanism for the base-catalyzed isomerization of 5-methyl-2-cyclopentenone to 2-methyl-2-cyclopentenone. (Hint. A third isomeric unsaturated ketone is a required intermediate.)

Answer: (a) The enolate is resonance stabilized, with charge distributed between two secondary carbon atoms. Protonation can occur to give the original ketone or the isomer. This isomeric ketone is not resonance stabilized by conjugation of the ketone with the double bond, so it is less stable than the original conjugated ketone. Nevertheless, at equilibrium there is a low concentration of the less stable 2-methyl-3-cyclopentenone.

Answer: (a)

CH_3

$CH_3CH_2O^-$ / CH_3CH_2OH

$CH_3CH_2O^-$ / CH_3CH_2OH

Answer: (b) A proton can be abstracted from either C-2- or C-5 of the nonconjugated product, 2-methyl-3-cyclopentenone. Loss of a proton from C-5 generates the resonance stabilized enolate of the above reaction, and leads to formation of the original 5-methyl-2-cyclopentenone. However, abstraction of a proton from C-2 gives a different resonance stabilized enolate with negative charge distributed between C-2 and C-4.

432

Answer: (b)

Answer: (c) Protonation can occur to give the original 2-methyl-3-cyclopentenone or an isomeric conjugated ketone. 2-Methyl-2-cyclopentenone is the most stable of the three isomers in equilibrium in this series of reactions because it is both conjugated and the more highly substituted ketone. Thus, 5-methyl-2-cyclopentenone is isomerized by the base into 5-methyl-3-cyclopentenone, which in turn is isomerized into 2-methyl-2-cyclopentenone. Although the number of the methyl group changes, it does not rearrange. The numbering is controlled by the location of the double bond relative to the ketone.

22.25 Write a mechanism that explains why a solution of (R)-2-methyl-1-pheny1-1-pentanone in ethanol containing sodium ethoxide gradually loses optical activity, but a solution of (R)-3-methyl-1-phenyl-1-pentanone does not.

Answer: (a) The chiral center of (R)-3-methyl-1-phenyl-1-pentanone is at the β-carbon atom relative to the ketone, so its hydrogen atom is not sufficiently acidic to be abstracted by ethoxide ion. Thus, its stereochemistry is not affected in the basic solution. The chiral center of the isomeric (R)-2-methyl-1-phenyl-1-pentanone is at the α carbon atom relative to the ketone, so its hydrogen atom is abstracted to form an enolate. Reprotonation can give either the (R) or (S) isomer.

Answer: (b)

22.26 Predict the change in the optical activity of each of the following in a solution of sodium ethoxide in ethanol.
(a) (R)-2-methylcyclohexanone   (b) (R)-3-methylcyclohexanone   (c) (R)-2-ethyl-2-methylycyclohexanone

(a)

(R)-2-methylcyclohexanone

(b)

(R)-3-methylcyclohexanone

(c)

(R)-2-ethyl-2-methylcyclohexanone

Answer: (a) The chiral center of (R)-2-methylcyclohexanone is at the α carbon atom relative to the ketone, and it is abstracted to form an enolate. Reprotonation can give either the (R) or (S) isomer, so the solution gradually loses its optical activity.

Answer: (b) The chiral center of (R)-3-methylcyclohexanone is at the β carbon atom relative to the ketone, and its hydrogen atom is not sufficiently acidic to be abstracted by ethoxide ion. Thus, its stereochemistry is not affected in the basic solution.

Answer: (c) There is no hydrogen atom at the chiral center of (R)-2-ethyl-2-methylcyclohexanone, so an enolate can form only at C-6, which does not affect the chirality of the molecule.

Deuterium Exchange

22.27   (a) Explain why 7-bicyclo[2 2 1]heptanone does not undergo an exchange reaction using sodium hydroxide in D_2O, but 2-bicyclo[2.2.1]heptanone readily reacts. (b) Which hydrogen atoms are exchanged'?

7-bicyclo[2.2.1]heptanone   2-bicyclo[2.2.1]heptanone

Answer: (a) The α carbon atom of 7-bicyclo[2 2.1]heptanone is at a bridgehead position. The carbanion if formed could not form the carbon–carbon double bond of the enolate because the orbital of the carbanion and the orbitals of the carbonyl group are perpendicular to one another.

Answer: (b) However, the proton at C-3 of 2-bicyclo[2.2.1]heptanone is acidic because an enolate ion can form and is resonance stabilized. The proton at the bridgehead position is not acidic in this compound either. Deuterium exchange occurs at C-3.

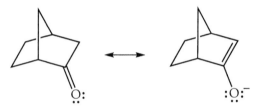

22.28   Explain why 3,3-dimethyl-2-bicyclo[2.2.1]heptanone does not undergo an exchange reaction using sodium hydroxide in D_2O.

3,3-dimethyl-2-bicyclo[2.2.1]heptanone

Answer: The α positions in this molecule are C-3 and C-1. There is no proton at C-3 of 3,3-bicyclo[2.2.1]heptanone. The proton at C-1 is at a bridgehead position, and is not acidic because a resonance-stabilized enolate would require a geometrically impossible double bond at the bridgehead position.

22.29 Explain how the following isomeric ketones could be distinguished using the base-catalyzed exchange reaction with deuterium.

I II

Answer: Compound I has two methylene carbon atoms at the α positions, so four hydrogen atoms can be exchanged by four deuterium atoms. Compound II has two C—H bonds at an α methylene carbon atom and one α C—H bond at a tertiary center. Thus, only three deuterium atoms can be incorporated into compound II.

22.30 Explain how 2-pentanone and 3-pentanone could be distinguished using the base-catalyzed exchange reaction with deuterium.

Answer: 2-Pentanone has a methyl carbon atom and a methylene carbon atom at the α positions, so five hydrogen atoms can be exchanged by five deuterium atoms. 3-Pentanone has two methylene carbon atoms at the α positions, so only four hydrogen atoms can be exchanged by deuterium atoms.

22.31 3-Methyl-2,4-pentanedione rapidly exchanges one hydrogen using sodium hydroxide and D_2O. After a long time, a total of seven hydrogen atoms are eventually exchanged. Explain these observations.

Answer: (a) The most acidic proton of 3-methyl-2,4-pentanedione is at C-3 because the enolate formed is resonance stabilized by conjugation with the second carbonyl group. Thus, this proton is exchanged rapidly.

Answer: (b) The protons at C-1 and C-5 can also be exchanged, but they are not as acidic as the proton at C-3 because the enolate is not resonance stabilized. Eventually, all six hydrogen atoms of the two equivalent methyl groups are exchanged by deuterium atoms.

22.32 After a long time, 3-methyl-2-cyclohexenone exchanges a total of eight hydrogen atoms. (a) Identify the hydrogen atoms exchanged and (b) write a step showing the transfer of deuterium to an enolate that gives exchange at each of the required sites.

Answer: The two hydrogen atoms are C-6 are exchanged in the first enolate shown. The hydrogen atom at C-2 and those of the methyl group are exchanged in the second enolate. The two hydrogen atoms at C-4 are exchanged in the third enolate.

α-Halogenation Reactions

22.33 Reaction of 3-methyl-2,4-pentanedione with bromine under acidic conditions rapidly yields a monobromo derivative. (a) Write the structure of the product and (b) explain how it forms.

Answer: The most stable enol is produced by protonating one of the two equivalent carbonyl oxygen atoms and forming a double bond between C-2 and C-3. Bromination of the enol thus occurs at C-3.

22.34 Reaction of 3-methyl-2-butanone with bromine under acidic conditions yields a mixture of two monobromo derivatives in a 95:5 ratio. (a) Write the structure of the products and (b) explain why the high ratio of isomers occurs.

Answer: There are two possible enols. The one with a double bond between C-2 and C-3 is more stable because it is more highly substituted. This enol gives the major product. The enol with a double bond between C-1 and C-2 accounts for the minor product.

22.35 Which of the following compounds will give a positive iodoform test when treated with iodine in a basic solution?

Answer: The formation of CHI_3 in an iodoform reaction requires a methyl group bonded to a carbonyl carbon atom. Only (c) and (d) have such structures.

22.36 Write the structure of a compound with molecular formula $C_8H_{14}O_2$ that gives adipic acid when reacted with excess bromine in a basic solution.

Answer: Adipic acid is hexanedioic acid, a six-carbon dicarboxylic acid. The product has two fewer carbon than the reactant, and a dicarboxylic acid. This means that the reactant was a diketone with methyl groups bonded to each carbonyl carbon atom before the bromination reaction.

adipic acid (hexanedioic acid)

22.37 (a) Explain why the indicated hydrogen atom at the bridgehead carbon of the following compound is not replaced by bromine in basic solution. (b) What competing reactions may occur?

Answer: (a) An enolate cannot form because a double bond cannot be located at the bridgehead carbon atom. The orbital containing the electron pair of a possible bridgehead carbanion is approximately perpendicular to the plane of the p bond of the carbonyl groups, which is not suitable for π bond formation. Either of the two methylene groups that are α to the two carbonyl groups could be brominated.

22.38 Predict the structure of the dibromo derivative obtained from the following ketone in basic solution.

Answer: Bromination occurs at the site of the most acidic hydrogen atom, which in this case is the methylene group. It is secondary; the bridgehead carbon is tertiary and therefore cannot be brominated twice. The second bromination occurs at a faster rate than the first and occurs at the site of the first bromination.

22.39 Bromination of 4-*tert*-butylcyclohexanone under acidic conditions yields a mixture of two isomeric monobromo derivatives in approximately equal amounts. (a) Write the structures of the products and (b) explain why the ratio of the two compounds is approximately one.

Answer: The methylene groups at C-2 and C-6 are equivalent. The enol formed under acid conditions has a double bond that may be attacked by the electrophilic bromine from either side of the ring; *cis*- and *trans*-2-bromo-4-*tert*-butylcyclohexanone compounds are produced.

22.40 Write the structures of the four isomeric monobromo products that could result from bromination of the following ketone in acidic solution.

Answers:

Reactions at the α Carbon Atom

22.41   Write the structure of the product obtained by the reaction of 2,2-dimethyl-3-pentanone with sodium hydride followed by addition of 1-iodobutane.

Answer: Alkylation occurs at C-4, whose hydrogen atom is abstracted by base to form an enolate ion. The alkyl iodide is primary and undergoes substitution without significant competing elimination reaction.

Answer:

2,2-dimethyl-3-pentanone

22.42   Explain why reaction of cyclohexanone with LDA followed by the addition of 2-bromopropane gives only the original ketone upon aqueous workup.

Answer: The alkyl bromide is secondary, so it undergoes an elimination reaction in the presence of the strong conjugate base formed abstraction of the proton at C-2 of cyclohexanone. Protonation of the conjugate base gives the original ketone.

22.43   The enolate derived from reaction of LDA with 4-*tert*-butylcyclohexanone reacts with ethyl iodide to give a mixture of two monoalkylated products in approximately equal amounts. (a) Write the structures of the products. (b) Explain why the ratio of the two compounds is approximately one.

Answer:

Answer: The methylene groups at C-2 and C-6 are equivalent. The enolate formed under basic conditions has electron density at C-2, which acts as a nucleophile that displaces iodide ion from iodoethane. The trigonal pyramidal carbanion can invert, so displacement can occur with the newly formed bond on either side of the ring. *Cis-* and *trans* isomers therefore form.

22.44   Write the structures of the four isomeric monobromo products that could result from bromination of the following ketone in acidic solution.

Answer:

Answer: There are two possible enolates because two nonequivalent methylene groups are located α to the carbonyl group. Each can react with methyl iodide, and the methyl group may bond to either side of the ring.

22.45 Reaction of 6-bromo-3,3-dimethyl-2-hexanone with LDA gives a product with the molecular formula $C_8H_{14}O$. Write its structure.

Answer: The enolate has some negative charge at C-1, which acts as a nucleophile to displace bromide ion at C-6 in an intramolecular reaction.

Answer:

22.46 Reaction of the following ketone with a sterically hindered strong base gives a product with the molecular formula $C_{10}H_{14}O$. Write its structure.

Answer:

Answer: The enolate has some negative charge, which acts as a nucleophile to displace bromide ion in an intramolecular reaction.

22.47 Trimethylchlorosilane, $(CH_3)_3SiCl$, reacts with enolates exclusively at the oxygen atom to give trimethylsilyl enol ethers. When heated with triethylamine and trimethylchlorosilane, the silyl ethers I and II derived from 2-methylcyclohexanone occur in a 1:3 ratio. Which is more stable? Why is it more stable?.

Answer: Compound II is more stable because the double bond is more highly substituted. Compound II is the major product because the enolates formed are in equilibrium with the weak base. The major product thus results from reaction of the major enolate in solution.

22.48 Using the data in Exercise 22.47, predict the structure of the maim product of the reaction of 2-pentanone with triethylamine and trimethylchlorosilane.

Answer:

Answer: There are two isomeric enolates. The enolate derived from abstraction of a proton at C-3 is more highly substituted than the enolate derived from abstraction of a proton at C-1. The major product is the isomer with the double bond at C-2.

22.49 The reaction of 2-methylcyclohexanone with LDA in 1,2-dimethoxyethane at 0 °C yields a solution that, when subsequently reacted with trimethylchlorosilane and triethylamine, yields I and II in a 99:1 ratio. Explain why the indicated ratio occurs.

Answer: The strong base removes a proton from the more acidic secondary carbon atom at a faster rate than from the tertiary carbon atom. Because the base is very strong, neither enolate reverts to the carbonyl compound, and the two are not in equilibrium. The product is the result of kinetic control.

22.50 Based on the data in Exercise 22.49, predict the structure of the major product of the reaction of 2-pentanone with LDA in 1,2-dimethoxyethane at 0 °C followed by trimethylchlorosilane and triethylamine.

Answer:

Answer: The strong base removes a proton from the more acidic primary carbon atom at a faster rate than from the secondary carbon atom. Because the base is very strong, neither enolate reverts to the carbonyl compound and the two are not in equilibrium. The major product, which has a double bond at C-1, is the result of kinetic control.

22.51 The reaction of 2-methylcyclohexanone with LDA in 1,2-dimethoxyethane at 0 °C yields a solution that reacts with benzyl bromide to give 2-benzyl-6-methylcyclohexanone and 2-benzyl-2-methylcyclohexanone in a 12:1 ratio. Explain why the indicated ratio occurs.

Answer: The strong base removes a proton from the more acidic secondary carbon atom at a faster rate than from the tertiary carbon atom. Because the base is very strong, neither enolate reverts to the carbonyl compound and the two are not in equilibrium. The major alkylation product is derived from the substitution at C-6 in the original compound. The product is named 2-benzyl-6-methylcyclohexanone because the benzyl group take alphabetic preference over the methyl group.

Answer:

442

22.52   What experimental conditions would favor formation of 2-benzyl-2-methylcyclohexanone by alkylation of 2-methylcyclohexanone with benzyl bromide?

Answer: Reaction of 2-methylcyclohexanone with benzyl bromide using triethylamine as a base generates two isomeric enolates. They exist in equilibrium with the ketone in the weak base. The major enolate is the more highly substituted compound with a double bond at C-2. This enolate reacts with benzyl bromide gives 2-benzy1-2- methylcyclohexanone as the major product.

22.53   Explain why the reaction of the following ketone with LDA in THE at −60 °C yields the indicated bicyclic ketone.

Answer: The strong base removes a proton from the more acidic primary carbon atom at a faster rate than from the tertiary carbon atom. Intramolecular attack of the primary carbanion at the primary bromoalkane center forms a seven-membered ring. Although formation of seven-membered rings is generally not favored, this process occurs because four of the bonds are restricted by the five-membered ring, and there is a larger probability of ring closure.

22.54   The ketone shown in Exercise 22.53 reacts with potassium *tert*-butoxide in *tert*-butyl alcohol to give a constitutional isomer of the bicyclic ketone shown above. (a) Write its structure and (b) explain its origin.

Answer: Potassium *tert*-butoxide in *tert*-butyl alcohol is a weaker base than LAD. It gives two enolates that are in equilibrium with the ketone, so two products can result. The enolate with a double bond to the atom of the five-membered ring can undergo an intramolecular reaction to produce a second five-membered ring.

Answer:

Aldol Condensations

22.55   Draw the structure of the product of the self-condensation of each of the following aldehydes in the presence of a catalytic amount of sodium hydroxide.
(a) 2-methylpropanal   (b) phenylethanal   (c) octanal

Answers:

22.56   What reactants are required to give the following compounds by a mixed aldol reaction?

(a)

(b)

(c)

(d)

Answers: (a) cyclohexanone and ethanal   (b) cyclohexanone and 2,4-pentanedione
            (c) acetophenone and 3-pentanone (d) benzaldehyde and acetophenone

22.57   Pseudoionone, a component of some perfumes, has the molecular formula $C_{13}H_{20}O$. It can be prepared by a mixed aldol reaction of citral and acetone. Write the structure of pseudoionone.

citral

Answer:

pseudoionone

22.58   A mixed aldol reaction between citral and 2-butanone yields two products. Write their structures.

Answers: Either C-1 or C-3 of 2-butanone can form an enolate. Subsequent reaction of each enolate at the carbonyl carbon of citral gives an unsaturated product.

Answer:

22.59   2,2-Dimethyl-l,3-propanediol can be synthesized by reduction of a mixed aldol product using sodium borohydride. (a) What is the aldol? (b) What two carbonyl compounds are required to produce it?

Answer: (a) The product is 3-hydroxy-2,2-dimethylpropanal. (b) It is the product of a mixed aldol condensation of 2-methylpropanal and methanal (formaldehyde).

3-hydroxy-2,2-dimethylpropanal

22.60   Suggest a synthesis of the following compound starting from acetophenone.

Answer: A mixed aldol condensation of acetophenone and methanal (formaldehyde) gives a product that can undergo a mixed aldol condensation two more times.

22.61   The favored products of the intramolecular aldol condensation of 2,5-hexanedione and 2,6-heptanedione are given in Section 22.9. (a) Write an alternative isomeric structure for each product. (b) Explain why it is not formed.

Answer: Formation of an enolate by abstraction of a proton at C-3 of 2,5-hexanedione followed by ring closure would give a cyclopropane ring. This process doesn't occur due to ring strain. Also, subsequent dehydration is disfavored because a double bond in the ring would further increase the ring strain. Formation of an enolate by abstraction of a proton at C-3 of 2,6-hexanedione followed by ring closure would give a cyclobutane. This process doesn't occur for the same reasons.

22.62 The intramolecular aldol condensation of 2,6-octanedione could yield two possible six-membered unsaturated products. (a) Write their structures. (b) Predict which isomer would be the major product.

2,6-octanedione

Answer: (a) Enolates can form as the result of proton abstraction from any of four nonequivalent α carbon atoms. However, six-membered rings result only from the enolates derived from abstraction of protons at C-1 and C-7. All of the enolates are in equilibrium with the diketone. Thus, the products formed are the result of favorable rates of ring closure and their individual stability. Reaction of the enolate derived from C-1 with the C-6 carbonyl group gives the first of the two products listed below. Reaction of the enolate derived from C-7 with the C-2 carbonyl group gives the second product.

Answer: (b) Formation of the first product requires nucleophilic attack at the more hindered carbonyl carbon atom. Also, the dehydration product has the less substituted double bond. Thus, the second product is the major product.

22.63 (a) What diketone will yield the following as a product of an intramolecular aldol condensation? (b) What isomeric bicyclic compound could also form, but in smaller amount?

Answers:

(a)   (b)

22.64 What reactant could yield the following product from an intramolecular aldol condensation?

Answer:

446

Conjugate Addition Reactions

22.65 Amines react with α,β-unsaturated ketones to give a conjugate addition products. Write the structure of the product for each of the following combinations of reactants.
(a) 2-cyclohexenone and $CH_3CH_2NH_2$
(b) 3-butenone and $(CH_3)_2NH$
(c) 4-methyl-3-penten-2-one and CH_3NH_2

Answers:

22.66 The conjugate addition of HCN to α,β-unsaturated ketones can be done using diethylaluminum cyanide, $(C_2H_5)_2Al—CN$, followed by acid workup. Write the structure of the addition product for each of the following reactants.

Answers:

22.67 (a) Write the structure of the addition product of 2-cyclohexenone with ethylmagnesium bromide after hydrolysis. (b) Do the same for the addition product of 2-cyclohexenone with lithium diethylcuprate.

Answer: (a) Addition of an ethyl Grignard reagent occurs 1,2 at the carbonyl group to give a tertiary alcohol. (b) Addition of lithium diethylcuprate gives a conjugate addition product.

Answers:

22.68 What combination of an α,β-unsaturated ketone and a Gilman reagent is required to synthesize each of the following compounds?
(a) 3-phenylcycloheptanone
(b) 2-hexanone
(c) 3-vinylcyclohexanone

Answers: (a)   2-cycloheptenone and lithium diphenylcuprate
(b)   3-buten-2-one and lithium diethylcuprate
(c)   2-cyclohexenone and lithium divinylcuprate

Michael Addition and Robinson Annulation Reactionss

22.69   Write the structure of the product of the Michael addition reaction of 2-methyl-1,3-cyclopentanedione with 3-buten-2-one followed by Robinson annulation.

Answer:

22.70   What combination of α,β-unsaturated ketone and a ketone is required to synthesize each of the following compounds by a Michael addition followed by Robinson annulation?

Answers:

Acidity of a Hydrogen Atom of Acid Derivatives

22.71   The cyano group is more deactivating in electrophilic aromatic substitution than a carboethoxy group. Which compound is more acidic, ethyl β-cyanoacetate or diethyl malonate?

Answer: Ethyl β-cyanoacetate is more acidic because the cyano group is more electron withdrawing than the carboethoxy group. Therefore, the cyano group stabilizes the conjugate base.

22.72   The pK_a of nitromethane is 11. Which compound is more acidic, nitroacetone or ethyl acetoacetate?

Answer: Nitromethane is more acidic than methyl acetate (pK_a = 25). Thus, the nitro group is more electron withdrawing than a carbomethoxy group as indicated by its stabilization of the conjugate base. The same effect is observed in nitroacetone, which is a stronger acid than ethyl acetoacetate.

22.73   The pK_a of malonitrile, $CH_2(CN)_2$, is 11. Calculate the equilibrium constant for the acid–base reaction of malonitnle with sodium ethoxide.

Answer: The K_a of malonitrile is 1×10^{-11}, and it is a stronger acid than ethanol, whose K_a is approximately 1×10^{16}. Thus, the equilibrium constant for the reaction of nitromethane with sodium ethoxide is 1×10^8.

22.74 The equilibrium constant for the reaction of ethyl 2-cyanoacetate with sodium ethoxide is approximately 10^7. What is the pK_a of ethyl 2-cyanoacetate?

Answer: Because $K > 1$, the K_a of ethyl 2-cyanoacetate is larger than the K_a of ethanol, which is approximately 1×10^{-16}. The equilibrium constant indicates the factor by which the K_a values of the two acids differ, which is 10^7. Thus, the K_a of ethyl 2-cyanoacetate is 1×10^{-9}, so the pK_a is 9.

Enolates of Acid Derivatives

22.75 Write the resonance forms of the conjugate base of (a) malonitrile and (b) dinitromethane.

Answers:

22.76 Write the resonance forms of the conjugate base of ethyl 2-cyanoacetate

Answer:

22.77 Ethyl acetoacetate reacts with two equivalents of LDA to give a dianion. Draw the structure of the dianion.

Answer:

22.78 Draw the resonance contributors of the anion formed by deprotonation of the most acidic hydrogen atom of each of the following compounds.

(a)   (b)   (c)

Answers:

(a)

Answers:

(b)

(c)

Reactions at the α Carbon Atom

22.79 What products result from the reaction of each of the following isomeric esters with sodium ethoxide in CH_3CH_2OD?
(a) ethyl pentanoate
(b) ethyl 2-methylbutanoate
(c) ethyl 3-methylbutanoate

Answers: (a) $CH_3CH_2CH_2CD_2CO_2CH_2CH_3$

(b) $CH_3CH_2CH_2\overset{\overset{\displaystyle CH_3}{|}}{C}DCO_2CH_2CH_3$

(c) $CH_3\overset{\overset{\displaystyle CH_3}{|}}{C}HCH_2CD_2CO_2CH_2CH_3$

22.80 Ethyl acetoacetate reacts rapidly with sodium ethoxide in CH_3CH_2OD to give a product incorporating two deuterium atoms. After a longer period of time, an additional three deuterium atoms are incorporated. Explain why.

Answer: The first two hydrogen atoms that are exchanged are the most acidic ones ($pK_a = 11$) at C-2 that are between the two carbonyl groups. The hydrogen atoms of the methyl group are less acidic ($pK_a = 19$) and are exchanged at a slower rate.

$$CH_3-\overset{\overset{\displaystyle O}{\|}}{C}-CD_2-\overset{\overset{\displaystyle O}{\|}}{C}-OCH_2CH_3 \qquad CD_3-\overset{\overset{\displaystyle O}{\|}}{C}-CD_2-\overset{\overset{\displaystyle O}{\|}}{C}-OCH_2CH_3$$

22.81 Write the equations for the synthesis of each of the following compounds starting from butanoic acid.

(a) $CH_3CH_2\overset{\overset{\displaystyle Br}{|}}{\underset{\underset{\displaystyle H}{|}}{C}}-CO_2H$ (b) $CH_3CH_2\overset{\overset{\displaystyle Br}{|}}{\underset{\underset{\displaystyle H}{|}}{C}}-\overset{\overset{\displaystyle O}{\|}}{C}-Br$

(c) $CH_3CH_2\overset{\overset{\displaystyle NH_2}{|}}{\underset{\underset{\displaystyle H}{|}}{C}}-CO_2H$ (d) $CH_3CH_2\overset{\overset{\displaystyle H}{|}}{\underset{\underset{\displaystyle CN}{|}}{C}}-\overset{\overset{\displaystyle O}{\|}}{C}-CO_2H$

Answers:

(a) $CH_3CH_2CH_2CO_2H$ $\xrightarrow[\text{Br}_2]{\text{PBr}_3}$
$CH_3CH_2-\overset{\overset{\displaystyle H}{|}}{\underset{\underset{\displaystyle Br}{|}}{C}}-\overset{\overset{\displaystyle O}{||}}{C}-OH$

(b) $CH_3CH_2CH_2CO_2H$ $\xrightarrow[\text{Br}_2]{\text{PBr}_3}$
$CH_3CH_2-\overset{\overset{\displaystyle H}{|}}{\underset{\underset{\displaystyle Br}{|}}{C}}-\overset{\overset{\displaystyle O}{||}}{C}-Br$

(c) $CH_3CH_2-\overset{\overset{\displaystyle H}{|}}{\underset{\underset{\displaystyle Br}{|}}{C}}-\overset{\overset{\displaystyle O}{||}}{C}-OH$ $\xrightarrow{\text{NH}_3}$
$CH_3CH_2-\overset{\overset{\displaystyle H}{|}}{\underset{\underset{\displaystyle NH_2}{|}}{C}}-\overset{\overset{\displaystyle O}{||}}{C}-OH$

(d) $CH_3CH_2-\overset{\overset{\displaystyle H}{|}}{\underset{\underset{\displaystyle Br}{|}}{C}}-\overset{\overset{\displaystyle O}{||}}{C}-OH$ $\xrightarrow{\text{NaCN}}$
$CH_3CH_2-\overset{\overset{\displaystyle H}{|}}{\underset{\underset{\displaystyle CN}{|}}{C}}-\overset{\overset{\displaystyle O}{||}}{C}-OH$

22.82   Draw the structure of the product of each of the following reactions.

(a) $CH_3CH_2CO_2H$ $\xrightarrow{\text{Br}_2\text{ / PBr}_3}$ $\xrightarrow{\text{CH}_3\text{NH}_2}$

(b) CH_3CO_2H $\xrightarrow[\text{PBr}_3\text{ / (1 eq)}]{\text{Br}_2}$ $\xrightarrow{\text{CN}^-}$

(c)
$-OH + BrCH_2CO_2H$ $\xrightarrow[\text{2. H}_3\text{O}^+]{\text{1. OH}^-}$

Answers:

(a) $CH_3-\overset{\overset{\displaystyle H}{|}}{\underset{\underset{\displaystyle Br}{|}}{C}}-\overset{\overset{\displaystyle O}{||}}{C}-OH$ then $H_3C-\overset{\overset{\displaystyle H}{|}}{\underset{\underset{\displaystyle NHCH_3}{|}}{C}}-\overset{\overset{\displaystyle O}{||}}{C}-OH$

(b) $CH_3-\overset{\overset{\displaystyle H}{|}}{\underset{\underset{\displaystyle Br}{|}}{C}}-\overset{\overset{\displaystyle O}{||}}{C}-Br$ then $CH_3-\overset{\overset{\displaystyle H}{|}}{\underset{\underset{\displaystyle CN}{|}}{C}}-\overset{\overset{\displaystyle O}{||}}{C}-Br$

(c)
$-O-CH_2CO_2H$

22.83   Draw the structure of the product resulting from reaction of each of the following esters with LDA followed by reaction of the enolate with the second reactant.
(a)   *tert*-butyl-2-methylpropanoate and benzoyl chloride
(b)   ethyl 2-methylpropanoate and ethyl iodide
(c)   *tert*-butyl-2-methylpropanoate and one equivalent of 1-bromo-3-chloropropane

Answers:

(a) CH$_3$—C(CH$_3$)—CO$_2$CH$_2$CH$_3$ with CH$_3$ and C=O attached to phenyl ring

(b) CH$_3$—C(CH$_3$)(CH$_2$CH$_3$)—CO$_2$CH$_2$CH$_3$

(c) CH$_3$—C(CH$_3$)(CH$_2$CH$_3$)—CO$_2$CH$_2$CH$_3$

22.84 (a) Explain why diethyl 2-phenylmalonate cannot be prepared by arylation of diethyl malonate. (b) Suggest a method of synthesis starting from ethyl 2-phenylacetate.

Answer: An S$_N$2 displacement reaction cannot occur at the sp²-hybridized carbon atom of an aryl halide such as bromobenzene. The compound can be made by reaction of the enolate of ethyl 2-phenylacetate with diethyl carbonate.

Ph—CH$_2$CO$_2$CH$_2$CH$_3$ CH$_3$CH$_2$O—C(=O)—OCH$_2$CH$_3$

22.85 Outline a synthesis of the herbicide 2,4-D using acetic acid as one of the reactants.

2,4-dichlorophenoxy structure: Cl (2-position), Cl (4-position), O—CH$_2$CO$_2$H

Answers: 1. Prepare ethyl 2-bromoacetate from acetic acid.
2. Then, react it with 2,4-dichlorophenol using ethoxide ion as a base to convert the phenol to a phenoxide; the phenoxide displaces bromide ion to give the ether.
3. Hydrolysis of the ethyl ester gives the product.

22.86 Outline a synthesis of valproic acid, a compound used in treatment of epilepsy, using ethyl acetate as one of the reactants.

CH$_3$—CH$_2$—CH$_2$ and CH$_3$—CH$_2$—CH$_2$—C(H)—CO$_2$H

Answers: 1. Prepare the enolate of ethyl acetate using LDA.
2. Then, alkylate 1-bromopropane.
3. Isolate the product. Prepare its enolate using LDA again. Then, alkylate a second time using 1-bromopropane.
4. Hydrolyze the dialkylated product to give valproic acid.

452

Answers:

1. CH_3—$\overset{\overset{\textstyle O}{\|}}{C}$—$OCH_2CH_3$ $\xrightarrow{\text{LDA}}$ $:\bar{C}H_2$—$\overset{\overset{\textstyle O}{\|}}{C}$—$OCH_2CH_3$

2. $:\bar{C}H_2$—$\overset{\overset{\textstyle O}{\|}}{C}$—$OCH_2CH_3$ $\xrightarrow{CH_3CH_2CH_2Br}$ $CH_3CH_2CH_2$—CH_2—$\overset{\overset{\textstyle O}{\|}}{C}$—$OCH_2CH_3$

3. $CH_3CH_2CH_2$—CH_2—$\overset{\overset{\textstyle O}{\|}}{C}$—$OCH_2CH_3$ $\xrightarrow[\text{2. } CH_3CH_2CH_2Br]{\text{1. LDA}}$ $CH_3CH_2CH_2$—$\overset{\overset{\textstyle H}{|}}{\underset{\underset{\textstyle CH_2CH_2CH_3}{|}}{C}}$—$\overset{\overset{\textstyle O}{\|}}{C}$—$OCH_2CH_3$

4. $CH_3CH_2CH_2$—$\overset{\overset{\textstyle H}{|}}{\underset{\underset{\textstyle CH_2CH_2CH_3}{|}}{C}}$—$\overset{\overset{\textstyle O}{\|}}{C}$—$OCH_2CH_3$ $\xrightarrow{H_3O^+}$ $CH_3CH_2CH_2$—$\overset{\overset{\textstyle H}{|}}{\underset{\underset{\textstyle CH_2CH_2CH_3}{|}}{C}}$—$\overset{\overset{\textstyle O}{\|}}{C}$—$OH$

valproic acid

Claisen Condensations

22.87   Draw the structure of the product of the self-condensation of each of the following esters in the presence of a molar equivalent of sodium methoxide.
(a) methyl propanoate
(b) methyl 3-phenylbutanoate
(c) methyl 2-cyclohexylethanoate

Answers:

(a) CH_3—CH_2—$\overset{\overset{\textstyle O}{\|}}{C}$—$\overset{\overset{\textstyle H}{|}}{\underset{\underset{\textstyle CH_3}{|}}{C}}$—$CO_2CH_3$

(b)

(c)

453

22.88   Draw the structure of the product of each of the following reactions.

(a)   [structure: CH2 chain with CO2H at top and CO2H at bottom]   + HCO2CH2CH3   —CH3CH2O⁻→

(b) C6H5—CH2—CO2CH2CH3 + CH3CH2O—C(=O)—C(=O)—OCH2CH3 —CH3CH2O⁻→

(c) CH3CO2CH3 + CH3O—C(=O)—C(=O)—OCH3 —CH3O⁻→

Answers:

(a)
$$H—\overset{O}{\underset{}{C}}—\overset{H}{\underset{CH_2CO_2CH_2CH_3}{C}}—CO_2CH_2CH_3$$

(b)
[benzene ring]—$\overset{H}{\underset{CO_2CH_2CH_3}{C}}$—CO2CH2CH3

(c)
$$CH_3O—\overset{O}{C}—\overset{O}{C}—\overset{H}{\underset{H}{C}}—CO_2CH_3$$

22.89   Explain why the following keto ester reacts with sodium ethoxide in ethanol to yield ethyl 2-methylpropanoate.

Answer: This compound is a product that cannot be obtained by a Claisen condensation of ethyl 2-methylpropanoate because it does not have an α hydrogen atom to be abstracted by a base in the final step that drives the reaction to completion. As a result, in a solution containing ethoxide ion, the reverse reaction occurs. The ethoxide ion attacks the ketone to form a tetrahedral intermediate, which then releases the enolate of ethyl 2-methylpropanoate and ethyl 2-methylpropanoate.

CH3—C(H)(CH3)—C(=O)—C(CH3)2—CO2CH2CH3   —CH3CH2O⁻→   CH3—C(H)(CH3)—C(=O)—OCH2CH3 + (CH3)2C=C(OCH2CH3)(:Ö:⁻)

22.90   Explain why the equilibrium constant for the following reaction is greater than one.

CH3—C(=O)—C(CH3)2—CO2CH3 —CH3O⁻ / CH3OH→ CH3—C(=O)—OCH3 + H—C(CH3)2—CO2CH3

Answer: The reverse of the given reaction would be a mixed Claisen condensation of methyl 2-methylpropanoate and methyl ethanoate to give the ketoester. This reaction is unfavorable because the ketoester does not have an α hydrogen atom to be abstracted by base in the final step that drives the reaction to completion. As a result, in a solution containing methoxide ion, the forward reaction occurs. The methoxide ion attacks the ketone to form a tetrahedral intermediate, which then releases the enolate of methyl 2 methylpropanate and methyl ethanoate.

454

22.91 What reactant is required to synthesize each of the following structures using a Dieckmann condensation?

(a)

(b)

(c)

Answers:

(a)

(b)

(c)

22.92 Each of the following pairs of compounds undergoes a "double" Claisen condensation in methanol and sodium methoxide to form a structure containing a second fused ring. Draw the structure of the product of each reaction.

(a)

(b)

(c)

Answers:

(a)

(b) CH₃CO₂ ... O ... OCH₃ ... CO₂CH₃ ... CO₂CH₃ ... O ... O ... CO₂CH₃

(c) ... OCH₃ ... CO₂CH₃ ... O ... O ... CO₂CH₃ ... O

22.93 What esters are required to give the following mixed Claisen products?

(a) $CH_3-\underset{\underset{CH_3}{|}}{\overset{\overset{CH_3}{|}}{C}}-\overset{\overset{O}{||}}{C}-\underset{\underset{CH_2CH_3}{|}}{\overset{\overset{H}{|}}{C}}-CO_2CH_2CH_3$

(b) $C_6H_5-\overset{\overset{O}{||}}{C}-\underset{\underset{CH_3}{|}}{\overset{\overset{H}{|}}{C}}-CO_2CH_3$

(c) $CH_3-O-\overset{\overset{O}{||}}{C}-\overset{\overset{H}{|}}{C}-CO_2CH_3$ (with phenyl substituent)

Answers: (a) ethyl 2,2-dimethylpropanoate and ethyl butanoate
(b) methyl benzoate and methyl propanoate
(c) dimethyl carbonate and methyl 2-phenylethanoate

22.94 There are two possible Dieckmann condensation products for each of the following compounds. Which product forms in each case?

(a) (aromatic ring with) CH₂CO₂CH₃ and CH₂CH₂CO₂CH₃

(b) $CH_3O-\overset{\overset{O}{||}}{C}-CH_2CH_2CH_2-\underset{\underset{CH_3}{|}}{\overset{\overset{H}{|}}{C}}-\overset{\overset{O}{||}}{C}-OCH_3$

Answers: (a) A six-membered ring results from either of the two possible Dieckmann condensation reactions. The acidity of the protons of the methyene groups are quite different. The protons of the methylene group of the side chain at the upper position are more acidic because the conjugate base is resonance stabilized by the aromatic ring. The larger quantity of this enolate at equilibrium should favor formation of the first product listed below. It results from a cyclization reaction with the carbonyl group of the side chain at the lower position. The second product results from abstraction of the α hydrogen atom of the side chain in the lower position, followed by a cyclization reaction with the carbonyl group of the side chain at the upper position.

Answers:

(a)

(b) A five-membered ring results from either of the two possible Dieckmann condensations. First, abstraction of a proton from the methylene group α to the carbon group on the left, followed by cyclization with the carbonyl group on the right gives the compound on the left, below. Second, abstraction of a proton from the carbon atom α to the carbonyl group on the right, followed by cyclization with the carbonyl group on the left could give the compound on the right, below. However, this product is unlikely because it does not have an acidic site to react with methoxide ion to stabilize the product.

Answers:

(b)

22.95   2-Methylcyclohexanone is treated with one molar equivalent of LDA to form an enolate. Draw the structure of the product of the reaction of the enolate with diethyl oxalate.

Answer: The major product results from abstraction of the more acidic hydrogen atom, located at C-6, followed by attack at one of the two equivalent carbonyl carbon atoms of diethyl oxalate. Loss of an ethoxide ion gives the product.

22.96   A mixture of cyclohexanone and diethyl carbonate is allowed to react in a solution of ethanol containing sodium ethoxide. Write the structure of the product.

Answer: Abstraction of the acidic hydrogen atom, located at C-2, gives an enolate, which then attacks the carbonyl carbon atom of diethyl carbonate. Loss of an ethoxide ion gives the product.

Reduction of Acyl Derivatives

22.97   Draw the structure of the product of each of the following combinations of reagents in a reaction using one equivalent of sodium ethoxide.
(a) p-nitrobenzaldehyde and diethyl malonate
(b) cyclopentanone and ethyl acetoacetate
(c) cyclooctanone and diethyl succinate

Answers:

(a)

(b)

(c)

22.98   Explain why the reaction of ethyl 2-bromopropanoate with zinc and acetophenone gives a pair of isomers.

Answer: The Reformatskii reaction gives a secondary benzylic alcohol that dehydrates to give n unsaturated carboxylic acid whose double bond is conjugated with the aromatic ring. Two geometric isomers result.

22.99   What reactants are required to prepare the following compounds using the Knoevenagel condensation?

(a)

(b)

(c)

Answers: (a) butanal and ethyl acetoacetate
(b) benzaldehyde and diethyl malonate
(c) cyclohexanone and ethyl 2-cyanoacetate

22.100 The product of a Reformatskii reaction can be dehydrated to give an α,β-unsaturated acid or ester. What reactants are required to synthesize each of the following products using the Reformatskii reaction as one of the steps?

(a) $CH=CH-CO_2H$

(b) $CH-CO_2CH_2CH_3$

(c)

Answers:

(a) CHO $BrCH_2CO_2CH_3$

(b) $BrCH_2CO_2CH_2CH_3$

(c) CHO CO_2CH_3 / Br

Synthesis Using β Dicarbonyl Compounds

22.101 Outline the synthesis of each of the following compounds using diethyl malonate as one of the reactants.
(a) 2-methyl-4-pentenoic acid   (b) 3-propylpentanoic acid (c) 2-benzylbutanoic acid
(d) 3-phenylpropanoic acid   (e) 2-ethyl-4-pentynoic acid

Answers: (a) Dialkylate diethyl malonate by first reacting the enolate with 3-bromopropene and then reacting the enolate of this product with iodomethane. Hydrolyze the diester and decarboxylate.
(b) Alkylate the enolate of diethyl malonate with 3-bromohexane. Hydrolyze the diester and decarboxylate.
(c) Dialkylate diethyl malonate by first reacting the enolate with benzyl bromide and then reacting the enolate of this product with iodoethane. Hydrolyze the diester and decarboxylate.
(d) Alkylate the enolate of diethyl malonate with benzyl bromide. Hydrolyze the diester and decarboxylate.
(e) Dialkylate diethyl malonate by first reacting the enolate with 3-bromopropyne, and then react the enolate of this product with iodoethane. Hydrolyze the diester an decarboxylate.

22.102 Outline the synthesis of each of the following compounds using ethyl acetoacetate as one of the reactants.
(a) 4-phenyl-2-butanone   (b) 5-hexene-2-one   (c) 5-methyl-2-hexanone   (d) 3-propyl-5-hexene-2-one
(e) 4-cyclopentyl-3-methyl-2-butanone

Answers: (a) Alkylate the enolate of ethyl acetoacetate with benzyl chloride. Hydrolyze the ketoester and decarboxylate.
(b) Alkylate the enolate of ethyl acetoacetate with 3-bromopropene. Hydrolyze the ketoester and decarboxylate.
(c) Alkylate the enolate of ethyl acetoacetate with 1-bromo-2-methylpropane. Hydrolyze the ketoester and decarboxylate.
(d) Dialkylate ethyl acetoacetate by first reacting the enolate with 3-bromopropene; then react the enolate of this product with iodopropane. Hydrolyze the ketoester and decarboxylate.
(e) Dialkylate ethyl acetoacetate by react the enolate with bromomethyl-cyclopentane. Then, react the enolate of this product with iodomethane. Hydrolyze the ketoester and decarboxylate.

22.103 The malonic acid synthesis can be used to prepare cycloalkanecarboxylic acids by a double alkylation using a dihalide. Write the structure of the product of each of the steps in the following sequence.

Answer: The enolate of dimethyl malonate displaces a bromide ion from the primary carbon atom. Adding a second mole of base forms an enolate of the product, which displaces a chloride ion from the secondary carbon atom, and forms a four-membered ring. Hydrolysis and decarboxylation occur in the last steps give a carboxylic acid.

22.104 What is the product of a malonic acid synthesis if the final step uses aqueous hydroxide ion in a base-catalyzed hydrolysis reaction (saponification) followed by careful neutralization with aqueous HCl rather than an acid-catalyzed hydrolysis reaction?

Answer: Saponification yields a dicarboxylate salt which cannot decarboxylate. Careful hydrolysis yields an alkylated malonic acid.

22.105 Ethyl acetoacetate reacts with one molar equivalent of sodium ethoxide followed by the addition of 2,2-dimethyloxirane to give a cyclic compound of molecular formula $C_8H_{12}O_3$. Draw its structure.

Answer: The enolate of ethyl acetoacetate attacks the primary carbon atom of 2,2-dimethyloxirane to give a ring-opened primary alkoxide. The alkoxide ion attacks the carbonyl carbon atom of the ester in an intramolecular process to give a tetrahedral intermediate that subsequently loses an ethoxide ion.

460

22.106 The malonic acid synthesis can be used to prepare cycloalkane carboxylic acids by a double alkylation using a dihalide. Write the structure of the product of each of the steps in the following sequence.

(a)

$$\xrightarrow[\text{2. CH}_3\text{CH}_2\text{I}]{\text{1. CH}_3\text{O}^-}$$

cyclooctanone with CO$_2$CH$_3$

(b)

$$\xrightarrow[\text{2. CH}_2\text{=CHCH}_2\text{Br}]{\text{1. NaH}}$$

cyclohexanone with CHO

(c)   NC—CH$_2$—CO$_2$CH$_2$CH$_3$ $\xrightarrow[\text{2. CH}_3\text{—C}\equiv\text{CCH}_2\text{CH}_2\text{OTs}]{\text{1. CH}_3\text{CH}_2\text{O}^-}$

Answers:

(a)

CO$_2$CH$_3$
CH$_2$CH$_3$

(b)

CHO
CH$_2$CH=C(CH$_3$)$_2$

(c)   CH$_3$—C≡CCH$_2$CH$_2$—C(—CN)(H)—CO$_2$CH$_2$CH$_3$

Michael Addition Reactions

22.107 Which member of each of the following pairs is more reactive as an acceptor in the Michael addition reaction?

(a)   CH$_2$=CH—C(=O)—OCH$_3$ or   CH$_2$=CH—C(=O)—CH$_3$

(b)   C$_6$H$_5$—CH=CH—C(=O)—OCH$_3$ or   CH$_3$—CH=CH—C(=O)—OCH$_3$

(c)   CH$_2$=CH—C(=O)—OCH$_3$ or   CH$_2$=C(CO$_2$CH$_3$)(CO$_2$CH$_3$)

Answers: (a) The ketone is the more reactive because the intermediate enolate is more stable. We recall that the α carbon atoms of ketones are more acidic than those of esters because the ketone carbonyl group stabilizes the negative charge better than the carbonyl group of an ester.
(b) The second compound is the more reactive because its β carbon atom is less sterically hindered for attack by a nucleophile. The charge of the intermediate cannot be delocalized by the aromatic ring of the first compound.
(c) The second compound is the more reactive because the intermediate enolate is more resonance stabilized.

461

22.108 What reactants are required to synthesize each of the following compounds using a Michael addition reaction in one of the steps?

(a)

(b) $CH_3O-\overset{\overset{O}{\|}}{C}-\overset{\overset{H}{|}}{\underset{\underset{CN}{|}}{C}}-CH_2CH_2-\overset{\overset{H}{|}}{\underset{\underset{OCH_3}{|}}{C}}-OCH_3$

(c)

Answers:

(a)

(b) $CH_3O-\overset{\overset{O}{\|}}{C}-\overset{\overset{H}{|}}{\underset{\underset{CN}{|}}{C}}-H$ $CH_2{=}CH-\overset{\overset{O}{\|}}{C}-H$

(c)

22.109 Write the reaction sequence required to synthesize the following structures using a Michael addition reaction as one of the steps.

(a)

$$CH_3-\underset{\underset{CH(CN)_2}{\overset{\overset{H}{|}}{|}}}{C}-CH_2-C\equiv N$$

(b)

$$CH_3-\underset{}{\overset{\overset{O}{\|}}{C}}-\underset{\underset{CH_3}{\overset{\overset{H}{|}}{|}}}{C}-CH_2CH_2-\overset{\overset{O}{\|}}{C}-CH_3$$

(c)

$$CH_3O-\overset{\overset{O}{\|}}{C}-CH_2CH_2-\underset{\underset{CO_2CH_3}{\overset{\overset{CN}{|}}{|}}}{C}-CH_2-\overset{\overset{O}{\|}}{C}-OCH_3$$

Answers:

(a)

$$\underset{H}{\overset{CH_3}{\diagdown}}C=C\underset{CN}{\overset{H}{\diagup}}\qquad N\equiv C-CH_2-C\equiv N$$

(b)

$$CH_3-\overset{\overset{O}{\|}}{C}-\underset{\underset{CH_3}{\overset{\overset{H}{|}}{|}}}{C}-CO_2CH_2CH_3 \qquad CH_2=CH-\overset{\overset{O}{\|}}{C}-CH_3$$

(c)

$$CH_3O-\overset{\overset{O}{\|}}{C}-CH=CH_2 \qquad H-\underset{\underset{CO_2CH_3}{\overset{\overset{CN}{|}}{|}}}{C}-CH_2-\overset{\overset{O}{\|}}{C}-OCH_3$$

463

23 AMINES AND AMIDES

KEYS TO THE CHAPTER 23.1 Organic Nitrogen Compounds

Because nitrogen has five valence shell electrons, it can form three covalent bonds in neutral compounds, leaving one nonbonding electron pair. These bonds can be three single bonds as in amines and amides, one double bond and a single bond in imines, or a triple bond in nitriles.

Nitrogen is found in many biologically important compounds that have a wide range of physiological properties. However, once a nitrogen-containing functional group is identified, its chemical reactions can often be predicted since the functional groups in these compounds, whose structures are often complex, have the characteristic reactivities of much simpler compounds.

23.2 Bonding and Structure of Amines

Amines are pyramidal at the nitrogen atom, with approximately tetrahedral bond angles to all bonded atoms. The nitrogen atom in amines is sp³ hybridized. However, the configuration of an amine is not static. Amines undergo nitrogen inversion to give mixtures of mirror images. The process occurs via a planar transition state. The energy barrier to inversion is low, and nitrogen inversion is rapid so that amines with chiral nitrogen atoms cannot be isolated.

23.3 Classification and Nomenclature of Amines

Amines are classified according to the number of alkyl or aryl groups bonded to the nitrogen atom. Primary, secondary, and tertiary amines have 1, 2, and 3 groups bonded, respectively. Amides are classified the same way, with the acyl group counting as one of the carbon groups bonded to the nitrogen atom. Abbreviations for the classes of amines and amides are 1°, 2°, and 3°.

The common names of simple amines are based on the identity of the alkyl or aryl groups bonded to the nitrogen atom. The names of the alkyl groups are written in alphabetical sequence as one word, followed by the word amine.

The common name of a complex amine is based first on identifying the longest continuous chain containing an attached nitrogen atom. The chain is numbered to assign the lowest number to the carbon atom bonded to the nitrogen atom. The nitrogen atom may be contained in an amino group ($-NH_2$), an *N*-alkylamino group ($-NHR$), or an *N,N*-dialkylamino group ($-NR_2$). Aryl groups may be present in place of alkyl groups, and the same procedure is followed. In naming alkylamino or arylamino groups, the prefix *N*-indicates that the alkyl or aryl group is attached to the nitrogen atom, and not to the parent chain.

The IUPAC name of an amine is also based on the longest continuous chain containing an attached nitrogen atom. The -e ending of the parent alkane is changed to -amine. The chain is numbered to give the lowest number to the carbon atom bearing the nitrogen atom. Alkyl groups attached to nitrogen are designated with *N*-, but they are named along with other substituents on the parent chain.

Heterocyclic aromatic amines have rings that are numbered using a selected nitrogen atom as the number one atom. The Chemical Abstract System (CAS) of heterocyclic ring nomenclature has been accepted by the IUPAC. We use CAS names in this text.

23.4 Physical Properties of Amines

Amines may be gases, liquids, or solids depending on their molecular weight and structure. The boiling points of primary and secondary amines are higher than those for alkanes of similar molecular weight because these amines form intermolecular hydrogen bonds. Tertiary amines have lower boiling points than isomeric primary and secondary amines because tertiary amines do not have an N—H bond to form intermolecular hydrogen bonds. The lower molecular weight amines are soluble in water as a result of hydrogen bonding to water molecules.

Organic Chemistry Study Guide: Key Concepts, Problems, and Solutions http://dx.doi.org/10.1016/B978-0-12-801889-7.00023-6

23.5 Basicity of Amines

The basicity of amines of different classes do not follow a simple pattern because the number of groups bonded to nitrogen affects the electron density at the nitrogen atom. And, the stability of the conjugate acid in the solvent has a major affect on basicity. Thus, the basicity of amines can be explained only for amines with similar structures at the nitrogen atoms.

The basicity of an amine is increased by electron-donating groups and decreased by electron-withdrawing groups. Aryl amines are less basic than alkyl-substituted amines because some electron density provided by the nitrogen atom is distributed throughout the aromatic ring. Basicity is expressed using K_b values measured from the reaction of the amine with water. An alternate indicator of basicity is pK_b, which is $-\log K_b$. A strong base has a large K_b and a small pK_b. The basicity of amines is also expressed by the acidity of their conjugate acids. A strong base has a weak conjugate acid, as given by a small value of K_a and a large pK_a.

The basicity of heterocyclic amines depends on the location of the electron pair of the nitrogen atom, its hybridization, and whether or not resonance stabilization is possible. In pyrrole, the electron pair is part of the aromatic system. As a result, pyrrole is a very weak base. Pyridine is a weaker base than saturated amines of similar structure because its electron pair is in an sp²-hybridized orbital, and the electron pair is more tightly held by the atom. Protonation of a similar nitrogen atom in pyrimidine is more favorable because the charge is delocalized to the second nitrogen atom.

23.6 Solubility of Ammonium Salts

Formation of the conjugate acid of an amine gives a quaternary ammonium ion, an ionic substance that is more soluble in water than the original amine. Amines may be separated from other organic compounds that are not basic by adding acid to dissolve the amine. After physical separation and subsequent neutralization with base, the free amine separates from water.

23.7 Synthesis of Amines by Substitution Reactions

Ammonia and amines are nucleophiles, and they displace halide ion from haloalkanes to give more highly substituted amines. However, multiple substitution reactions are possible since each product successively acts as a nucleophile. Therefore, the reaction is not synthetically useful in the synthesis of a single product.

The Gabriel synthesis is used to convert a haloalkane into an amine in which the amino group replaces the halogen. However, only primary amines can be prepared because of competing elimination reactions in one of the steps. The steps required are:
1.   Convert phthalimide into its conjugate base
2.   Add a primary haloalkane to form an alkylated phthalimide
3.   Release the amine using either strong base or hydrazine

23.8 Synthesis of Amines by Reduction

Any functional group containing nitrogen in a higher oxidation state can be reduced to give an amine, which contains nitrogen in its lowest oxidation state. These include azides, imines, nitriles, amides, and nitro compounds.

Azides are prepared by the S_N2 displacement of a halide from a haloalkane by the azide ion. Reduction of the azide by either hydrogen and platinum as catalyst, or lithium aluminum hydride in ether gives a primary amine.

The double bond of an imine can be reduced by hydrogen and Raney nickel, or by a metal hydride. Sodium borohydride reduces imines, so lithium aluminum hydride is seldom used. The **reductive amination** reaction forms an imine, by reaction of a primary amine with either an aldehyde or ketone, which is reduced immediately in the reaction by hydrogen in the presence of a nickel catalyst. Secondary amines also react by formation of an iminium ion followed by reduction.

Nitriles are reduced to primary amines using lithium aluminum hydride. The nitrile can be made by displacement of a halide ion from a haloalkane by cyanide ion. This method allows the formation of primary amines having one additional carbon atom.

Reduction of amides using lithium aluminum hydride is the most versatile way of producing amines. Amides are easily prepared by reaction of an acyl chloride and an amine.

However, the reaction is most versatile because primary, secondary, and tertiary amines can be synthesized using primary, secondary, and tertiary amides, respectively.

Reduction of nitroaromatic compounds is used to produce anilines. Tin and HCl is the usual reducing agent.

23.9 Hofmann Rearrangement

The Hofmann rearrangement occurs when a primary amide reacts with a basic solution of a halogen such as chlorine or bromine. In this process, the carboxyl carbon atom is lost as carbonate ion and a primary amine results. The rearrangement occurs when an alkyl group is transferred from the carboxyl carbon atom to the nitrogen atom in one of the several intermediates involved in the reaction mechanism. The rearrangement occurs with retention of configuration of the alkyl group.

23.10 Conjugate Addition Reactions

The reactions of amines are distinctly different than the reactions of alcohols. Amines are substantially stronger bases than alcohols. Amines are sufficiently basic to exist to some degree as the conjugate acid in water. Alcohols require strong acids to form the conjugate acid. Amines are much less acidic than alcohols—the pK_a values of amines and alcohols are 35 and 16, respectively.

Within a period of the periodic table, the nucleophilicity decreases from left to right for the elements in compounds of similar structure. Ammonia is a distinctly better nucleophile than water because nitrogen is less electronegative (EN 3.0) than oxygen (EN 3.5). Likewise, amines are better nucleophiles than alcohols. Usually, it is necessary to convert an alcohol to its alkoxide ion to make it sufficiently nucleophilic to displace a leaving group such as a halide ion from a haloalkane. The neutral amine is sufficiently nucleophilic for this type of displacement reaction.

Substitution reactions to replace oxygen as a leaving group occur if the oxygen is protonated to provide for water as a leaving group. The NH_2^- ion is a much stronger base than the hydroxide ion and is a much poorer leaving group. Even protonation of the amine to give an ammonium ion doesn't allow for the loss of ammonia as a leaving group. Elimination reactions of amines are similarly less likely than elimination reactions of alcohols.
Nitrogen moieties are more basic than the oxygen analogs and are poorer leaving groups.

Amines can be oxidized, but the resulting imines, which correspond to the carbonyl groups obtained by the oxidation of alcohols, are much more sensitive to further reaction.

23.11 Enamines

Enamines and enols are structural analogs. Enamines are formed by the reaction of secondary amines with carbonyl compounds. Common secondary amines used to form enamines include pyrrolidine, piperidine, and morpholine.

Enamines react as nucleophiles, resulting in alkylation at the position equivalent to the α carbon atom of the original carbonyl compound. The reaction is restricted to displacement of halide ion from primary alkyl halides. The product is an alkylated imine. Upon hydrolysis, it gives an α alkylated carbonyl compound.

One advantage of using enamines to alkylate carbonyl compounds is that strong base is not required. The second advantage is the formation of a singly alkylated product. In contrast, multiple alkylation of ketones occurs because proton exchange generates the enolate of the alkylated product.

23.12 Sulfonamides

Sulfonyl chlorides and acyl halides react with amines in much the same way. The resulting sulfonamides are more acidic than amides because the sulfonyl group is more electron withdrawing than an acyl group.

The **Hinsberg test** can be used to classify amines. Tertiary amines do not react with benzenesulfonyl chloride in the presence of base. There are no signs of a reaction. Secondary amines give a water insoluble sulfonamide because there are no acidic N—H bonds. Primary amines give a soluble conjugate base of the sulfonamide. Addition of acid results in protonation of the conjugate base, and the insoluble sulfonamide forms.

23.13 Quaternary Ammonium Salts

The reaction of an amine with a haloalkane does not stop after one step and eventually gives a quaternary ammonium ion. Reaction of an amine with methyl iodide to give a quaternary ammonium ion is termed **exhaustive methylation**.

Quaternary ammonium salts containing one long carbon chain are invert soaps because their polar end is positively charged in contrast to a negative charge on soaps. These compounds are effective against bacteria and are used in hospitals.

Quaternary ammonium hydroxide salts formed by exhaustive methylation followed by exchange of the halide ion by hydroxide undergo an elimination reaction called the **Hofmann elimination**. The elimination occurs by an *anti* periplanar transition state and gives the least substituted alkene. In cases where both E and Z isomers are possible, the E isomer predominates.

23.14 Spectroscopy of Amines

Infrared spectroscopy is usually not used to confirm the presence of the C—N bond because the stretching vibrations occur in a region complicated by other absorptions. The N—H stretching vibration of amines is easily seen as a broad absorption, similar to that found for the O—H vibration of alcohols, on the "left" of the spectrum in the 3200–3375 cm^{-1} region. Primary amines give two absorptions; secondary amines give one absorption.

In NMR spectra, the chemical shift of hydrogen atoms bonded to the carbon atom bearing the nitrogen atom of amines occurs in the 2–3 δ region. The N—H group has a variable chemical shift due to rapid exchange among various hydrogen bonding species whose identities are concentration dependent.

The α carbon atom of an amine has a ^{13}C chemical shift that reflects the smaller deshielding effect of the nitrogen atom relative to the more electronegative oxygen atom. The carbon absorptions are in the 30–50 δ region.

Summary of Reactions

1. Synthesis of Amines by Substitution Reactions of Haloalkane

$$\text{(cyclohexyl)}-CH_2-Br \xrightarrow[\text{excess}]{NH_3} \text{(cyclohexyl)}-CH_2-NH_2$$

2. Gabriel Synthesis

$$\text{phthalimide} \xrightarrow[\text{ethanol}]{KOH} \text{phthalimide anion}$$

$$:N^-\ K^+ + CH_2{=}CH-CH_2-Cl \xrightarrow{DMF} :N-CH_2-CH{=}CH_2 + KCl$$

$$:N-CH_2-CH{=}CH_2 + NH_2NH_2 \xrightarrow{\text{ethanol}} \text{phthalhydrazide} + \ :N-CH_2-CH{=}CH_2$$

3. Synthesis of Amines by Reductive Methods

$$\text{(4-bromocyclohexene)} \xrightarrow[\substack{\text{1. NaN}_3 \\ \text{2. LiAlH}_4 \\ \text{3. H}_3O^+}]{} \text{(4-aminocyclohexene)}$$

$$\text{acetophenone} \xrightarrow[\substack{\text{1. (CH}_3)_2NH \\ \text{2. H}_2 / Ni}]{} \text{N(CH}_3)_2 \text{ enamine}$$

$$\text{1-cyclohexyl ketone} \xrightarrow[\text{NaBH}_3CN]{\text{CH}_3NH_2} \text{NHCH}_3 \text{ product}$$

469

CH$_3$CH$_2$—C(H)(CH$_3$)—CH$_2$—C≡N $\xrightarrow[\text{2. H}_3\text{O}^+]{\text{1. LiAlH}_4}$ CH$_3$CH$_2$—C(H)(CH$_3$)—CH$_2$—CH$_2$NH$_2$

(4-methoxyphenyl)C(=O)—NHCH$_3$ $\xrightarrow[\text{2. H}_3\text{O}^+]{\text{1. LiAlH}_4}$ CH$_3$O—(phenyl)—CH$_2$—NHCH$_3$

CH$_3$O—(phenyl)—NO$_2$ $\xrightarrow{\text{Sn / HCl}}$ CH$_3$O—(phenyl)—NH$_2$

4. Hofmann Rearrangement

CH$_3$—(cyclohexane)—C(=O)NH$_2$ $\xrightarrow{\text{Br}_2 \text{ / NaOH}}$ CH$_3$—(cyclohexane)—NH$_2$

5. Alkylation of Enamines

(1-tetralone) + pyrrolidine (N–H) $\xrightarrow{-\text{H}_2\text{O}}$ (1-pyrrolidino-3,4-dihydronaphthalene)

(1-pyrrolidino-3,4-dihydronaphthalene) $\xrightarrow[\text{2. H}_2\text{O}]{\text{1. CH}_2=\text{CHCH}_2\text{Cl}}$ (2-allyl-1-tetralone)

6. Hofmann Elimination

SOLUTIONS TO EXERCISES

Bonding and Structure

23.1   Which compound has the greater N—H bond length, pyrrole or pyrrolidine?

Answer: The N—H bond of pyrrolidine is longer because the nitrogen atom is sp³ hybridized. The nitrogen atom of pyrrole is sp² hybridized. Bond length decreases with increasing percent s character.

23.2   Which compound has the larger activation energy for the nitrogen inversion, *tert*-butyldimethylamine or trimethylamine?

Answer: The hybridization of the nitrogen atom changes from sp³ in the ground state to sp² in the transition state for pyramidal inversion. Thus, groups bonded to the nitrogen atom are less sterically crowded in the transition state where the bond angles between groups is 120°. *tert*-Butyldimethylamine is more sterically hindered in the ground state than trimethylamine. Since it loses this steric strain in the transition state, *tert*-butyldimethylamine inverts at a faster rate.

Classification of Amines

23.3   Classify each of the following amines.

Answers:
(a) 2°, (b) 3°

(c) 2°, (d) 1°

(a)   CH₃—N(H)—CH₂CH₃   (b)   CH₃CH₂—N(CH₂CH₃)—CH₂CH₃

(c)   CH₃—N(H)—C(H)(CH₃)—CH₃   (d)   CH₃—C(CH₃)(H)—CH₂—NH₂

23.4   Classify each of the following amines.

Answers:
(a) 2°, (b) 2°

(c) 3°, (d) 2°

(a) piperidine N—CH₃   (b) diphenylamine N—H

(c) cyclohexyl-N(CH₃)—CH₃   (d) phenyl-N(H)—CH₂CH₃

23.5   Classify the nitrogen-containing functional group in each of the following structures.
(a) methadone, a heroin substitute used in treating addicts

Answer:
3° amine

(a) The nitrogen atom located on the right in the structure has two methyl groups and a complex alkyl group bonded to it. The amine is tertiary.

472

(b) coniine, the hemlock poison that was used to execute Socrates

Answer:
2° amine

coniine

(b) The nitrogen atom located in the ring of the structure has two carbon atoms as part of the ring bonded to it. The amine is secondary.

(c) pantothenic acid, vitamin B₅

Answer:
2° amide

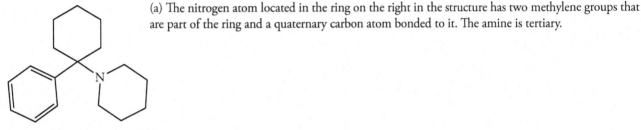

pantothenic acid, vitamin B₅

(c) The nitrogen atom is bonded an alkyl group and to a carbonyl group, so the functional group is a 2° amide.

23.6   Classify the nitrogen-containing functional group in each of the following structures.
(a) phencyclidine, a hallucinogen

Answer:
3° amine

phencyclidine

(a) The nitrogen atom located in the ring on the right in the structure has two methylene groups that are part of the ring and a quaternary carbon atom bonded to it. The amine is tertiary.

(b) encainide, an antiarrhythmic drug

Answer:
3° amine
2° amide

encainide

(b) The nitrogen atom bonded to the aromatic ring is also bonded to a carbonyl group, so the functional group is a 2° amide. The nitrogen atom located in the ring on the right in the structure has two methylene groups that are part of the ring and a methyl group bonded to it. The amine is tertiary.

(c) practolol, an antihypertensive drug

Answer:
2° amine
2° amide

practolol

(c) The nitrogen atom located on the right of the structure has two alkyl groups and a hydrogen atom bonded to it. The amine is secondary. The nitrogen atom bonded to the aromatic ring is also bonded to a carbonyl group, so the functional group is a 2° amide.

473

Nomenclature

23.7   Give the IUPAC name for each of the following compounds.

(a)
$$\underset{\displaystyle CH_3CH_2\underset{\textstyle |}{C}HCH_2CH_2CH_3}{\overset{NH_2}{}}$$

(b)
$$\underset{\displaystyle CH_3CH_2CH_2CH_2{-}\underset{\textstyle |}{N}{-}CH_3}{\overset{CH_3}{}}$$

Answers:
(a) 3-hexanamine (b) *N,N*-dimethy1-1-butanamine

23.8   Give the IUPAC name for each of the following compounds.

(a) —CH₂CH₂NH₂ (b) —NH₂

(c) —CHCH₃ (with NH₂) (d) —N(CH₃)₃

Answers:
(a) 2-cyclohexy1-1-ethanamine (b) 3-cyclohexenamine
(c) 1-cyclohexy1-1-ethanamine (d) *N,N*-dimethylcycloheptanamine

23.9   An antidepressant drug is named *trans*-2-phenylcyclopropylamine. Draw its structure.
Answer:

23.10   Tranexamic acid is a drug that aids blood clotting. Its IUPAC name is *trans*-4-(aminomethyl)cyclohexanecarboxylic acid. Draw its structure.
Answer:

23.11   Draw the structure of each of the following compounds.
(a) 2-ethylpyrrole (b) 3-bromopyridine (c) 2,5-dimethylpyrimidine (d) 2,6,8-trimethylpurine
Answers:

474

23.12 Name each of the following compounds.

(a)

(b)

(c)

(d)

Answers:

(a) 3-chloropyrrole   (b) 4-ethylpyrimidine   (c) 3,5-dimethylpyridine   (d) 1,2-dimethylindole

Molecular Formulas of Amines

23.13 (a) What is the general molecular formula for a saturated amine? (b) What is the general molecular formula for a saturated cyclic amine?

Answer: A saturated amine has one more hydrogen atom than an alkane with the same number of carbon atoms. The general formula is C_nH_{2n+3}. A cyclic amine has two fewer hydrogen atoms, so the general formula is C_nH_{2n+1}.

23.14 How many isomers are possible for each of the following molecular formulas?
(a) C_2H_7N (b) C_3H_9N (c) C_3H_7N

Answers:
(a) two; they are ethylamine and dimethylamine
(b) four; they are propylamine, isopropylamine, ethylmethylamine, and trimethylamine
(c) eight; however, three are enamines

23.15 Draw the isomers of the primary amines with molecular formula $C_4H_{11}N$.

Answer: Four alkyl groups can be part of the isomeric primary amines. The alkyl groups are butyl, *sec*-butyl, isobutyl, and *tert*-butyl.

23.16　Draw the isomers of the tertiary amines with molecular formula $C_5H_{13}N$.

Answer: The five carbon atoms must be distributed among three alkyl groups. If two of the groups are methyl, then the remaining three carbon atoms may be in either a propyl or an isopropyl group. The only other possible distribution of atoms is among two ethyl groups and a methyl group. Thus, there are three isomeric tertiary amines with molecular formula $C_5H_{13}N$.

Properties of Amines

23.17　The boiling points of the isomeric compounds propylamine and trimethylamine are 49 and 3.5 °C, respectively. Explain this large difference.

Answer: Propylamine can form hydrogen bonds between an N—H bond and an electron pair on a neighboring molecule. Trimethylamine has no N—H bond and therefore cannot form hydrogen bonds. Hydrogen bonding increases the boiling point of propylamine.

23.18 The boiling point of 1,2-diaminoethane is 116 °C. Explain why this compound boils at a much higher temperature than propylamine (49 °C).

Answer: Two sites in 1,2-diaminoethane two sites can form hydrogen bonds. Both nitrogen atoms provide hydrogen bond donor sites and hydrogen bond acceptor sites. The extent of hydrogen bonding is greater than for propylamine, which has only one site where hydrogen bonds may form.

23.19 The boiling points of pentane (36 °C) and 1-butanamine (78 °C) differ by 42 °C. Explain why the difference between the boiling points of nonane (151 °C) and dibutylamine (159 °C) is smaller.

Answer: Hydrogen bonds form more readily at the nitrogen atom of 1-butanamine because the site is sterically unhindered. In dibutylamine, the nitrogen atom is secondary, and the site is more crowded. Therefore, hydrogen bonding does not occur as easily.

23.20　Explain why the boiling points of tertiary amines are close to the boiling points of the structurally related alkanes.

Answer: Tertiary amines have no N—H bond and cannot form hydrogen bonds. Amines are slightly more polar than, but the dipole–dipole forces are small.

Basicity of Amines

23.21　The pK_a values for cyclohexylamine and triethylamine are 3.34 and 2.99, respectively. Which compound is the stronger base?

Answer: The stronger base has a smaller pK_b value. Thus, triethylamine is the stronger base.

23.22　The K_a values For dimethylamine and diethylamine are 4.7×10^{-4} and 3.1×10^{-4}, respectively. Which compound is the stronger base?

Answer: The stronger base has a larger K_b value. Thus, dimethylamine is the stronger base.

23.23　Explain the difference between the pK_a values of the conjugate acids of the following bases.

$$N{\equiv}C-CH_2-CH_2-NH_2 \qquad\qquad N{\equiv}C-CH_2-NH_2$$

$$\text{I } (pK_a\ 7.8) \qquad\qquad\qquad \text{II } (pK_a\ 5.3)$$

Answer: The smaller pK_a of the conjugate acid of the second compound means that it is the stronger acid. Thus, the second amine is the weaker base. The cyano group is inductively electron withdrawing, and it decreases the electron density at the nitrogen atom of the amine. In the second compound, the distance between the nitrogen atom and the cyano group is smaller, so the inductive effect is more pronounced.

23.24 Explain the difference between the pK_a values of the conjugate acids of the following bases.

$CH_3(CH_2)_3NH_2$ $CH_3O(CH_2)_3NH_2$

I (pK_a 10.6)   II (pK_a 9.9)

Answer: The smaller pK_a of the conjugate acid of compound II means that it is the stronger acid. Thus, the second amine is the weaker base. The methoxy group is inductively electron withdrawing, and it decreases the electron density at the nitrogen atom of the amine. Note, however, that the effect is relatively small because the methoxy group is far from the nitrogen atom.

23.25 Explain the order of acidity of the following generalized structures. Why is the difference between the pK_a values of the conjugate acids of an imine and a nitrile much larger than the difference between those of the conjugate acids of an amine and an imine?

$R_3\overset{+}{N}-H$ $R_2C{=}\overset{+}{N}H_2$ $RC{\equiv}\overset{+}{N}H$

pK_a 9 pK_a 3 pK_a −9

Answer: The acidity of the N—H bond depends on the hybridization of the nitrogen atom. The tendency to hold electrons closer to the nitrogen atom increases with increasing % s character of the bond, which is in the order $sp^3 < sp^2 < sp$. The % s character changes from 25% to 33% to 50% in this series. The difference in % s character is larger between sp^2 and sp than between sp^3 and sp^2.

23.26 Explain why the basicity of guanidine is comparable to that of an alkoxide ion. Where does protonation occur?

Answer: Based on information of Exercise 23.25, we would expect the sp^3-hybridized nitrogen atoms to be more basic than the imine nitrogen atom. However, the amine groups are part of an electron-delocalized system with the multiple bond of the imine.

Answer: As a consequence, protonation at either amino group destabilizes the conjugate acid because the number of resonance contributors is reduced. Protonation at the imine nitrogen atom gives a conjugate acid that is resonance stabilized. Note that all three contributors are equivalent.

Answer: Protonation at the imine nitrogen atom gives a conjugate acid that is resonance stabilized. Note that all three contributors are equivalent.

23.27   Explain why oxazole is a weaker base than thiazole.

oxazole   thiazole

Answer: The site of protonation is at the sp² orbital of nitrogen in oxazole because nitrogen is more basic than oxygen. The oxygen atom of oxazole decreases the electron density at the nitrogen atom by an inductive effect. Thus, the oxazole is a weaker base than thiazole. Sulfur is less electronegative than oxygen and inductively is a weaker electron withdrawing atom.

23.28   Explain why pyrazole is a weaker base than imidazole.

pyrazole   imidazole

Answer: The sp² orbital of nitrogen is the site of protonation in both pyrazole and imidazole. The nitrogen atom bonded to hydrogen is not basic because its lone pair of electrons is part of the aromatic ring system. However, that nitrogen atom withdraws electrons inductively from the atom where protonation occurs. The inductive effect is larger in pyrazole because the two atoms are closer. Inductive effects decrease with distance separating an electron withdrawing group and a reaction center.

Synthesis of Amines

23.29   Write the steps required for the synthesis of each of the following compounds starting from 1-pentanol.
(a) 1-butanamine   (b) 1-pentanamine   (c) 1-hexanamine

Answers:

(a) $CH_3CH_2CH_2CH_2CH_2OH \xrightarrow[H_2SO_4]{CrO_3} CH_3CH_2CH_2CH_2CO_2H \xrightarrow[pyridine]{SOCl_2}$

$$CH_3CH_2CH_2CH_2\overset{\overset{O}{\|}}{C}-Cl \xrightarrow{NH_2} CH_3CH_2CH_2CH_2\overset{\overset{O}{\|}}{C}-NH_2$$

$\downarrow$ Br₂ / NAOH

$$CH_3CH_2CH_2CH_2NH_2$$

(b) $CH_3CH_2CH_2CH_2CH_2OH \xrightarrow{PBr_3} CH_3CH_2CH_2CH_2CH_2Br \longrightarrow$

phthalimide $\xrightarrow{NaOH}$ phthalimide anion $N:^-$

$\downarrow$ $CH_3CH_2CH_2CH_2CH_2Br$

$CH_3CH_2CH_2CH_2CH_2NH_2 \xleftarrow{NaOH / H_2O}$ N—$CH_2CH_2CH_2CH_2CH_3$

(c) $CH_3CH_2CH_2CH_2CH_2OH$ $\xrightarrow{PBr_3}$ $CH_3CH_2CH_2CH_2CH_2Br$ $\xrightarrow[\text{DMF}]{\text{NaCN}}$

$CH_3CH_2CH_2CH_2CH_2C\equiv N$ $\xrightarrow[\text{2. } H_3O^+]{\text{1. LiAlH}_4}$ $CH_3CH_2CH_2CH_2CH_2CH_2NH_2$

23.30 Write the steps required for the synthesis of each of the following compounds starting from 3-methyl-l-butanol.

(a) 2-methyl-l-propanamine   (b) 3-methyl-1-butanamine   (c) 4-methyl-1-pentanamine

Answers:

(a) $\underset{\underset{CH_3CHCH_2CH_2OH}{\vert}}{CH_3}$ $\xrightarrow[\text{H}_2\text{SO}_4]{\text{CrO}_3}$ $\underset{\underset{CH_3CHCH_2CO_2H}{\vert}}{CH_3}$ $\xrightarrow[\text{pyridine}]{\text{SOCl}_2}$

$\underset{CH_3CHCH_2\overset{O}{\overset{\|}{C}}-Cl}{\overset{CH_3}{|}}$ $\xrightarrow{\text{NH}_3}$ $\underset{CH_3CHCH_2\overset{O}{\overset{\|}{C}}-NH_2}{\overset{CH_3}{|}}$ $\xrightarrow[\text{NaOH}]{\text{Br}_2}$ $\underset{CH_3CHCH_2NH_2}{\overset{CH_3}{|}}$

(b) $\underset{CH_3CHCH_2CH_2OH}{\overset{CH_3}{|}}$ $\xrightarrow{PBr_3}$ $\underset{CH_3CHCH_2CH_2Br}{\overset{CH_3}{|}}$

Phthalimide $\xrightarrow{\text{NaOH}}$ Phthalimide anion $\xrightarrow{\underset{CH_3CHCH_2CH_2Br}{\overset{CH_3}{|}}}$

$\underset{CH_3CHCH_2CH_2NH_2}{\overset{CH_3}{|}}$ $\xleftarrow{\text{NaOH / H}_2\text{O}}$ N-substituted phthalimide $(N-CH_2CH_2\overset{CH_3}{\overset{|}{C}HCH_3})$

(c) $\underset{CH_3CHCH_2CH_2OH}{\overset{CH_3}{|}}$ $\xrightarrow{PBr_3}$ $\underset{CH_3CHCH_2CH_2Br}{\overset{CH_3}{|}}$ $\xrightarrow[\text{DMF}]{\text{NaCN}}$

$\underset{CH_3CHCH_2CH_2C\equiv N}{\overset{CH_3}{|}}$ $\xrightarrow[\text{2. } H_3O^+]{\text{1. LiAlH}_4}$ $\underset{CH_3CHCH_2CH_2CH_2NH_2}{\overset{CH_3}{|}}$

23.31　Suggest a synthesis of *cis*-4-methylcyclohexylamine starting from each of the following compounds.
　　　(a) *trans*-4-methylcyclohexanol　　　(b) *cis*-4-methylcyclohexanecarboxylic acid
　　　(c) *trans*-4-bromomethylcyclohexane (d) *cis*-4-methylcyclohexanol

Answers:
(a) Convert the alcohol to the *trans* chloro compound using thionyl chloride in *dioxane*, which occurs with retention of configuration to give *cis*-1-chloro-4-methylcyclohexane. Prepare the conjugate base of phthalimide and react it with the haloalkane. The reaction occurs with inversion of configuration. Hydrolyze the product to obtain *cis*-4-methylcyclohexylamine.

(b) Convert the carboxylic acid to an amide by reacting it with thionyl chloride in pyridine followed by ammonia. Treat the amide with bromine and sodium hydroxide in the Hofmann rearrangement, which occurs with retention of configuration to give *cis*-4-methylcyclohexylamine.

(c) Prepare the conjugate base of phthalimide and react it with the halogen compound. The reaction occurs with inversion of configuration. Hydrolyze the product.

(d) Convert the alcohol to the *trans* chloro compound using thionyl chloride in *pyridine*, which will invert the configuration. Prepare the conjugate base of phthalimide and react it with the halogen compound. This reaction also occurs with inversion of configuration, so the methyl group and the phthalimide group are *cis*. Hydrolyze the product.

23.32　Outline the steps required to convert each reactant into the indicated product.
　　　(a) benzoic acid into *N*-ethylbenzylamine
　　　(b) benzyl chloride into 2-phenylethanamine
　　　(c) 1,4-dibromobutane into 1,6-diaminohexane

Answers:
(a) Convert the acid into an acid chloride using thionyl chloride and then react it with ethylamine to give *N*-ethylbenzamide. Reduce the product with $LiAlH_4$.
(b) React the chloro compound with sodium cyanide in dimethylformamide. Reduce the product with lithium aluminum hydride.

(c) React the dibromo compound with two equivalents of sodium cyanide in dimethylformamide. Reduce the product with lithium aluminum hydride.

23.33　Reaction of (*R*)-2-butyl tosylate with azide ion gives an alkyl azide. Subsequent reduction by lithium aluminum hydride gives an amine. Write the structure of the product.

Answer:
The displacement of tosylate by azide ion occurs with inversion of configuration. Subsequent reduction of the azide does not affect the configuration of the stereogenic center. Therefore, product has the (*S*) configuration.

23.34　Cyclohexene oxide reacts with azide ion to give an azido alcohol. Subsequent reduction by lithium aluminum hydride gives an amino alcohol. Write the structure of the product showing its stereochemistry.

Answer:
The ring opening of the epoxide by azide ion occurs with inversion of configuration at the site of attack by azide ion. The configuration of the carbon atom of the alkoxide ion and subsequent alcohol is retained. Thus, the azido alcohol is the *trans* isomer. Subsequent reduction of the azide does not affect the configuration of the stereogenic center. The product is the *trans* isomer.

23.35 Write the structure of the product of each of the following reactions.

(a) $CH_3-\overset{\overset{O}{\|}}{C}-CH_2CH_3$ + $CH_3CH_2NH_2$ $\xrightarrow{\text{NaBH}_3\text{CN}}$ $CH_3CH_2NH\overset{\overset{CH_3}{|}}{C}HCH_2CH_3$

(b)

$\xrightarrow[\text{2. H}_3\text{O}^+]{\text{1. LiAlH}_4}$

(c)

$\xrightarrow[\text{2. H}_3\text{O}^+]{\text{1. LiAlH}_4}$ $NH_2CH_2-$$-CH_2NH_2$

23.36 Write the structure of the product of each of the following reactions.

Answers:

(a) $CH_3-\overset{\overset{O}{\|}}{C}-CH_2CH_2CH_3$ + $-CH_2NH_2$ $\xrightarrow{\text{NaBH}_3\text{CN}}$ $-CH_2NH\overset{\overset{CH_3}{|}}{C}HCH_2CH_2CH_3$

(b) $N\equiv C-$ $\xrightarrow{\text{LiAlH}_4}$ $NH_2CH_2-$$-CH_2NH_2$

(c) $CH_3-\overset{\overset{O}{\|}}{C}-CH_2CH_2CH_3$ + CH_3NH_2 $\xrightarrow[\text{Ni}]{\text{H}_2}$ $CH_3NH\overset{\overset{CH_3}{|}}{C}HCH_2CH_2CH_3$

23.37 Write the product of reduction of each of the following compounds by lithium aluminum hydride.

Answers:

(a) $CH_3-\overset{\overset{O}{\|}}{C}-\overset{\overset{}{\underset{\underset{H}{|}}{N}}}-$$-OCH_2CH_3$ $\xrightarrow[\text{2. H}_3\text{O}^+]{\text{1. LiAlH}_4}$ $CH_3CH_2NH-$$-OCH_2CH_3$

(b) $-\overset{\overset{O}{\|}}{C}-\overset{\overset{}{\underset{\underset{H}{|}}{N}}}-\overset{\overset{CH_3}{|}}{\underset{\underset{CH_3}{|}}{C}}-CH_3$ $\xrightarrow[\text{2. H}_3\text{O}^+]{\text{1. LiAlH}_4}$ $-\overset{\overset{H}{|}}{\underset{\underset{H}{|}}{C}}-\overset{\overset{}{\underset{\underset{H}{|}}{N}}}-\overset{\overset{CH_3}{|}}{\underset{\underset{CH_3}{|}}{C}}-CH_3$

481

(c)

(d)

23.38 Write the product of reduction of each of the following lactams by lithium aluminum hydride.

Answers:

(a)

(b)

(c)

23.39 Write the structure of the final product of each of the following sequences of reactions.

(a) $CH_3(CH_2)_4CO_2H$ $\xrightarrow[\text{2. NH}_3]{\text{1. SOCl}_2}$ $\xrightarrow{\text{LiAlH}_4}$ $CH_3(CH_2)_4CH_2NH_2$

(b) ▷—CH_2CH_2Br $\xrightarrow{CN^-}$ $\xrightarrow{\text{LiAlH}_4}$ ▷—$CH_2CH_2CH_2NH_2$

(c) $CH_3(CH_2)_4CO_2CH_3$ $\xrightarrow[\text{2. H}_3O^+]{\text{1. LiAlH}_4}$ $\xrightarrow[\text{2. NH}_3 \text{ (excess)}]{\text{1. PBr}_3}$ $CH_3(CH_2)_3CH_2NH_2$

Answers:
(a) Reaction with thionyl chloride yields an acid chloride, which is converted into an amide by the reaction with ammonia. Reduction of the amide gives a primary amine.
(b) Reaction with cyanide displaces the bromide ion to give a nitrile. Reduction of the nitrile gives a primary amine.
(c) Reaction with lithium aluminum hydride gives an alcohol. The subsequent reaction with phosphorus tribromide gives a bromoalkane. Excess ammonia favors formation of a primary amine.

23.40   Write the structure of the final product of each of the following sequences of reactions.

(a)
$$\text{CH}_3\overset{\overset{\displaystyle \text{OH}}{|}}{\text{CH}}\text{CH}_2\text{CH}_2\text{CH}_3 \xrightarrow{\text{PCC}} \xrightarrow[\text{2. CH}_3\text{NH}_2]{\text{1. Ni / H}_2} \text{CH}_3\overset{\overset{\displaystyle \text{NHCH}_3}{|}}{\text{CH}}\text{CH}_2\text{CH}_2\text{CH}_3$$

(b) ⬡—CH$_2$CO$_2$H $\xrightarrow[\text{2. NH}_3]{\text{1. SOCl}_2}$ $\xrightarrow{\text{LiAlH}_4}$ ⬡—CH$_2$CH$_2$NH$_2$

(c) HOCH$_2$(CH$_2$)$_4$CH$_2$OH $\xrightarrow[\text{(excess)}]{\text{HBr}}$ $\xrightarrow[\text{(excess)}]{\text{CN}^-}$ $\xrightarrow[\text{H}_3\text{O}^+]{\text{1. LiAlH}_4 \text{ (excess)}}$ NH$_2$CH$_2$CH$_2$(CH$_2$)$_4$CH$_2$CH$_2$NH$_2$

Answers:
(a) Reaction with PCC gives a ketone. Subsequent reaction with methylamine under reductive amination conditions gives a secondary amine.
(b) Reaction with thionyl chloride yields an acid chloride, which is converted into an amide by the reaction with ammonia. Reduction of the amide gives a primary amine. (c) Reaction with excess HBr gives a dibromo compound. Reaction with excess cyanide ion gives a dinitrile, which reacts with lithium aluminum hydride to give a diamine.

23.41   Devise a synthesis of Apetinil, an appetite suppressant, from each of the following starting materials.
   (a) 1-phenyl-2-bromopropane   (b) 1-phenyl-2-propanone
   (c) 1-phenyl-2-propanamine   (d) 2-methyl-3-phenylpropanoic acid

Apetinil

Answers:
(a) React the 1-phenyl-2-bromopropane with excess aminoethane. Multiple alkylation is less likely for this process because the secondary amine product is moderately hindered.
(b) React the 1-phenyl-2-propanone with aminoethane under conditions of reductive amination. Use sodium cyanoborohydride in methanol as the solvent.
(c) React the 1-phenyl-2-propanamine with acetyl chloride to produce an amide. Reduce the amide with lithium aluminum hydride in ether as the solvent.
(d) Prepare the amide of 2-methyl-3-phenylpropanoic acid by reacting it with SOCl$_2$ and then reacting the acid chloride with ammonia. React the amide with bromine and base under the conditions of the Hofmann rearrangement to give 1-phenyl-2- propanamine. Then proceed as in (c).

23.42   Devise a synthesis of tetracaine, a spinal anesthetic, from *p*-nitrobenzoic acid.

Tetracaine

Answer:
Form the ester by reaction with 2-(*N,N*-dimethyl)ethanol. Reduce the nitro group using tin and HCl. React the amine with butanal in the presence of hydrogen and Raney nickel.

Enamines

23.43   Identify the compounds used to prepare each of the following enamines.

(a)

(b)

(c)

Answers:

(a)

(b)

(c)

23.44   Draw the structure of the enamine prepared from each of the following combinations of reagents.
(a) cyclooctanone and piperidine   (b) 2-pentanone and pyrrolidine   (c) butanal and morpholine

Answers:

(a)

(b)

(c)

23.45 Draw the structure of the product obtained in the reaction of each of the following combination of reactants.

(a)

(b)

(c)

Answer: In each reaction, the enamine acts as a nucleophile. Subsequent hydrolysis of the iminium ion obtained gives the alkylated product.

23.46 (a) Explain why a single enamine forms from the following combination of reactants. (b) Draw its structure. (c) Draw the structure of the product of the reaction of this enamine and allyl bromide.

Answers: (a) The enamine with a double bond in conjugation with the aromatic ring is formed. The allyl group alkylates the enamine.

(b)

(c)

485

Amides

23.47   What amine and acid derivative are required to form the amides contained in each of the following compounds?
(a) acetaminophen, an analgesic

Answers:

(a) acetaminophen

(b) encainide, an antiarrhythmic drug

encainide

(c) nubucaine, a local anesthetic

23.48   What amine and acid derivative are required to form the amides contained in each of the following compounds?
(a) DEET, an insect repellent.

(b) crotamiton, used to treat scabies

(c) bupivacaine, a local anesthetic

Answer:

Sulfonamides

23.49   Explain how the following isomeric amines can be distinguished by the Hinsberg test.

I   II   III

Answer:
Although the Hinsberg reaction is a classical qualitative test for distinguishing primary, secondary, and tertiary amines that has been supplanted by IR spectral analysis. However, these reactions reveal chemical properties of amines. Neither I nor III give evidence of a reaction. Compound II, which is a secondary amine, gives an insoluble sulfonamide. Adding acid to the solution of III yields a precipitate of a water insoluble sulfonamide. Compound I, a tertiary amine, remains in solution even after acidification because it did not react with the reagent.

23.50   What observations would be made for the Hinsberg test for each of the following amines?

I   II   III

Answer:
Compound I reacts with the benzenesulfonyl chloride to give a base-soluble compound. Adding acid gives a precipitate. Compound II, a secondary amine, gives an insoluble sulfonamide. Compound III does not react with the benzenesulfonyl chloride, and no precipitate forms when acid is added.

23.51   Explain why sulfamethazine is a weaker acid than sulfadiazine.

Answer:
The acidity of an acid increases if its conjugate base is stabilized by electron withdrawing substituents. In the case of sulfamethazine, the methyl groups donate electron density, and their effect is the opposite of that required to stabilize the conjugate base.

23.52   Explain why sulfadiazine is a stronger acid than sulfapyridine.

sulfadiazine

sulfapyridine

Answer:

The acidity of an acid increases if its conjugate base is stabilized by electron withdrawing substituents. In the case of sulfadiazine, which has one more nitrogen atom than structure sulfapyridine causes greater withdrawal of electron density from the nitrogen atom of the sulfonamide. Therefore, sulfadiazine is a stronger acid than sulfapyridine.

Quaternary Ammonium Salts

23.53   Propose a synthesis of hexafluorenium bromide, a neuromuscular blocking agent, using 1,6-dibromohexane.

Answer:

React 1,6-dibromohexane with two moles of the following amine. Then, exhaustive methylation with methylamine places two methyl groups on each nitrogen atom.

23.54   Draw the structure of the primary amine required to synthesize benzethonium chloride, an antimicrobial agent in first aid antiseptics.

benzethonium chloride

Answer:

There are two possible primary amines. One is benzylamine, $C_6H_5CH_2NH_2$. The structure of the second is shown below. This compound is the more likely candidate because it is relatively easy to alkylate amines with benzylchloride, $C_6H_5CH_2Cl$, which reacts readily with amines.

Hofmann Elimination

23.55 The following quaternary ammonium ion undergoes a Hofmann elimination to give a mixture of two products. The structurally related bromoalkane reacts with sodium ethoxide to give a mixture of the same two products. One of the two reactants gives a 6:4 ratio of the two alkenes. The other gives a 1:12 ratio of the same two alkenes. What are the structures of the two alkenes, and which product ratio corresponds to each reactant?

Answer:
The major product of the Hofmann elimination is the less substituted alkene, derived from elimination of a hydrogen atom from either of the two methyl groups bonded to nitrogen. The Hofmann is regioselective, resulting in a 12:1 product ratio in this case. The β elimination of the bromoalkane favors the more substituted product by a 6:4 ratio.

β elimination product   Hoffman elimination product

23.56 Explain why the following quaternary ammonium ion undergoes a Hofmann elimination to give propene and 1-butene in a 2:1 ratio.

Answer:
The conformation required to form propene has a gauche arrangement of the methyl group of the propyl group and the dibutylpropylamino group. This conformation is less sterically hindered than the conformation required to form 1-butene. It has a gauche arrangement of an ethyl group and the butyldipropylamino group.

23.57 Exhaustive methylation of Apetinil, the ingredient of a "diet pill," followed by the Hofmann elimination reaction gives a mixture of alkenes. (a) Draw their structures. (b) Which alkene is the major product?

Answer:

Ethylene forms in the largest amount because it is the least substituted alkene and there is no steric hindrance in the conformation required for the elimination reaction. There are also two alkenes containing a phenyl ring, which form in lesser amounts. Although the alkene derived from loss of hydrogen from the methyl group is less substituted, the alkene derived from loss of hydrogen at the methylene group is conjugated with the phenyl ring, so it forms in a greater amount.

ethene is major product

minor products

23.58 Draw the structure of the product, indicating the stereochemistry around the double bond, when the following compound undergoes exhaustive methylation and subsequent Hofmann elimination.

Answer:

To visualize the product, rewrite the Fischer projection structure as an eclipsed "sawhorse" conformation. Then rotate about the carbon–carbon bond to form a staggered conformation with a hydrogen atom and a nitrogen atom in an *anti* periplanar arrangement. The quaternary ammonium ion derived from this amine is used to predict the stereochemistry of the product alkene, in which the phenyl groups are *cis* to one another.

eclipsed conformation   staggered conformation

23.59 What unsaturated compound is obtained by exhaustive methylation of each of the following amines followed by a Hofmann elimination?

(a) (b) (c)

Answers:

(a) $CH_2{=}CH_2$ (b) (c)

23.60 Draw the structure of the unsaturated compound obtained by exhaustive methylation followed by Hofmann elimination for quinuclidine.

Answer:

quinuclidine

1. CH_3I / OH^-
heat

23.61 What unsaturated compound is obtained by exhaustive methylation of each of the following amines followed by Hofmann elimination?

(a) (b) (c)

Answers:

(a)
1. CH_3I (excess)/ OH^-
2. heat

(b)
1. CH_3I (excess)/ OH^-
2. heat

(c)
1. CH_3I (excess)/ OH^-
2. heat

23.62   A compound found in the venom of the red fire ant has the molecular formula $C_{17}H_{35}N$. It undergoes exhaustive methylation and Hoffman elimination to give the following mixture of products. Draw the structure of original amine.

$$\underset{N(CH_3)_2}{CH_2=CHCH_2CH_2CH_2\overset{|}{C}HCH_2(CH_2)_9CH_3}$$

$$\underset{N(CH_3)_2}{CH_3\overset{|}{C}HCH_2CH_2CH=CHCH_2(CH_2)_9CH_3}$$

$$\underset{N(CH_3)_2}{CH_3\overset{|}{C}HCH_2CH_2CH_2CH=CHCH_2(CH_2)_8CH_3}$$

Answer:

Answer:

There are three ways to obtain alkenes by exhaustive methylation of the above amine followed by Hoffman elimination. (a) The top compound results from elimination that occur by elimination beginning at the methyl group at C-6. (b) The middle compound arises from elimination from C-3. (c) The bottom compound arises from elimination beginning at the methylene group at C-2.

Spectroscopy of Amines

23.63   How can the following amines be distinguished using infrared spectroscopy?

Answers:

(a) There are two absorptions in the 3200–3375 cm^{-1} region due to the N—H bonds in the NH$_2$ group.
(b) There is one absorption in the 3200–3375 cm^{-1} region due to the single N—H bond.
(c) There are no absorptions in the 3200–3375 cm^{-1} region because there are no N—H bonds in the molecule.

23.64   Which of the isomeric compounds with molecular formula $C_4H_{11}N$ have the following characteristics in their infrared spectrum?
(a) no absorption in the 3200–3400 cm^{-1} region
(b) a single absorption in the 3200–3400 cm^{-1} region
(c) two absorptions in the 3200–3400 cm^{-1} region

Answers:

(a) Only ethyldimethylamine $(CH_3CH_2)N(CH_3)_2$ has no N—H bonds, and therefore no absorption in the 3200–3375 cm^{-1} region.
(b) Secondary amines have a single N—H bond in the molecule and give a single absorption in the 3200–3375 cm^{-1} region. There are three possible isomers whose structures are given below.

$$CH_3CH_2-NH-CH_2CH_3 \qquad CH_3CH_2CH_2-NH-CH_3 \qquad \underset{CH_3}{CH_3\overset{|}{C}H-NH-CH_3}$$

(c) Primary amines have two N—H bonds in the molecule and give two absorptions in the 3200–3375 cm⁻¹ region. There are four possible isomers whose structures are given below.

$$CH_3CH_2CH_2CH_2\!—\!NH_2 \quad CH_3CH_2\overset{\displaystyle CH_3}{\overset{|}{CH}}\!—\!NH_2 \quad CH_3\overset{\displaystyle CH_3}{\overset{|}{CH}}CH_2\!—\!NH_2 \quad (CH_3)_3C\!—\!NH_2$$

23.65   Deduce the structure of isomeric amines with molecular formula $C_6H_{15}N$ that have the following hydrogen NMR spectra.

(a)

Answer:
(a) The quartet and triplet pattern is typical of an ethyl group. The six carbon atoms are distributed in three equivalent ethyl groups in triethylamine. Note that the peak for the methylene groups appears at 2.5 δ, as expected for groups attached directly to the nitrogen atom.

(b)

Answer:
(b) The pattern is typical of an isopropyl group although the expected septet is such a weak signal that it often appears as a quintet. The six carbon atoms are distributed in two equivalent isopropyl groups in diisopropylamine.

23.66   Deduce the structure of each of the following diamines with the indicated molecular formulas and proton NMR spectra. The number of hydrogen atoms and the multiplicity of each resonance are given within parentheses.
(a) $C_5H_{14}N_2$; 2.25 δ (12H singlet), 2.7 δ (2H singlet)
(b) $C_3H_{10}N_2$; 1.1 δ (4H singlet), 1.6 δ (2H singlet), 2.75 δ (4H singlet)
(c) $C_4H_{12}N_2$; 1.2 δ (4H singlet), 1.1 δ (6H singlet), 2.5 δ (2H singlet)

Answers:

(a) The singlet of intensity 12H at 2.25 δ is due to 12 equivalent protons, a good indicator of four methyl groups; its position at 2.25 δ indicates that the protons are bonded to carbon atoms bonded to nitrogen. The singlet of intensity 2H is due to two equivalent protons; its position at 2.7 δ indicates that these protons are also bonded to a carbon atom bonded to nitrogen. The compound is shown below.

(b) The triplet of intensity 4H at 2.7 δ is due to four equivalent protons on carbon atoms bonded to nitrogen and split by a neighboring methylene group. The singlet of intensity 4H at 1.1 δ is due to four equivalent protons bonded to nitrogen. The quintet of intensity 2H at 1.6 δ is due to two equivalent protons split by a two neighboring methylene groups. The compound is shown below.

(c) The singlet of intensity 2H at 2.5 δ is due to two equivalent protons on a carbon atom that has no neighboring protons and is bonded to nitrogen. The singlet of intensity 6H at 1.1 δ is due to six equivalent protons on carbon atoms that have no neighboring protons and are not bonded to nitrogen. The singlet of intensity 4H at 1.2 δ is due to four protons bonded to nitrogen atoms. The compound is shown below. Note that a single peak is observed for the protons bonded to nitrogen although the nitrogen atoms are not equivalent. This is because these protons undergo such rapid exchange that a single "time-averaged" signal is obtained.

(a) CH_3—N—CH_2—N—CH_3 (b) $NH_2CH_2CH_2CH_2CH_2NH_2$ (c) CH_3—$\overset{\overset{\displaystyle CH_3}{|}}{\underset{\underset{\displaystyle CH_3}{|}}{C}}$—$CH_2NH_2$
(a) with CH_3 groups on each N

23.67 Using the data from the carbon NMR spectrum, deduce the structure of each of the following amines with the molecular formulas $C_6H_{15}N$ that have no absorption in the 3200–3400 cm^{-1} region of the infrared spectrum. The chemical shifts for the two compounds are:
        (a) 12.6 δ, 46.9 δ
        (b) 23.6 δ, 38.7 δ, 53.2 δ.

Answers:
The absence of an absorbance in the 3200–3600 cm^{-1} region of the IR spectrum indicates that the compounds are tertiary amines.
(a) This compound has six carbon atoms but only two signals, so in the tertiary amine, there must be two sets of three equivalent carbon atoms. The absorbance at 12.6 δ is due to methyl groups, but they are not bonded to a nitrogen atom. The triplet at 46.9 δ is due to methylene groups bonded to a nitrogen atom. The compound is triethylamine.

(b) This compound has six carbon atoms but only three absorbances, so there must be three sets of equivalent carbon atoms. The absorbance at 23.6 δ is due to one or more methyl groups, but they are not bonded to a nitrogen atom. The absorbance at 38.7 δ is due to one or more methyl groups which are bonded to a nitrogen atom. The absorbance at 53.2 δ is due to a carbon atom bonded to a nitrogen atom. The compound is *tert*-butyldimethylamine.

23.68 Using the data from the carbon NMR spectrum, deduce the structure of each of the following amines with the molecular formula $C_6H_{15}N$ that have a single absorption in the 3200–3400 cm^{-1} region of the infrared spectrum. The chemical shifts for the two compounds are:
        (a) 23.7 δ, 45.3 δ
        (b) 12.0 δ, 23.9 δ, 52.3 δ

Answers:
The presence of a single absorbance in the 3200–3600 cm^{-1} region of the IR spectrum indicates that the compounds are secondary amines.
(a)The compound has six carbon atoms but only two signals, so in the secondary amine, there must be two sets of equivalent carbon atoms. The absorbance at 23.7 δ is due to methyl groups, but they are not bonded to a nitrogen atom. The absorbance at 45.3 δ is due to a carbon atom bonded to a nitrogen atom and one hydrogen atom. The compound is diisopropylamine

(b) The compound has six carbon atoms but only three signals, so the secondary amine must have three sets of two equivalent carbon atoms. The absorbance at 12.0 δ is due to methyl groups, but they are not bonded to a nitrogen atom. The absorbance at 23.9 δ is due to methylene groups, but they are not bonded to a nitrogen atom either. The absorbance at 52.3 δ is due to methylene groups bonded to a nitrogen atom. The compound is dipropylamine.

24

ARYL HALIDES, PHENOLS, AND ANILINES

KEYS TO THE CHAPTER

This chapter is organized around the chemistry of halogens, hydroxyl groups, and amino groups when they are bonded to an aromatic ring. Most of the chapter focuses on the chemistry unique to those functional groups precisely because they are bonded to an aromatic ring. A comparison with the chemical reactions that we have encountered before for these functional groups illustrates the decreased reactivity that is a consequence of the aromatic ring.

24.1 Properties of Aromatic Compounds

Based on our discussions of electrophilic aromatic substitution in Chapter 13, we know that electronegative atoms with lone-pair electrons affect the electron density of the aromatic ring and therefore the reactivity of the ring toward electrophilic aromatic substitution. The electron withdrawing inductive effect of the electronegative atom, and its ability to donate its lone pair electrons by resonance are both important. We recall that chlorine inductively withdraws electron density, and is a poor donor of electrons by resonance. Both oxygen and nitrogen inductively withdraw electron density, with oxygen being the strongest because it is more electronegative. Both oxygen and nitrogen are electron donors by resonance, but nitrogen is the more effective because it is less electronegative. The net effect is that both oxygen and nitrogen are net donors of electron density with nitrogen being more effective. It follows that the properties of the chlorine, oxygen, and nitrogen atoms bonded to the aromatic ring are themselves changed.

The C—Cl bond in aryl chlorides is shorter than in alkyl chlorides because the sp²-hybridized bond of the aromatic ring carbon atom has a larger % s character. In addition, the bond energy is larger. The C—O and C—N bonds are also shorter in aromatic compounds than in saturated compounds for the same reason. In addition, the bonds are shorter because they both have some double bond character as a result of contributing resonance forms based on the donation of their lone pair electrons to the aromatic ring.

The bond polarity of the carbon bond to chlorine, oxygen, or nitrogen is affected by both the hybridization of the bond and the donation of electron density to the ring by resonance. The higher % s character to some degree withdraws electrons away from the heteroatom, resulting in some decrease in the polarity of the bond. This effect is seen in the aryl chloride. There is a further effect in the same direction as a result of donation of electrons by resonance. In both oxygen and nitrogen, there is a net reversal of polarity of the C—O and C—N bonds.

24.2 Acid–Base Properties

The pK_a values of phenols are smaller than the pK_a values of alcohols. This difference is the result of delocalization of the negative charge of the phenoxide ion by the benzene ring. Those groups that are electron-withdrawing, such as the nitro group, more effectively stabilize the negative charge in the conjugate base, and thus increase the acidity of the phenol. Therefore, their pK_a values decrease.

The pK_a values of anilines are larger than the pK_a values of saturated amines. In this case, the difference is the result of delocalization of the lone pair electron of the nitrogen atom in the aniline itself. Stabilization of the compound as a result of resonance diminishes the availability of the lone-pair electrons for protonation by an acid. Electron-withdrawing groups, such as the nitro group, more effectively delocalize the lone pair electrons in the aniline itself and thus further decrease the basicity of the aniline. Hence, the pK_a values increase.

24.3 Formation of Organometallic Reagents

The reaction to form aryl Grignard reagents is similar to the formation of alkyl Grignard reagents. However, in the case of the chloro compounds, it is necessary to use THF as the solvent rather than diethyl ether. Bromo compounds form Grignard reagents in ether. As a result, it is possible to prepare

Organic Chemistry Study Guide: Key Concepts, Problems, and Solutions http://dx.doi.org/10.1016/B978-0-12-801889-7.00024-8

the Grignard reagent from a C—Br bond without affecting a C—Cl bond. Once formed, both types of Grignard reagents react in the same way and add to the typical carbon–oxygen double bonds of carbonyl compounds, esters, and carbon dioxide.

Aryllithium compounds are prepared by direct reaction of aryl halides with lithium metal or by a transmetallation reaction between an aryl halide and an alkyllithium compound.

24.4 Nucleophilic Substitution

The typical S_N2 and S_N1 mechanisms of alkyl halides do not occur for aryl halides. Nucleophilic substitution does occur, but by two different mechanisms termed addition–elimination and elimination–addition reactions.

The addition–elimination reaction results from attack of a nucleophile at the carbon atom bearing a leaving group, forming a tetrahedral intermediate. We recall that an intermediate is not a transition state. Although the intermediate in this addition–elimination reaction may resemble the transition state structure in an S_N2 mechanism, however an intermediate has a lifetime that in some cases may allow for its isolation. The intermediate is not formed by attack of the nucleophile from the back of the carbon–halogen bond, but rather from "front" side.

The addition–elimination reaction, also known as nucleophilic aromatic substitution, occurs only if electron withdrawing groups are bonded to the ring to stabilize the negative charge of the cyclohexadienyl anion. The strongly electron withdrawing nitro group is most effective, but it must be *ortho* or *para* to the carbon–halogen bond for resonance stabilization to occur.

The second step in the addition–elimination reaction is the ejection of the halide ion as the leaving group. This step is not rate determining. Thus, the normal order of leaving group tendencies of the halide ions is not observed. Rather, the effect of the halogen atom in stabilizing the cyclohexadienyl anion is observed. As a result of inductive electron withdrawal, the fluorine atom is the most effective in stabilizing the anion and hence the transition state leading to that anion is lowest in energy for the fluoro compound.

The elimination–addition mechanism of aryl halides requires an amide ion, a very strong base, to abstract a proton from the position *ortho* to the halogen atom. An elimination reaction in occurs in which the halide ion is the leaving group. A very reactive intermediate called benzyne results. The two carbon atoms in the triple bond of benzyne are equivalent. Attack of the amide ion at either of carbon atom gives an aryl anion which is subsequently protonated by transfer of a proton from ammonia. This step regenerates an amide ion.

When substituents such as a methyl group are also present, the two carbon atoms of the benzyne intermediate are no longer structurally equivalent, and mixtures of anilines result. There is little regioselectivity, so mixtures of isomers usually form. Some regioselectivity may result if only the ring substituent inductively stabilizes charge. Because the charge is located in an sp^2-hybridized orbital, resonance stabilization of charge cannot occur.

24.5 Overview of the Reactions of Phenols

Phenols undergo some of the same reactions as alcohols. They form ethers and esters. Ethers are synthesized by the Williamson synthesis. Unlike the formation of alkoxides, which require a strong base such as sodium hydride, phenoxides are easily produced using hydroxide ion. The phenoxide ion then acts as a nucleophile and displaces a halide ion from an alkyl halide. Aryl methyl ethers can be obtained using dimethyl sulfate.

Esters of phenols cannot be obtained by the Fischer ester synthesis because the equilibrium constant for esters of phenols is even more unfavorable than for esters of alcohols. Either an acyl halide or an acid anhydride must be used.

Phenols undergo electrophilic aromatic substitution. However, the activation of the aromatic ring by the hydroxyl group makes monosubstitution difficult unless the electrophilic group introduced strongly deactivates the ring.

24.6 Reactions of Phenoxide Ions

Based on the resonance forms of the phenoxide ion, we can write structures with some negative charge on the carbon atoms *ortho* and *para* to the oxygen atom. This carbanion reacts with certain carbon–oxygen double bonds that resembles the reaction of the carbanion of Grignard reagents. Reaction with formaldehyde gives a hydroxymethyl derivative which can dehydrate to give a conjugate ketone, which can undergo conjugate addition reactions that resemble the aldol reaction.

Reaction of the phenolate ion at the *ortho* position with carbon dioxide resembles the reaction of a Grignard reagent with carbon dioxide in that a carboxylic acid is formed. The Kolbe synthesis is used to produce salicylic acid.

24.7 Quinones

Quinones are cyclohexadienediones that can be prepared by the reaction of diphenols. Quinones are oxidizing agents as they are reduced to regenerate the aromatic ring. The reduction potential of quinones depends on the stability of the aromatic ring formed and the electron-attracting characteristics of substituents. Quinones with several electron-withdrawing groups are sufficiently strong oxidizing agents to dehydrogenate some suitably substituted hydrocarbons.

24.8 Substitution Reactions of Aryldiazonium Salts

Primary amines form diazonium salts when they react with nitrous acid. Homolytic cleavage of the carbon–nitrogen bond readily occurs in the case of sp^3-hybridized carbon atoms, but less readily in for sp^2-hybridized carbon atoms. Thus, aryldiazonium ions are sufficiently stable to be used as intermediates for substitution reactions, giving compounds with groups such as halogen, cyanide, or hydroxide replacing the original nitrogen atom. We discussed these reactions in Section 13.8. A specialized reaction of secondary amines with nitrous acid give a nitroso compound. Recall from Chapter 22 that the NO^+ ion is the initial intermediate that is responsible for the reaction of amities with nitrous acid. The nitrosonium ion is an electrophile that attacks the aromatic ring of activated compounds.

24.9 Azo Compounds

The diazonium ion is an electrophile that can attack activated aromatic rings to give a nitrogen–nitrogen double bond of azo compounds. Formation of these compounds follows the pattern predicted by the concepts we developed for electrophilic aromatic substitution. Many azo compounds are colored and are used as dyes.

SUMMARY OF REACTIONS

1. Formation of Organometallic Reagents

2. Nucleophilic Aromatic Substitution (addition–elimination)

3. Nucleophilic Aromatic Substitution (elimination–addition)

4. Synthesis of Aryl Ethers

5. Synthesis of Esters of Phenols

pyridine

6. Kolbe Synthesis

1. NaOH / CO_2 (100 atm)

2. H_3O^+

7. Quinone Synthesis

$Na_2Cr_2O_7$

8. Synthesis Using Aryldiazonium Ions

1. $NaNO_2$ / H_2SO_4

2. CuCN

1. $NaNO_2$ / H_2SO_4

2. CuCl

1. $NaNO_2$ / H_2SO_4

2. H_3PO_2

1. NaNO$_2$ / H$_2$SO$_4$
2. H$_2$O

Answers to Exercises

Properties of Aromatic Compounds

24.1   The boiling points of the 1,2-, 1,3-, and 1,4-benzenediols are 245, 276, and 285 °C, respectively. Explain why the boiling point of the *ortho* isomer is significantly lower than those of the other two isomers.

Answer: The *ortho* isomer has the hydroxyl groups in a position to form an intramolecular hydrogen bond, as shown below. Intermolecular hydrogen bonding in the *meta* and *para* isomers results in an increased boiling point because there is extensive aggregation of these compounds.

24.2   The boiling points of the three isomeric hydroxyanisoles are 205, 243, and 244 °C, respectively. What is the structure of the compound corresponding to the lowest boiling point?

Answer: The compound with the lowest boiling point is the *ortho* isomer. It forms an intramolecular hydrogen bond between the hydrogen atom of the hydroxyl group and the oxygen atom of the ether. The *meta* and *para* isomers form intermolecular hydrogen bonds, resulting in a higher boiling point.

24.3   The dipole moments of toluene and chlorobenzene are 0.4 and 1.7 D, respectively. Predict the dipole moment of *p*-chlorotoluene.

Answer: The negative end of the bond moment for the methyl group bonded to the aromatic ring is toward the aromatic ring because the methyl group is an inductively electron donating group. The bonding electron pair of the carbon–carbon bond is polarized toward the sp^2-hybridized carbon atom of the aromatic ring. Recall that sp^2-hybridized atoms have a greater % s character, and as a result, bonding electrons are held more tightly than for spa-hybridized atoms. The negative end of the bond moment for the chlorine atom bonded to the aromatic ring is toward the chlorine atom. Thus, the two bond moments reinforce one another to give a net dipole moment of 2.1 D.

p-chlorotoluene
(2.1 D)

24.4   The dipole moments of toluene and phenol are 0.4 and 1.5 D, respectively. Predict the dipole moment of *p*-methylphenol.

p-methylphenol
(1.1 D)

Answer: The negative end of the bond moment for the methyl group bonded to the aromatic ring is toward the aromatic ring because the methyl group is an inductively electron donating group. The negative end of the bond moment for the hydroxyl group bonded to the aromatic ring is toward the aromatic ring mainly because oxygen donates electrons to the ring by resonance. Thus, the two bond moments are opposed, giving a net dipole moment of 1.1 D.

24.5 The dipole moments of two of the isomeric dichlorobenzenes are 1.72 and 2.50 D. Assign a structure to each value.

o-dichlorobenzene   m-dichlorobenzene

Answer: The *para* isomer has no dipole moment because the two bond moments are directly opposed at a 180° angle to give a net resultant of 0 D. The isomer with the larger dipole moment is the *ortho* compound because the two bond moments are at a smaller angle and reinforce each other more effectively than for the meta compound. Thus, the net dipole moment resulting from the two moments of the *ortho* isomer is larger than for the meta isomer.

24.6 The dipole moments of chlorobenzene and phenol are 1.7 and 1.5 D, respectively. Predict the dipole moment of *p*-chlorophenol.

Answer: The negative end of the bond moment for the chlorine atom bonded to the aromatic ring is toward the electronegative chlorine atom. The negative end of the bond moment for the hydroxyl group bonded to the aromatic ring is toward the aromatic ring. Thus, the two bond moments reinforce one another to give an estimated dipole moment of 3.2 D.

24.7 Which compound has the longer C—N bond length, *p*-methoxyaniline or *p*-cyanoaniline?

Answer: The cyano group withdraw electrons from the aromatic ring in resonance structures. The amino group can cooperatively donate electrons by resonance, so the double bond character of the C—N bond to NH_2 increases. As a result, the C—N bond of *p*-cyanoaniline will be shorter than C—N bond of *p*-methoxyaniline, which does not have such double bond character.

24.8 Which compound has the larger activation energy for the nitrogen inversion, cyclohexylamine or aniline?

Answer: The bonds to the nitrogen atom form a shallower pyramid in aniline compared to cyclohexylamine. As a consequence, the structure of aniline is closer to that of the planar structure at the transition state for inversion, so the activation energy is less.

24.9 The dipole moment of pyrrolidine is 1.57 D, and the negative end of the dipole is directed toward nitrogen. The dipole moment of pyrrole is 1.80 D, but the dipole is opposite that of pyrrolidine. Explain why.

pyrrolidine   pyrrole

Answer: The nitrogen atoms of both compounds contribute three electrons to σ bonds. In the case of pyrrolidine, the remaining two valence electrons are present as a nonbonded pair, and the dipole is directed away from the ring. In pyrrole, two valence electrons are incorporated into the aromatic π system. In resonance forms for pyrrole, there is a decrease in electron density at the nitrogen atom. In short, two valence electrons shown as a lone pair in the resonance structure on the left are part of the π system. They are drawn away from nitrogen toward the carbon atoms in the ring as a result.

24.10 The dipole moments of aniline, *p*-(trifluoromethyl)aniline, and (trifluoromethyl)benzene are 1.3, 4.3, and 2.9 D, respectively. Explain how these data are used to deduce the direction of the dipole moment of aniline.

aniline
(1.3 D)

p-trifluoromethylaniline
(4.3 D)

trifluoromethylbenzene
(2.9 D)

Answer: The negative end of the bond moment for the trifluoromethyl group bonded to the aromatic ring is away from the ring toward the carbon atom of the trifluoromethyl group. The observed dipole moment of *p*-(trifluoromethyl)aniline is larger than that of trifluoromethylbenzene. Thus, the bond moment of aniline must be in the direction that reinforces the bond moment of the trifluoromethyl group. The negative end of the bond moment for the amino group bonded to the aromatic ring is toward the aromatic ring.

Acid–Base Properties

24.11 Explain why *p*-hydroxybenzaldehyde (pK_a = 7.62) is a substantially stronger acid than phenol.

Answer: The aldehyde group can contribute to the delocalization of charge, stabilizing the conjugate base as a result. Resonance stabilization of the conjugate base favors ionization and increases the acid dissociation constant of *p*-hydroxybenzaldehyde compared to phenol.

24.12 Explain why 2,4-dinitrophenol (pK_a = 3.96) is a stronger acid than 3,5-dinitrophenol (pK_a = 6.73).

Answer: The nitro groups that are *ortho* and *para* to the site bearing the negatively charged ion can stabilize that charge by delocalization as indicated by contributing resonance forms. Such stabilization cannot occur when the nitro group is in a *meta* position.

24.13 Based on resonance structures, explain why 1-naphthol (pK_a = 9.31) is a stronger acid than 2-naphthol (pK_a = 9.55).

Answer: Five resonance forms with the negative charge on carbon atoms can be written for each conjugate base. However, in the case of 1-naphthol, two of them have a formal, more stable benzene ring, whereas in 2-naphthol, only one of them does. Thus, the conjugate base of 1- naphthol is more stable, so the K_a of this phenol larger.

Answer: 1-naphthol resonance forms.

Answer: 2-naphthol resonance forms.

24.14 Which of the following isomeric phenols is the more acidic?

I   II

Answer: Compound I is more acidic because the charged nitrogen atom of the ammonium ion is closer to the carbon atom bearing the hydroxyl group. Withdrawal of electron density stabilizes the conjugate base and therefore increases the acid dissociation constant of compound I.

24.15 Estimate K_b in each of the following compounds.

Answers:
(a) The compound is a primary amine with an alkyl group bonded to the nitrogen atom. The K_b should be approximately 5×10^{-4}.
(b) The compound is a secondary amine with only alkyl groups bonded to the nitrogen atom. The K_b should be approximately 5×10^{-4}.
(c) The compound is a tertiary amine with only alkyl groups bonded to the nitrogen atom. The K_b should be approximately 5×10^{-4}.
(d) The compound is a primary amine with an aromatic ring bonded to the nitrogen atom. The K_b should be approximately 4×10^{-10}, like that of aniline.

24.16 Estimate K_b in each of the following compounds.

Answers:
(a) The compound is a primary amine with an aromatic ring bonded to the nitrogen atom. The K_b should be approximately 4×10^{-10}, like that of aniline.
(b) The compound is a substituted aniline. The electron donating methyl group should increase the basicity above that of aniline, so the K_b should be larger than 4×10^{-10}.
(c) The compound is a tertiary amine with only alkyl groups bonded to the nitrogen atom. The K_b should be approximately 5×10^{-4}.
(d) The compound is a tertiary amine with only alkyl groups bonded to the nitrogen atom. The K_b should be approximately 5×10^{-4}.

24.17 Physostigmine is used in 0.1–1.0% solutions to decrease the intraocular pressure in treatment of glaucoma. Rank the three nitrogen atoms in the molecule in order of increasing basicity.

Answer: The amide-like nitrogen atom is the weakest base. Of the other two nitrogen atoms, the tertiary amine contained in the five-membered ring on the right is the most basic. The nitrogen atom bonded directly to the aromatic ring is intermediate in basicity compared to the other two.

24.18 Quinine is an antimalarial drug, and reserpine is an antihypertensive drug. Estimate the pK_b values of both nitrogen atoms in each drug.

quinine

reserpine

Answer: For quinine, the nitrogen contained within the aromatic ring should have pK_b = 8.7, like that of pyridine. The tertiary amine should have pK_b = 3.4. For reserpine, the nitrogen atom within a pyrrole-like ring is a very weak base and pK_b = 15. The tertiary amine contained in the six-membered ring should have pK_b = 3.4.

24.19 Explain why the pK_a values of the anilinium ions of *m*-cyanoaniline and *p*-cyanoaniline are 2.75 and 1.74, respectively.

Answer: The cyano group is an electron withdrawing group, both by resonance and by an inductive effect. In the *para* position, both resonance and inductive effects operate to decrease the electron density on the nitrogen atom of the aniline. As a consequence, the K_b is decreased and the K_a of the anilinium ion is larger. This effect is seen in the small pK_a. In the *meta* position, only an inductive effect operates, so the decrease in the electron density on the nitrogen atom of the aniline is less than for the *para* isomer. As a consequence, the K_b is larger and the K_a of the anilinium ion is smaller than for the *para* isomer. The effect is seen in the larger pK_a of the *meta* isomer.

24.20 Explain why the pK_a values of the anilinium ions of *m*-methoxyaniline and *p*-methoxyaniline are 4.2 and 5.3, respectively.

Answer: The methoxy group is an electron donating group by resonance, but electron withdrawing by an inductive effect. In the *meta* position, the inductive effect operates to decrease the electron density on the nitrogen atom of the aniline. As a consequence, the K_b is decreased and the K_a of the anilinium ion is larger. This effect is seen in the small pK_a.
    In the *para* position, the resonance effect counters the inductive effect and increases the electron density on the nitrogen atom of the aniline relative to the *meta* isomer. As a consequence, the K_b is larger and the K_a of the anilinium ion is smaller than for the *meta* isomer. The effect is seen in the larger pK_a of the *para* isomer.

24.21 Explain why morpholine (pK_b = 5.67) is a weaker base than piperidine (pK_b = 2.88).

morpholine

Answer: The oxygen atom is inductively electron withdrawing and therefore decreases the electron density on the nitrogen atom, which is reflected in its decreased basicity.

24.22 The acidities of benzoic acid and of acetic acid differ by a factor of about 4, whereas the acidities of the anilinium ion and of the methylammonium ion differ by about a factor of 10^6. Why does the aromatic ring have so little effect on the acidity of carboxylic acids and a large effect on the acidity of ammonium ions?

Answer: The resonance forms of the carboxylate ion have negative charge on the two oxygen atoms, but there is no delocalization of charge into the ring. Hence the acidity of carboxylic acids is affected only by the electron density on the carbon atom bearing the carboxyl group, which in turn is affected by the substituents on the ring.

Resonance structures of benzoate ion

Answer: In anilines, the electron pair of the nitrogen atom is delocalized into the aromatic ring, and as a result, there is a direct resonance interaction with ring substituents. Thus aniline is a very weak base, and the anilinium ion is a fairly strong acid compared to the methylammonium ion.

Nucleophilic Aromatic Substitution

24.23 Rank the following compounds in order of increasing reactivity toward sodium methoxide in methanol.

I   II   III   IV

Answer:
Only the nitro groups in *ortho* or *para* positions relative to the site of attack by a nucleophile affect the reactivity. Such groups can stabilize the negative charge of the intermediate by resonance. Groups in the *meta* position can stabilize the negative charge only by an inductive effect. Compound III is the most reactive. Compound IV is the next most reactive and is slightly more reactive than compound II as a consequence of the additional nitro group in the *meta* position. Compound I is very much less reactive.

24.24 Draw the product of each of the following reactions.

Answer: (a) The carboethoxy groups can stabilize the negative charge of the intermediate formed when the ethoxy ion displaces the bromide ion.

Answer: (b) The two nitro groups that are *ortho* and *para* to one of the bromine atoms can stabilize the negative charge of the intermediate formed when the thiol displaces that bromide ion. The other site containing bromine is far less reactive toward nucleophiles.

Answer: (c) The nitro and cyano groups that are *ortho* and *para* to the chlorine atom can stabilize the negative charge of the intermediate that forms when hydrazine displaces the chloride ion.

Answer: (d) The nitro group that is *para* to one of the chlorine atoms can stabilize the negative charge of the intermediate formed when the amine displaces that chloride ion. The other site containing chlorine is far less reactive toward nucleophiles.

24.25   Explain why 4-chloropyridine reacts with methoxide to give 4-methoxypyridine under conditions where 3-chloropyridine is unreactive.

Answer: The negative charge of the intermediate formed in the displacement of chloride from the C-4 atom of pyridine can be resonance stabilized by transfer of the charge to the electronegative nitrogen atom. In the case of 3-chloropyridine, there is no resonance stabilization possible, and as a result, the transition state for formation of the intermediate has a much higher energy. Therefore, the reaction rate is slow.

24.26   At one time, 2,4-dinitrofluorobenzene was used to form derivatives of peptides at the N-terminal amino acid. Write a general structure of this type of derivative. Explain why the reaction readily occurs.

Answer: The two nitro groups can stabilize the intermediate formed by nucleophilic attack by the amino group of the amino acid to displace fluoride. The fluorine atom is the most effective of the halogens in the stabilization of the intermediate.

24.27 Explain why hexafluorobenzene readily reacts with sodium methoxide in methanol at 75 °C to yield 2,3,4,5,pentafluoroanisole.

Answer: Three of the fluorine atoms that are *ortho* and *para* to the point of attack by methoxide ion are strongly electron withdrawing and can stabilize the negative charge.

24.28   2,3,4,5,6-Pentafluoronitrobenzene reacts with sodium methoxide in methanol at 25 °C to yield a mixture of two isomeric products with the molecular formula $C_7H_3F_4NO_3$. Draw their structures.

Answer: The nitro group can stabilize the intermediate formed by nucleophilic attack of methoxide ion at positions *ortho* and *para* to the nitro group. The products result from displacement of a fluoride ion by methoxide ion at either of these two positions.

24.29   The following compound reacts in basic solution to give a product with the molecular formula $C_8H_4F_4O_2$. Suggest a structure for this product.

Answer: In basic solution, an alkoxide ion forms that displaces a fluoride ion from the *ortho* position adjacent to the side chain in an intramolecular reaction. The intermediate is stabilized by fluorine atoms at positions *ortho* and *para* to the site of the displacement reaction.

24.30   1-Bromonaphthalene reacts slowly with piperidine at 230 °C to give compound I. The addition of sodium amide accelerates the reaction, which then occurs at 100 °C to give compounds I and II. (a) Explain the difference in the two reaction conditions. (b) Explain the product distribution.

Answer: Reaction with piperidine only occurs by a nucleophilic substitution mechanism in which the piperidine replaces a bromide ion at the C-1 atom. In the presence of sodium amide, the reactions proceeds via a benzyne intermediate. In this case, the piperidine can react at either C-1 or C-2 of naphthalene.

24.31   The two benzyne intermediates derived from reaction of sodium amide and 3-chlorotoluene are formed in approximately equal amounts. Calculate the composition of the mixture of aniline isomers formed in this reaction.

Answer: The benzyne intermediate with a triple bond between C-2 and C-3 can react to give equal amounts of 2-methyl- and 3-methylaniline. The benzyne intermediate with a triple bond between C-3 and C-4 can react to give equal amounts of 3-methyl- and 4-methylaniline. As a consequence, the 2-methyl, 3-methyl, and 4-methyl compounds are formed in a 1:2:1 ratio.

24.32   The reaction of 3-chloro(trifluoromethyl)benzene with sodium amide is regioselective. Which of the two possible isomeric benzynes is formed? Suggest a reason why the reaction is regioselective.

Answer: The benzyne intermediate with a triple bond between C-2 and C-3 forms because the proton at C-2 is more acidic than the proton at C-4. The trifluoromethyl group stabilizes the negative charge of the conjugate base by an inductive effect.

24.33   Draw the structures of the products formed in the reaction of the following deuterated chlorobenzene with sodium amide in liquid ammonia.

Answer: A benzyne intermediate is formed by elimination of DCl. The amide ion can then add to either of two carbon atoms, and two products result that differ in the location of the deuterium atom that was still present in the benzyne intermediate.

24.34   Reaction of 2-bromoanisole with sodium amide in liquid ammonia gives a high yield of 3-aminoanisole. Explain why the reaction of the benzyne intermediate is regioselective.

Answer: The benzyne intermediate has a triple bond between C-2 and C-3. Reaction of amide ion at C-3 occurs because the resulting charge at C-2 is stabilized by the inductive effect of the methoxy group.

24.35   Reaction of o-bromofluorobenzene with magnesium in tetrahydrofuran gives an intermediate that decomposes to yield benzyne. Write the mechanism of this reaction.

Answer: The carbon atom bonded to the magnesium atom of a Grignard has some carbanionic character. Transfer of those electrons to form a carbon–carbon bond results in the loss of fluoride ion from the adjacent carbon atom.

24.36   Reaction of 2-aminobenzoic acid with nitrous acid yields an intermediate that decomposes to yield benzyne. Write the mechanism of this reaction.

Answer: A diazonium ion forms that can lose nitrogen. The carboxyl group can exist as the carboxylate ion. Loss of carbon dioxide and nitrogen can occur in a concerted reaction to give benzyne.

24.37 Draw the structures of the products formed in the reaction of the following deuterated chlorobenzene with sodium amide in liquid ammonia.

Answer: (a) The reaction of this unactivated chlorobenzene occurs via a single benzyne intermediate, and as a result two isomeric phenols form.

Answer: (b) The reaction of this unactivated bromobenzene occurs via either of two benzyne intermediates, and as a result *three* isomeric substituted amines form.

Answer: (c) The reaction of this unactivated bromonaphthalene occurs via a single benzyne intermediate, and as a result two isomeric products form.

24.38 The following compound reacts with sodium amide in ether to give a product with the molecular formula $C_{10}H_{14}N_2$. Suggest a structure for this product.

Answer: Two benzyne intermediates are possible if the elimination occurs with amide ion. However, the conjugate base formed from the secondary amine by proton exchange with amide ion can also react. This intramolecular process is favored. The benzyne formed can react with the secondary amine to give a cyclic product.

24.39 Draw the product of each of the following reactions.

Answers:

(a)

(b)

(c)

(d)

24.40 Draw the product of each of the following reactions.

Answers:

(a)

(b)

(c) + Cl₂ (1 equiv.) ⟶

(d)

Reactions of Phenolate Ions

24.41 Could a polymer result from the reaction of phenolate ion with acetone?

Answer: No, a polymerization reaction does not occur because the carbonyl group of ketones is more stable than the carbonyl group of aldehydes and is less reactive toward nucleophiles.

24.42 Draw the structure of the product of the reaction of CO_2 with the phenolate ion of each of the following compounds.

Answers:

(a)

(b)

(c)

24.43   Based on the structures of ubiquinone and plastoquinone, which compound has the larger reduction potential?

ubiquinone (CoQ)   plastoquinone

Answer: Electron withdrawing groups increase the reduction potential; electron releasing groups decrease the reduction potential. The methoxy groups of ubiquinone are electron donating and therefore decrease the reduction potential. The methyl groups of plastoquinone should also reduce the reduction potential, but less than the methoxy groups of ubiquinone. Therefore, plastoquinone has a larger reduction potential.

24.44   Phenanthrene can be directly oxidized by vanadium(V) oxide to 9,10-phenanthraquinone. Based on the resonance forms of phenanthrene, explain why the central ring is oxidized rather than the other rings.

Answer: Four of the five resonance forms of phenanthrene have a double bond at the C-9 to C-10 atoms, as shown below, and therefore this bond has substantial double bond character. As a result, this bond is more susceptible to oxidizing agents.

24.45   Predict whether the following reaction has an equilibrium constant greater or less than 1.

Answer: The reduction potential for 2-methoxy-1,4-benzoquinone is lower than for 2-methyl-1,4-benzoquinone because the methoxy group is a stronger electron donating group than a methyl group. Thus, the reaction has an equilibrium constant greater than 1.0.

24.46 The $E°$ of the following reaction is 0.08 V. What is the reduction potential for 1,2-benzoquinone?

Answer: The reduction potential for 1,-2-benzoquinone must be 0.08 V more positive than the reduction potential for 1,4-benzoquinone (0.699 V). The value-for 1,2-benzoquinone is 0.78 V.

Quinones

24.47 Select the better nucleophile from each of the following pairs of amines.
(a) aniline and cyclohexylamine
(b) *p*-nitroaniline and *p*-methoxyaniline
(c) aniline and *N,N*-dimethylaniline

Answers:
(a) Cyclohexylamine is a better nucleophile because the nitrogen atom is more basic. The electron pair on nitrogen for aniline is delocalized by interaction with the π electrons of the benzene ring.

(b) *p*-Methoxyaniline is a better nucleophile because the methoxy group is electron donating by resonance and increases the electron density at the nitrogen atom. The nitro group is strongly electron withdrawing and decreases the electron density at the nitrogen atom.

(c) The methyl groups of *N,N*-dimethylaniline increase the electron density at the nitrogen atom and increase it nucleophilicity. However, these groups also increase the size of the nucleophile and therefore decrease its nucleophilicity. These effects more or less cancel, and the nucleophilicities of the two amine are about the same.

24.48 The amidine group (—N—C=N—) is a stronger base than amines. Determine the site of protonation in 1,5-Diazabicyclo[4.3.0]non-5-ene, DBN, a base used in organic reactions. Explain why DBN is a stronger base than an amine.

DBN

Answer: (a) Based on the hybridization of the nitrogen atoms, the site of protonation is the tertiary amino group because it is sp^3 hybridized. The lone-pair electrons of the imino group are sp^2-hybridized and are more tightly held to the nitrogen atom because the % s character of the orbital is greater. However, the effect of resonance stabilization of the base and its conjugate acid are also important. The basic form is resonance stabilized.

Answer: (b) Protonation at the amine group eliminates the possibility of delocalization of resonance stabilization and is therefore not favored.

Answer: (c) However, protonation of the nitrogen atom of the imino group gives a resonance-stabilized conjugate acid. That is why DBN is a stronger base than an amine. Resonance stabilization of this conjugate acid favors its formation compared to the alternate conjugate acid resulting from protonation of the tertiary amino group.

Synthesis Using Diazonium Compounds

24.49   Write a series of reactions required to prepare 2-bromo-4-methylphenol from toluene.

Answer: This synthetic sequence requires seven steps.

2-bromo-4-methylphenol

24.50   Starting from *m*-dimethylbenzene, write the products formed in each step of the following reaction sequence.

Answer:

1-bromo-2,4-dimethylbenzene

Azo Compounds

24.51   Explain why the *p*-nitrobenzenediazonium ion reacts faster than benzenediazonium ion with 2-naphthol.

Answer: The nitro group withdraws electrons from the aromatic ring and from the diazonium group, thus increasing its electrophilicity relative to the unsubstituted benzenediazonium ion.

24.52   Which member of each of the following pairs of aromatic compounds reacts faster with benzenediazoniumchloride?
   (a) aniline and *o*-bromoaniline
   (b) *p*-methylphenol and *p*-methylphenoxide
   (c) anisole and *N,N*-dimethylaniline

Answers:
(a) Aniline is more reactive because the bromine atom of *o*-bromoaniline deactivates the ring to a small degree toward electrophilic aromatic substitution.

(b) *p*-Methylphenoxide is more reactive because the electron pairs of oxygen are more available to the aromatic ring when it is attacked by an electrophile.

(c) *N,N*-Dimethylaniline is more reactive because nitrogen is a better donor of electrons by resonance than oxygen.

24.53   When *o*-aminobenzoic acid (anthranilic acid) is treated with $NaNO_2$ and HCl followed by addition of *N,N*-dimethylaniline to the solution, a dye called methyl red is formed. Draw its structure.

Answer:

methyl red

517

24.54 The following compound is a red dye used in some plastics. Write the structure of the amine needed to form the diazonium ion required to produce the compound. Outline a synthesis of this amine starting from toluene.

Answer: The starting compound is toluene. It is converted to the red dye shown above by the following steps. 1. Nitrate toluene and separate the *para* isomer from the reaction mixture. 2. Reduce the nitro group to give an aniline derivative. 3. Acetylate the amino group with acetyl chloride. 4. Nitrate the amide derivative to place a nitro group at the position *ortho* to the nitrogen atom. Hydrolyze the amide to obtain the amine.

24.55 What compounds are required to synthesize each of the following dyes?

Answers:

518

Answer:

(c) NH₂—⬡—⬡—NH₂   and

24.56   Reaction of 7-amino-2-naphthol with benzenediazonium ion at pH 9 results in substitution *ortho* to the hydroxyl group. At pH 5, the substitution occurs *ortho* to the amino group. Explain how the pH determines the ring in which substitution occurs. Remember that an amino group is a stronger *ortho–para* director than a hydroxyl group.

7-amino-2-naphthol

Answer: At pH 9, the phenol exists as a phenoxide, and this group is a much more effective donor of electrons by resonance. At pH 5, the phenol form exists, and the reaction is controlled by the stronger donor of electrons by resonance, which is the amino group in this case.

25

KEYS TO THE CHAPTER

PERICYCLIC REACTIONS

25.1 Concerted Reactions

In a concerted reaction, bond formation and bond cleavage occur simultaneously to give a product in a single step. **Pericyclic reactions** are concerted processes in which only changes in π and σ bonds occur via cyclic transition states. No intermediates form in these reactions.

  The energy required for a pericyclic reaction can be supplied by heat or light. The differences in the products formed by these two sources of energy reflect the molecular orbitals involved in the reaction. Thermal reactions occur using molecular orbitals of the ground state; photochemical reactions occur via excited states in which electrons have been promoted to higher-energy molecular orbitals.

25.2 Classification of Pericyclic Reactions

Pericyclic reactions are separated into two classes based on the number of electrons that change molecular orbitals in the transition state. Thus, we must account for either $4n$ or $4n+2$ π electrons where n is an integer. We must also know which molecular orbital is involved in the reaction and recognize its symmetry properties. Mechanistically, pericyclic reactions are divided into three classes: electrocyclic reactions, cycloaddition reactions, and sigmatropic rearrangements.

  Electrocyclic reactions interconvert polyenes and isomeric cyclic products that differ by one double bond. The ends of the polyene join to form the single bond of the ring. These reactions favor the cyclic product because it has two additional single bonds at the expense of losing only one double bond. The $\Delta H°$ for the reaction is thus negative. The reverse of the cyclization reaction is also considered an electrocyclic reaction. In general, the reverse reactions are not spontaneous unless there is ring strain in the product that destabilizes the cyclic isomer relative to the polyene, which contains one more double bond.

  Cycloaddition reactions join two components such as a conjugated diene and an alkene (or alkyne). In the Diels–Alder reaction, the ends of the diene join one each to the two carbon atoms of the alkene. The net result of this thermal reaction is the formation of four single bonds and the loss of two double bonds—a process with $\Delta H° < 0$. Although $\Delta S°$ for the cycloaddition is negative, the $\Delta H°$ is sufficiently negative to provide a negative $\Delta G°$. Cycloaddition reactions can occur between two alkenes in which there are $4n +2$ π electrons in the transition state occur thermally. Six π electrons are in the transition state for the Diels–Alder reaction. Cycloaddition reactions can also occur between two alkenes in which 4 π electrons are involved in the transition state, but the process occurs only photochemically. Cycloaddition reactions are designated by the number of π electrons in each of the two reactants contained in brackets and separated by a plus sign. For example, the Diels–Alder reaction is a [4 +2] cycloaddition reaction.

  Sigmatropic rearrangements transfer a σ-bonded group such as hydrogen from one end of a π system to the other: In the process, the positions of single and double bonds are interchanged. These rearrangements are conceptually the most difficult of the pericyclic reactions to recognize and understand.

  The stereochemistry of a pericyclic reaction depends on both the reaction conditions and the number of π electrons in the transition state. The stereochemistry of the reaction, which is based on the symmetry of the reacting molecular orbitals, is established by the stereochemistry of attached groups such as the methyl group. As the component atomic orbitals that contribute to the molecular orbital "rotate" to form bonds, the methyl groups can move in the same direction in a conrotatory process or in opposite directions in a disrotatory process.

25.3 Stereospecificity of Pericyclic Reactions and Molecular Orbitals

Pericyclic reactions occur stereospecifically because the symmetry of the molecular orbitals must be conserved for the reaction to occur. We can analyze these reactions using a frontier orbital method that considers the symmetry of only the highest occupied molecular orbital (HOMO) and lowest unoccupied molecular orbital (LUMO).

  Pericyclic reactions occur if the symmetries of π orbitals in the reactants and products match. These reactions are symmetry allowed. These reactions occur under relatively mild reaction conditions.

Organic Chemistry Study Guide: Key Concepts, Problems, and Solutions http://dx.doi.org/10.1016/B978-0-12-801889-7.00025-X

A molecular orbital is symmetric if the signs on each side of the vertical plane are the same. If there is a reversal of sign, the molecular orbital is antisymmetric. We can determine the symmetry of the molecular orbital by recognizing that π_1 is symmetric for all polyenes and that the symmetry of the molecular orbitals alternate between symmetric and antisymmetric for π_1, π_2, π_3, and so forth.

25.4 Electrocyclic Reactions

In an electrocyclic reaction, the terminal molecular orbitals rotate to form a σ bond by overlap of the p orbital lobes having the same sign. Rotation of the two orbitals in the same direction is conrotatory; rotation of the orbitals in opposite directions is disrotatory.

If the π_2 orbital of butadiene is involved in a thermal electrocyclic reaction, the rotation must occur in a conrotatory sense because the π_2 orbital is antisymmetric. However, in a photochemical electrocyclic reaction of butadiene, π_3, which is symmetric, is involved. As a result, rotation occurs in a disrotatory sense.

In 1,3,5-hexatriene, the results are reversed because the corresponding HOMO and LUMO molecular orbitals that react have different symmetries. The π_3 molecular orbital is required in the thermal electrocyclic reaction. Therefore, rotation must occur in a disrotatory sense because the orbital is symmetric. However, in a photochemical electrocyclic reaction of the triene, π_4, which is antisymmetric, is involved. As a result, rotation occurs in a conrotatory sense.

A quick way to remember this information is to learn one process; for example, the thermal electrocyclic reaction of butadiene. This compound is a $4n$ π system, and the thermal reaction a conrotatory process. If either the number of π electrons changes or the symmetry of the orbitals change compared to this reference reaction, then the rotational process changes. Thus, if the system has $4n+2$ π electrons, the "change" is to a disrotatory process. If we then change the reaction conditions from thermal to photochemical for this new conjugated system, then the direction changes again. In this case, a $4n+2$ π electron compound reacts photochemically by a conrotatory process (Table 25.1).

Once we know the direction of rotation of the orbitals, we also know the direction of movement of the attached alkyl groups. We thus need to know where groups are originally and where they will be located after the rotation occurs.

25.5 Cycloaddition Reactions

Cycloaddition reactions occur by joining the terminal atoms of one π system to the terminal atoms of another π system. We can view this process as the approach of one plane containing one set of π orbitals to another plane containing another set of π orbitals. The π orbitals interact in a suprafacial process in most cycloaddition reactions. For bond formation to occur, the symmetry of the appropriate molecular orbitals must be the same.

Antarafacial cycloaddition reactions are a bit more difficult to visualize because the ends of one π system must simultaneously bond to different sides of the plane of the second π system. Such processes are not as common because of geometric constraints due to insufficient numbers of carbon atoms to bridge from one side of the plane to the other.

The combination of orbitals for a cycloaddition reaction must include the HOMO of one π system and the LUMO of the other π system. The HOMO of the diene and the LUMO of the alkene are selected. Therefore, the diene is viewed as the donor of electrons for the reaction. Both the HOMO of the diene LUMO of the alkene are antisymmetric. The Diels–Alder reaction is thus a suprafacial [4+2] cycloaddition.

The photochemical cycloaddition of two moles of an alkene is a [2+2] reaction. It is visualized by the interaction of one excited state molecule with a ground-state molecule. The electrons in the "donor" excited state molecule are in π_2. The available orbital in the ground-state molecule is π_2. The symmetries of identical molecular orbitals are the same, so the reaction can occur suprafacially.

Table 25.2 summarizes the various possible symmetry-allowed cycloaddition reactions. As in the case of electrocyclic reactions, we should make sure that we know the results of one reaction. If any of the characteristics of that type of reaction are different, then reverse the stereochemical consequences for each difference. Remember that the Diels–Alder reaction is a thermal [4+2] process that occurs suprafacially. Changing from $4n+2$ π electrons to $4n$ π electrons or changing the conditions from thermal to photochemical changes the stereochemistry from suprafacial to antarafacial. Changing two of the conditions gets us back to a suprafacial reaction.

25.6 Sigmatropic Rearrangements

There are two common types of sigmatropic shifts, [1,5] and [3,3]. In each case, a group from one site of a π system migrates to another site at the other "end" of the π system. The thermal process occurs suprafacially in both cases, which are $4n+2$ π systems.

There are two difficulties in understanding the stereochemical consequences of sigmatropic shifts and why they occur. First, we must consider the molecular orbital over which the migrating group moves. Both the [1,5] and [3,3] systems have an odd number of electrons. Thus, the molecular orbitals are arranged differently than for polyenes. However, the lowest energy molecular orbital is still symmetric, and the symmetries of additional molecular orbitals in the energy diagram still alternate.

It is convenient to picture the reactant as two radicals, one of which separates at one point and is rejoined with the other radical at the other end of conjugated radical. (This approach does not mean that the process pictured constitutes the actual mechanism. Indeed, it does not, because the actual process is concerted and no radicals are formed.) For the pentadienyl radical, the HOMO has a single electron in π_3, and that molecular orbital is symmetric. Thus, the migrating group moves along a single face in going from one end of the π system to the other.

For [3,3] sigmatropic shifts, the reaction is visualized as the cleavage and rebonding of two allyl radicals. The HOMO of the allyl radical is π_2, which is antisymmetric. Because the two radicals are identical, the cleavage and rebonding can occur suprafacially.

Vitamin D_3 is activated by a highly unusual antarafacial [1,7] sigmatropic rearrangement.

SOLUTIONS TO EXERCISES

Classification of Pericyclic Reactions

25.1   Classify each of the following reactions as electrocyclic, cycloaddition, or sigmatropic.

Answers:

(a) Electrocyclic reaction

(a)

(b) Electrocyclic reaction

(b)

(c) Electrocyclic reaction

(c)

(d) Sigmatropic rearrangement

(d)

(e) Cycloaddition

(e)

25.2   Classify each of the following reactions as electrocyclic, cycloaddition, or sigmatropic.

Answers:

(a) Cycloaddition

(a)

(b) Sigmatropic rearrangement

(b)

(c) Electrocyclic reaction

(c)

524

(d) Cycloaddition

(e) Sigmatropic rearrangement

Molecular Orbitals

25.3 (a) What π molecular orbital would account for the thermal [1,7] sigmatropic shift in previtamin D_3? (b) What is the symmetry of the molecular orbital?

Answer: We saw earlier that previtamin D_3 is a heptatrienyl system. (a) The HOMO for a [1,7] sigmatropic shift is the seven-electron heptatrienyl radical. It is π_4. (b) This orbital is antisymmetric. This symmetry requires that the hydrogen atom migrate antarafacially, which is geometrically possible for a system of this size and flexibility.

25.4 (a) What π molecular orbital would account for the photochemical cyclization of an octatetraene to give a cyclooctatriene? (b) What is the symmetry of the molecular orbital?

Answer: The HOMO for the eight-electron octatetraene is π_4, so the molecular orbital into which an electron is promoted for a photochemical reaction is π_5, which is symmetric.

Symmetry-Allowed Reactions

25.5 Describe the stereochemistry associated with each of the following symmetry-allowed reactions.
    (a) a thermal [4 + 6] cycloaddition
    (b) a photochemical [2 + 6] cycloaddition
    (c) a thermal [1,7] sigmatropic rearrangement
    (d) a photochemical [1,3] sigmatropic rearrangement

Answer: Refer to Tables 25.2 and 25.3 to relate the number of electrons and type of reaction to the stereochemistry.
    (a) $4 + 6 = 10 = 4n + 2$ π electrons, so the stereochemistry is suprafacial for a thermal cycloaddition.
    (b) $2 + 6 = 8 = 4n$ π electrons, so the stereochemistry is suprafacial for a photochemical cycloaddition.
    (c) $1 + 7 = 8 = 4n$ π electrons, so the stereochemistry is antarafacial for a thermal sigmatropic rearrangement.
    (d) $1 + 3 = 4 = 4n$ π electrons, so the stereochemistry is suprafacial for a photochemical sigmatropic rearrangement.

25.6 Describe the motion that occurs in each of the following symmetry-allowed reactions.
    (a) thermal ring closure of a triene to a cyclohexadiene
    (b) photochemical ring closure of a diene to a cyclobutene
    (c) thermal ring opening of a cyclic triene to a tetraene
    (d) photochemical ring opening of a cyclic diene to a triene

Answer: Refer to Table 25.1 to relate the number of electrons and type of reaction to the motion of the orbitals as they close or open a ring.
    (a) A triene has 6π electrons $(4n + 2)$, so the motion is disrotatory for a thermal ring closure.
    (b) A diene has 4π electrons $(4n)$, so the motion is disrotatory for a photochemical ring closure.
    (c) The tetraene product has 8π electrons $(4n)$, so the motion is conrotatory for a thermal ring opening.
    (d) The triene product has 6π electrons $(4n + 2)$, so the motion is conrotatory for a photochemical ring opening.

25.7   (a) Classify the following thermal cycloaddition reaction. Is the reaction symmetry allowed? (b) What π molecular orbitals are required to explain your answer? (c) What are their respective symmetries?

Answer: (a) The reaction is a [4 + 6] cycloaddition. (b) This is a $4n+2$ system, which is symmetry allowed. (c) The LUMO of the diene is π_3, which is symmetric. The HOMO of the triene is π_3, which is also symmetric.

25.8   (a) Show why the following thermal reaction is regarded as a [2 + 8] cycloaddition. (b) What are their respective symmetries? (c) What π molecular orbitals are required to determine if the reaction is symmetry allowed?

Answer: (a) One of the two π bonds of the acetylene group adds to the methylene carbon atom and C-2 of the ring. Therefore, the entire system contains four pairs of π electrons. (b) The HOMO of the triene is π_4, which is antisymmetric. The LUMO of the reacting double bond of the acetylene is π_2, which is antisymmetric. (c) Thus, since the symmetries of the two orbitals are the same, the reaction is symmetry allowed and can occur suprafacially.

25.9   The following diene does not undergo a Diels–Alder reaction with maleic anhydride. Explain why this symmetry-allowed reaction does not occur.

Answer: The reaction cannot occur because that double bonds are in an s-*trans* arrangement, so the two "ends" of the π system cannot be bridged by the atoms of a dienophile.

25.10 The ring-opening reaction of a cyclobutene ring fused to an eight-membered ring occurs at temperatures below 200 °C. Even though a structurally related compound with a cyclobutene ring fused to a five-membered ring is more strained, this ring-opening reaction requires temperatures near 300 °C. Explain why.

Answer: For a concerted process, the thermal ring opening of a *cis*-dialkyl-substituted cyclobutene occurs by a conrotatory process to give an *E, Z* diene. In the first compound, this configuration can be accommodated within the cyclic product. 1,3-Cycloheptadiene cannot exist with an *E, Z* configuration. The product has the *Z, Z* configuration, so it cannot be the product of a concerted electrocyclic process. The reaction must occur via a multistep process that requires more energy.

25.11 Photochemical [1,3] sigmatropic shifts occur in some allylic systems. Draw the structures of the products resulting from the following alkene. Which product should predominate?

Answer: There are two nonequivalent allylic sites from which the hydrogen atom could move. The predominant product should be the compound with the more stable double bond, which is the first of the following two products because its double bond is more highly substituted.

25.12 Explain why the following compound cannot undergo a [1,7] sigmatropic rearrangement, whereas previtamin D can. A photochemical [1,7] sigmatropic rearrangement occurs. Draw the structure of the first product formed.

Answer: The thermal [1,7] sigmatropic rearrangement occurs antarafacially in previtamin D_3 because the open conformation allows the hydrogen atom to transfer from the "top" of one end of the system to the "bottom" of the other end. In the phenyl-substituted cycloheptatriene, the "ends" of the system are bonded, and a hydrogen atom cannot move from one side of the ring to the other. A photochemical rearrangement occurs suprafacially, which is geometrically possible. The resulting product is a resonance-stabilized triene.

Electrocyclic Reactions

25.13   Draw the structure of the product of each of the following reactions, and indicate the stereochemistry of each product.

(a)   heat ⟶

(b)   light ⟶

(c)   light ⟶

Answer: Refer to Table 25.1 to relate the number of electrons and type of reaction to the motion of the orbitals as they close or open a ring. The structures resulting from each reaction are given below.

(a) For a thermal ring closure with $4n$ π electrons, the motion is conrotatory.
(b) For a photochemical ring closure with $4n$ π electrons, the motion is disrotatory.
(c) For a photochemical ring closure with $4n$ π electrons, the motion is disrotatory.

25.14 Draw the structure of the product of each of the following reactions and indicate the stereochemistry of each product.

(a) heat

(b) light

(c) heat

Answer: Refer to Table 25.1 in the text to relate the number of electrons and type of reaction to the motion of the orbitals as they close or open a ring. The structures resulting from each reaction are shown below.

(a) For a thermal ring closure with $4n + 2$ π electrons, the motion is disrotatory.
(b) For a photochemical ring closure with $4n$ π electrons, the motion is disrotatory.
(c) For a thermal ring opening with $4n$ π electrons, the motion is conrotatory.

(a) heat

(b) light

(c) heat

25.15 Explain why the thermal ring opening of *trans*-3,4-dimethylcyclobutene could yield two isomeric 2,4-hexadienes. Explain why only one isomer forms.

Answer: The conrotatory process could occur in a clockwise or counterclockwise manner to give isomeric dienes. However, the process that moves two hydrogen atoms toward one another is less sterically hindered that the process that moves two methyl groups toward one another. The first of the following two products is the preferred product.

(a)

(2*E*,4*E*)-hexadiene
(preferred, less steric hindrance)

(b)

(2*Z*,4*Z*)-hexadiene
(not favored, more steric hindrance)

25.16 Which of the following compounds cannot undergo a thermal electrocyclic ring closure?
(a) (2*E*,4*Z*,6*E*)-2,4,6-octatriene
(b) (2*E*,4*E*,6*E*)-2,4,6-octatriene
(c) (2*E*,4*Z*,6*Z*)-2,4,6-octatriene

Answer: Compound (b) has an *E* configuration at C-4 to C-5 atoms which geometrically prevents the "ends" of the triene system from bonding in an electrocyclic ring closure. Compounds (a) and (c) can react.

Electrocyclic Reactions

25.17 Draw the structure of the product of each of the following reactions.

Answers:

530

(b)

(c)

25.18   What reactants are required to produce each of the following compounds via a Diels–Alder reaction?

Answers:

(a)

(b)

(c)

25.19   Explain why 1,3-cyclopentadiene reacts faster than 1,3-butadiene with maleic anhydride.

Answer: The two "ends" of the diene system in 1,3-cyclopentadiene are held in a *cis* arrangement by the ring. The reaction of 1,3-butadiene requires conversion of the more stable s-*trans* conformation into an s-*cis* conformation via rotation about the C-2 to C-3 bond before reaction with maleic anhydride can occur.

25.20 Explain why (2Z,4Z)-hexadiene does not react with maleic anhydride even though the diene has two methyl groups that increase the electron density of the diene.

Answer: The s-*cis* conformation is sterically hindered by the terminal methyl groups.

(2Z,4Z)-hexadiene

25.21 One equivalent of 1,3-butadiene reacts with the following quinone to give a single product. Draw the structure and explain why it forms.

Answer: The dienophile has two nonequivalent double bonds. The more reactive double bond is the one with electron withdrawing groups. In this case, the methyl group is more electron donating than hydrogen, so the double bond lacking the methyl group is more reactive.

25.22 One equivalent of (E)-1,3-pentadiene reacts with the following quinone to give a mixture of two products. Draw their structures and explain why they form.

Answer: The dienophile has two nonequivalent double bonds. Both the methoxy group and the methyl group can stabilize their respective double bonds. The methoxy group is more effective in donation of electrons by resonance. Thus, the double bond with the methyl group is the more reactive one. The diene can add in two ways to give isomeric adducts.

25.23 Draw the structure of the Diels–Alder product of the following combination of reactants.

Answer:

532

25.24　What reactants are required to produce each of the following compounds via a Diels–Alder reaction?

Answer:

Sigmatropic Rearrangements

25.25　Explain why *cis*-1,2-divinylcyclobutane undergoes a [3,3] sigmatropic rearrangement faster than 1,5-hexadiene.

Answer: The two "ends" of the diene that must form a bond are held in a *cis* arrangement by the ring in *cis*-1,2-divinylcyclobutane. 1,5-Hexadiene is conformational flexible about three sigma bonds. The reaction of this compound requires a conformation that has an eclipsed arrangement around the C-3 to C-4 bond.

25.26　(a) What type of reaction occurs in the conversion of allyl vinyl ether into 4-pentenal? (b) Why does the reaction have a large equilibrium constant?

Answer: The reaction is a [3,3] sigmatropic shift. The driving force for the reaction is the formation of a stable carbonyl group compared to the less stable carbon–carbon double bond in the reactant.

25.27　Draw the structure of the product of the Claisen rearrangement of each of the following compounds.

Answers:

25.28 Write a mechanism to account for the following rearrangement reaction.

Answer: The reaction occurs in two steps. First, a Claisen-type rearrangement occurs. This is a [3,3] sigmatropic rearrangement. It gives an unstable intermediate that undergoes a tautomerization reaction.

25.29 The following compound undergoes a Claisen-type rearrangement to yield a ketone. Draw the structure of the product.

Answer: The reaction occurs by a [3,3] sigmatropic rearrangement in which a carbonyl group forms rather than a carbon–carbon double bond. The driving force for the reaction is the stability of the carbonyl group.

25.30 What reactant is required to yield the following aldehyde by a Claisen-type rearrangement?

Answer: The reaction occurs by a [3,3] sigmatropic rearrangement. The driving force of the reaction is the stability of the tetrasubstituted carbon–carbon double bond and the carbonyl group.

25.31 (a) Draw the product of a thermal [5,5] sigmatropic rearrangement of the following compound. (b) Does the reaction occur by a suprafacial or antarafacial process?

Answer: (a) The reaction is a [5,5] sigmatropic rearrangement that involves $4n + 2$ π electrons. (b) It occurs thermally by a suprafacial process.

25.32 The following ether undergoes a thermal [5,5] sigmatropic rearrangement. Draw the structure of the product.

Answer: This [5,5] sigmatropic rearrangement involves two of the π bonds of the aromatic ring as the side chain diene bonds to the *para* position. To visualize how this occurs, imagine the aromatic ring lying flat and the diene side chain curved above it. The rearrangement occurs in a concerted process in which a π bond forms between oxygen and C-1 of the ring, a π bond breaks between oxygen and C-5 of the side chain, and a sigma bond forms between C-1 of the side chain and C-4 of the diene. The unstable intermediate that forms then tautomerizes to give a phenol.

Multiple Pericyclic Reactions

25.33 The following thermal isomerization reaction occurs by two similar sequential pericyclic reactions. Identify them and draw the structure of the intermediate compound.

Answer: The reactions are two electrocyclic reactions. The first reaction forms a cyclobutene ring fused to the cyclobutane ring. The double bond is between the two carbon atoms at the points of fusion. Thus, the double bond "belongs" to both rings, and either one can be considered the cyclobutene ring. The second reaction could be the opening of the newly formed ring to give the original diene or an opening of the original ring to give the isomeric diene.

25.34 The following thermal isomerization occurs by two similar, sequential pericyclic reactions. Identify them and draw the structure of the intermediate compound.

Answer: The product results from two electrocyclic reactions. The first reaction forms a cyclobutene ring. The second opens the ring. The ring closure and ring opening reactions are both conrotatory and can occur in either a clockwise or a counterclockwise direction to give the same product.

25.35 The following compound can lose its optical activity when heated. Two similar, sequential pericyclic reactions are required to account for this result. Identify them and draw the structure of the intermediate compound.

Answer: A [1,5] sigmatropic shift involving the hydrogen atom at the chiral center gives an achiral isomer. Reversal of the reaction can occur by transferring either of the two hydrogen atoms from a methylene group. The transfer can occur from the "bottom" in the case of one hydrogen, and from the "top" for the other hydrogen, so the product is a racemic mixture.

25.36   The following isomerization reactions occur by related sequential pericyclic processes. Draw the structure of the intermediate
        compound involved in each reaction.

Answers: A sequence of [1,5] sigmatropic rearrangements leads from the reactant to the products. Migration of either a hydrogen atom
or a deuterium atom gives similar structures that do not have the resonance stabilization of the benzene ring. The first of the following two
structures results from migration of a hydrogen atom; the second results from migration of a deuterium atom.

Transfer of a hydrogen atom from the first intermediate to either of the equivalent sites in a [1,5] sigmatropic shift gives one of the isomers.

Transfer of a hydrogen atom from the second intermediate to the site that does not contain deuterium gives the other isomer.

26 CARBOHYDRATES

KEYS TO THE CHAPTER

26.1 Carbohydrates in the Biosphere

Carbohydrates are polyhydroxy aldehydes and ketones or compounds that produce them in hydrolysis reactions. Carbohydrates are a major source of stored chemical energy. That energy is incorporated into carbohydrates in their formation from carbon dioxide and water in photosynthesis. The energy is released in metabolic reactions in microorganisms, plants, and animals.

26.2 Classification of Carbohydrates

Monosaccharides are carbohydrates that cannot be hydrolyzed to simpler carbohydrates. These compounds may be aldehydes or ketones, or acetals or ketals which yield an alcohol and a monosaccharide when hydrolyzed. **Oligosaccharides** contain **glycosidic linkages** (acetal or ketal) that release two or more monosaccharide units upon hydrolysis. One monosaccharide serves as the acetal or ketal center that reacts with the hydroxyl group of the next monosaccharide. **Disaccharides** are oligosaccharides that contain two monosaccharide units. **Polysaccharides** contain a large number of monosaccharide units bonded to each other by a series of glycosidic bonds. Each unit serves both as the acetal (or ketal) center to form one glycosidic bond and as an alcohol to form the other glycosidic bond to neighboring monosaccharides.

A monosaccharide is classified as an **aldose** if the carbonyl group is an aldehyde and as a **ketose** if the carbonyl group is a ketone. The number of carbon atoms is commonly between three and six. Both pieces of information are used in classifying monosaccharides. The term aldohexose indicates that the compound is an aldehyde and that it contains six carbon atoms. The term ketopentose indicates that the compound is a ketone and that it contains five carbon atoms.

26.3 Chirality of Monosaccharides

Figures 26.1 and 26.3 show the Fischer projection structures of the D series of monosaccharides as for aldoses and ketoses, respectively. The terms D and L refer to the configuration at the chiral center farthest away from the carbonyl group, which is an aldehyde group in aldoses and a ketone group in ketoses. The nomenclature used for each of these monosaccharides is a name that conveys the internal configuration at each center with respect to the carbon atom establishing the D configuration assigned by Fischer. Because the relative configuration of all chiral centers is given by the name, the configuration of *every* center of an L-monosaccharide with the same name is reversed compared to the D-monosaccharide.

Each isomeric monosaccharide of the D-series is a diastereomer of every other isomer. The enantiomer of a particular D compound is the L compound. These diastereomers have different physical properties.

26.4 Isomerization of Monosaccharides

Inversion of one or more of the stereogenic centers of a monosaccharide gives diastereomers. Such reactions generally require specialized chemistry that involves protection of one or more of the functional groups. Compounds with multiple stereogenic centers that differ in configuration at one stereogenic center are **epimers**. Epimerization at selected centers, as in galactose and glucose, occurs in biochemical reactions catalyzed by enzymes. Epimerization at the α carbon atom occurs readily via an enediol intermediate and can in an enzyme-catalyzed reaction. Both aldoses and ketose are interconverted in this equilibrium reaction.

Organic Chemistry Study Guide: Key Concepts, Problems, and Solutions http://dx.doi.org/10.1016/B978-0-12-801889-7.00026-1

Monosaccharides form intramolecular hemiacetals or hemiketals containing five-membered rings called **furanoses** and six-membered rings called **pyranoses**. These structures are the predominant form of monosaccharides. Only a very small fraction of the open chain form exists in solution.

The Haworth projection formulas show the cyclic form of carbohydrates in a flat five- or six-membered ring drawn as if perpendicular to the plane of the page. The ring oxygen atom is drawn in the back of the ring in a furanose (five-membered ring). In a pyranose (six-membered ring), the oxygen atom is shown in the back right position in the plane of the ring. The hemiacetal or hemiketal carbon atom is on the right and lies on the line where the plane of the ring and the plane of the page intersect. The hydroxyl group at this center may be below or above the plane of the ring, designated as α or β, respectively. The hydroxyl groups on the right in the open chain form are drawn below the plane of the ring in the Haworth projection formula; hydroxyl groups on the left are drawn above the plane of the ring.

The —CH$_2$OH group is "up" in Haworth projections of D-monosaccharides. The β anomer has its OH group "up" in a Haworth projection.

When a cyclic hemiacetal or hemiketal forms, a new stereogenic center is created. The configuration at this center is the only difference between the α and β forms of a monosaccharide. The α and β isomers are diastereomers called **anomers**. Mixtures of anomers of both pyranoses and furanoses exist at equilibrium in aqueous solution. Table 26.1 gives composition of these mixtures.

Haworth projection formulas convey stereochemical information but do not realistically picture the location of the functional groups in space. Cyclohexane-like conformations are much more accurate. There is a standard way of orienting the ring. The "foot" of the chair is on the right, and the back of the chair is on the left. The substituents in pyranoses can be axial or equatorial. Reversing the equatorial and axial positions of a substituent generates a different diastereomer, that is, an epimer. Even drawing the chair forms of cyclohexanes does not disclose the full picture; for that we need molecular models.

Since the α and β anomers are diastereomers, they have different optical activities. The change in optical activity when these anomers interconvert is called **mutarotation**. Mutarotation is a result of the reversible opening and closing of a hemiacetal or hemiketal center in the cyclic forms of a carbohydrate. When the ring is open, rotation about the bond between C-1 and C-2 can occur, changing the position of the hydroxyl group. When the ring closes again, the α anomer may have changed to β or vice versa. Eventually, a point of equilibrium is attained between the α and β forms. The α and β anomers of a monosaccharide are diastereomers, so the optical rotation values of the two anomers are not reversed as they are for enantiomers. The optical rotation of a mixture of anomers has an intermediate value determined by the percent composition of the two anomers.

26.6 Reduction and Oxidation of Monosaccharides

Any hemiacetal or hemiketal of a carbohydrate can exist in equilibrium with its open-chain form. The resultant aldehyde or ketone is reduced to an **alditol** by reaction with NaBH$_4$. An alditol is a polyalcohol obtained when the carbonyl group is reduced to a —CH$_2$OH group. Thus, to write the structure of the reduction product, simply write the open-chain structure of the monosaccharide, replacing —CHO with —CH$_2$OH.

As a result of the equilibrium of an hemiacetal of a monosaccharide with its open-chain form, the aldehyde of an aldose will reduce Tollens's or Benedict's reagent. The ketone of a ketose will undergo the same reactions because it undergoes tautomerization to the aldehyde form. The products, which have a carboxylic acid at C-1, are **aldonic acids**. All monosaccharides that are oxidized to aldonic acids are called **reducing sugars** because they reduce the reagent.

Aldaric acids result from oxidation at both terminal positions of an aldose to give a dicarboxylic acid. Special enzyme-catalyzed reactions convert the highest numbered carbon atom of an aldose into a carboxylic acid without affecting the aldehydic carbon atom. These compounds are **uronic** acids.

26.7 Glycosides

The hemiacetal and hemiketal forms of monosaccharides react with alcohols to form acetals and ketals called glycosides. The C—O bond formed is a glycosidic bond, and the —OR from the alcohol is called an **aglycone**. The formation of glycosides occurs by a mechanism identical to the one we described in Section 19.5 for simple hemiacetals and hemiketals. A carbocation is formed that is stabilized by the lone pair electrons of a directly bonded oxygen atom. As a result, a mixture of anomeric glycosides results.

Any carbohydrate that exists as an acetal or ketal cannot revert to an aldehyde or ketone in the presence of the basic solution of Tollens' or Benedict's reagent. Thus, no reaction occurs, and the compound is called a **non reducing sugar**.

26.8 Disaccharides

Disaccharides are glycosides formed from two monosaccharides that can be either aldoses or ketoses. One of the —OR groups is provided by the original cyclization to give the hemiacetal or hemiketal. The second —OR group is derived from an aglycone; it is a second monosaccharide that provides the alcohol functional group of the glycosidic bond. The common disaccharides are maltose, cellobiose, lactose, and sucrose. To identify a disaccharide it is necessary to carry out the following steps, which form a kind of recipe for reading the structure.

1. Identify the constituent monosaccharides.
2. Identify the ring form of each monosaccharide.
3. Identify which monosaccharide serves as the acetal center or ketal center and its configuration.
4. Identify the configuration of the hemiacetal center of this monosaccharide.
5. Identify which hydroxyl group of the monosaccharide serving as the aglycone is bonded to the acetal center.

The first two steps listed above are straightforward. In the third step, we must locate the acetal center. It is the one with two oxygen atoms bonded to a common carbon atom, and each of them is bonded to a second carbon atom. For step 4, check the hemiacetal center at the carbon atom bearing both an —OH and —OR group. Step 5 is straightforward because the aglycone is derived from the —OR group that is not part of the furanose or pyranose ring of the monosaccharide that is providing the acetal or ketal center.

26.9 Polysaccharides

Polysaccharides that consist of only one type of monosaccharide are **homopolysaccharides**. Starch and cellulose are both homopolysaccharides of glucose. They differ only in the stereochemistry of the glycosidic linkage, which is 1,4′ both cases. In starch the linkage is α; in cellulose it is β. The ability of organisms to hydrolyze α or β-glycosidic bonds determines what they can use as food sources. Most animals can only digest α-linked polysaccharides of glucose. Cattle and other herbivores can digest β-linked polysaccharides because they harbor microorganisms which have the necessary enzymes.

26.10 Chemical Determination of Monosaccharide Structures

A monosaccharide can be shown to be an aldose or a ketose based on the products of oxidation by periodate. Aldoses give an equivalent of formic acid, whereas ketoses give an equivalent of carbon dioxide. Each secondary alcohol site also gives an equivalent of formic acid.

Oxidation by nitric acid may provide information about the relationship of the stereogenic centers in an aldose. If the secondary hydroxyl groups are symmetrically arranged, the aldaric acid will be an optically inactive *meso* compound. That fact decreases the number of possible arrangements to consider for the stereogenic centers.

Formation of an osazone identical with that of a second monosaccharide establishes the identity of the unknown monosaccharide. All stereogenic centers other than the one a to the carbonyl group have the same configuration.

Both the chain extension and chain degradation methods provide information about the configuration of the stereogenic centers of an unknown monosaccharide. Chain degradation indicates the configuration of all but one of the original stereogenic centers, and thus the unknown may be one of two possible compounds. Chain elongation gives two diastereomeric products whose configuration at the C-3 atom and beyond is the same as for the original monosaccharide.

26.11 Determination of Ring Size

The size of a monosaccharide ring is determined by first preparing a glycoside with an aglycone such as a methyl group. Subsequent oxidation by periodate gives a dialdehyde whose structure can be used to deduce the ring size.

An alternate method of determining the ring size involves complete methylation of the monosaccharide, followed by hydrolysis of the acetal. Subsequent treatment using a strong oxidizing agent not only oxidizes the aldehyde group but also cleaves a carbon–carbon bond at the carbon atom bearing the oxygen atom of the original ring, and gives a second carboxylic acid site.

26.12 Structure of Disaccharides

Hydrolysis of disaccharides establishes the identity of its component monosaccharides. However, three questions remain.

1. What is the configuration at the acetal or ketal center?
2. What are the sites of the glycosidic linkage?
3. Are the component monosaccharides furanoses or pyranoses?

Stereospecific enzymatic hydrolysis identifies the configuration of the acetal or ketal center, and combinations of methylation and hydrolysis identify the sites of glycosidic linkage and ring size.

26.13 Blood Group Antigens

Human blood cells contain antigenic determinants that divide blood into three major types, designated A, B, and O. The blood group oligosaccharides contain the monosaccharides galactose, *N*-acetylgalactosamine, and *N*-acetylglucosamine. They also contain 6-deoxyl-α-L-galactose, which has the common name α-L-fucose.

Summary of Reactions

1. Reduction of Monosaccharides

CHO
H——OH
H——OH
H——OH
CH₂OH

D-ribose

→ NaBH₄ →

CH₂OH
H——OH
H——OH
H——OH
CH₂OH

D-ribitol

CH₂OH
C=O
H—C—OH
H—C—OH
CH₂OH

D-ribulose

→ NaBH₄ →

CH₂OH
H——OH
H——OH
H——OH
CH₂OH

D-ribitol

+

CH₂OH
HO——H
H——OH
H——OH
CH₂OH

D-arabinitol

2. Oxidation of Monosaccharides

CHO
H——OH
HO——H
H——OH
CH₂OH

D-xylose

→ Ag(NH₃)₂⁺ →

CO₂H
H——OH
H——OH
H——OH
CH₂OH

D-xylonic acid

+ Ag(s)

CHO
HO——H
H——OH
H——OH
CH₂OH

D-arabinose

→ HNO₃ →

CO₂H
HO——H
H——OH
H——OH
CO₂H

D-arabinaric acid

3. Isomerization of Monosaccharides

D-ribose   D-ribulose   D-arabinose

4. Formation of Glycosides

5. Periodate Oxidation

vicinal diol at
1° alcohol

an aldehyde

formaldehyde

(an aldose)

formic acid

(an aldehyde)

(a ketose)

(a carboxylic acid)

formaldehyde

6. Formation of Osazones

$$\begin{array}{c}
\text{CHO} \\
| \\
\text{H}-\text{C}-\text{OH} \\
| \\
(\text{CH}_2\text{OH})_n \\
| \\
\text{CH}_2\text{OH}
\end{array}
\quad \xrightarrow{\text{NH}_2\text{NHC}_6\text{H}_5} \quad
\begin{array}{c}
\text{CH}=\text{NHNHC}_6\text{H}_5 \\
| \\
\text{C}=\text{NHNHC}_6\text{H}_5 \\
| \\
(\text{CH}_2\text{OH})_n \\
| \\
\text{CH}_2\text{OH}
\end{array}$$

$\qquad$ (an aldose) $\qquad\qquad\qquad\qquad\qquad$ (an osazone)

$$\begin{array}{c}
\text{CH}_2\text{OH} \\
| \\
\text{C}=\text{O} \\
| \\
(\text{CH}_2\text{OH})_n \\
| \\
\text{CH}_2\text{OH}
\end{array}
\quad \xrightarrow{\text{NH}_2\text{NHC}_6\text{H}_5} \quad
\begin{array}{c}
\text{CH}=\text{NHNHC}_6\text{H}_5 \\
| \\
\text{C}=\text{NHNHC}_6\text{H}_5 \\
| \\
(\text{CH}_2\text{OH})_n \\
| \\
\text{CH}_2\text{OH}
\end{array}$$

$\qquad$ (a ketose) $\qquad\qquad\qquad\qquad\qquad$ (an osazone)

7. Aldose Chain Elongation: The Kiliani–Fischer Synthesis

Step 1

New stereogenic centers

$$\begin{array}{c}
\text{CHO} \\
| \\
\text{H}-\text{C}-\text{OH} \\
| \\
\text{R}
\end{array}
\quad \xrightarrow{\text{HCN}} \quad
\begin{array}{c}
\text{CN} \\
| \\
\text{H}-\text{C}-\text{OH} \\
| \\
\text{H}-\text{C}-\text{OH} \\
| \\
\text{R}
\end{array}
\quad + \quad
\begin{array}{c}
\text{CN} \\
| \\
\text{HO}-\text{C}-\text{H} \\
| \\
\text{H}-\text{C}-\text{OH} \\
| \\
\text{R}
\end{array}$$

Step 2

$$\begin{array}{c}
\text{CN} \\
| \\
\text{H}-\text{C}-\text{OH} \\
| \\
\text{H}-\text{C}-\text{OH} \\
| \\
\text{R}
\end{array}
\quad \xrightarrow[\text{Pd / BaSO}_4]{\text{H}_2} \quad
\begin{array}{c}
\text{C}=\text{NH} \\
| \\
\text{HO}-\text{C}-\text{H} \\
| \\
\text{H}-\text{C}-\text{OH} \\
| \\
\text{R}
\end{array}
\quad \xrightarrow{\text{H}_2\text{O}} \quad
\begin{array}{c}
\text{CHO} \\
| \\
\text{HO}-\text{C}-\text{H} \\
| \\
\text{H}-\text{C}-\text{OH} \\
| \\
\text{R}
\end{array}$$

$\quad$ (one diastereomer) $\qquad\qquad$ (one imine intermediate) $\qquad\qquad$ (an extended aldose)

8. Chain Shortening of Aldoses: The Wohl Degradation

Steps 1 and 2

$$\begin{array}{c}
\text{CHO} \\
| \\
\text{H}-\text{C}-\text{OH} \\
| \\
\text{H}-\text{C}-\text{OH} \\
| \\
\text{R}
\end{array}
\quad \xrightarrow{\text{NH}_2\text{OH}} \quad
\begin{array}{c}
\text{CH}=\text{NOH} \\
| \\
\text{H}-\text{C}-\text{OH} \\
| \\
\text{H}-\text{C}-\text{OH} \\
| \\
\text{R}
\end{array}
\quad \xrightarrow{\text{Ac}_2\text{O}} \quad
\begin{array}{c}
\text{C}\equiv\text{N} \\
| \\
\text{H}-\text{C}-\text{OAc} \\
| \\
\text{H}-\text{C}-\text{OAc} \\
| \\
\text{R}
\end{array}$$

$\quad$ (one diastereomer) $\qquad\qquad\qquad$ (an oxime)

Steps 3 and 4

$$\begin{array}{c}
\text{C}\equiv\text{N} \\
| \\
\text{H}-\text{C}-\text{OAc} \\
| \\
\text{H}-\text{C}-\text{OAc} \\
| \\
\text{R}
\end{array}
\quad \xrightarrow{\text{CH}_3\text{O}^-} \quad
\begin{array}{c}
\text{C}\equiv\text{N} \\
| \\
\text{H}-\text{C}-\text{OH} \\
| \\
\text{H}-\text{C}-\text{OH} \\
| \\
\text{R}
\end{array}
\quad \xrightarrow{\text{CH}_3\text{O}^-} \quad
\begin{array}{c}
\text{H} \diagdown \quad \diagup \text{O} \\
\quad \text{C} \\
| \\
\text{H}-\text{C}-\text{OH} \\
| \\
\text{R}
\end{array}
\quad + \text{CH}_3\text{OH} + \text{CN}^-$$

9. Permethylation: Reaction with Dimethyl Sulfate

β-D-glucopyranose

(CH₃)₂SO₄

methyl 2,3,4,6-tetra-O-methyl-β-D-glucopyranoside

SOLUTIONS TO EXERCISES

Classification of Monosaccharides

26.1   Classify each of the following monosaccharides.

```
                              CHO           CH2OH
                    CHO       |             |
              CHO   |      H—C—OH         C=O
        CHO   |   H—C—OH   |           HO—C—H
        |   H—C—OH   |     HO—C—H        |
(a) H—C—OH (b) H—C—OH (c) H—C—OH   (d) H—C—OH
        |     |           |             |
      H—C—OH H—C—OH     H—C—OH         H—C—OH
        |     |           |             |
      CH2OH  CH2OH       CH2OH         CH2OH
```

Answer: The classes of carbohydrates are designated by aldo if they contain an aldehyde group or keto if they contain a keto group. The length of the carbon chain is indicated by terms *tetr*, *pent*, and *hex*, prefixes we have seen many times. The ending of a carbohydrate name is *ose*.

  (a) aldotetrose   (b) aldopentose   (c) aldohexose   (d) ketohexose

26.2   Classify each of the following monosaccharides.

```
                    CHO
        CHO       HO—C—H
        |           |              CH2OH
        C=O       H—C—OH           |
(a) H—C—OH  (b)  H—C—OH    (c)    C=O
        |           |           H—C—OH
      H—C—OH      H—C—OH          |
        |           |            CH2OH
      CH2OH       CH2OH
```

Answers: (a) ketopentose (b) aldohexose   (c) ketotetrose

26.3   Classify each of the following monosaccharides as D or L.

```
                    CHO           CH2OH
                  HO—C—H          |
        CHO       HO—C—H          C=O
      H—C—OH        |           H—C—OH
(a) HO—C—H  (b)  H—C—OH  (c) HO—C—H
      H—C—OH     HO—C—H       HO—C—H
        |           |             |
      CH2OH       CH2OH         CH2OH
```

Answer: The highest numbered stereogenic center is used to assign D and L configurations. If the —OH group at that point is on the right in the projection formula, it is D; if the —OH group is on the left, it is L. The chains are numbered to give the lower number to the most highly oxidized center. Thus, (b) and (c) are L carbohydrates; (a) is a D carbohydrate.

26.4   Classify each of the following monosaccharides as D or L.

(a)
```
        CHO
        |
        C=O
        |
   HO—C—H
        |
   HO—C—H
        |
       CH₂OH
```

(b)
```
       CH₂OH
        |
   HO—C—H
        |
        C=O
        |
   HO—C—H
        |
    H—C—OH
        |
       CH₂OH
```

(c)
```
       CH₂OH
        |
        C=O
        |
   HO—C—H
        |
       CH₂OH
```

Answer: The highest numbered stereogenic center is used to assign D and L configurations. If the —OH group at that point is on the right in the projection formula, it is D; if the —OH group is on the left, it is L. The chains are numbered to give the lower number to the most highly oxidized center. Thus, (a) and (c) are L carbohydrates; (b) is a D carbohydrate.

Fischer Projection Formulas

26.5   Draw the Fischer projection formulas of the isomeric D-3-ketopentoses.

Answer: There are two equivalent stereogenic centers in a 3-ketopentose. An enantiomeric pair of compounds and a *meso* compound are possible. The first structure, (a), has the D configuration. The second, (b), is its enantiomer, and has the L configuration. The third is *meso* a compound.

(a)
```
       CH₂OH
        |
   HO—C—H
        |
        C=O
        |
    H—C—OH
        |
       CH₂OH
```
a D-3-ketopentose

(b)
```
       CH₂OH
        |
    H—C—OH
        |
        C=O
        |
   HO—C—H
        |
       CH₂OH
```
an L-3-ketopentose

(c)
```
       CH₂OH
        |
    H—C—OH
        |
        C=O
        |
    H—C—OH
        |
       CH₂OH
```
a *meso* 3-ketopentose

26.6   Draw the Fischer projection formulas of the isomeric D-3-ketohexoses.

Answer: There are three nonequivalent stereogenic centers, so a total of eight stereoisomers are possible. Four have a D configuration, as shown below.

3-Ketohexoses

```
       CH₂OH
        |
    H—C—OH
        |
        C=O
        |
    H—C—OH
        |
    H—C—OH
        |
       CH₂OH
```

```
       CH₂OH
        |
   HO—C—H
        |
        C=O
        |
    H—C—OH
        |
    H—C—OH
        |
       CH₂OH
```

```
       CH₂OH
        |
    H—C—OH
        |
        C=O
        |
   HO—C—H
        |
    H—C—OH
        |
       CH₂OH
```

```
       CH₂OH
        |
   HO—C—H
        |
        C=O
        |
   HO—C—H
        |
    H—C—OH
        |
       CH₂OH
```

26.7 Draw the Fischer projection formula of each of the following monosaccharides.
(a) L-xylose (b) L-erythrose (c) L-galactose (d) L-ribose (e) L-fructose

Answers:

(a)

CHO
HO———H
H———OH
HO———H
CH$_2$OH

L-xylose

(b)

CHO
HO———H
HO———H
CH$_2$OH

L-erythrose

(c)

CHO
HO———H
H———OH
H———OH
HO———H
CH$_2$OH

L-galactose

(d)

CHO
HO———H
HO———H
HO———H
CH$_2$OH

L-ribose

(e)

CH$_2$OH
C═O
H—C—OH
HO—C—H
HO—C—H
CH$_2$OH

L-fructose

26.8 Draw the Fischer projection formula of each of the following monosaccharides.
(a) 6-deoxy-L-galactose   (b) 3-deoxy-D-ribose   (c) 2,6-dideoxy-D-allose   (d) 6-deoxy-L-mannose

Answers:

(a)

CHO
HO———H
H———OH
H———OH
HO———H
CH$_3$

6-deoxy-L-galactose

(b)

CHO
H—C—OH
H—C—H
H—C—OH
CH$_2$OH

3-deoxy-D-ribose

(c)

CHO
H—C—H
H—C—OH
H—C—OH
H—C—OH
CH$_3$

2,6-dideoxy-D-allose

(d)

CHO
H———OH
H———OH
HO———H
HO———H
CH$_3$

6-deoxy-L-mannose

Haworth Projection Formulas

26.9 Draw the Haworth projection formula of the hemiacetal of 5-hydroxyhexanal.

Answer: Two isomers are possible at the hemiacetal center.

549

26.10   Draw the Haworth projection formula of the hemiketal of 5-hydroxy-2-hexanone.

Answer: Two isomers are possible at the hemiacetal center.

26.11   Draw the Haworth projection formula of the pyranose form of each of the following compounds
(a) α-D-mannose   (b) β-D-galactose   (c) α-D-glucose   (d) α-D-galactose

Answers:

(a)

α-D-mannose

(b)

β-D-galactose

(c)

α-D-glucose

(d)

α-D-galactose

26.12   Draw the Haworth projection formula of the pyranose form of each of the following compounds
(a) α-D-fructose   (b) β-D-fructose   (c) α-D-ribulose   (d) β-D-xylulose

Answers:

(a)

α-D-fructose

(b)

β-D-fructose

(c)

α-D-ribulose

(d)

β-D-xylulose

26.13   Identify the monosaccharide represented by each of the following structures. Name each compound.

Answers: (a) α-D-galactopyranose   (b) β-D-lyxopyranose   (c) α-D-ribopyranose

26.14   Identify the monosaccharide represented by each of the following structures. Name each compound.

Answers: (a) α-D-xylulofuranose   (b) α-D-fructopyranose   (c) β-D-ribulofuranose

Conformations of Monosaccharides

26.15   Draw the chair conformation of β-galactopyranose and β-mannopyranose and compare the number of axial hydroxyl groups in each compound.

Answer: The C-4 hydroxyl group of β-galactopyranose is axial. The C-2 hydroxyl group of β-mannopyranose is axial. They have the same number of axial hydroxyl groups.

β-D-galactopyranose   β-D-mannopyranose

26.16   Draw the standard chair conformation of β-talopyranose and β-allopyranose and compare the number of axial hydroxyl groups in each compound.

Answer: The C-2 and C-4 hydroxyl groups of β-talopyranose are axial. The C-3 hydroxyl group of β-allopyranose is axial. Thus, β-talopyranose has two axial hydroxyl groups and β-allopyranose has only one.

β-D-talopyranose   β-D-allopyranose

26.17   Identify each of the following monosaccharides.

(a)   (b)   (c)

Answer: The rings of both (b) and (c) must be flipped into an alternate chair conformation to give the standard representation of the pyranose ring.

(a) β-D-6-deoxygalactose   (b) α-D-arabinose   (c) β-D-glucose

26.18   Identify each of the following monosaccharides.

(a)   (b)   (c)

Answer: (a) This compound is β-D-lyxose. Flip ring (c) into an alternate chair conformation to give the standard representation of the pyranose ring. In either conformation, the hydroxyl groups on C-2, C-3, and C-4 are up, down, and up, respectively, which corresponds to left, right, left, respectively, in the Fischer projection. The compound is the mirror image of D-xylose, so it is L-xylose. Compound (c) has a methyl group at C-6; therefore, it is a 6-deoxypyranose.

(a)   β-D-lyxose   (b) β-D-6-deogallose   (c) α-L-xylose

26.19   Write the conventional chair conformation of α-D-idose in which the oxygen atom is at the upper right hand corner of the pyranose ring. Convert it to an alternate conformation by a chair–chair interconversion. Determine which conformation is more stable.

Answer: All hydroxyl groups of α-idopyranose, shown in the structure on the left, are axial. A chair–chair interconversion places all four hydroxyl groups in equatorial positions and the —CH₂OH group in an axial position in the structure on the right. The 1,3-diaxial interaction of two hydroxyl groups on each side of the ring is larger than the 1,3-diaxial interaction of the axial —CH₂OH group with two hydrogen atoms. Therefore, the conformation in which all hydroxyl groups are equatorial is more stable.

two sets of unfavorble   more stable conformation
1,3 diaxial interactions

26.20   Identify and name the following aldohexose.

α-D-glulose

Answer: The ring must be flipped into an alternate chair conformation to give the standard representation of the pyranose ring. The compound is α-D-gulose.

552

Mutarotation

26.21   Which of the following compounds can mutarotate?

Answer: Compound (a) is a hemiacetal and (b) is a hemiketal. Therefore, they can mutarotate. Compound (c) is a ketal and cannot mutarotate.

26.22   Which of the following compounds can mutarotate?

Answer: Compound (a), which has a hemiacetal center on the *right* ring, can mutarotate. Compound (b) has acetal centers in both rings and cannot mutarotate.

26.23   The $[\alpha]_D$ values of the α and β anomers of D-galactose are +150.7° and +52.8°, respectively. In water, mutarotation of D-galactose results in a specific rotation of +80.2°. Which anomer predominates?

Answer: The specific rotation, which is a weighted average of the rotation the two anomers, is closest to that of the β anomer. Thus, the β anomer predominates.

26.24   The $[\alpha]_D$ values of the α and β anomers of D-mannose are +20.3° and −17.0°, respectively. In water, mutarotation of D-mannose results in a specific rotation of +14.2. Disregarding the furanose forms present (less than 1%), calculate the percent of the α anomer.

Answer: The specific rotation, which is a weighted average of the rotation the two anomers, is closest to that of the a anomer. Thus, the a anomer predominates. Let x equal the mole fraction of the α anomer. Then, the observed rotation is set equal to the specific rotations of each anomer multiplied by their respective mole fractions. The percent of the α anomer is 83.6%.

$$(20.3)\, x + (-17.0)\, (1 - x) = 14.2$$
$$37.3\, x = 31.2$$
$$x = 0.836$$

The α anomer accounts for 83.6% of anomeric mixture.

26.25   In solution, D-ribose forms an equilibrium mixture containing 6% α-furanose, 18% β-furanose, 20% α-pyranose, and 56% β-pyranose. Explain why β-pyranose forms predominates at equilibrium.

Answer: Disregarding the substituents, the pyranose ring is more stable than a furanose ring. There is torsional strain in a five-membered ring and none in the six-membered pyranose. In ribose, the C-2 and C-4 hydroxyl groups are equatorial in the pyranose. In the β anomer, the C-1 hydroxyl group is also equatorial, but in the α anomer, it would be axial. Thus, the β anomer, with three equatorial hydroxyl groups, predominates over the α anomer, with two.

26.26 Suggest a season why D-glucose, D-mannose, D-galactose, and D-aldose all have larger percentages of the pyranose form than the other four diastereomeric aldohexoses.

Answer: Disregarding the substituents, the pyranose ring is more stable than a furanose ring. There is torsional strain in a five-membered ring, and none in the six-membered pyranose. In glucose, the C-2, C-3, and C-4 hydroxyl groups and the C-6—CH_2OH is also equatorial. In mannose, galactose, and allose, there are two equatorial hydroxyl groups; the C-6—CH_2OH group is also equatorial. The other diastereomers would have a majority of hydroxyl groups in axial positions in the standard representation. Thus, the higher energies of these conformations become comparable to that of the furanose form.

β-D-gluopyranose
(all substituents equatorial)

β-D-mannopyranose
(C-2 epimer of glucose)

β-D-allopyranose
(C-3 epimer of glucose)

β-D-galactopyranose
(C-3 epimer of glucose)

Reduction of Monosaccharides

26.27 Draw the Fischer projections of the alditols of D-erythrose and D-threose. One compound is optically active, and the other is a *meso* compound. Explain why.

Answer: Reduction with sodium borohydride converts the —CHO group of D-erythrose and D-threose into a —CH_2OH group. In the product from D-erythrose, the two hydroxyl groups are on the same side in the Fischer projection formula, and on opposite sides of a plane of symmetry placed perpendicular to the formula and bisecting the bond between C-2 and C-3. Therefore, the alditol shown on the left is a *meso* compound. In the product from D-threose, there is no plane of symmetry, so the alditol shown on the right is optically active.

plane of symmetry, a
meso compound

chiral, stereogenic centers
at C-2 and C-3.

26.28   Which of the alditols of the D-pentoses are optically inactive? Explain why.

Answer: Conversion of the —CHO group into a —CH$_2$OH group in an aldopentose makes the C-1 and C-5 atoms equivalent. The potential plane of symmetry to be considered bisects C-3 and its substituents. Only ribose and xylose have the C-2 and C-4 hydroxyl groups arranged in a mirror image relationship as required for a meso compound, so their alditols will be optically inactive.

```
      CH2OH              CH2OH              CH2OH              CH2OH
H ——— OH        HO ——— H          H ——— OH          HO ——— H
H ——— OH        H ——— OH          HO ——— H          HO ——— H
H ——— OH        H ——— OH          H ——— OH          H ——— OH
      CH2OH              CH2OH              CH2OH              CH2OH
      ribitol          D-arabinitol         xylitol           D-lyxitol
(optically inactive)                   (optically inactive)
```

26.29   Reduction of D-fructose with sodium borohydride yields a mixture of two alditols. Explain why. Name the two alditols.

Answer: Reduction of the carbonyl group gives a mixture of diastereomers that differ at C-2. One of these isomers has the same configuration as the alditol of glucose, glucitol. The other is the same as the alditol of mannose, mannitol.

```
      CH2OH                           CH2OH              CH2OH
      C=O                       H —— C —— OH        HO —— C —— H
HO —— C —— H      NaBH4        HO —— C —— H        HO —— C —— H
H —— C —— OH      ———————>     H —— C —— OH    +   H —— C —— OH
H —— C —— OH                   H —— C —— OH        H —— C —— OH
      CH2OH                           CH2OH              CH2OH
    D-fructose                     D-glucitol          D-mannitol
```

26.30   Reduction of D-tagatose with sodium borohydride yields a mixture of galactitol and talitol. What is the structure of D-tagatose?

Answer: Galactitol and talitol differ only in the configuration at C-2, so it must be the carbonyl carbon atom of tagatose. The configuration of the remaining stereogenic centers of tagatose are the same as in galactose and talose.

```
      CH2OH                           CH2OH              CH2OH
      C=O                       H —— C —— OH        HO —— C —— H
HO —— C —— H      NaBH4        HO —— C —— H        HO —— C —— H
HO —— C —— H      ———————>     HO —— C —— H    +   H —— C —— OH
H —— C —— OH                   H —— C —— OH        H —— C —— OH
      CH2OH                           CH2OH              CH2OH
    D-tagatose                      galactitol          D-talitol
```

26.31   What relationship exists between the reduction products of D-galactose and L-galactose?

Answer: Both enantiomers give the same *meso* alditol, galactitol.

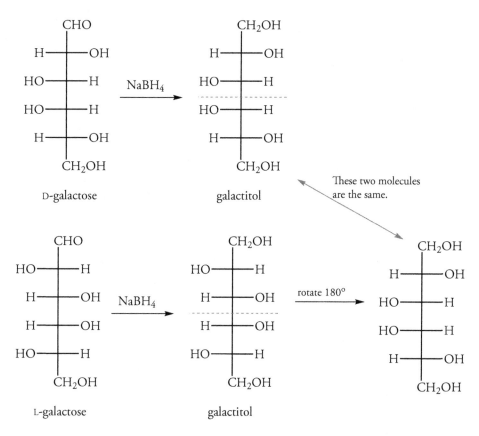

26.32   Explain why the alditol of D-glucose is identical to the alditol of L-gulose.

Answer: The same product is obtained from both compounds. To show this relationship, rotate the structure of D-gulitol by 180° and compare it to the structure of D-glucitol. They are the same.

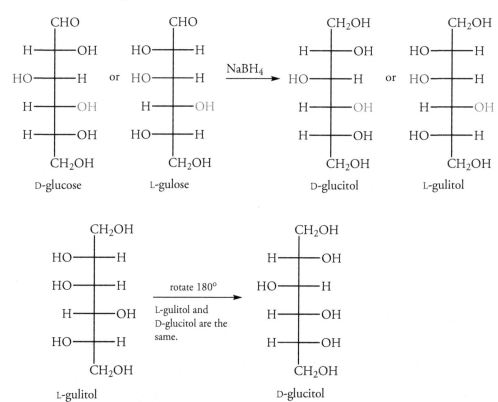

26.33   Draw the structures of one aldose and one ketose that can exist in equilibrium with D-ribose in basic solution.

Answer:

```
     CHO                    CH2OH                  CHO
 H——|——OH                 ═══O                HO——|——H
 H——|——OH      ⇌      H——|——OH     ⇌      H——|——OH
 H——|——OH             H——|——OH             H——|——OH
    CH2OH                  CH2OH                  CH2OH
   D-ribose               D-ribulose             D-arabinose
```

26.34   Draw the structures of two aldoses that can exist in equilibrium with D-xylulose in basic solution.

Answer:

```
     CHO                    CH2OH                  CHO
  H——|——OH                 ═══O                HO——|——H
 HO——|——H       ⇌      HO——|——H     ⇌      H——|——OH
  H——|——OH             H——|——OH             H——|——OH
    CH2OH                  CH2OH                  CH2OH
   D-xylose               D-xylulose             D-arabinose
```

26.35   Explain why an equilibrium mixture of dihydroxyacetone phosphate and D-glyceraldehyde 3-phosphate contains the two compounds in a 96:4 ratio.

Answer: In Chapter 17, we saw that the carbonyl groups of ketones are generally more stable than those of aldehydes as the result of inductive effects of the two alkyl groups compared to a hydrogen atom and an alkyl group of an aldehyde. Thus, dihydroxyacetone phosphate is more stable than D-glyceraldehyde-3-phosphate.

```
        CHO                      CH2OH
         |                        |
  H——C——OH                      C═O
         |                        |
      CH2OPO3^{2-}             CH2OPO3^{2-}
 D-glyceraldehyde-3-phosphate   dihydroxyacetone-phosphate
```

26.36   Although ketones are more stable than isomeric aldehydes by approximately 12 kJ mole^{-1}, fructose 6-phosphate is less stable than glucose 6-phosphate by approximately 1.7 kJ mole^{-1}. Explain why.

Answer: Glucose-6-phosphate and fructose-6-phosphate do not exist in solution in open chain form, but as cyclic hemiacetal and hemiketal derivatives. Thus, the relative stabilities reflect other differences between the cyclic derivatives, such as ring structure and steric interactions.

Glycosides

26.37 Draw the Haworth projection formulas of the two glycosides derived from each of the following pairs of components.
  (a) the pyranose form of D-glucose and ethanol
  (b) the furanose form of D-fructose and phenol
  (c) the pyranose form of D-ribose and methanol
  (d) the furanose form of D-arabinose and benzyl alcohol

Answers:

(a)

ethyl α-D-glucopyranoside ethyl β-D-glucopyranoside

(b)

phenyl α-D-fructofuranoside phenyl β-D-fructofuranoside

(c)

methyl α-D-ribopyranoside methyl β-D-ribopyranoside

(d)

benzyl α-D-arabinofuranoside benzyl β-D-arabinofuranoside

26.38 The individual, isomeric methyl acetals of D-glucose can be prepared only by a series of special reactions. Explain why each compound cannot be prepared by direct reaction of D-glucose with methanol.

Answer: The reaction proceeds by an oxocarbocation in which C-1, the anomeric carbon, is achiral (Section 26.7). Thus, a mixture of diastereomers must form when D-glucose reacts directly with methanol.

558

26.39 Linamarin is found in manioc, a yam found in Brazil. Explain why this compound is toxic.

linamarin

Answer: The aglycone of linamarin is the cyanohydrin of acetone. The cyanohydrin is unstable, and its hydrolysis reverses releases HCN, which is poisonous.

26.40 Peonin, a red pigment found in the red peony, has the following structure. What are the monosaccharide products of the hydrolysis of peonin. Draw the structure of the aglycone. Explain why knowing the structure of the aglycone is not sufficient to determine the structure of peonin.

peonin

Answer: Both monosaccharide units are D-glucopyranose. The aglycone has three hydroxyl groups, so three isomeric diacetal derivatives could give this compound.

26.41 Vanillin is found as the β-glycoside of D-glucose. Draw the structure of the glycoside.

vanillin

Answer: The structure of the β-glycoside of D-glucopyranose and vanillin is shown below.

26.42 Arbutin, an antibiotic used for urinary tract infections, is methylated to give a pentamethyl derivative. The derivative is hydrolyzed to give a tetramethylglucose and *p*-methoxyphenol. What is the aglycone of arbutin?

Answer: There are only four methyl groups in the carbohydrate portion of the antibiotic. The fifth methyl group had to be introduced in the aglycone portion of the molecule. Because hydrolysis gives β-methoxyphenol, the aglycone portion must be hydroquinone, whose chemistry we discussed in Chapter 23.

arbutin

26.43 Salicin is found in the bark of several species of fruit trees. Upon hydrolysis, it yields glucose and 2-(hydroxymethyl)phenol. The mild oxidation of salicin followed by hydrolysis of the oxidation product yields glucuronic acid and salicylic acid (2-hydroxymethyl benzoic acid). What are the possible structures of salicin?

Answer: The hydrolysis products identify the carbohydrate part of salicin as glucose and the aglycone as 2-(hydroxymethyl)phenol. There are two hydroxyl groups in the aglycone that could be bonded to the acetal center of glucose in salicin. The oxidation reaction converts the —CH_2OH of the carbohydrate and the —CH_2OH of the aglycone. Because 2-hydroxybenzoic acid is obtained, the acetal must have had a bond between C-1 of glucose and the phenolic oxygen atom.

salicin

26.44 Phlorizin is a glycoside found in the root bark of certain fruit trees. Hydrolysis of phlorizin yields the following phenolic material. How many possible structures are possible for the glycoside? Explain how methylation of the glycoside using dimethyl sulfate followed by hydrolysis of the product with dilute acid can establish the structure of the glycoside.

Answer: Two of the four hydroxyl groups in the aglycone are equivalent. They are *ortho* to the carbonyl group. Thus, the glycoside could be any of three possible isomers. Methylation of the glycoside followed by hydrolysis would give a trimethyl ether of the aglycone. Determination of its structure would indicate the position of the free hydroxyl group, which must have formed the acetal linkage of the glycoside.

26.45   Suggest a reason why methyl a-D-glucofuranoside (I) is more slowly hydrolyzed than methyl a-D-fructofuranoside (II) at the same pH.

Answer: Both compounds give an oxocarbocation in the hydrolysis reaction. The anomeric carbon atom bearing the charge in compound I (an acetal) has a hydrogen atom bonded to it, whereas the related carbon atom in compound II (a ketal) has a carbon atom bonded. Thus, the carbocation of compound II is more stable and the transition state leading to its formation is lower in energy than that for compound I.

26.46   Suggest a reason why methyl β-D-2-deoxyglucopyranoside (1) is hydrolyzed faster than methyl β-D-glucopyranoside (II) at the same pH.

Answer: Both compounds give an oxocarbocation in the hydrolysis reaction. The anomeric carbon atom bearing the charge in compound I (an acetal) has a hydrogen atom bonded to it, whereas the related carbon atom in compound II (a ketal) has a carbon atom bonded. Thus, the carbocation of compound II is more stable and the transition state leading to its formation is lower in energy than that for compound I.

26.47   The carbohydrate daunosamine is contained in the antibiotic Adriamycin. Is daunosamine a D or L carbohydrate?

daunosamine

Answer: Daunosamine is an L carbohydrate because the C-6 methyl group is "down." A ring flip to give the standard chair representation would place the C-6 methyl group in an axial position, which corresponds to the L configuration. We recall that D carbohydrates always have a —CH₂OH group "up" in the Haworth projection, and that it is equatorial in the standard chair representation.

26.48   The melting point of methyl α-D-glucopyranoside is 166 °C. Explain why the melting point of methyl β-D-glucopyranoside, 105 °C, is different. Predict the melting point of methyl β-L-glucopyranoside.

Answer: The α and β anomers are diastereomers. Since they differ in configuration at one stereogenic center (C-1), they have different physical properties. Methyl β-L-glucopyranoside is the enantiomer of methyl β-D-glucopyranoside. Enantiomers has the same physical properties, so their melting points are the same, 105 °C.

Disaccharides

26.49 Identify the component monosaccharides of each of the following compounds and describe the type of glycosidic linkage in each.

Answers:
(a) A β linkage from C-1 of D-2-deoxyribose with a β linkage from C-1 of L-glucose
(b) A β linkage from C-1 of D-glucose to C-3 of D-galactose.
(c) A β linkage from C-1 of D-galactose to C-4 of D-mannose

26.50 Identify the component monosaccharides of each of the following compounds and describe the type of glycosidic linkage in each.

Answers:
(a) There is no glycosidic linkage. The compound is an ether formed between C-2 of D-xylose and the C-6 atom of D-galactose.
(b) An α linkage from C-1 of D-glucose to C-1 of D-ribose
(c) A β linkage from C-1 of D-fructose to C-3 of D-glucose

Structure Determination of Monosaccharides

26.51 What are the products of the periodate oxidation of each of the following monosaccharides?
    (a) ribose   (b) ribulose   (c) galactose   (d) erythrulose
Answers:
(a) Periodate oxidation of ribose gives one mole of formaldehyde and four moles of formic acid.
(b) Periodate oxidation of ribulose gives two moles of formaldehyde, two moles of formic acid, and one mole of CO_2.
(c) Periodate oxidation of galactose gives one mole of formaldehyde and five moles of formic acid.
(d) Periodate oxidation of erythrulose gives two moles of formaldehyde, one mole of formic acid, and one mole of CO_2.

26.52 What are the products of the periodate oxidation of each of the following monosaccharides?
    (a) xylose   (b) sorbose   (c) erythrose   (d) idose
Answers:
(a) Periodate oxidation of xylose gives one mole of formaldehyde and four moles of formic acid.
(b) Periodate oxidation of sorbose gives two moles of formaldehyde, three moles of formic acid, and one mole of CO_2.
(c) Periodate oxidation of erythrose gives one mole of formaldehyde and three moles of formic acid.
(d) Periodate oxidation of idose gives one mole of formaldehyde and five moles of formic acid.

26.53   What are the products of the periodate oxidation of each of the following monosaccharides?

(a)
```
        CH2OH
         |
   HO—C—H
         |
   (a)  C=O
         |
    H—C—OH
         |
        CH2OH
```

(b)
```
        CHO
         |
   H—C—CH2OH
         |
  HO—C—H
         |
   H—C—OH
         |
        CH2OH
```

(c)
```
        CHO
         |
    H—C—H
         |
    H—C—OH
         |
  HO—C—H
         |
        CH3
```

Answers:

(a) Periodate oxidation of compound (a) gives two moles of formaldehyde, two moles of formic acid, and one mole of CO_2.

(b) Periodate oxidation of compound (b) gives one mole of formaldehyde, one mole of formic acid, and the following dialdehyde, which containing the original C-1, C-2, and C-3 atoms.

```
        CHO
         |
   H——————CH2OH
         |
        CHO
```

(c) Periodate oxidation of compound (c) gives acetaldehyde from the original C-4 and C-5 atoms, and the following dialdehyde containing the original C-1, C-2, and C-3 atoms.

```
        CHO
         |
   H——————H
         |
        CHO
```

26.54   What are the products of the periodate oxidation of each of the following monosaccharides?

(a)
```
        CH2OH
         |
   HO—C—H
         |
   (a)  C=O
         |
    H—C—OH
         |
        CH3
```

(b)
```
        CHO
         |
  HO—C—CH3
         |
  HO—C—H
         |
   H—C—OH
         |
        CH2OH
```

(c)
```
        CHO
         |
   H—C—OH
         |
   H—C—OH
         |
  HO—C—CH2OH
         |
        CH3
```

Answers:

(a) Periodate oxidation of compound (a) gives one mole of formaldehyde, one mole of formic acid, one mole of CO_2, and one mole of acetaldehyde from the original C-4 and C-5 atoms.

(b) Periodate oxidation of compound (b) gives one mole of formaldehyde, three moles of formic acid, and one mole of acetic acid.

(c) Periodate oxidation of compound (c) gives one mole of formaldehyde, three moles of formic acid, and one mole of acetic acid.

26.55   There are eight diastereomeric D-aldoses, but they yield only four diastereomeric osazones? Explain why.

Answers:

Formation of the osazone decreases the number of stereogenic centers from four to three and cuts the number of possible stereoisomers in half. Each of the following pairs of D-aldoses are C-2 epimers: D-Allose and D-altrose form the same osazone, D-glucose and D-mannose form the same osazone, D-gulose and D-idose form the same osazone, D-galactose and D-talose form the same osazone.

26.56 Draw the structure of the product of the reaction of 2-deoxy-D-ribose with phenylhydrazine.

Answer: A simple phenylhydrazone derivative forms because there is no α-hydroxy group.

CHO
H—C—H
H—C—OH
H—C—OH
CH₂OH

2-deoxy-D-ribose

+ NHNH₂ (phenyl) ⟶

CH=NHNH—(phenyl)
H—C—H
H—C—OH
H—C—OH
CH₂OH

26.57 A Wohl degradation is done on a monomethyl ether of D-idose. The product, when oxidized with nitric acid, gives an optically inactive compound. At what position of idose is the methyl ether located?

Answer: The Wohl degradation of D-idose gives D-xylose. Oxidation of D-xylose gives an optically inactive aldaric acid. The position of the methyl ether in the oxidation product described must be at C-3, which is where the plane of symmetry would be located. If the methyl ether were at C-2 or C-4, the oxidation product would be chiral. The methyl ether of the original idose was at C-4.

CHO
HO—C—H
H—C—OH
CH₃O—C—H
H—C—OH
CH₂OH

3-O-methyl-D-idose

⟶

CHO
H—C—OH
CH₃O—C—H
H—C—OH
CH₂OH

3-O-methyl-D-xylose

⟶

CO₂H
H—C—OH
·CH₃O—C—H·····
H—C—OH
CO₂H

3-O-methyl-D-xylaric acid
(optically inactive)

26.58 A Kiliani–Fischer chain extension is done on a monomethyl ether of D-glucose. One of the products, when oxidized with nitric acid, gives an optically inactive compound. At what position of glucose is the methyl ether located?

Answer: The position of the methyl ether in the chain-extended oxidation product described must be at C-4, which is where the plane of symmetry would be located. Any other position for the methyl ether would lead to an optically active product. The methyl ether of the original glucose was at C-3.

CHO
H—C—OH
CH₃O—C—H
H—C—OH
H—C—OH
CH₂OH

3-O-methyl-D-glucose

⟶

CHO
H—C—OH
H—C—OH
CH₃O—C—H
H—C—OH
H—C—OH
CH₂OH

⟶

CO₂H
H—C—OH
H—C—OH
··CH₃O—C—H····
H—C—OH
H—C—OH
CO₂H

(optically inactive)

26.59 An aldohexose is methylated using dimethyl sulfate and then treated with a mild acid. The resulting product, when subjected to a strong oxidizing agent, gives an optically inactive dicarboxylic acid containing five carbon atoms. What are the possible structures of the aldohexose?

Answer: The aldohexose exists in a pyranose form, so methylation occurs at the hydroxyl groups at C-1, C-2, C-3, C-4, and C-6. Treatment with acid hydrolyzes the acetal, restoring the —OH group at C-1. After methylation and hydrolysis, the "open chain" isomer in equilibrium with the methylated pyranose is oxidized at C-1, giving a carboxylic acid, and at C-5 giving a ketone. Vigorous oxidation cleaves the C-5 to C-6 bond, and the product is a dicarboxylic acid. To give an optically inactive product, the methylated sites at C-2, C-3, and C-4 have to be symmetrically arranged about a plane through C-3. The aldohexoses that have related symmetrical arrangements are allose, talose, glucose, and idose. The reaction for allose is given as an example.

1,2,3,4,6-penta-O-methyl
β-D-allopyranoside

2,3,4,6-tetra-O-methyl-
β-D-allopyranose

2,3,4,6-tetra-O-methyl-D-allose

2,3,4,6-tetra-O-methyl-D-allose   2,3,4-tri-O-methyl-D-xylose   2,3,4-tri-O-methylxylaric acid (optically inactive)

26.60 An aldohexose is converted to a methyl glycoside and then is oxidized by periodate to yield the following product. Is the aldohexose a pyranose or a furanose? What is the configuration of the anomeric center of the glycoside? What are the possible structures of the aldohexose?

Answer: The aldohexose is in the furanose form as evidenced by the aldehyde group on the left side of the molecule. This C-5 atom was originally bonded to C-6 in a —CH₂OH group. Oxidation by periodate released the C-6 atom as formaldehyde, leaving C-5 in an aldehyde group. The acetal center is located on the right side of the molecule as evidenced by the methoxy group, and it has the α configuration. The ring oxygen atom of the glycoside is derived from the hydroxyl group at C-4 of the aldohexose and is located on the right in the Fischer projection formula. If the aldohexose has the D configuration, then allose, altrose, glucose, and mannose are possible compounds.

565

Structure Determination of Disaccharides

26.61   Hydrolysis of the disaccharide primeverose yields D-xylose and D-glucose. Methylation of primeverose using dimethyl sulfate followed by mild acid hydrolysis yields 2,3,4-tri-*O*-methyl-D-xylose and 2,3,4-tri-*O*-methyl-D-glucose. What features of the structure are determined by these data?

Answer: The location of the three methoxy groups in the xylose derivative means that the C-5 oxygen atom was part of a pyranose ring, and the compound was linked as a glycoside at C-1. The location of the three methoxy groups in the glucose derivative means that the C-5 oxygen atom was part of a pyranose ring. The compound must have been linked to the glycosidic center of the xylose by the C-6 oxygen atom. The compound is 6-*O*-β-D-xylopyranosyl-β-D-glucopyranose, otherwise known as primeverose.

6-*O*-β-D-xylopyranosyl- β-D-glucopyranose   β-D-xylopyranose   β-D-glucopyranose

primeverose   (CH₃)₂SO₄

HCl (aq)   2,3,4-tri-*O*-methyl-D-xylose   2,3,4-tri-*O*-methyl-D-glucose

26.62   Hydrolysis of the disaccharide trehalose yields only D-glucose. Methylation of trehalose using dimethyl sulfate followed by mild acid hydrolysis yields 2,3,4-tri-*O*-methyl-D-glucose and 2,3,4,6-tetra-*O*-methyl-D-glucose. What features of the structure are determined by these data?

Answer: The location of the four methoxy groups in one glucose derivative means that the C-5 oxygen atom was part of a pyranose ring, and the compound was linked as a glycoside at its C-1 atom. The location of the three methoxy groups in the other glucose derivative means that the C-5 oxygen atom was part of a pyranose ring. This compound must have been linked to the glycosidic center of the first glucose by the C-6 oxygen atom.

HCl (aq)   D-glucopyranose

6-*O*-α-D-glucopyranosyl β-D-glucopyranose

(CH₃)₂SO₄ → [reaction of tetrol disaccharide to permethylated disaccharide]

$(CH_3)_2SO_4$

HCl (aq) →

2,3,4,6-tetra-*O*-methyl-D-glycose + 2,3,4-tri-*O*-methyl D-glucose

26.63   Hydrolysis of the disaccharide turanose yields D-fructose and D-glucose. Methylation of turanose using dimethyl sulfate followed by mild acid hydrolysis yields 1,4,5-tri-*O*-methyl-D-fructose and 2,3,4,6-tetra-*O*-methyl-D-glucose. What features of the structure are determined by these data?

Answer: The location of the four methoxy groups in one glucose derivative means that the C-5 oxygen atom was part of a pyranose ring, and the compound was linked as a glycoside at C-1. The location of the three methoxy groups in the other glucose derivative means that the C-5 oxygen atom was part of a pyranose ring. This compound must have been linked to the glycosidic center of the first glucose by the C-3 oxygen atom. Turanose is α-D-glucopyranosyl-(1→3)-α-D-fructofuranose.

26.64   Hydrolysis of a trisaccharide yields two equivalents of glucose and one equivalent of galactose. Methylation, using dimethyl sulfate followed by mild acid hydrolysis, yields 3,6-di-*O*-methyl-D-glucose as one of the products. What features of the structure are determined by these data?

Answer: The location of the two methoxy groups in the glucose derivative could mean that either the C-4 or the C-5 oxygen atom was part of a furanose or pyranose ring, respectively. If this portion of the structure is a furanose, then the C-2 and C-5 oxygen atoms were part of glycosidic linkages to the other monosaccharide units. If this portion of the structure is a pyranose, then the C-2 and C-4 oxygen atoms were part of glycosidic linkages to the other monosaccharide units. For each possible case, one oxygen atom must be linked to the acetal center of another glucose molecule and the second to the acetal center of a galactose molecule.

27

Amino Acids, Peptides, and Proteins

KEYS TO THE CHAPTER

This chapter illustrates the bridge between organic chemistry and biochemistry. We can understand the reactions of amino acids, peptides, and proteins based on the chemistry of the amide functional groups and the component amines and carboxylic acids.

27.1 Structures of α-Amino Acids

Amino acids contain both an amino group and a carboxylic acid group. A group of 20 amino acids with the amino group at the α carbon atom is the subject of this chapter. With the exception of glycine, which is not chiral, these amino acids have an *S* configuration and are designated as L using the Fischer configurational system. (Priority rules for the *R/S* nomenclature dictate that cysteine has an *R* configuration.) It is an L amino acid in the Fischer configurational system.

$$
\begin{array}{c}
\text{CO}_2\text{H} \\
| \\
\text{NH}_3\text{—C}_\alpha\text{—H} \\
| \\
\text{R}
\end{array}
$$

L-α-amino acid

Amino acids are classified according to their side chains. If the side chain does not contain an acidic or basic functional group, it is neutral, although it may contain polar or nonpolar side chains. An amino acid that has a basic functional group in its side chain is a basic amino acid; an amino acid that has an acidic functional group in its side chain is an acidic amino acid.

27.2 Acid–Base Properties of α-Amino Acids

Near a neutral pH, a neutral amino acid exists in an electrically neutral form. However, the amino acid has a negatively charged carboxylate group and a positively charged amino group (NH_3^+). The dipolar ion is called a **zwitterion**.

$$
\begin{array}{c}
\text{CO}_2^- \\
| \\
^+\text{NH}_3\text{—C—H} \\
| \\
\text{CH}_3
\end{array}
$$

dipolar ion or zwitterion
at neutral pH

If base is added to a neutral solution, it removes a proton from the ammonium group and a free amino group results. There is still a charge at the carboxylate group, so the amino acid has a negative charge.

$$
\begin{array}{c}
\text{CO}_2^- \\
| \\
\text{NH}_2\text{—C—H} \\
| \\
\text{CH}_3
\end{array}
$$

basic pH, negatively charged

If acid is added to a neutral solution, the carboxylate group is protonate, and a carboxylic acid group results. There is still a charge at the ammonium group, so the amino acid has a positive charge.

$$
\begin{array}{c}
\text{CO}_2\text{H} \\
| \\
^+\text{NH}_3\text{—C—H} \\
| \\
\text{CH}_3
\end{array}
$$

acidic pH, positively charged

Organic Chemistry Study Guide: Key Concepts, Problems, and Solutions http://dx.doi.org/10.1016/B978-0-12-801889-7.00027-3

27.3 Isoionic Point and Titration of α-Amino Acids

A zwitterion has a negative carboxylate ion and ammonium ion. The pH at which the concentration of the zwitterion predominates is the **isoionic point**. For neutral amino acids, the isoionic point lies halfway between the pK_a values of the carboxylic acid group and the ammonium group. See Table 27.1.

For neutral amino acids, the isoionic point is near 7. Acidic amino acids have isoionic points smaller than 7. For acidic amino acids, the side chain carboxylic acid group of aspartate and glutamic acid exist as a carboxylate ion at a pH near 7. To protonate that carboxylate ion, the solution must be acidic. Basic amino acids have isoionic points larger than 7. For basic amino acids, the basic functional group in the side chain tends to be protonated at a pH near neutrality. Thus, there are two ammonium ions and only one carboxylate ion. In order to remove a proton from the ammonium ion, solution must be basic. Table 27.2 list the isoionic points of the amino acids.

27.4 Synthesis of α-Amino Acids

Amino acids can be prepared by the nucleophilic substitution of a halide ion of α-halocarboxylic acids by ammonia. The α-halocarboxylic acids are prepared by the Hell–Volhard–Zelinsky reaction.

The Strecker synthesis occurs by the addition of ammonia to an aldehyde to give an imine which then adds HCN across the carbon–nitrogen double bond to give an α-amino nitrile. The nitrile is then hydrolyzed in a second step to give the α-amino carboxylic acid.

Reductive amination of an α-keto acid using ammonia yields an α-imino acid which is then reduced by hydrogen and a palladium catalyst. The entire reductive amination is carried out in a single step with all reagents present.

The acetamidomalonate synthesis uses a malonate ester-type synthesis. Diethyl acetamidomalonate has an acidic α hydrogen atom. Treating the compound with base converts it to a carbanion. The carbanion displaces a halide ion from a haloalkane to form an alkylated product. This alkyl group corresponds to the side chain of the amino acid. Acid-catalyzed hydrolysis cleaves the acetyl group of the amide and also hydrolyzes the ester of the malonate. Under the reaction conditions, the malonic acid undergoes decarboxylation, yielding an α amino acid.

27.5 Chiral Synthesis α-Amino Acids

We recall that a chiral ruthenium catalyst with binaphthyl ligands can be employed to carry out chiral hydrogenation reactions (Section 17.8). Many chiral ligands have been developed. One that is widely used for the chiral synthesis of amino acids is called **degphos**. It can be prepared with either an (R,R) or an (S,S) configuration. Since the ligand is chiral, the transition states leading to either an R or S amino acid are diastereomers. Therefore, there energies differ, and the rate of formation of one enantiomer is favored over the other. As a result, chiral α-amino acids are readily synthesized with high enantiomeric purity.

27.6 Reactions of α-Amino Acids

The individual functional groups of amino acids each have their own characteristic reactivity. Carboxylic acids are converted to esters by reaction of an alcohol under acid-catalyzed conditions. Ethyl and benzyl esters are commonly used for this reactions, which protects the carboxyl group so that additional reactions can be carried out at other functional groups. Acid-catalyzed hydrolysis readily reverses the process to deprotect the carboxylic acid. Benzyl esters are cleaved under neutral conditions using hydrogen and a palladium catalyst. The process, called hydrogenolysis, gives toluene as the by-product

The α-amino group of an amino acid can be converted to an amide. This reaction protects the amino group so that other functional groups can be modified. For example, benzyl chloroformate, which is similar to an acyl chloride in its reactivity, converts an α-amino group into a benzyloxycarbonyl (**Cbz**) derivative. Di-*tert*-butyldicarbonate is an anhydride that is also used to protect α-amino groups by converting them into a *tert*-butoxycarbonyl (**Boc**) derivative. Either derivative is easily hydrolyzed using trifluoroacetic acid to give a carbamic acid which decarboxylates to give the free α-amino group. The Cbz derivative also reacts with hydrogen and palladium in a hydrogenolysis reaction that gives a carbamic acid which decarboxylates.

27.7 Peptides

The combination of two amino acids joined by an amide bond is a **dipeptide**. Larger numbers of amino acids are **tripeptides**, **tetrapeptides**, etc. Peptides have two ends. The ends containing the free α-amino group and the free carboxylic group are the N-terminal and C-terminal amino acids, respectively. Peptides are named from the N-terminal amino acid in sequence toward the C-terminal amino acid. The shorthand three-letter representations of the amino acids are used in the name.

The number of possible isomeric peptides with n different amino acids is equal to $n!$, where $n!$ is equal to the product of all numbers from 1 to n.

Many peptides are hormones that function as cellular signaling molecules. The functions of hormonal peptides depend on their amino acid composition and their sequence. Small changes in either result in functional differences. Thus, a change from a neutral to an acidic or basic amino acid often has a large difference in function. Many peptide hormones act by binding to membrane proteins called guanine-nucleotide receptor proteins (GCRP),

27.8 Peptide Synthesis

Peptide synthesis requires precise amide bond formation between specific amino acids. Because each amino acid is both an amine and a carboxylic acid, direct reaction between two amino acids A and B can form the four dipeptides A–A, B–B, A–B, and B–A. To obtain a desired product such as B–A, it is necessary to protect the amino group of B and the carboxylic acid group of A. In this way, only the carboxylic acid group of B and the amino group of A can react with each other.

Synthesis of higher peptides such as C–B–A is accomplished by removing the protecting group of B in the dipeptide B–A and reacting it with the amino acid C, which is protected at the amino group. Then, the free carboxylic acid group of C can react with the free amino group of B–A, which is protected at the carboxylic acid group of A.

The carboxyl group of an amino acid is protected by forming a benzyl ester. The amino group is protected by forming a Boc derivative. These two protected amino acids are joined in an amide bond using dicyclohexylcarbodiimide (DCCI). Continued extension of the peptide is accomplished by hydrolysis of the Boc group using trifluoroacetic acid. The free amino group is now available for reaction with another amino acid protected at the amino group. In the last step, the polypeptide is deprotected by hydrolysis of both protecting groups using basic hydrolysis.

27.9 Solid-Phase Peptide Synthesis

The efficiency of polypeptide synthesis is improved by attaching the C-terminal amino acid to a solid polymer. As a result, the products of either peptide bonds or the hydrolysis step in the last reaction of a synthetic sequence remain attached to the polymer. The polymer with the growing polypeptide chain is easily handled by physical methods without mechanical loss. The result is a net yield that approaches 100%. The peptide chain is liberated from the polymer by hydrolysis with anhydrous hydrogen fluoride in the very last synthetic step. Solid-phase peptide synthesis is an automated process that is carried out by a peptide synthesizer. Many companies in the biotechnology industry routinely carry out custom peptide synthesis. Some university laboratories also provide this service.

27.10 Determination of the Amino Acid Compositions of Proteins

The composition of peptides or proteins is determined by complete hydrolysis using 6 M HCl. The identity and number of each amino acid is determined by chromatographic methods. The composition is written using subscripts located on the three letter abbreviation for the amino acid separated by commas between each of the component amino acids. This process, too, has been automated. Commercial instruments called amino acid analyzers carry out the necessary reactions.

27.11 Determination of the Amino Acid Sequences of Proteins

The N-terminal amino acid of a peptide is identified by reaction with phenyl isothiocyanate (the Edman reagent) followed by hydrolytic release of a heterocyclic compound called a phenylthiohydantoin. In these compounds, the "R" group of the N-terminal amino acid is attached to a ring carbon atom located between the carbonyl carbon atom and the nitrogen atom of the original amino acid. The amino acid is identified by comparing the phenylthiohydantoin obtained to those of known amino acids. The Edman degradation may be repeated sequentially to identify the amino acids one by one from the N-terminal end. Automation of this process makes it possible to determine the sequence of peptide that contains 50 or so amino acid residues using very small samples.

Polypeptide chains are cleaved at methionine residues by cyanogen bromide to give homoserine lactone products at the carboxyl group of methionine residues. Large polypeptides can also be cleaved by enzymes to give smaller fragments. The products of an enzyme-catalyzed hydrolysis depend on the enzyme chosen and the amino acid sequence in the peptide. Chymotrypsin catalyzes the hydrolysis of peptide bonds at the carboxyl end of the aromatic amino acids. Trypsin catalyzes the hydrolysis of peptide bonds at the carboxyl end of the basic amino acids lysine and arginine. After the structures of the peptide fragments are determined, the manner in which they were originally joined may be determined by aligning the peptides produced by enzymatic cleavage.

27.12 Bonding in Proteins

The covalent bonds in proteins are the peptide bonds that link amino acids and disulfide bonds between cysteine residues that cross-link the polypeptide chain. Intramolecular hydrogen bonds occur between amide groups often give rise to regularly repeating structures called α-helices and β-pleated sheets. These regularly repeating sequences define the secondary structure of the protein. Hydrogen bonds can also occur between polar side chains which have groups that can act as hydrogen bond donors and acceptors. Hydrophobic interactions between nonpolar side chains also stabilize the folded structure of the protein. Acidic and basic amino acids can also interact by forming ion pairs, which are sometimes called "salt bridges."

27.13 Protein Structure

The biological function of a protein is observed only in its native conformation. The entire structure is viewed at various levels:

1. The **primary structure** of a protein is the linear sequence of amino acid residues and the location of the disulfide bonds.
2. The **secondary structure** of a protein consists of regularly repeating sequences called α-helices and β-pleated sheets.
3. The the three-dimensional conformation of the protein chain is called the tertiary structure.
4. Quaternary structure is the association of several protein chains or subunits into a closely packed arrangement.

27.14 Oxygen Storage and Transport: Myoglobin and Hemoglobin

Humans and other vertebrates transport oxygen in red blood cells called **erythrocytes**. Hemoglobin transports oxygen throughout the body; myoglobin stores oxygen in cardiac and skeletal muscle until it is consumed during metabolism.

Most of the amino acid residues in myoglobin are in α-helices. The oxygen binding site in myoglobin is a **heme** group. Heme contains Fe^{2+}, and O_2 is the ligand that binds it.

Hemoglobin has a quaternary structure. It consists of two pairs of different proteins, designated the α and the β chains. Each subunit is linked covalently to a molecule of heme. Thus, hemoglobin binds four O_2 molecules. The subunits of hemoglobin do not act independently. When one subunit binds O_2, its conformation changes. When a change in conformation at one site of an oligomeric protein is caused by a change in a spatially separated site of the oligomer, the change is called an **allosteric** effect, and the protein is called an **allosteric protein**. Hemoglobin is an allosteric protein. When one heme group in hemoglobin binds oxygen, it is easier for successive oxygen molecules to bind at the remaining three sites.

In some people, glutamate 6 of the β chain has undergone a mutation to valine. This mutation changes the charge on the surface of hemoglobin. The mutant protein is called sickle cell hemoglobin, HbS. Valine residue 6 of the β chain of deoxy HbS lies on the surface of the protein. This hydrophobic residue, present in each β chain, forms a hydrophobic contact with a pocket in a neighboring β chain of another hemoglobin molecule. The mutation that replaces a glutamate residue by a valine residue decreases the solubility of deoxy-HbS. Deoxy-HbS precipitates and forms long fibers that impart a sickle shape of erythrocytes. People who have the mutant hemoglobin, HbS, have impaired oxygen transport and suffer from a condition called sickle cell anemia.

Solutions to Exercises

Amino Acids

27.1 A D-glutamic acid residue is found in some bacterial cell walls. Draw its Fischer projection formula.

Answer: Note that the α-amino group and hydrogen have "switched places," converting the L-isomer to its enantiomer. In the Fischer projection formula, the amino group is on the right for an amino acid with the D configuration.

$$CO_2H$$
$$H—C—NH_3^+$$
$$CH_2CH_2CO_2H$$

D-glutamatic acid, conjugate acid

27.2 Gramicidin S is a cyclic peptide antibiotic that contains a D-phenylalanine residue. Draw the projection formula of D-phenylalanine.

Answer: In the Fischer projection formula, the amino group is on the right for an amino acid with the D configuration.

$$CO_2^-$$
$$H—C—NH_3^+$$
$$CH_2$$

D-phenylalanine, dipolar ion

27.3 The following amino acid is present in collagen. From what amino acid is it derived?

$$NH_2—CH_2—CH—CH_2—CH_2—\overset{\overset{\displaystyle H}{|}}{C}—\overset{\overset{\displaystyle O}{\|}}{C}—OH$$
$$\underset{HO}{\quad} \qquad \underset{NH_2}{\quad}$$

Answer: The amino acid is derived from lysine. It is γ-hydroxylysine.

27.4 The following antibacterial agent, called allin, is present in garlic. From what amino acid might it be derived?

$$CH_2=CH—CH_2—\overset{\overset{\displaystyle O}{\|}}{S}—CH_2—\overset{\overset{\displaystyle H}{|}}{C}—\overset{\overset{\displaystyle O}{\|}}{C}—OH$$
$$\underset{NH_2}{\quad}$$

alliin

Answer: The amino acid is derived from cysteine. It has an allyl group in place of the hydrogen of the cysteine S—H group. Also, the sulfur is in an oxidized state known as a sulfoxide.

27.5 The following compound is an amino acid that acts as a neurotransmitter. Classify this amino acid and give its IUPAC and common name. (This neurotransmitter is universally known by its common name.)

$$NH_2—CH_2—CH_2—CH_2—CO_2H$$

Answer: The amino acid is derived biochemically from glutamic acid. It has one carboxylic acid group and one amino group. Thus, it is a neutral amino acid. Its IUPAC name is 4-aminobutanoic acid; its common name is γ-aminobutyric acid. It is commonly known by its abbreviation, GABA.

27.6   The following compound is one of the amino acids formed in the biosynthesis of penicillin. Classify this amino acid and determine its common name.

$$HO-\underset{\underset{O}{\parallel}}{C}-CH_2-CH_2-CH_2-CH_2-\underset{\underset{NH_2}{|}}{\overset{\overset{H}{|}}{C}}-\underset{\underset{O}{\parallel}}{C}-OH$$

Answer: The compound has two carboxylic acid groups and one amino group. Thus, it is an acidic amino acid. The common name of the dicarboxylic acid is adipic acid. The amino group is on an α carbon atom. The common name is α-aminoadipic acid. Biochemically, it is derived from lysine, whose ε-amino group has been replaced by a carboxylic acid group.

Acid–Base Properties of Amino Acids

27.7   Draw the structures of alanine and glutamic acid at pH = 1 and pH = 12.

Answer: At pH 1, both compounds exist as their respective conjugate acids. At pH 12, both compounds exist as their conjugate bases. At pH 12, both carboxyl groups of glutamic acid are carboxylate ions.

alanine   glutamic acid

27.8   Draw the structures for the dipolar ions (zwitterions) of serine and valine.

Answer: The dipolar ions have a negative charge on the carboxylate ion and a positive charge on the nitrogen atom of the ammonium ion.

serine   valine
dipolar ions (zwitterions)

27.9   How could you distinguish between aqueous solutions of asparagine and aspartic acid?

Answer: Aspartic acid is an acidic amino acid, so its solution has a lower pH than a solution of asparagine, which is a neutral amino acid.

aspartic acid   asparagine
acidic amino acid   neutral amino acid

27.10   Would you expect an aqueous solution of lysine at pH 7 to be neutral, acidic, or basic? Explain.

Answer: Lysine is a basic amino acid, so its solutions will be basic.

lysine
basic amino acid

27.11 One of the pK_a values of tyrosine is 9.11. What functional group is responsible for this acidic hydrogen atom?

Answer: The pK_a of the —NH_3^+ group of tyrosine is 9.11. The phenolic hydroxyl group of tyrosine has a pK_a of 10.07. The pK_a of phenol is 10.

$$p K_a \; 9.11 \quad H_3N^+ - \overset{\displaystyle CO_2^-}{\underset{\displaystyle CH_2-}{C}} - H \qquad \text{(phenol ring)} - OH \quad p K_a \; 11.07$$

tyrosine

27.12 One of the pK_a values of cysteine is 8.33. What functional group is responsible for this acidic hydrogen atom?

Answer: The pK_a value of 8.33 is for the —SH group. The pK_a values of thiols are approximately 8.

$$H_3N^+ - \overset{\displaystyle CO_2^-}{\underset{\displaystyle CH_2SH}{C}} - H \quad p K_a \; 8.33$$

cysteine

27.13 Explain why the pK_a for the —NH_3^+ group of tyrosine is slightly smaller than the corresponding pK_a of phenylalanine.

Answer: The decrease in pK_a means that the "group" bonded to the α—NH_3^+ group withdraws electron density inductively. This effect increases the acidity of the α—NH_3^+ group. The hydroxyl group of tyrosine group is inductively more electron withdrawing than a hydrogen atom in the phenyl group of phenylalanine.

$$p K_a \; 9.91 \quad H_3N^+ - \overset{\displaystyle CO_2^-}{\underset{\displaystyle CH_2-}{C}} - H \qquad \text{(ring)} - OH \quad \longrightarrow$$

tyrosine

$$p K_a \; 9.24 \quad H_3N^+ - \overset{\displaystyle CO_2^-}{\underset{\displaystyle CH_2-}{C}} - H \qquad \text{(ring)}$$

phenylalanine

27.14 Explain why the pK_a for the side chain —CO_2H group of aspartic acid (3.92) is smaller than the corresponding pK_a of glutamic acid (4.32).

Answer: The decrease in pK_a means that the "group" bonded to the carboxyl group withdraws electron density inductively. This effect increases the acidity of carboxylic acids. In the case of aspartic acid, the electron withdrawing group is the ammonium group, which is closer to the acidic site than the ammonium group in glutamic acid. The pK_a scale is logarithmic. Thus, a difference of 0.36 pK_a units means that glutamic acid is approximately twice as strong an acid as aspartic acid.

$$H_3N^+ - \overset{\displaystyle CO_2H}{\underset{\displaystyle CH_2CO_2H}{C}} - H \quad p K_a \; 3.96$$

aspartic acid

$$H_3N^+ - \overset{\displaystyle CO_2H}{\underset{\displaystyle CH_2CH_2CO_2H}{C}} - H \quad p K_a \; 4.32$$

glutamic acid

27.15 Explain why the difference between the pK_a values of aspartic acid and asparagine is larger than the difference between the pK_a values of glutamic acid and glutamine.

Answer: The difference in the pK_a values for the α-ammonium and α-carboxyl groups of aspartic acid and the structurally related asparagine reflects the difference in the inductive effect of a carboxylate group and an amide group on the ionization sites. The effect is smaller for glutamic acid compared to glutamine because the groups are further removed from the ionization sites. As the chain between the groups and the reaction site increases, the difference between the effects of the groups decreases until eventually there is no effect.

27.16 Consider the amino and imino nitrogen atoms of the side chain of arginine. Which one would be protonated in acid solution? How can resonance stabilization account for the site of protonation?

Answer: Based on the hybridization of the nitrogen atoms, we would expect the site of protonation to be the amino group because it is sp^3 hybridized. The lone-pair electrons of the imino group are sp^2 hybridized. They are more tightly held to the nitrogen atom because the % s character of the orbital is greater. However, the effect of resonance stabilization of the base and its conjugate acid is very important. The basic form of the side chain group is resonance stabilized.

Protonation at the amino group eliminates the possibility of resonance stabilization. However, protonation of the nitrogen of the imino group gives a resonance-stabilized conjugate acid with two equivalent groups bonded to carbon. Resonance stabilization of this conjugate acid favors its formation.

Isoionic Points of Peptides and Proteins

27.17   Estimate the isoionic points of the following tripeptides.
(a) Ala-Val-Gly   (b) Ser-Val-Asp   (c) Lys-Ala-Val

Answers: (a) It is close to 7, because the component amino acid residues are all neutral amino acids.
(b) It is less than 7, because there is an acidic amino acid (Asp) in the tripeptide.
(c) It is greater than 7, because there is a basic amino acid (Lys) in the tripeptide.

27.18   Estimate the isoionic points of the following tripeptides.
(a) Glu-Val-Ala   (b) Arg-Val-Gly   (c) His-Ala-Val

Answers: (a) It is less than 7, because there is an acidic amino acid (Glu) in the tripeptide.
(b) It is greater than 7, because there is a basic amino acid (Lys) in the tripeptide.
(c) It is close to 7, because the side chain of the basic amino acid (His) in the tripeptide has a pK_a of 6.0.

27.19   Examine the structures of oxytocin and vasopressin in Section 27.7. Which one has the higher isoionic point?

Answer: Vasopressin has an arginine residue, so its isoionic point is higher. Oxytocin contains only neutral amino acids.

27.20   Examine the structure of the enkephalin whose sequence is shown below and estimate its isoionic point.
Ala-Gly-Phe-Leu-Gly

Answer: It is close to 7 because the component amino acid residues—Ala, Gly, Phe, and Leu—are all neutral amino acids.

27.21   The isoionic point of hen egg white lysozyme is 10.8. What does this value indicate about its amino acid composition?

Answer: An isoionic point greater than 7 implies that basic residues in the protein outnumber acidic ones. In fact, the 129 residue protein contains 10 acidic residues and 18 basic residues. Therefore, its isoionic point is greater than 7.

27.22   The isoionic point of pepsin is 1.1. What does this value indicate about its amino acid composition?

Answer: An isoionic point less than 7 implies that acidic residues in the protein outnumber basic ones. In fact, human pepsin protein contains 71 acidic residues and 3 basic residues. Therefore, its isoionic point is less than 7.

Amino Acid Synthesis

27.23 What haloalkane is required to synthesize isoleucine by the acetamidomalonate method? What side reaction might decrease the yield?

Answer: 2-Bromobutane is required to supply a *sec*-butyl group for the synthesis. Secondary halides competitively undergo dehydrohalogenation in syntheses designed to substitute the halide by a nucleophile. The major elimination by product is *trans-2-butene*.

isoleucine

27.24 What reactants are required to synthesize phenylalanine by reductive amination?

Answer: The reaction requires the structurally related α-keto acid and ammonia.

27.25 3-Aminopropanoic acid, sometimes called β-alanine, is a nonsteroidal anti-inflammatory agent used in veterinary medicine. It is prepared by a conjugate addition reaction using ammonia and acrylonitrile (CH_2=CH—CN). The resulting nitrile is then hydrolyzed to give the product. Why is conjugated addition favored?

Answer: Conjugate addition is favored because the reaction proceeds through a resonance-stabilized intermediate that is analogous to conjugate addition to enolates.

27.26 Methionine can be prepared from propenal in a multistep sequence. (a) Explain how is the thiomethyl group introduced? (b) Why is the carbon chain length increased by one carbon atom?

Answer: Conjugate addition of methylthiol gives a compound that introduces the thiomethyl group. Subsequent reaction with sodium cyanide and ammonium chloride (Strecker synthesis) adds a carbon atom to the chain and gives methionine.

27.27 The structure of alliin is shown in Exercise 27.4. Propose a synthesis of alliin starting from an amino acid.

Answer: Reaction of cysteine with allyl bromide introduces the allyl group via an S_N2 reaction of bromide by the thiol group. Subsequent oxidation by hydrogen peroxide gives the sulfoxide group of alliin.

27.28 One of the amino acids in the blood-clotting protein prothrombin is shown below. It was difficult to detect because it decomposes under hydrolysis conditions. (a) What reaction occurs and (b) what is the product?

Answer: The two carboxylic acid groups on the left of the structure can be regarded as components of malonic acid. Decarboxylation occurs in aqueous acid to give glutamic acid.

(a substituted malonic acid)   glutamic acid

Peptides

27.29 Write the bond line structure for alanylserine at pH 7.

Answer:

alanylserine

27.30 How does glycylserine differ from serylglycine?

Answer: The C-terminal amino acid of glycylserine is serine, whereas the C-terminal amino acid of serylglycine is glycine. Peptides are always named starting with the N-terminal amino acid.

glycylserine   serylglycine

27.31 Which amino acids can form peptides with carboxylic acid groups or carboxylate groups at internal positions in the peptide chain?

Answer: Aspartic acid and glutamic acid have carboxylic acid groups as part of their side chains that can exist as an acid or its conjugate base in peptides.

27.32 Which amino acids can form peptides with amino groups or ammonium groups at internal positions in the peptide chain?

Answer: Arginine, lysine, and histidine have side chains that contain groups that can exist as a base or its conjugate acid in peptides.

27.33   Identify the amino acids contained in the following tripeptide. Name the compound.

Answer: serylglycylalanine

$$NH_3^+ \!-\! C \!-\! C \!-\! NH \!-\! C \!-\! C \!-\! NH \!-\! C \!-\! C \!-\! OH$$

seryl (CH_2OH)   glycyl (H)   alanine (CH_3)

27.34   Identify the amino acids contained in the following tripeptide. Name the compound.

Answer: glycylcysteinylvaline

$$NH_3^+ \!-\! C \!-\! C \!-\! NH \!-\! C \!-\! C \!-\! NH \!-\! C \!-\! C \!-\! OH$$

glycyl (H)   cysteinyl (CH_2SH)   valine ($CH(CH_3)_2$)

27.35   Thyrotropin-releasing hormone (TRH) causes the release of thyrotropin from the pituitary gland, which then stimulates the thyroid gland. Examine its structure and comment on one unusual structural feature.

Answer: TRH is a tripeptide containing glutamic acid, histidine, and proline. Proline is the C-terminal amino acid and it exists as an amide. The N-terminal amino acid is glutamic acid, but the γ carboxyl group in its side chain exists as a cyclic amide (lactam) by reaction with the α amino group.

27.36   The tripeptide glutathione, which is important in detoxifying metabolites, has an unusual structural feature. Identify it.

Answer: Glutathione is a tripeptide that contains glutamic acid, cysteine, and glycine residues. Glutamic acid forms an amide with the amino group of cysteine using the γ carboxyl group of its side chain rather than the α carboxyl group.

glu   cys   gly

glutathione

27.37   How many peptide isomers with the composition Gly_2,Ala_2 are possible?

Answer: Six isomers are possible.

Gly-Gly-Ala-Ala   Gly-Ala-Gly-Ala   Gly-Ala-Ala-Gly
Ala-Ala-Gly-Gly   Ala-Gly-Ala-Gly   Ala-Gly-Gly-Ala

27.38   How many peptide isomers with the composition Gly$_2$,Ala,Leu are possible?

Answer: Twelve isomers are possible.

Gly-Gly-Ala-Leu	Gly-Ala-Gly-Leu	Gly-Ala-Leu-Gly
Ala-Leu-Gly-Gly	Ala-Gly-Leu-Gly	Ala-Gly-Gly-Leu
Gly-Gly-Leu-Ala	Gly-Leu-Gly-Ala	Gly-Leu-Ala-Gly
Leu-Ala-Gly-Gly	Leu-Gly-Ala-Gly	Leu-Gly-Gly-Ala

Peptide Hydrolysis and Primary Structure Determination

27.39   Assuming that only dipeptides are formed by partial acid hydrolysis, what is the minimum number that must be identified to establish the amino acid sequence of a pentapeptide?

Answer: Four dipeptides must be identified. For the general formula A—B—C—D—E, the necessary dipeptide are A—B, B—C, C—D, and D—E.

27.40   Assuming that only tripeptides are formed by partial hydrolysis, what is the minimum number that must be identified to establish the amino acid sequence of an octapeptide?

Answer: For the general formula A—B—C—D—E—F—G—H, four tripeptides must be identified. One combination of tripeptides is A—B—C, C—D—E, E—F—G, and F—G—H.

27.41   The tetrapeptide tuftsin is hydrolyzed to produce Pro-Arg and Thr-Lys. Does this information establish the structure of tuftsin?

Answer: No, because there are two ways that the two dipeptides may have been combined in the tetrapeptide. They are Pro-Arg-Thr-Lys and Thr-Lys-Pro-Arg.

27.42   Partial hydrolysis of the octapeptide angiotensin II produces Pro-Phe, Val-Tyr-Ile, Asp-Arg-Val, and Ile-His-Pro. What is its amino acid sequence?

Answer: Align the hydrolysis products to list common amino acids in a vertical column. Overlapping the sequences of the hydrolysis products gives the complete sequence.

```
Asp-Arg-Val
        Val-Try-Ile
                Ile-His-Pro
                        Pro-His
Asp-Arg-Val-Try-Ile-His-Pro-His
```

27.43   Treatment of somatostatin with the Edman reagent gives a derivative of alanine. Partial hydrolysis of the polypeptide gives the following oligopeptides. Write the structure of the polypeptide.
I: Phe-Trp   II: Lys-Thr   III: Thr-Ser-Cys   IV: Thr-Phe-Thr-Ser-Cys
V: Asn-Phe-Phe-Trp-Lys   VI: Ala-Gly-Cys-Lys-Asn-Phe

Answer: Treatment with the Edman reagent identifies alanine as the N-terminal amino acid. Align the hydrolysis products, starting with alanine in a vertical column.

```
Ala-Gly-Cys-Lys-Asn-Phe
                Asn-Phe-Phe-Trp-Lys
                        Phe-Trp
                            Lys-Thr
                                Thr-Phe-Thr-Ser-Cys
                                        Thr-Ser-Cys
```

These correspond to the following peptide sequence.

Ala-Gly-Cys-Lys-Asn-Phe-Phe-Trp-Lys-Thr-Phe-Thr-Ser-Cys

27.44   The amino acid composition of the peptide is (Arg$_2$,Gly,Phe$_2$,Pro$_3$,Ser). Treatment of bradykinin with the Edman reagent gives the PTC-derivative of arginine. Partial hydrolysis yields several fragments that include the following oligopeptides. What is the amino acid sequence of bradykinin?

   I: Gly-Phe-Ser   II: Arg-Pro-Pro-Gly   III: Phe-Arg-Ser-Pro-Phe

Answer: Treatment with the Edman reagent identifies arginine as the N-terminal amino acid. Align the hydrolysis products, starting with arginine in a vertical column.

                    Gly-Phe-Ser
            Arg-Pro-Pro-Gly
                            Ser-Pro-Phe-Arg

The amino acid sequence of the polypeptide is

        Arg-Pro-Pro-Gly-Phe-Ser-Pro-Phe-Arg

Enzymatic Hydrolysis of Peptides

27.45   Which of the following tripeptides will be cleaved by trypsin? If cleavage occurs, name the products.
   (a) Arg-Gly-Tyr   (b) Glu-Asp-Gly   (c) Phe-Trp-Ser   (d) Ser-Phe-Asp

Answer: Trypsin cleaves peptides at the C-terminal side of the basic amino acids lysine and arginine.
   (a) Arg and Gly-Tyr   (b) no reaction occurs (c) no reaction occurs   (d) no reaction occurs

27.46   Which of the following tripeptides will be cleaved by trypsin? If cleavage occurs, name the products.
   (a) Asp-Lys-Ser   (b) Lys-Tyr-Cys   (c) Asp-Gly-Lys   (d) Arg-Glu-Ser

Answer: Trypsin cleaves peptides at the C-terminal side of the basic amino acids lysine and arginine.
   (a) Ser and Asp-Lys   (b) Lys and Tyr-Cys   (c) no reaction occurs   (d) Arg and Glu-Ser

27.47   Indicate which of the tripeptides in Exercise 26.45 will be cleaved by chymotrypsin and name the products.

Answer: Chymotrypsin cleaves peptides at the C-terminal side of the aromatic amino acids phenylalanine, tyrosine, and tryptophan.
   (a) no reaction occurs (b) no reaction occurs (c) Ser, Trp, and Ser   (d) Asp and Ser-Phe

27.48   Indicate which of the tripeptides in Exercise 26.46 will be cleaved by chymotrypsin and name the products.

Answer: Chymotrypsin cleaves peptides at the C-terminal side of the aromatic amino acids phenylalanine, tyrosine, and tryptophan.
   (a) no reaction occurs   (b) Cys and Lys-Tyr   (c) no reaction occurs   (d) no reaction occurs

27.49   The tetrapeptide tuftsin is hydrolyzed by trypsin to produce Pro-Arg and Thr-Lys. Does this information establish the amino acid sequence of tuftsin?

Answer: No, it does not. Trypsin cleaves peptides at the C-terminal side of the basic amino acids lysine and arginine. Both dipeptides have a C-terminal basic amino acid. The tetrapeptide could be either of the following two structures. The site of cleavage is indicated by bold face.

        Pro-**Arg**-Thr-Lys   Thr-**Lys**-Pro-Arg

27.50   The pentapeptide met-enkephalin is hydrolyzed by chymotrypsin to give Met, Tyr, and Gly-Gly-Phe. Does this information establish the amino acid sequence of met-enkephalin?

Answer: No, it does not. Chymotrypsin cleaves peptides at the C-terminal side of the aromatic amino acids phenylalanine, tyrosine, and tryptophan. The hydrolysis products have two aromatic amino acids and they can result from either of the following two structures. The sites of cleavage are indicated by bold face.

        Tyr-Gly-Gly-**Phe**-Met   Gly-Gly-**Phe**-**Tyr**-Met

27.51   The nonapeptide known as the sleep peptide is hydrolyzed by chymotrypsin to produce Ala-Ser-Gly-Glu and Ala-Arg-Gly-Tyr and Trp. What two amino acid sequences are possible for the sleep peptide?

Answer: Chymotrypsin cleaves peptides at the C-terminal side of the aromatic amino acids phenylalanine, tyrosine, and tryptophan. The hydrolysis products have two aromatic amino acids, and they can result from either of the following two structures.

   Ala-Arg-Gly-**Tyr**-**Trp**-Ala-Ser-Gly-Glu

   Trp-Ala-Arg-Gly-**Tyr**-Ala-Ser-Gly-Glu

27.52   The sleep peptide is hydrolyzed by trypsin to produce Gly-Tyr-Ala-Ser-Gly-Glu and Trp-Ala-Arg. What is the amino acid sequence of the sleep peptide?

Answer: Trypsin cleaves peptides at the C-terminal side of the basic amino acids lysine and arginine. Only one of the peptides has a C-terminal basic amino acid. The structure of the sleep peptide is the second of the possible structures given in Exercise 26.51.

   Trp-Ala-Arg-Gly-Tyr-Ala-Ser-Gly-Glu

27.53   Feline gastrin, a hormone that stimulates secretion of gastric juice in cats, has the amino acid composition (Ala$_2$,Asp,Gly$_2$,G1u$_5$,Leu,Met,Phe,Pro,Trp2,Tyr).

   End group analysis shows that the C-terminal and N-terminal amino acids are Phe and Glu, respectively. Hydrolysis with chymotrypsin yields the following four peptides. Write two possible amino acid sequences of feline gastrin.

   I: Gly-Trp   II: Met-Asp-Phe   III: Glu-Gly-Pro-Trp   IV: Leu-Glu-Glu-Glu-Glu-Ala-Ala-Tyr

Answer: Chymotrypsin cleaves peptides at the C-terminal side of the aromatic amino acids phenylalanine, tyrosine, and tryptophan. Four of the five glutamic acid residues are in one octapeptide. Thus, based on the end group analysis and fragment Ill, the sequence of amino acids from the N-terminal position is Glu-Gly-Pro-Trp. Based on the end group analysis and fragment II, the sequence of amino acids at the C-terminal position is Met-Asp-Phe, because there is only one phenylalanine in the peptide. These two peptides can combine with the octapeptide and dipeptide I in two ways. The two possible locations of the dipeptide fragment I are shown in bold face.

   Glu-Gly-Pro-Trp-**Gly**-**Trp**-Leu-Glu-Glu-Glu-Glu-Ala-Ala-Tyr-Met-Asp-Phe

   Glu-Gly-Pro-Trp-Leu-Glu-Glu-Glu-Glu-Ala-Ala-Tyr-**Gly**-**Trp**-Met-Asp-Phe

27.54 Corticotropin, a pituitary hormone, stimulates the adrenal cortex. Hydrolysis by chymotrypsin yields six peptides:

I: Arg-Trp   II: Ser-Tyr   III: Ser-Met-Glu-His-Phe   IV: Pro-Leu-Glu-Phe
V: Pro-Asp-Ala-Gly-Glu-Asp-Gln-Ser-Ala-Glu-Ala-Phe
VI: Gly-Lys-Pro-Val-Gly-Lys-Lys-Arg-Pro-Val-Lys-Val-Tyr

Hydrolysis by trypsin produces lysine, arginine, and five peptides:
I: Trp-Gly-Lys   II: Pro-Val-Gly
III: Pro-Val-Gly-Lys   IV: Ser-Tyr-Ser-Met-Glu-His-Phe-Arg
V: Val-Tyr-Pro-Asp-Ala-Gly-Glu-Asp-Gln-Ser-Ala-Glu-Ala-Phe-Pro-Leu-Glu-Phe

What is the amino acid sequence of corticotropin?

Answer: Both peptide IV and peptide V from hydrolysis by chymotrypsin are contained with peptide V from hydrolysis by trypsin. Both peptide II and peptide III from hydrolysis by trypsin are contained in peptide VI from hydrolysis by chymotrypsin. Both peptide III and peptide II from hydrolysis by chymotrypsin are contained in peptide IV from hydrolysis by trypsin. Therefore, the following three peptides contain the majority of the amino acids.

     A: Val–Tyr–Pro–Asp–Ala–Gly–Glu–Asp–Gln–Ser–Ala–Glu–Ala–Phe–Pro–Leu–Glu–Phe
     B: Gly–Lys–Pro–Val–Gly–Lys–Lys–Arg–Pro–Lys–Val–Tyr
     C: Ser–Tyr–Ser–Met–Glu–His–Phe–Arg

Tryptophan occurs in only two fragments. Thus, the dipeptide I from hydrolysis by chymotrypsin and peptide I from hydrolysis by trypsin can be conbined to give an amino acid sequence Arg–Trp–Gly–Lys. Part of this sequence is duplicated in the two amino acids at the N-terminal positions of peptide B. The appropriate overlap gives

Arg–Trp–Gly–Lys–Pro–Val–Gly–Lys–Lys–Arg–Pro–Lys–Val–Tyr

The N-terminal amino acid of this fragment is common with the arginine of peptide C, which does not match the arginine of peptide B. The resulting peptide with appropriate overlap is

Ser–Tyr–Ser–Met–Glu–His–Phe–Arg–Trp–Gly–Lys–Pro–Val–Gly–Lys–Lys–Arg–Pro–Lys–Val–Tyr

The two amino acids at the C-terminal end of this peptide match the two amino acids at the N-terminal positions of peptide A. The resulting joined peptide with appropriate overlap gives the structure of corticotropin.

Ser–Tyr–Ser–Met–Glu–His–Phe–Arg–Trp–Gly–Lys–Pro–Val–Gly–Lys–Lys–Arg–Pro–Lys–Val–Tyr–Pro–Asp–Ala–Gly–Glu–Asp–Gln–Ser–Ala–Glu–Ala–Phe–Pro–Leu–Glu–Phe

End Group Analysis

27.55 Edman degradation of tuftsin yields Thr as the N-terminal amino acid. Using the information in Exercise 27.49, what is the structure of tuftsin?

Answer: The two possible tetrapeptides based on information given in Exercise 26.49 are
    Pro-Arg-Thr-Lys   Thr-Lys-Pro-Arg

Hydrolysis by the Edman reagent frees the N-terminal amino acid as a PTC-threonine derivative. Because threonine forms, the second tetrapeptide with an N-terminal threonine is the correct structure.

27.56 Hydrolysis of met-enkephalin with CNBr yields the homoserine lactone derivative of methionine and a tetrapeptide. Using the information in Exercise 27.50, what is the structure of met-enkephalin?

Answer: The two possible pentapeptides based on information given in Exercise 26.50 are
    Tyr-Gly-Gly-Phe-Met   Gly-Gly-Phe-Tyr-Met

Cleavage with CNBr releases the C-terminal amino acids the HSL derivative of methionine. Because methionine is the C-terminal amino acid in both cases, the structure is still not established.

27.57 Explain why structure determination of insulin using the Edman method yields two phenylthiohydantoin products.

Answer: There are two peptide chains in insulin joined by disulfide linkages. One chain has glycine as the N-terminal amino acid. The other chain has phenylalanine as the N-terminal amino acid.

27.58 Cholecystokinin, a peptide that contains 33 amino acids, plays a role in reducing the desire for food, and its production is stimulated by food intake. Its N-terminal amino acid is lysine. Draw the structure of the phenylthiohydantoin product.

Answer:

27.59 Reaction of angiotensin II with the Edman reagent yields the following product. What information has been established?

Answer: The N-terminal amino acid is aspartic acid.

27.60 Reaction of angiotensin II with the Edman reagent yields the following product. What information has been established?

Answer: The N-terminal amino acid is serine.

Protein Structure

27.61 Which of the following amino acids are likely to exist in the interior of a protein dissolved in an aqueous solution?
(a) glycine   (b) phenylalanine   (c) glutamic acid   (d) arginine

Answer: Amino acids without polar side chains exist in the interior of the protein. Glycine and phenylalanine do not have polar side chains. Glycine, however, is an ambiguous case, and it is sometimes found on a protein's surface.

26.62 Which of the following amino acids are likely to exist in the interior of a protein dissolved in an aqueous solution?
(a) proline   (b) cysteine   (c) glutamine   (d) aspartic acid

Answer: Amino acids with polar side chains seldom exist in the interior of the protein. Proline and cysteine tend to be buried, glutamine and aspartic acid are usually found on the surface of a protein.

26.63 If a protein is embedded in a hydrophobic lipid bilayer of a biological membrane, which of the amino acids listed in Exercise 26.61 will be in contact with the interior of the bilayer?

Answer: Phenylalanine is very hydrophobic, and its side chain will be in contact with a lipid bilayer; Glu and Arg have very polar side chains and will not be in contact with the lipid bilayer. Once again, however, glycine is an ambiguous case and can be in contact with a lipid bilayer.

26.64   If a protein is embedded in a lipid bilayer, which of the amino acids listed in Exercise 26.62 will be in contact with the interior of the bilayer?

Answer: Cysteine and proline will be in contact with the bilayer; glutamine and aspartic acids are polar and will not be in contact with the bilayer.

26.65   Noting that proline is a secondary amine, explain how proline can disrupt the α helix of a protein.

Answer: Proline does not have an amide hydrogen; therefore, it cannot be a hydrogen bond donor in an α helix.

26.66   Examine the structures of valine and glutamic acid and suggest a reason why human hemoglobin is affected by the substitution of valine for glutamic acid at position 6 in the β chain.

Answer: Valine is a neutral amino acid, whereas glutamic acid is an acidic amino acid. The structure and properties of the side chains of these two amino acids differ significantly. Valine lies on the surface of the mutant HbS molecule and forms hydrophobic contacts with other hemoglobin molecules that result in the formation of long fibers that precipitate when HbS releases oxygen.

28

SYNTHETIC POLYMERS

KEYS TO THE CHAPTER

In this chapter, we focus on chemical reactions that we have discussed in previous chapters and understand in principle, but which we have considered only in the context of the synthesis of lower-molecular-weight compounds. In this chapter, we extend those principles by considering the reactions of low, molecular, weight compounds, called monomers, that can react over and over again with one another to give polymers.

28.1 Natural and Synthetic Macromolecules

Large molecules, known as **macromolecules**, occur naturally in carbohydrates and proteins. The chemical industry has learned how to produce similarly structured macromolecules with properties designed for specific uses. The creativity of polymer chemists has resulted in an avalanche of new polymers with remarkable properties are a ubiquitous in our environment. We restrict our discussion to the most commonly encountered types of polymers and the methods for their synthesis.

28.2 Physical Properties of Polymers

The physical properties of polymers result from the types of intermolecular attractive forces between chains of the polymer. The functional groups in the polymer are formed from those in the monomers from which it is made. Hydrogen bonds, dipole–dipole forces, and London forces all play a role in the interactions of polymer chains.

The types of bonds formed in the polymerization process constitute the primary structure of the polymer. Polymerization gives mixtures of polymers with different molecular weights, and the properties of the polymer depend on the length of the chains. Reactions that form bonds between chains, known as **cross-linking**, are used to change the properties of polymers.

London forces, although the weakest of the intermolecular forces, are exceedingly important in polymers because of the length of the chain, which allows many "points of contact," between neighboring chains. Branching of chains changes the density of the polymers of ethylene. Polymers containing aromatic rings allow fewer conformations within the chain, and these rings are more polarizable as well. As a result, polymers of aromatic compounds have strong intermolecular forces.

Hydrogen bonding between functional groups such as amides in polyamides gives significant strength to these synthetic polymers. The cumulative effect of a large number of hydrogen bonds also helps to stabilize the folded conformations of proteins.

Crystalline regions that are the result of regular alignment of polyethylene chains are called crystallites. These regions have strong intermolecular attractive forces. Crystallites also occur in polymers that form hydrogen bonds. In each case, it is the cumulative effect of numerous individual interactions that stabilizes the crystallite and hence affects the property of the polymer.

28.3 Classification of Polymers

There are three types of polymers: elastomers, plastics, and fibers. **Elastomers** regain their original shape after being distorted in a physical process. The individual chains in elastomers are coiled and can be lengthened by stretching the material. However, after the stress is released, the molecule resumes its original coiled conformation. Rubber is an elastomer. It consists of unsaturated units separated by sp^3-hybridized carbon atoms that give the polymer its flexibility.

Plastics are polymers that harden upon cooling. Plastics that soften when heated are **thermoplastics** and can be molded while warm. **Thermosetting plastics** cannot be softened by heating—they are "set." Polyethylene is a thermoplastic. Thermosetting plastics usually have extensively cross-linked chains.

Fibers are thermoplastics that can be spun into materials that resemble natural fibers. One method of generating fibers is to pass molten thermoplastics through tiny pores in a die, after which the material hardens. A second method passes thermoplastics dissolved in a volatile solvent through tiny pores in a die, during which time the solvent evaporates.

Organic Chemistry Study Guide: Key Concepts, Problems, Synd Solutions http://dx.doi.org/10.1016/B978-0-12-801889-7.00028-5

28.4 Polymerization Methods

Addition polymerization is the successive addition of alkene monomers to one another. The addition reaction may occur by way of radical, cationic, or anionic intermediates. **Condensation polymerization** is a reaction that joins two functional groups such as an alcohol and a carboxylic acid and forms a second small molecule such as water.

Addition polymers are **chain growth polymers** because each intermediate adds another monomer unit one at a time. Condensation polymers are **step-growth polymers** because condensation may occur between two smaller molecular weight chains. Thus, the joining of oligomers results in a substantial increase in molecular weight in a single step.

28.5 Addition Polymerization

The reaction conditions control the length of the hydrocarbon chain of polyethylene and the degree of its branching. **Termination steps** stop the growth of the polymer chain by destruction of radicals. Two such steps are dimerization, which joins two radicals, and disproportionation, which transfers a hydrogen atom between radicals, resulting in an alkane and an alkene.

The regulation of chain length is accomplished with **chain transfer agents**. These substances react by transferring a hydrogen atom to the radical end of a developing polymer chain. To continue polymerization, the resulting radical of the chain transfer agent must be sufficiently reactive to initiate another polymerization process by reacting with the monomer.

Inhibitors react with growing polymer chains to transfer a hydrogen atom and give a stabilized radical derived from the inhibitor. This less reactive radical does not allow chain propagation steps to continue.

There are two chain branching processes. **Short-chain branching** produces butyl side chains to the polymer chain as a result of intramolecular hydrogen transfer by a transition state that contains five carbon atoms and a hydrogen atom. In this process, a primary radical is converted into a secondary radical. **Long-chain branching** occurs by random transfer of a hydrogen atom between the reacting primary radical site and some other site within the hydrocarbon skeleton.

28.6 Copolymerization of Alkenes

Mixtures of monomeric alkenes can form **copolymers** that contain varying amounts and varying sequences of each monomer. The degree to which the polymer has a random arrangement of monomers or a regular alternation of monomers depends on the reactivity of the radical derived from one monomer toward either itself or the other monomer.

28.7 Cross-linked Polymers

Bonds between polymer chains known as cross-links are formed in two different ways. Some monomers used in copolymerization processes have more than one site for addition reactions to occur. In such cases, one site is used in forming the polymer chain and the other site is used to link to another chain by a similar polymerization step. In this process, the cross-links are produced while the polymer forms. The second method involves adding a substance after the polymer chain has formed. This material finds reactive sites along the individual chain and chemically reacts to link one chain to another.

28.8 Stereochemistry of Addition Polymerization

Polymerization of substituted alkenes gives polymers with stereogenic centers. The relationship of these stereogenic centers to one another affects the physical properties of the polymer. For example, in polypropylene the methyl groups are on the same side of the chain in an **isotatic** polymer. The regular alternation of methyl groups on each side of the chain is a **syndiotatic**, polymer. A random distribution of methyl groups occurs in **atactic** polymers. Ziegler–Natta catalysts are used to control the type of polymer formed.

Diene monomers react to give polymers with one double bond per monomer unit. The stereochemistry of the double bond may be *E* or *Z*. The properties of these two polymers differ markedly. Ziegler–Natta catalysts are used to control the type of polymer formed.

28.9 Condensation Polymers

Any of the reactions used to join two functional groups together and form a second smaller molecule such as water are candidates for condensation polymerization reactions. Two functional groups per monomer unit are required. Two monomers, each containing two units of one of the two possible functional groups, such as dicarboxylic acids and diols, are commonly used in condensation reactions. Monomers containing one of each type of functional group are more difficult to synthesize and yet prevent from polymerizing.

28.10 Polyesters

Polyesters are produced from dicarboxylic acids or their derivatives and a diol. Polyethylene terephthalate, or PET, is made by a transesterification reaction of dimethyl terephthalate and ethylene glycol. Cyclic anhydrides such as maleic anhydride or phthalic anhydride are also used to form condensation polymers. Glycerol provides polymer that are cross-linked by reactions of the third hydroxyl group with the acid derivative. Glyptal is a cross-linked polymer of glycerol and phthalic anhydride.

28.11 Polycarbonates

Carbonic acid is an unstable diprotic acid. Its structurally related "diacid chloride," known as phosgene, or the "diester" diethyl carbonate react with alcohols to form polycarbonates. The reaction with phosgene resembles the reaction of an alcohol with an acid chloride. The reaction of an alcohol with diethyl carbonate is a transesterification process.

28.12 Polyamides

Polyamides can be made by direct reaction of a dicarboxylic acid and a diamine. The resulting salt is heated, and the polyamide forms by loss of water. Nylon is a polyamide of a six-carbon diacid and a six-carbon diamine. Certain lactams, among them ε-caprolactam, can polymerize by a ring opening followed by condensation of the resulting amino acid.

28.13 Phenol-Formaldehyde Polymers

Phenol-formaldehyde polymers result from addition of a carbanion, pictured as one of the resonance forms of the phenoxide ion. Addition of the carbanion to the carbon–oxygen double bond of formaldehyde gives a hydroxymethyl derivative, which can dehydrate to give a conjugated ketone, which can undergo conjugate addition reactions that resemble the aldol reaction. As a consequence, a polymeric product known as Bakelite results.

28.14 Polyurethanes

A urethane is an ester of a carbamic acid. Because carbamic acids are unstable, the direct esterification is not possible. Urethanes are made by reaction of alcohols with isocyanates. Polyurethanes are then made by using diisocyanates and diols. The diols may be simple compounds such as ethylene glycol or oligomers.

SOLUTIONS TO EXERCISES

Properties of Polymers

28.1   Explain why the polymer of 2-methylpropene is a sticky elastomer with few crystalline domains.

Answer: The elastomer has alternating quaternary carbon atoms that prevent the polymer chain from packing closely in a regular array to give a crystalline-like region.

28.2   How would the properties of the polymer of the following diamine and adipic acid differ from those of nylon 6,6?

$$CH_3—NH—(CH_2)_6—NH—CH_3$$

Answer: The secondary amine would give a polymer containing tertiary amide functional groups. This polymer could not form the hydrogen bonds that are responsible for the strong intermolecular forces in polyamides such as nylon 6,6.

28.3   Explain how 1,2,4,5-benzenetetracarboxylic acid dianhydride could be used to make a thermosetting polyester.

1,2,4,5-benzenetetracarboxylic acid dianhydride

Answer: Reaction of the dianhydride at one of the anhydride sites with a diol gives a polyester that retains an anhydride unit which could react with added monomers to give bridged links between chains. Extensive cross-linking is responsible for the properties of thermosetting polyesters.

590

28.4 How would the properties of the copolymer of 1,4-butanediol with terephthalic acid differ from those of PET?

Answer: The additional two methylene units of the diol result in a decrease in London forces and give a polymer that is more flexible. It should have less tensile strength.

28.5 Why is neoprene less susceptible to oxidation than polyisoprene?

Answer: The neoprene is a polymer of 2-chloro-1,3-butadiene, whereas polyisoprene is a polymer of 2-methyl-1,3-butadiene. The electronegative chlorine atom of neoprene reduces the availability of electrons to oxidizing agents, and the large chlorine atom reduces the accessibility of the C—H bonds to oxidizing agents.

Neoprene

28.6 Explain why teflon, a polymer of tetrafluoroethylene, is not sensitive to oxidation.

Answer: The electronegative fluorine atom reduces the availability of electrons to oxidizing agents. There are no C—H bonds to be attacked by oxidizing agents.

Addition Polymers

28.7 Vinyl acetate is used to make a polymer used in chewing gum. Draw a bond-line representation of the polymer.

vinyl acetate

Answer: In the following structure, the symbol OAc represents the acetate group.

vinyl acetate   poly(vinyl) acetate

28.8 Draw a bond-line structure of polyvinyl alcohol. Explain why the polymer is prepared by the hydrolysis of polyvinyl acetate.

vinyl alcohol   poly(vinyl) alcohol   poly(vinyl) acetate

Answer: The required monomer would have to be vinyl alcohol, which does not exist because vinyl alcohol tautomerizes to ethanal (acetaldehyde). Hydrolysis of polyvinyl acetate, however, does produce polyvinyl alcohol.

28.9 What monomer is required to prepare the following polymer?

$$-CFCl-CF_2-CFCl-CF_2-CFCl-CF_2-CFCl-CF_2-$$

Answer: There is a repeating unit every two carbon atoms: Thus, the polymer is formed from a substituted ethylene compound. One carbon atom has a chlorine and fluorine atom, and the other carbon atom has two fluorine atoms.

28.10 Hexafluoropropylene is a monomer used to prepare a polymer called Viton. Draw a representation of the polymer.

Answer: The formula tells us that there are no hydrogen atoms in this propene, whose formula is C_3H_6.

28.11 Draw the structure of the ozonolysis product of natural rubber under oxidative workup conditions.

natural rubber (polyisoprene, all *cis*)

Answer: We recall that ozonolysis followed by oxidative workup gives a dicarboxylic acid if there are no substituents on the double bond, or an aldehyde or ketone if there are substituents of the double bond.

28.12 The polymer formed from a compound with molecular formula C_6H_{10} undergoes ozonolysis to give 2,5-hexane-dione. What is the structure of C_6H_{10}?

Answer: The structure of the polymer shown below gives 2,5-hexanedione. The diene required to form the polymer is 2,3-butadiene.

2,6-hexanedione

2,3-dimethyl-1,3-butadiene

Chain Transfer Reactions

28.13   Draw the structure of the branch formed by a short-chain transfer reaction in the formation of polystyrene.

Answer: The short-chain branching reaction has a step in which intramolecular abstraction of a hydrogen atom occurs. The step produces a new radical site where polymerization continues. The branch formed is 2,4-diphenylbutyl.

(2,4-diphenylbutyl branch)

28.14   Explain why formation of a polymer of 1-hexene under free radical conditions would produce some molecules with methyl groups bonded to the main chain.

Answer: The side chain on the polymer is a butyl group, and the reaction center is a secondary radical. Abstraction of a hydrogen atom from the methyl group could occur via a six-membered transition state, but a primary radical would result. Abstraction of a hydrogen atom of a methylene group can occur through a five-membered transition state to give a secondary radical. Continued reaction of this radical gives a product with a methyl group at that point.

Copolymers

28.15   Draw a representation of an alternating polymer of isoprene and 2-methylpropene.

Answer:

isoprene   +   2-methylpropene

28.16   Styrene and 1,3-butadiene form a random polymer. What is the probability that a 1,3-butadiene unit will react with a growing polymer chain with styrene at its end?

Answer: If the polymer is completely random, then any reactive end has a 50:50 chance of reacting with either of the two monomers.

28.17 Some hair sprays contain a solution of a copolymer made from the following monomers. Draw a representation of the polymer. Why does the copolymer hold hair in place?

Answer: If the polymer is completely random, then any reactive end has a 50:50 chance of reacting with either of the two monomers.

28.18 Saran is a copolymer of vinylidene chloride (CH_2=CCl_2) and a smaller amount of vinyl chloride. Draw a representation of the polymer.

Answer:

28.19 Nitrile rubber, which is used to make automotive hoses, has the following structure. What monomers are used to produce the polymer?

nitrile rubber

Answer: The monomers are 1,3-butadiene and acrylonitrile, CH_2=$CHCN$.

nitrile rubber

28.20 Draw a section of a copolymer of acrylonitrile, styrene, and 1,3-butadiene, which is used as a synthetic rubber.

Answer:

Cross-Linked Polymers

28.21 What is the difference between the number of cross-links in the rubber used in tires and the rubber used in gloves?

Answer: The number of cross-links in rubber gloves is less because much greater flexibility is required in gloves than in the rubber of tires.

28.22   Draw a representation of the polyester formed from butenedioic anhydride (maleic anhydride) and 1,2-propane-diol. Explain how this polymer could be cross-linked by reacting it with styrene.

Answer:

Answer: There are double bonds in the polymer that can react with styrene and form cross links containing aromatic rings.

Stereochemistry of Polymerization

28.23   Which of the following alkenes can be polymerized to give isotactic and syndiotactic structures?
(a) 1-chloroethene (b) 1,1-dichloroethene (c) 2-methylpropene (d) styrene

Answer: Only (a) and (d) can give isotactic and syndiotactic polymers. Both 1,1-dichloroethene and 2-methylpropene give polymers with two equivalent atoms or groups on a carbon atom.

28.24   Are syndiotactic or isotactic forms of polypropylene optically active?

Answer: Neither one is optically active. Both have a mirror plane of symmetry perpendicular to the zigzag chain and passing through the center of the polymer chain.

28.25   *S*-Methyl-1-pentene reacts with a Ziegler–Natta catalyst to give an isotactic polymer. What relationship exists between the alkyl groups on the polymer chain?

Answer: Each tertiary carbon atom of the chain contains a *sec*-butyl group. Each alkyl group is on the same side of the backbone of the zigzag chain.

28.26   Ethylene and *cis*-2-butene form a syndiotactic copolymer in a reaction catalyzed by a vanadium catalyst. Draw a representation of the polymer.

Answer:

Condensation Polymers

28.27   What monomers are required to prepare the following polymers?

(a)

Answer:

(b)

Answer:

(c)

Answer:

(d)

Answer:

$NH_2(CH_2)_9NH_2$

28.28   What monomers are required to prepare the following polymers?

(a)

Answer:

(b)

Answer:

(c)

Answer:

(d)

Answer:

Polyesters

28.29   A homopolymer of lactic acid can be used to make body implants. Write a bond-line representation of the polymer.

Answer: Lactic acid is $CH_3CH(OH)CO_2H$.

28.30   A polymer of β-propiolactone is obtained by using a catalytic amount of hydroxide ion. Draw the structure of the polymer. Why does the polymerization reaction continue?

β-propiolactone

Answer: A nucleophile attacks the carbonyl carbon atom giving a tetrahedral intermediate, and an alkoxide ion "leaves" but remains attached to the reacting unit at the β-position as an alcohol. The alkoxide ion attacks another molecule of the lactone and forms an ester while releasing another alkoxide ion. The reaction continues because the ring strain of the lactone is released.

28.31   Kodel is a polymer of terephthalic acid and *trans*-di-1,4-(hydroxymethyl)cyclohexane. Draw a representation of the polymer.
Answer:

28.32   What monomers are used to prepare the following polyester? Identify an unusual feature of this polyester.

Answer: A 2-methyl-1,3-propanediol, phthalic acid, and *trans*-butenedioic acid. The copolymer is produced using a mixture of dicarboxylic acids and that is the unusual feature of this polymer.

Polyamides

28.33   Draw a representation of each of the following polymers.
         (a) nylon 6,10 (b) nylon 11 (c) nylon 4,6
Answer:

nylon 6,10

Answer:

nylon 11

Answer:

nylon 4,6

28.34 The following structure represents a group of polyamides called Qiana. The value of x is 8, 10, or 12. What are the component monomers? What is the significance of the value of x?

Answer: The diamine portion of the polyamide is relatively rigid. The flexibility of the carbon chain of the dicarboxylic acid affects the properties of the polymer. The structure of the diamine monomer and the dicarboxylic acid with $x = 10$ are shown.

28.35 Is it likely that nylon 11 could be prepared from a lactam?

Answer: A polyamide could be made by ring opening of the lactam. However, the lactam cannot be easily made because of the size of the ring. In fact, if an attempt were made to synthesize the lactam using the following amino acid, it would polymerize.

28.36 A polyamide contains the following structural unit, which is prepared from the reaction of a lactam. Draw the structure of the lactam.

Answer:

Polyethers

28.37 Carbowax is a polyether named polyethylene glycol. Why is ethylene oxide used to prepare this polymer?

Answer: The ring strain of the epoxide allows the polymerization to occur readily. A nucleophile opens the epoxide ring to give an alkoxide, which in turn opens another epoxide ring.

28.38 Polymerization of (S)-2-methyloxirane catalyzed by a Lewis acid in ether yields an optically inactive polymer.

Answer: Hydroxide ion attacks C-3, which is primary, to form a chiral alkoxide. This alkoxide in turn opens another epoxide ring.

Acid catalysis occurs by addition of a Lewis acid (shown below with a proton) to the oxygen atom and subsequent ring opening, giving a carbocation that is achiral. Attack of the carbocation on the oxygen atom of another epoxide continues the polymerization.

28.39 A polyether oligomer of tetramethylene glycol is produced by an acid-catalyzed ring opening of tetrahydrofuran. Write the steps that account for the formation of the oligomer.

28.40 Poly(vinylbutyral) is used in automobile windshield glass. How is this polymer prepared starting from polyvinyl acetate

poly(vinyl butyral)

Answer: Hydrolyze the polyvinyl acetate to give a poly alcohol. Form the cyclic acetal from diol units and butanal under acidic conditions.

Polyurethanes

28.41 Explain why the addition of glycerol to the polymerization of toluene diisocyanate and ethylene glycol produces a stiffer foam.

Answer: Glycerol provides a third hydroxyl group that can react with additional toluene diisocyanate to give a cross-linked polymer.

28.42 An oligomer of tetramethylene glycol reacts with toluene diisocyanate to form a polyurethane called Lycra. Draw a representation of the polyurethane.

tetramethylene glycol

INDEX

Printed in the United States
By Bookmasters